Engineering Materials

This series provides topical information on innovative, structural and functional materials and composites with applications in optical, electrical, mechanical, civil, aeronautical, medical, bio- and nano-engineering. The individual volumes are complete, comprehensive monographs covering the structure, properties, manufacturing process and applications of these materials. This multidisciplinary series is devoted to professionals, students and all those interested in the latest developments in the Materials Science field, that look for a carefully selected collection of high quality review articles on their respective field of expertise.

Indexed at Compendex (2021) and Scopus (2022)

Lalita Ledwani · Saurabh Srivastava
Editors

Experimental and Computational Advances in Materials

Springer

Editors
Lalita Ledwani
Department of Chemistry
Manipal University Jaipur
Jaipur, Rajasthan, India

Saurabh Srivastava
Department of Chemistry
Manipal University Jaipur
Jaipur, Rajasthan, India

ISSN 1612-1317 ISSN 1868-1212 (electronic)
Engineering Materials
ISBN 978-981-95-4179-9 ISBN 978-981-95-4180-5 (eBook)
https://doi.org/10.1007/978-981-95-4180-5

This Springer imprint is published by the registered company Springer Nature Singapore Pte Ltd.
The registered company address is: 152 Beach Road, #21-01/04 Gateway East, Singapore 189721, Singapore

Preface

The Book *Experimental and Computational Advances in Materials* highlights the recent advances made in the vast field of Materials Sciences and related computational techniques used to explore various type of materials. Since the field of materials science is an interdisciplinary field where the researchers from all the domain of sciences such as Physical Sciences, Chemical Sciences, Biological Sciences, and Mathematical and Computer Application Sciences work towards the development of novel materials, this book provides a common platform to present their work with a comprehensive discussions and methods to explore materials development and the properties of novel and existing materials. The book also includes the research work in computational techniques developed by mathematical and Machine Learning models to understand and design various devices and their applications and properties as an integral part of materials science.

This book consists of 9 parts covering different areas of materials science across all the domains. The first part, **Advanced Materials for Environmental Applications**, focuses on the environmental applications of the novel and existing materials. The part covers five chapters on the use of nanoparticles, nanomaterials, and nano biosurfactants for the purification and treatment of contaminants in water and soil. A chapter on the biochar-enhanced fertilization in this part discusses the role of biochar materials in the plant growth and disease control in the plants.

Second part of the book, **Energy Storage Materials**, includes three chapters focusing on the energy storage applications of the materials. The need to balance the demand and supply related to energy consumption can only be fulfilled by novel materials supporting energy storage and providing a sustainable environment at the same time. The chapters in this part discuss not only the popular perovskite and photocatalytic materials but also the carbon extraction from tire waste as an energy storage material.

A very important field in the development of modern materials is the advent of **Multifunctional Nanomaterials** which constitute third part of the book. Multifunctional materials show more than one application of the same material in different conditions. The single chapter in this part discusses the metal-based biogenic nanoparticles which show catalytic behavior as well as the anti-microbial activity.

Another aspect of materials includes coatings and surface engineering to develop corrosion resistance and anti-microbial surfaces for various applications. Fourth part of this book, **Surface Functionalized Materials**, enlightens the development of different materials used for coatings, surface engineering, and modifications to improve the functions of the metals and other polymers. This part consists of 3 chapters discussing antibacterial paints and coatings, epoxy polymers, and textile surface modifications.

Fifth part of this book, **Sensors and Actuators**, insights into the role of materials, especially biomaterials as sensors for environmental monitoring and remediation. The part consists of one chapter about Bioengineered Endophyte-Integrated Biosensors. The advancements in this area could be very beneficial for the selective and rapid detection of the environmental contaminants.

Another area of the materials involves the development of **Composite Materials**, the Sixth part of this book. With the advancements in the field of composite manufacturing, various materials have been developed to constitute the matrix and fillers of the composites. The chapter in this part overviews the development of future technologies through innovative composites.

The discussions about the computational advances in materials are divided into two parts, as Seventh part, **Computational Modelling for Materials**, primarily focuses on the development of computational techniques to study bio-materials such as proteins, enzymes, and the development of novel drugs, whereas Eighth part, **Data Science and Artificial Intelligence in Materials**, delves into the role of data science, and various machine learning models to study different materials or their industrial production viability.

In view of the contents of the book, the last part is included to provide an overview of the **Experimental and Computational Methodologies in Materials**. This part includes a chapter which incorporates an overview of most of the modern experimental and theoretical approaches available for the analysis of the existing and novel materials.

We hope that the book provides a comprehensive vision of the developments related to various materials science fields and the readers will benefit immensely from the collection of works in various branches of the materials science and engineering.

We would like to thank all our contributors without whom the book could not have been completed and our publishing partner Springer Nature for providing all the support to publish the book.

Jaipur, India

Dr. Lalita Ledwani
Dr. Saurabh Srivastava

Contents

Contributors

Gurpreet Kaur Bamrah Department of Chemistry, Computational and Theoretical Chemistry Laboratory, Manipal University Jaipur, Jaipur, Rajasthan, India

Rishabh Basrur Catalytic Applications Laboratory, Department of Chemistry, School of Physical and Biological Sciences, Faculty of Science, Technology and Architecture, Manipal University Jaipur, Jaipur, Rajasthan, India

Sanjay Singh Bisht Department of Chemistry, Amity Institute of Applied Sciences, Amity University, Noida, Uttar Pradesh, India

Anand G. Chakinala Chemical Reaction Engineering Laboratory, Department of Chemical Engineering, Manipal University Jaipur, Jaipur, Rajasthan, India

Ishita Chopra Department of Chemistry, Manipal University Jaipur, Jaipur, India

Abhishek Dadhich Department of Biosciences, Manipal University Jaipur, Jaipur, Rajasthan, India

Hemen Dave School of Pharmacy, National Forensic Sciences University, Gandhinagar, Gujarat, India

Mousumi Debnath Department of Biosciences, Manipal University Jaipur, Jaipur, Rajasthan, India

Irra Dhar Department of Biosciences, Manipal University Jaipur, Jaipur, Rajasthan, India

Veena Dhayal Department of Chemistry, Manipal University Jaipur, Jaipur, India

Saumya Dubey Structural Biology and Bioinformatics Laboratory, Department of Biosciences, Manipal University Jaipur, Jaipur, Rajasthan, India

Seema Garg Department of Chemistry, Amity Institute of Applied Sciences, Amity University, Noida, Uttar Pradesh, India

Varsha Goyal Department of Chemistry, IIS Deemed to be University, Jaipur, India

Kanchan Guru Solar Energy Conversion and Nanomaterials Laboratory, Department of Chemistry, Manipal University Jaipur, Jaipur, Rajasthan, India

Rohit Jain Department of Biosciences, Manipal University Jaipur, Jaipur, India

Grijesh Jaiswal Structural Biology and Bioinformatics Laboratory, Department of Biosciences, Manipal University Jaipur, Jaipur, Rajasthan, India

Sivasankar Kakku Waste to Resources Laboratory, Department of Biotechnology and Chemical Engineering, Manipal University Jaipur, Jaipur, India

Renu Khangarot Cancer Research Laboratory, Department of Biosciences, Manipal University Jaipur, Jaipur, Rajasthan, India

Ashish Kumar Department of Mathematics and Statistics, Manipal University Jaipur, Jaipur, India

Sanjay Kumar Functional Ceramics and Smart Materials Lab, Department of Physics, Manipal University Jaipur, Jaipur, India

Ankita Kumawat Department of Biosciences, Manipal University Jaipur, Jaipur, Rajasthan, India

Lalita Ledwani Department of Chemistry, Manipal University Jaipur, Jaipur, India;
Nanotechnology and Catalysis Research Centre, University of Malaya, Kuala Lumpur, Malaysia

Rashi Nathawat Functional Ceramics and Smart Materials Lab, Department of Physics, Manipal University Jaipur, Jaipur, India

Debbie Ndlovu School of Pharmacy, National Forensic Sciences University, Gandhinagar, Gujarat, India

Saraswati Kumari Ola Department of Chemistry, B K Birla Institute of Higher Education, Pilani, Jhunjhunu, Rajasthan, India

Akanksha Pandey School of Doctoral Studies and Research, National Forensic Sciences University, Gandhinagar, Gujarat, India

Diya Patel School of Pharmacy, National Forensic Sciences University, Gandhinagar, Gujarat, India

Shruti Patel School of Doctoral Studies and Research, National Forensic Sciences University, Gandhinagar, Gujarat, India

Himanshu Priyadarshi Department of Electrical Engineering, Manipal University Jaipur, Jaipur, Rajasthan, India

Smita Purohit Department of Botany, IIS Deemed to be University, Jaipur, India

Sumaira Rasool Department of Mathematics and Statistics, Manipal University Jaipur, Jaipur, India

Rajnandani Rathore Department of Biosciences, Manipal University Jaipur, Jaipur, India

Satyapal S. Rathore Department of Physics, Cluster University of Jammu, Jammu, India

Nikita Rawat Department of Biosciences, Manipal University Jaipur, Jaipur, India

Sriparna Ray Catalytic Applications Laboratory, Department of Chemistry, School of Physical and Biological Sciences, Faculty of Science, Technology and Architecture, Manipal University Jaipur, Jaipur, Rajasthan, India

Suresh Sagadeven Manipal University Jaipur, Dehmi Kalan, Jaipur, Rajasthan, India

Himanshi Saini Department of Chemistry, Manipal University Jaipur, Jaipur, India

Monika Saini Department of Mathematics and Statistics, Manipal University Jaipur, Jaipur, India

Abhishek Sharma Waste to Resources Laboratory, Department of Biotechnology and Chemical Engineering, Manipal University Jaipur, Jaipur, India; Department of Chemical Engineering, BITS Pilani K. K. Birla Goa Campus, Goa, India

Anchal Sharma Department of Biosciences, Manipal University Jaipur, Jaipur, Rajasthan, India

Lakshika Sharma S.S. Jain Subodh P.G. College, Jaipur, India

Madan Mohan Sharma Department of Biosciences, Manipal University Jaipur, Jaipur, Rajasthan, India

Rakesh Kumar Sharma Department of Life Sciences, Chhatrapati Shahu Ji Maharaj University, Kanpur, India

Abhijeet Singh Cancer Research Laboratory, Department of Biosciences, Manipal University Jaipur, Jaipur, Rajasthan, India

Jai Singh Structural Biology and Bioinformatics Laboratory, Department of Biosciences, Manipal University Jaipur, Jaipur, Rajasthan, India

Mumal Singh Department of Chemistry, Manipal University Jaipur, Jaipur, India

Nitish Kumar Singh Structural Biology and Bioinformatics Laboratory, Department of Biosciences, Manipal University Jaipur, Jaipur, Rajasthan, India

Tanisha Singh Department of Biosciences, Manipal University Jaipur, Jaipur, Rajasthan, India

Ashish Somwanshi Functional Ceramics and Smart Materials Lab, Department of Physics, Manipal University Jaipur, Jaipur, India

Ananya Srivastava Structural Biology and Bioinformatics Laboratory, Department of Biosciences, Manipal University Jaipur, Jaipur, Rajasthan, India

Sandeep Kumar Srivastava Structural Biology and Bioinformatics Laboratory, Department of Biosciences, Manipal University Jaipur, Jaipur, Rajasthan, India

Saurabh Srivastava Department of Chemistry, Computational and Theoretical Chemistry Laboratory, Manipal University Jaipur, Jaipur, Rajasthan, India

Praveen K. Surolia Solar Energy Conversion and Nanomaterials Laboratory, Department of Chemistry, Manipal University Jaipur, Jaipur, Rajasthan, India

Sandesh P. Telang Structural Biology and Bioinformatics Laboratory, Department of Biosciences, Manipal University Jaipur, Jaipur, Rajasthan, India

Indu Tomar Department of Biosciences, Manipal University Jaipur, Jaipur, Rajasthan, India

Triveni Department of Chemistry, Smt. Indramani Mandelia Shiksha Niket, Pilani, Jhunjhunu, Rajasthan, India

Mona Vajpayee Department of Biosciences, Manipal University Jaipur, Jaipur, India

Abbreviations

•OH	Hydroxyl radicle
3D	Three Dimensional
3-NHPAA	3-Nitro-4-hydroxy-phenylarsonic acid
5-FU	5-Fluorouracil
16S rRNA	16S rRNA sequencing
ABX3	General formula for perovskite materials
AC	Activated Carbon
ACC	1-Aminocyclopropane-1-Carboxylic Acid
ACE	Angiotensin-Converting Enzyme
ADMET	Absorption, Distribution, Metabolism, Excretion, and Toxicity
ADR	Adenoma Detection Rate
AFM	Atomic Force Microscopy
AFRP	Aramid Fiber Reinforced Polymer
AFS	Atomic fluorescence spectrometry
AgNPs	Silver Nanoparticles
AI	Artificial Intelligence
AIMD	Ab Initio Molecular Dynamics
AJCC	American Joint Committee on Cancer
ALD	Atomic Layer Deposition
ALL	Acute Lymphoblastic Leukemia
AMBER	Assisted Model Building with Energy Refinement
AMPs	Antimicrobial Peptides
AMR	Adenoma Miss Rate
AOCs	Aromatic organoarsenic compounds
AOPs	Advance Oxidation Processes
aOR	Adjusted Odds Ratio
AP4	Activator Protein 4
APC	Adenomatous Polyposis Coli
APIs	Active Pharmaceutical Ingredients
Ar	Argon

As(III)	Arsenite
As(V)	Arsenate
ASIR	Age-Standardized Incidence Rate
ASTM E2180	Standard Test Method for Determining the Activity of Incorporated Antimicrobial Agent(s) in Polymeric or Hydrophobic Materials
ATR-FTIR	Attenuated total internal reflectance Fourier transform infrared spectroscopy
AUC	Area Under the Curve
AuNPs	Gold Nanoparticles
B. subtilis	*Bacillus subtilis*
B3LYP	Becke, 3-parameter, Lee–Yang–Parr functional
BC	Biochar
BD	Bulk Density
BET	Brunauer–Emmett–Teller
b-HLH-LZ	Basic Helix-Loop-Helix Leucine Zipper
bioMEMs	Biomedical micro-electro-mechanical systems
BITS	Birla Institute of Technology and Science
BJH	Barrett–Joyner–Halenda
BMC AI	BioMed Central Artificial Intelligence
BMW	Bayerische Motoren Werke (automobile company)
BRAFV600E/ERK1/2	B-Rapidly Accelerated Fibrosarcoma/Extracellular signal-Regulated Kinase.
BTEX	Benzene, Toluene, Ethylbenzene and Xylenes
C_0	Initial concentration
CAPOX	Capecitabine + Oxaliplatin
CAR-T	Chimeric Antigen Receptor T-cell Therapy
CAS9	CRISPR-associated protein 9
CCS	Carbon Capture and Storage
CCSD	Coupled Cluster with Singles and Doubles
C_e	Final concentration
CEA	Carcinoembryonic Antigen
CEC	Cation Exchange Capacity
CF_4	Carbon Tetrafluoride
CFRP	Carbon Fiber Reinforced Polymer
CFU	Colony-Forming Unit
CFUs	Colony-Forming Units
CH_4	Methane
CHARMM	Chemistry at Harvard Advanced Region Molecular Mechanics
CHARMM-GUI	Chemistry at Harvard Macromolecular Mechanics-Graphical User Interface
ChIP-seq	Chromatin Immunoprecipitation
CHT	Conventional Hydrothermal
CI	Confidence Interval

CLEAR-box	Coordinated Lysosomal Expression and Regulation
CMC	Ceramic Matrix Composite
CMC	Critical micelle concentration
CMS1–4	Consensus Molecular Subtypes 1–4
CNSPI	Cesium–Sodium mixed-cation Perovskite Iodide (Cs$_x$Na$_y$PbI$_3$)
CNT	Carbon Nano Tube
CNTs	Carbon Nanotubes
CO	Carbon Monoxide
CO_2	Carbon Dioxide
COMMAD	Coloboma, osteopetrosis, microphthalmia, macrocephaly, albinism, and deafness
Conc	Concentration
COSMO	Conductor-like Screening Model
CPIC	Clinical Pharmacogenetics Implementation Consortium
CPS	Conducting Polymers
CQDs	Carbon Quantum Dots
CRC	Colorectal Cancer
CRC	CRC Press (Publisher)
CRISPR	Clustered Regularly Interspaced Short Palindromic Repeats
CSCF	Carbon Supercritical Fluid
CSR	Corporate Social Responsibility
CTCL	Cutaneous T-cell Lymphoma
ctDNA	Circulating Tumor DNA
CTE	Coefficient of Thermal Expansion
$CuNO_3$	Copper(II) nitrate
CuO	Copper(II) oxide
CVD	Chemical Vapor Deposition
CVI	Chemical Vapor Infiltration
DFT	Density Functional Theory
DJ phase	Dion–Jacobson phase
DL	Deep Learning
DMA	Dimethyl arsenate
DMF	Dimethylformamide
DNA	Deoxyribonucleic Acid
DNAPLs	Dense Non-Aqueous Phase Liquids
DNT	Dinitrotoluene
DOS	Density of States
DRS	Diffuse reflectance spectroscopy
DSC	Differential Scanning Calorimetry
DST	Department of Science and Technology (India)
DWNTs	Double-Walled Nanotubes
E. coli	*Escherichia coli*
E_{ads}	Adsorption Energy

EC	Electrical Conductivity
ECHA	European Chemicals Agency
EDAX	Energy Dispersive X-ray Analysis
EDM	Electrical Discharge Machining
EDS	Energy Dispersive Spectroscopy
EDX	Energy-Dispersive X-ray Spectroscopy
EELS	Electron Energy Loss Spectroscopy
EGFR	Epidermal Growth Factor Receptor
EHPs	Electron–Hole Pair
EMA	European Medicines Agency
ENCORI	Encyclopedia of RNA Interactomes
EPA	Environmental Protection Agency
EPR	Enhanced Permeability and Retention
EPSS	Extracellular Polymeric Substances
EQE	External Quantum Efficiency
EUA	Emergency Use Authorization
EZVI	Emulsified Zero Valent Iron
FA	Formamidinium
FAP	Familial Adenomatous Polyposis
FBDD	Fragment-Based Drug Discovery
FDA	Food and Drug Administration
Fe(110)	Iron (110) Surface
Fe_2O_3	Iron(III) oxide
Fe_3O_4 NPs	Magnetite nanoparticles
$FeCl_2.H_2O$	Ferrous Chloride Hydrate
$FeCl_3$	Iron(III) Chloride
$FeCl_3.6H_2O$	Ferric Chloride Hexahydrate
FESEM	Field Emission Scanning Electron Microscopy
FIT	Fecal Immunochemical Test
FO	Forward Osmosis
FOLFOX	5-Fluorouracil + Oxaliplatin
FREI-39	Endophytic fungal isolate of Eupenicillium shearii
FRET	Fluorescence Resonance Energy Transfer
FTIR	Fourier Transform Infrared spectroscopy
Fukui −	Fukui Function for Electrophilic Attack
Fukui +	Fukui Function for Nucleophilic Attack
GANs	Generative Adversarial Networks
GC-MS	gas chromatography–mass spectrometry
GDP	Gross Domestic Product
GFRP	Glass Fiber Reinforced Polymer
GGA	Generalized Gradient Approximation
GHG	Greenhouse Gas
GI Genius	AI-assisted colonoscopy system (Medtronic)
GO	Graphene Oxide
GQDs	Graphene Quantum Dots

GRAMMDOCK	Global RAnge Molecular Matching
GROMACS	GROningen MAchine for Chemical Simulations
GROMOS	GROningen MOlecular Simulation
GST	Glutathione S Transferase
H&E	Hematoxylin and Eosin
H_2O_2	Hydrogen Peroxide
HAIs	Hospital-Acquired Infections
HCl	Hydrochloric acid
HDAC	Histone Deacetylase inhibitor
HDI	Human Development Index
HEA	High-Entropy Alloys
HEBM	High-Energy Ball Milling
HEPs	High-Entropy Perovskites
HER2	Human Epidermal Growth Factor Receptor 2
HF	Hartree–Fock
HG-AFS	Hydride generation atomic fluorescence spectrometry
HGCNS	Hierarchical Graphitic Carbon Nanosheets
HGP	Human Genome Project
HIPIC	High-Performance Integrated Composite
HOMO	Highest Occupied Molecular Orbital
HPLC	High-performance liquid chromatography
HR	Hazard Ratio
HRTEM	High-resolution transmission electron microscopy
HTS	High-Throughput Screening
HTS	High-Temperature Superconductor
IAA	Indole-3-Acetic Acid
IBM WFO	IBM Watson for Oncology
IC	Internal Combustion (Engine)
ICP-AES	Inductively coupled plasma atomic emission spectrometer
ICP-MS	Inductively Coupled Plasma Mass Spectrometry
ICP-OES	Inductively coupled plasma optical emission spectrometry
IEEE	Institute of Electrical and Electronics Engineers
II	Industrial Institute/Section reference (context dependent)
IR	Infrared Spectroscopy
ISO 22196	Measurement of Antibacterial Activity on Plastics and Other Non-Porous Surfaces
ISO	International Organization for Standardization
JEOL	Japan Electron Optics Laboratory (manufacturer)
JIS Z 2801	Antimicrobial Products—Test for Antimicrobial Activity and Efficacy (Japanese Industrial Standard)
JOM	Journal of Metals (TMS publication)
JSM	JEOL Scanning Microscope series
K. pneumonia	*Klebsiella pneumoniae*
kb	Kilobase
KMnO4	Potassium permanganate

LBDD	Ligand-Based Drug Design
LbL	Layer by Layer
LC	Liquid chromatography
LC/Q-TOF/MS	Liquid chromatography/quadrupole time-of-flight mass spectrometry
LCA	Life Cycle Assessment
LC-MS/MS	Liquid chromatography–mass spectrometry/mass spectrometry
LDA	Local Density Approximation
LDpred2	LD-based Polygenic Risk Score method
LEBM	Low-Energy Ball Milling
LEDs	Light-Emitting Diodes
LINCS	Linear Constraint Solver
LLTO	$Li_{0.5}La_{0.5}TiO_3$
LMICs	Low- and Middle-Income Countries
LNPs	Lipid Nanoparticles
LUMO	Lowest Unoccupied Molecular Orbital
Lynch	Lynch Syndrome (Hereditary Nonpolyposis CRC)
MA	Methylammonium
MALDI-TOF	Matrix Assisted Laser Desorption/Ionization (MALDI)
MAPI	Methylammonium Lead Iodide (CH$_3$NH$_3$PbI$_3$)
MAX	MYC-associated factor X
MC	Monte Carlo
MD	Molecular Dynamics
MDR	Multidrug-Resistant
MHT	Microwave-assisted Hydrothermal
MIC	Minimum inhibitory concentration
MIL	Military Standard
MIPS	Molecularly Imprinted Polymers
miRNAs	MicroRNAs
MiT	Microphthalmia-associated Transcription
MITF	Microphthalmia-associated transcription factor
ML	Machine Learning
MMC	Metal Matrix Composite
MMR	Mismatch Repair
MNMG	magnetite nanoparticles modified graphite
MOF	Metal–Organic Framework
MOFs	Metal–Organic Frameworks
MoS	Molybdenum Sulfide
MP2	Møller–Plesset (second order)
MRI	Magnetic Resonance Imaging
MSI	Microsatellite Instability
MSI-H	Microsatellite Instability-High
MSNs	Mesoporous Silica Nanoparticles

MSS	Microsatellite Stable
MTBF	Mean Time Between Failures
MTTF	Mean Time to Failures
MWCNTs	Multi-Walled Carbon Nanotubes
MWNTs	Multi-Walled Nanotubes
MYC	MYeloCytomatosis
N_2	Nitrogen
NaOH	Sodium hydroxide
NASA	National Aeronautics and Space Administration
NBDDS	Nanoparticle-Based Drug Delivery Systems
NCEs	Novel Chemical Entities
NCT	National Clinical Trial (identifier)
NF	Nanofiltration
NHS	National Health Service (UK)
NMR	Nuclear Magnetic Resonance
NMs	Nanomaterials
NO	Nitric Oxide
NOM	Natural Organic Matter
NPs	Nanoparticles
O_2	Oxygen
P. aeruginosa	*Pseudomonas aeruginosa*
P. amboinicus	*Plectranthus amboinicus*
P. mirabilis	*Proteus mirabilis*
PAH	Polycyclic Aromatic Hydrocarbon
PAHs	Polycyclic Aromatic Hydrocarbons
PAIP	Pathology AI Platform
p-ASA	p-arsanilic acid
PCE	Power Conversion Efficiency
PCM	Polarizable Continuum Model
PCR	Polymerase Chain Reaction
PD-1	Programmed Death-1
PDI	Polydispersity index
PE	Power Electronics
PEEK	Polyether Ether Ketone
PEI	Polyethyleneimine
PeLEDs	Perovskite Light-Emitting Diodes
PET	Polyethylene Terephthalate
pH	Hydrogen ion concentration
PhAs and PAA	Phenylarsonic acid
PI	Polyimide
PLA	Polylactic Acid
PMC	Polymer Matrix Composite
PME	Particle Mesh Ewald
PMP	Paper Manufacturing Plants
PNIPAM	poly-N-isopropylacrylamide

POps	Persistent Organic Pollutants
PP	Polypropylene
ppb	Per Billion
PPIs	Protein-Protein Interactions
PROTACs	Proteolysis targeting chimera
PRS	Polygenic Risk Scores
PS	Pulping Systems
PSCs	Perovskite Solar Cells
PXRD	Powder X-Ray Diffraction
QACs	Quaternary Ammonium Compounds
QBL	Quadruple Bottom Line
QCancer 10	Cancer Risk Prediction Model (UK)
QDs	Quantum Dots
QM	Quantum Mechanics
QSAR	Quantitative Structure Activity Relationship
QSPR	Quantitative Structure Property Relationship
QuantCRC	Quantitative Colorectal Cancer Algorithm (Mayo Clinic)
QUV	Accelerated Weathering Tester (used for UV and weather resistance testing)
RAM	Reliability, Availability, and Maintainability
RAMAN	Raman spectroscopy
RAMD	Reliability, Availability, Maintainability and Dependability
rCDI	Recurrent *Clostridium difficile* infection
RCSB	Research Collaboratory for Structural Bioinformatics
RDF	Radial Distribution Function
REACH	Registration, Evaluation, Authorisation and Restriction of Chemicals
rGO	Reduced Graphene Oxide
RL	Reinforcement Learning
RMSD	Root Mean Square Deviation
RMSF	Root Mean Square Fluctuation
RNNs	Recurrent Neural Networks
RO	Reverse Osmosis
ROS	Reactive Oxygen Species
ROX	Roxarsone
RP phase	Ruddlesden–Popper phase
RPE	Retinal Pigment Epithelium
RTM	Resin Transfer Molding
S	Chemical Softness
S. aureus	*Staphylococcus aureus*
S. epidermidis	*Staphylococcus epidermidis*
S. pneumonia	*Streptococcus pneumoniae*
S. typhi	*Salmonella typhi*
SAD	Selected Area Diffraction

SAHA	Suberoylanilide hydroxamic acid
SAR	Structure-Activity Relationship
SAXS	Small-Angle X-ray Scattering
SBDD	Structure-Based Drug Design
SDG	Sustainable Development Goals (also listed as SDG1 … SDG17 individually)
SEM	Scanning Electron Microscopy
SEM-EDX	Scanning Electron Microscopy-Energy Dispersive X-ray Spectroscopy
SiNPs	Silicon Dioxide Nanoparticles
SKCM	Skin Cutaneous Melanoma
SLNs	Solid Lipid Nanoparticles
SMA	Spinal Muscular Atrophy
SMARTLAB	Rigaku SmartLab XRD System
SMILES	Simplified Molecular Input Line Entry System
SOD	Superoxide Dismutase
SPAAC	Strain-promoted azide–alkyne cycloaddition
SPIONs	Superparamagnetic Iron Oxide Nanoparticles
SPR	Surface Plasmon Resonance
SSA	Specific Surface Area
SSR	Solid-State Reaction
STA	Simultaneous Thermal Analysis
STP	Standard Temperature and Pressure
SVM	Support Vector Machine
SWCNT	Single-Walled Carbon Nanotubes
SWNTs	Single-Walled Nanotubes
TAD	Transactivation Domain
TCE	Trichloroethylene
TCGA	The Cancer Genome Atlas
TCO	Tetrazine–Trans-Cyclooctene
TEM	Transmission Electron Microscopy
TFE3	Transcription Factor E3
TFEB	Transcription Factor EB
TFEC	Transcription Factor EC
TGA	Thermogravimetric analysis
TILs	Tumor-Infiltrating Lymphocytes
TiO_2	Titanium Dioxide
TMSS	Topochemical Molten Salt Synthesis
TNFR1	Tumor Necrosis Factor Receptor 1
TNT	Trinitrotoluene
TOC	Total organic carbon analyzer
TOC	Total organic content
TOF-SIMS	Time-of-flight secondary ion mass spectrometry
TP53	Tumor Protein p53
TPH	Total Petroleum Hydrocarbon

TPMT	Thiopurine Methyltransferase
TRP-1	Tyrosine-Related Protein
TSCA	Toxic Substances Control Act
UHMWPE	Ultra-High-Molecular-Weight Polyethylene
UHPC	Ultra-High-Performance Concrete
UK Biobank	United Kingdom Biobank
UNEP	United Nations Environment Programme
UPLC	Ultra-performance liquid chromatography
UPLC-QTOF-MS	Ultraperformance liquid chromatography quadrupole time-of-flight mass spectrometry
USY	Ultra Stable Y-type (zeolite catalyst)
UV	Ultraviolet
UV-A	Ultraviolet A (320–400 nm)
UV-B	Ultraviolet B (280–320 nm)
UV-C	Ultraviolet C (200–280 nm)
UVM	Uveal Melanoma
UV-Vis	Ultraviolet–visible spectroscopy
UV-VIS-NIR	Ultraviolet Visible Near Infrared
VAEs	Variational Autoencoders
VMD	Visual Molecular Dynamics
VOC	Volatile Organic Compound
VOCs	Volatile Organic Compounds
VSM	Vibrating sample magnetometer
WDS	Wavelength-Dispersive X-ray Spectroscopy
WFO	Watson for Oncology
WHC	Water Holding Capacity
WHO	World Health Organization
Wnt/β	Wingless-related integration site/Beta
WSA2	Waardenburg type 2A
WSI	Whole Slide Image
XGBoost	Extreme Gradient Boosting
XPS	X-Ray Photoelectron Spectroscopy
XRD	X-Ray Diffraction
ZA43	Zinc-Aluminum Alloy (ZA-43)
ZIF	Zeolitic Imidazolate Framework
ZnO NPs	Zinc Oxide Nanoparticles
ZnO	Zinc(II) oxide
ZVI	Zero Valent Iron
α-CD	Alpha-Cyclodextrin
α-Fe_2O_3	Hematite
β-CD	Beta-Cyclodextrin
γ-CD	Gamma-Cyclodextrin
γ-Fe_2O_3	Maghemite
ΔE	Energy Gap (LUMO—HOMO)
η	Chemical Hardness

μ	Chemical Potential
χ	Electronegativity
ω	Electrophilicity Index

Chemical Groups (Functional Groups)

–C=O–	Carbonyl group
–CF, $–CF_2$, $–CF_3$	Fluorocarbon groups
–CN–	Cyano group
–COOH–	Carboxyl group
$–NH_2–$	Amine group
–NO–	Nitroso group
–OH–	Hydroxyl group
^{14}C	Carbon-14 isotope

Advanced Materials for Environmental Applications

Adsorptive Removal and Degradation of *p*-Arsanilic Acid Using Magnetite Nanoparticles and Fenton Reagent: A Sustainable Approach

Akanksha Pandey, Shruti Patel, Debbie Ndlovu, Diya Patel, and Hemen Dave

Abstract *P*-arsanilic acid (*p*-ASA), an organoarsenic compound widely used as a veterinary feed additive in poultry and swine industries, is an emerging pollutant and one of the potential sources of arsenic in the environment. This study opens up a novel and sustainable approach for the adsorption, desorption, and Fenton reagent-based subsequent degradation of *p*-arsanilic acid, using green magnetite nanoparticles (Fe_3O_4 NPs) as an adsorbent. An aqueous extract prepared from the leaves of milkweed (*Calotropis procera*) plant was used to produce magnetic Fe_3O_4 NPs. The properties of nanoparticles were further studied using advanced characterization tools, including Zeta Potential Analysis, Fourier-Transform Infrared Spectroscopy, Scanning Electron Microscopy-Energy Dispersive X-ray Spectroscopy and X-ray diffraction. Adsorption of *p*-ASA from aqueous solutions on the surface of Fe_3O_4 NPs was studied and optimized. Being magnetic, the Fe_3O_4 NPs could be easily separated from the aqueous solution after the adsorption and mixed with distilled water for desorbing *p*-ASA into the wash solution, which was further degraded using Fenton's reagent. Furthermore, the Fe_3O_4 NPs, after desorption, retained their adsorption efficiency, enabling them to be recycled multiple times for *p*-ASA's adsorption and desorption. The green Fe_3O_4 NPs exhibited an adsorption efficiency of 76%, even during the eighth cycle of adsorption. Thus, the study presents a sustainable and recyclable nanoparticle-based system for eliminating *p*-ASA for effective treatment and purification of water.

Keywords Aromatic organoarsenic compounds · *p*-arsanilic acid · Magnetite nanoparticles · Fenton treatment · Arsenic

A. Pandey · S. Patel
School of Doctoral Studies and Research, National Forensic Sciences University, Gandhinagar, Gujarat, India

D. Ndlovu · D. Patel · H. Dave (✉)
School of Pharmacy, National Forensic Sciences University, Gandhinagar, Gujarat, India
e-mail: hemen.dave@nfsu.ac.in

L. Ledwani and S. Srivastava (eds.), *Experimental and Computational Advances in Materials*, Engineering Materials, https://doi.org/10.1007/978-981-95-4180-5_1

1 Introduction

Emerging environmental contaminants pose a high risk to the aquatic ecosystems, human health, as well as general environmental quality. AOCs, like *p*-ASA, are one of these contaminants that have gained attention because of their excessive use in the swine and poultry industries in the form of veterinary feed additives [1]. *p*-ASA eventually gets released into the environment via animal excrement, contaminated water, and agricultural soils [2]. This contamination is a possible source of non-geogenic inorganic arsenic, a known toxic and carcinogenic element [3]. Thus, developing sustainable degradation/removal techniques to eliminate and degrade *p*-ASA from soil and water is the need of the hour [4]. Traditional methods of eliminating organoarsenic contaminants, such as filtering, coagulation, and precipitation, frequently have demerits in terms of cost, effectiveness, and environmental sustainability [5].

Multiple research works have been carried out regarding the removal and degradation of *p*-ASA [6]. For instance, Yang et al. applied AOPs for degrading *p*-ASA, which turned out to be highly efficient in mineralizing the compound under optimal conditions [7]. In a study, biochar produced through valorized agricultural waste was used as an adsorbent for removing *p*-ASA, showing this material's potential for cost-effective and sustainable treatment [8]. Another study by Vega et al. deals with the degradation (photocatalytic) of *p*-ASA using titanium dioxide NPs that exhibited high efficiency in degrading the contaminant under visible light [9]. In recent times, nanotechnology has emerged as an appealing alternative owing to the various physicochemical characteristics of NPs, including their substantial surface area, reactivity, as well as magnetic attributes [10–12]. Among the various nanomaterials used for environmental remediation, NPs of iron oxides have drawn a lot of attraction because of their easy availability, low toxicity, and versatility. Maghemite (γ-Fe_2O_3), hematite (α-Fe_2O_3), and magnetite (Fe_3O_4) are the most studied phases; their unique physicochemical characteristics affect their efficiency in removing myriad pollutants [13]. γ-Fe_2O_3 consists entirely of Fe^{3+} ions. It offers good stability under oxidative conditions and has demonstrated effective adsorption of pollutants such as dyes and heavy metals [14, 15]. α-Fe_2O_3 has been used as a photoctalyst for degrading dye [16, 17]. Fe_3O_4 is the iron oxide phase that performs best in removing pollutants because of its strong redox activity and magnetic recoverability [18].

The current study has investigated the utilization of green Fe_3O_4 NPs as a new and sustainable method for the degradative removal of *p*-ASA via adsorption and degradation [19]. An aqueous extract prepared from the leaves of *Calotropis procera* (milkweed) was used as a natural reducing and capping agent for the green synthesis, in contrast to conventional and hazardous chemical synthesis methods [20]. The plant *Calotropis procera* is renowned for its ability to grow in challenging environmental conditions of arid and semi-arid areas. Its medicinal and phytochemical properties have been acknowledged for a long time because of the presence of phenolic chemicals, flavonoids, and alkaloids. These bioactive ingredients work well as natural stabilizing and reducing agents in green synthesis procedures. Using *Calotropis procera* as a renewable and biodegradable resource for material synthesis is coherent with green chemistry concepts [20].

A sustainable and recyclable nanoparticle-based system for the degradation and removal of *p*-ASA has been presented in this work, addressing a primary environmental concern. Unlike most previous research works, which were mainly concentrated on the single-stage removal processes or relied on chemically synthesized NPs that may be environmentally concerning, the present study incorporates a multi-step approach by merging adsorption, desorption, and degradation. This ensures the effective removal of *p*-ASA along with dealing with its degradation products. Moreover, the use of green Fe_3O_4 NPs makes the process environmentally friendly and sustainable. This environmentally friendly synthesis route strengthens the biocompatibility of the NPs, reducing the adverse effects on the environment [21]. The study examined the adsorption of *p*-ASA onto the Fe_3O_4 NPs from aqueous solutions with optimization of process parameters. The magnetic nature of the NPs enabled their easy separation from the solution post adsorption [22]. Fenton's reagent was subsequently utilized to degrade the concentrate of *p*-ASA which desorbed from the nanoparticles [23]. This two-step adsorption–desorption procedure provides a sustainable and inexpensive method to remove *p*-ASA [24]. Notably, the study demonstrated that even after multiple cycles of adsorption and desorption, the NPs maintained their adsorption efficiency. The system proved to be economically feasible for large-scale applications because of its reusability, which significantly lowers the operating expenses related to adsorbent substitution [25]. Furthermore, amalgamation of green synthesis with advanced oxidation processes such as Fenton like reactions, photocatalysis or ozonation can produce hybrid treatment systems that tackle a wide range of pollutants in complicated matrices.

By incorporating the adsorption, desorption, and degradation processes, the approach provides a reliable and eco-friendly alternative for removing organoarsenic pollutants. This study emphasizes combining cutting-edge material science with green chemistry concepts to successfully and sustainably solve urgent environmental issues.

2 Experimental Section

2.1 Materials

p-ASA ($C_6H_8AsNO_3$, 98%), Ferric Chloride Hexahydrate ($FeCl_3.6H_2O$, 98%), Hydrogen Peroxide (H_2O_2, 30%), Sulphuric acid (H_2SO_4, 98%) were purchased from Central Drug House (P) Ltd. India. Sodium hydroxide ($NaOH$, 97%), Ferrous chloride hydrate ($FeCl_2.H_2O$, 98%), Ferrous Sulphate Heptahydrate ($FeSO_4.7H_2O$, 99.5%), Acetone (C_3H_6O, 99.9%) were purchased from Sisco Research Laboratories Pvt. Ltd. India. The concentration of the *p*-ASA stock solution was 1000 ppm.

Fig. 1 Image showing the stages from the collection of leaves to the preparation of *Calotropis procera* leaf extract

2.2 *Synthesis of Green Fe_3O_4 NPs*

2.2.1 Preparation of *Calotropis Procera* Leaf Extract

Calotropis leaves were collected, rinsed with tap and distilled water, and then allowed to air dry for ten days. The dried leaves were then crushed into fine powder and preserved in an airtight container for further use. To prepare the leaf extract, 100 mL of distilled water was added to a round-bottom flask containing 5 g of the powdered leaves, and heating was provided for 15 min at 60 °C. It was allowed to cool and centrifuged at 6000 RPM for 10 min. The supernatant, upon centrifugation, was filtered on Whatman No. 1 filter paper to remove residual solids. The filtered leaf extract was preserved in a refrigerator at 4 °C for further use [26] (Fig. 1).

2.2.2 Preparation of Green Fe_3O_4 NPs

In the present study, Fe_3O_4 NPs were synthesized by a green synthesis procedure using ferric chloride hexahydrate ($FeCl_3 \cdot 6H_2O$) and ferrous chloride tetrahydrate ($FeCl_2 \cdot 4H_2O$) in a 2:1 molar ratio. Salts were dissolved in 100 mL of distilled water. The mixture was heated to 80 °C with gentle stirring for 10 min. After removing the heating source, an aqueous leaf extract of Calotropis was added to this, followed by agitation for 10 min. Afterwards, 1N NaOH was added at a controlled rate, which

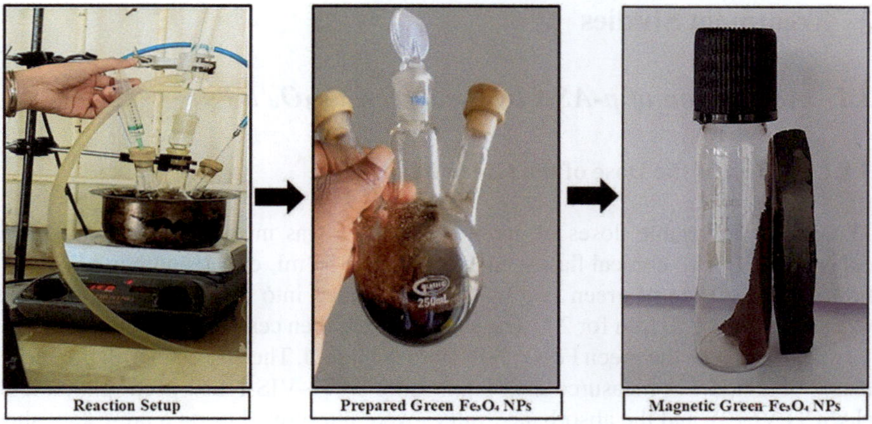

Fig. 2 Synthesis of green Fe_3O_4 NPs

allowed green Fe_3O_4 NPs to precipitate out. The mixture was continuously stirred for 30 min, followed by cooling to room temperature. The precipitate obtained was separated via centrifugation for 15 min at 7000 rpm. Further, it was washed with absolute acetone and distilled water for complete purification. Then, the NPs were kept in the oven at 100 °C for 24 h followed by storage in an airtight container for further use [27] (Fig. 2).

2.3 *Characterization of Green Fe_3O_4 NPs*

Zeta sizer (Zeta Sizer Zen3600, Malvern) was used for characterization of the particle size, zeta potential and polydispersity index (PDI) of Fe_3O_4 NPs. SEM–EDX (Zeiss Gemini 300) assisted in studying the morphology, size, and surface texture of the green Fe_3O_4 NPs. The SEM images of green Fe_3O_4 NPs before and after adsorption of *p-ASA* were collected at different magnifications (463 × and7.2 KX) and (576 × and 5.45 KX), respectively. ATR-FTIR (JASCO 4700) was employed to identify the functional moieties and chemical bonds in the NPs [28]. Spectral data were collected over a range of 500 cm^{-1} to 4000 cm^{-1} (32 Scans) at a resolution of 4 cm^{-1}. XRD (D8 Discover, Bruker) was used to analyze (Scanning range: 10°–100°) the crystalline structure and phase composition of the Fe_3O_4 NPs.

3 Treatment Studies

3.1 *Adsorption of p-ASA by the Green Fe_3O_4 NPs*

3.1.1 Effect of the Dose of the Green Fe_3O_4 NPs

The effect of variable doses of green Fe_3O_4 NPs was investigated using a batch experiment. In two conical flasks, each containing 50 mL of a 10-ppm *p*-ASA solution, 10 and 50 mg of green Fe_3O_4 NPs were added into the solutions. The flasks were agitated at 250 rpm for 2 h. The samples were then centrifuged at 5000 rpm for 10 min to separate the green Fe_3O_4 NPs from the liquid. The absorbance of the supernatant obtained was measured at 250 nm using a UV–VIS-NIR Spectrophotometer (JASCO V670), and the absorbance values were interpreted using a pre-established calibration curve to find the final concentration of *p*-ASA. The percentage removal of *p*-ASA was determined using the following formula:

$$\%Removal = \left(\frac{CO-Ce}{CO}\right) * 100 \qquad (1)$$

where,

C_0 = initial concentration of *p*-ASA.
C_e = final concentration of *p*-ASA.

3.1.2 Effect of pH

Four conical flasks were prepared for the batch experiments, each containing 50 mL of a 10-ppm *p*-ASA solution. The pH of the solutions was adjusted to pH 4 and 6 using 0.1 M HCl. In the two flasks, 10 mg of green Fe_3O_4 NPs were added, while in the other two, 50 mg of NPs were added. The flasks were agitated at 250 rpm for 2 h. After 2 h, the samples were centrifuged at 5000 rpm for 10 min. The resulting supernatant was analyzed using a UV–Vis-NIR spectrophotometer. Absorbance values obtained were interpreted with a pre-established calibration curve to determine the final concentration of *p*-ASA in the solution. The percentage removal of *p*-ASA was then calculated as per Eq. 1.

3.2 *Degradation of p-ASA Through Fenton Reagent*

In the Fenton reaction, the catalytic reaction between H_2O_2 and $Fe2^+$ ions at the acidic pH forms hydroxyl (•OH) radicals. The hydroxyl radicals possess high oxidative potential and can degrade almost all kinds of organic pollutants into innocuous end products like H_2O, CO_2, and simple inorganic salts [29]. The fundamental mechanism is represented as follows:

$$Fe^{2+} + H_2O_2 \rightarrow Fe^{3+} \bullet OH + OH^-$$

$$Fe^{3+} + H_2O_2 \rightarrow Fe^{2+} + HOO\bullet + H^+$$

$$2H_2O_2 \rightarrow \bullet OH + HOO\bullet + H_2O$$

Fe^{2+} serves as a catalyst in this process, generating hydroxyl radicals that target the $-AsO(OH)_2$ and $-NH_2$ functional groups of *p*-arsanilic acid, causing it to oxidatively degrade and transform into *p*-aminophenol and arsenite. Arsenite is immediately oxidized to arsenate. Oxidation of *p*- aminophenol leads to formation of intermediate products which eventually converted into carbon dioxide, water and ammonia (NH_4^+). The generated Fe^{3+} can be reduced back to Fe^{+2} by cyclic reactions to maintain the catalytic cycle [30, 31].

Experiments were performed using 50 ml of *p*-ASA in a concentration range of 2–10 ppm with the Fenton Reagent in a molar ratio of Fe^{2+} to H_2O_2 as 1:5 and 1:10 to study and optimize the degradation of *p*-ASA by the Fenton reagent. Degradation of *p*-ASA was studied using UV–visible Spectroscopy, and the release of inorganic arsenic was qualitatively analyzed by the molybdate-based method described by [32]. In this method, the oxidizing solution contains potassium iodate, while the reducing solution is a mixture of sodium metabisulfite, sodium thiosulfate, and sulfuric acid in the ratio of 2:2:1. The color reagent consists of ascorbic acid, potassium antimonyl tartrate, ammonium molybdate, and sulfuric acid in the ratio of 2:2:1:5. The procedure involves the formation of arsenomolybdate complex (yellow colored), which is afterwards reduced to molybdenum blue. The development of the blue coloration indicates the release of arsenate from the degradation of *p*-ASA in the solution. There is no possibility of phosphate interference since we used only synthetic solution of *p*-ASA and Fenton Reagent. All the solutions were prepared using double distilled water.

The experiments found that the molar ratio of $Fe2^+$ to H_2O_2 of 1:5 is more efficient and hence used for the degradation of *p*-ASA desorbed from the magnetic Fe_3O_4 NPs [30]. The chosen ratio could ensure sufficient hydrogen peroxide for the effective generation of hydroxyl radicals to enhance the efficiency of *p*-ASA degradation.

3.3 Recovery of Green Fe_3O_4 NPs

The Fe_3O_4 NPs, being magnetic, can be retrieved easily from the aqueous solution after the adsorption cycle. The *p*-ASA was able to desorb into the wash solution when the nanoparticles were stirred rigorously with 5 mL of distilled water for 15 min with ultrasonication, and after desorption, the Fe_3O_4 NPs recovered their adsorption ability. The washed Fe_3O_4 NPs were then dried in an oven at a regulated temperature to ensure that all remaining moisture had evaporated. The dried NPs were utilized for additional adsorption cycles, exhibiting their reusability during

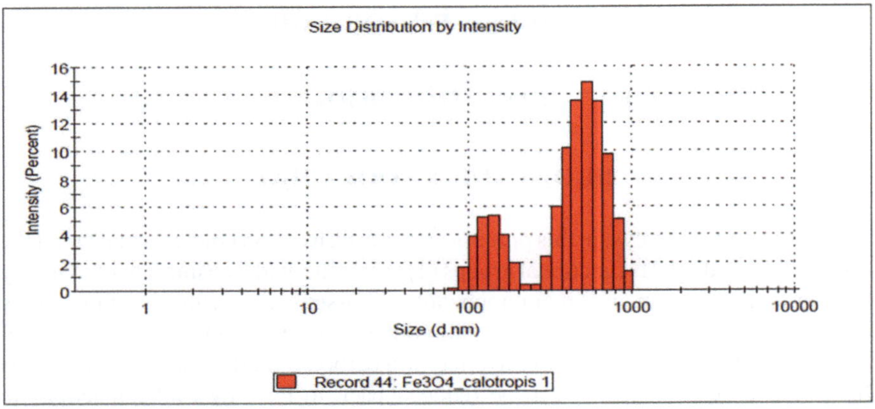

Fig. 3 Hydrodynamic size of the green Fe_3O_4 NPs

eight successive adsorption–desorption cycles. The ultrasonication resulted in almost complete desorption of *p*-ASA into the wash solution. The desorbed *p*-ASA in the wash solution was degraded using Fenton's reagent. For that, wash the solution (5 ml) of each of the adsorption cycles mixed (i.e., 40 ml from 8 cycles) and degraded with the Fenton reagent, the molar ratio of Fe^{2+} to H_2O_2 as 1:5.

4 Results and Discussion

4.1 Characterization

4.1.1 Zetasizer

The size and zeta potential of the green Fe_3O_4 NPs were determined using a zeta sizer. The average size of the green Fe_3O_4 NPs was 610 nm, which was relatively high and might be due to the particle aggregations and inclusion of the hydrated layer. Due to this, the apparent size measured by the zeta sizer is higher than that of the actual one found by the SEM (Fig. 3).

The zeta potential of green Fe_3O_4 NPs was determined by −29.7 mV, showing excellent stability in aqueous solution. The negative value suggests a net negative charge on the surface of green synthesized NPs (Fig. 4).

4.1.2 Attenuated Total Reflectance Fourier-Transform Infrared (ATR-FTIR)

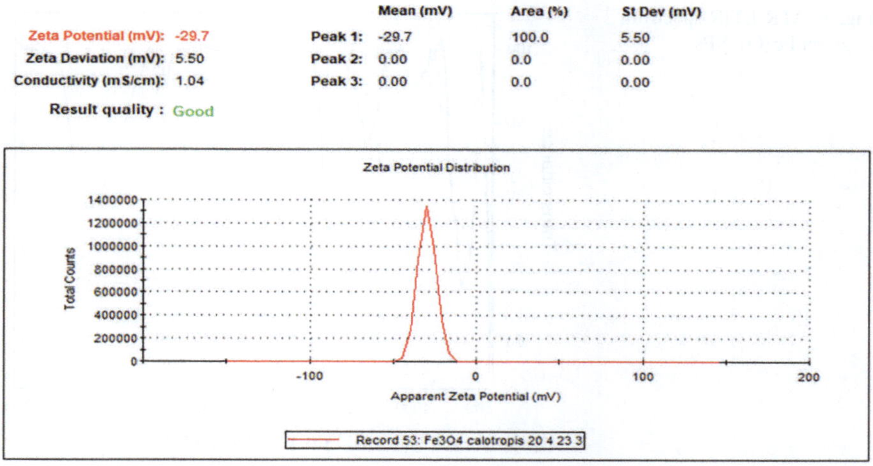

Fig. 4 Zeta potential of the green Fe_3O_4 NPs

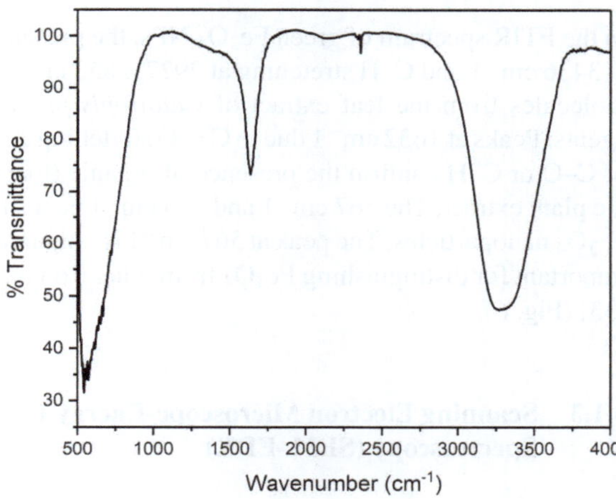

Fig. 5 ATR-FTIR spectrum of *Calotropis procera* leaf extract

ATR-FTIR of *Calotropis Procera* Leaf Extract

The FTIR spectrum of *Calotropis procera* leaf extract is presented in Fig. 5. The observed peaks at 3269 cm^{-1} and 1635 cm^{-1} are due to hydroxyl (OH) and carbonyl ($C = O$) groups, respectively. Such functional moieties indicate towards the phenolic compounds, flavonoids, and other bioactive metabolites present in the leaf extracts [26].

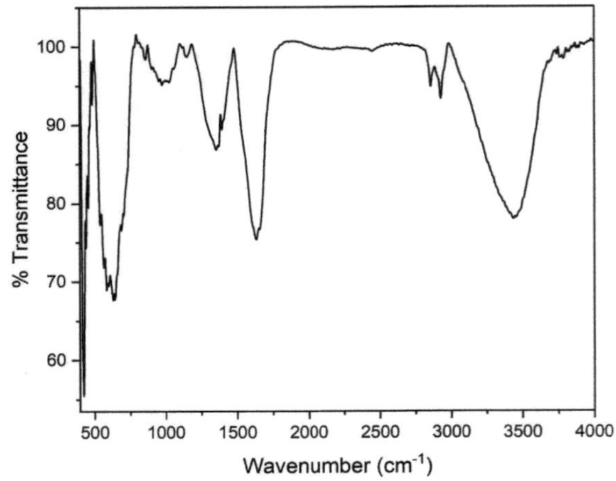

Fig. 6 ATR-FTIR spectrum of green Fe_3O_4 NPs

ATR-FTIR of Green Fe_3O_4 NPs

In the FTIR spectrum of green Fe_3O_4 NPs, the presence of O–H stretching vibration at 3436 cm^{-}1 and C–H stretching at 2927, 2853 cm^{-}1 confirms the role of bioactive molecules from the leaf extract of *Calotropis procera* as capping and stabilizing agents. Peaks at 1632 cm^{-}1 due to C = O stretching and at 1353 cm^{-}1 due to bending of C–O or C–H confirm the presence of organic (like phenolics or flavonoids) from the plant extract. The 567 cm^{-}1 and 423 cm^{-}1 peaks indicate the formation of green Fe_3O_4 nanoparticles. The peak at 567 cm^{-}1 is a signature peak for Fe_3O_4; this peak is important for distinguishing Fe_3O_4 from other iron oxides, such as hematite (Fe_2O_3) [33] (Fig. 6).

4.1.3 Scanning Electron Microscope-Energy Dispersive X-ray Spectroscopy (SEM–EDX)

SEM images of the green Fe_3O_4 NPs (before adsorption) are presented in Fig. 7. The images indicate that the nanoparticles show a relatively homogeneous size distribution with irregular shape and morphology. Aggregation to a certain degree is also observed; this often occurs for magnetic nanoparticles since they have strong magnetic interactions. The green Fe_3O_4 NPs surface shows a smooth surface texture. Figure 9 represents SEM images of green Fe_3O_4 NPs after the adsorption of *p*-ASA. No notable changes or morphological alterations were observed in the SEM Images of green Fe_3O_4 NPs after the adsorption of *p*-ASA. However, elemental composition analysis by EDX before and after the adsorption of *p*-ASA confirms the adsorption of *p*-ASA due to the appearance of arsenic peaks in the post-adsorption spectra. The EDX analysis in green Fe_3O_4 NPs indicated that all pre-adsorption spectra had no arsenic. In contrast, the after-adsorption spectra are consistent, and the presence of as further confirms the sorption of *p*-ASA on the green Fe_3O_4 NPs. Also, trace elements found in the spectrums are Na, Mg, Al, Si, Cl, Ca, and Cr; these will probably be due to the *Calotropis procera* leaf extract or impurities arising in the synthesis process (Figs. 8, 9 and 10; Tables 1 and 2).

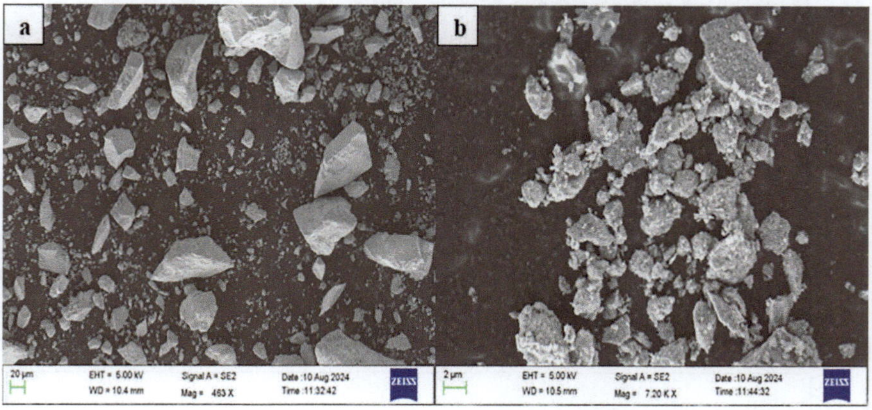

Fig. 7 SEM images of Green Fe_3O_4 NPs before adsorption of *p*-ASA at different magnification

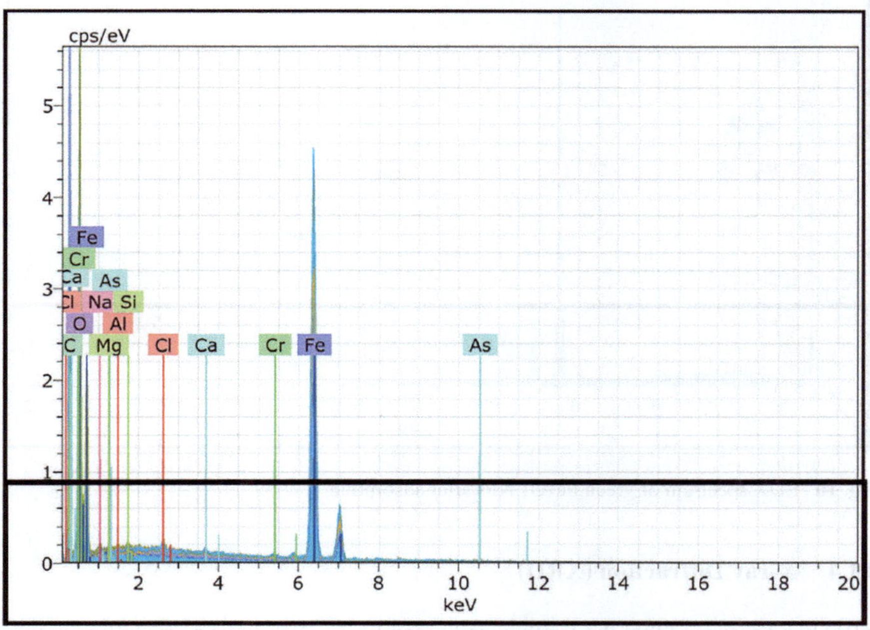

Fig. 8 EDX spectrum of green Fe_3O_4 NPs before adsorption

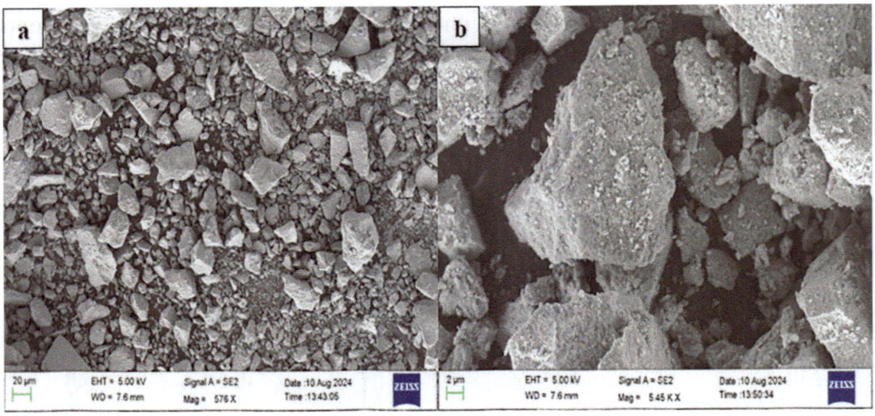

Fig. 9 SEM images of green Fe_3O_4 NPs after adsorption of *p*-ASA at the different magnification

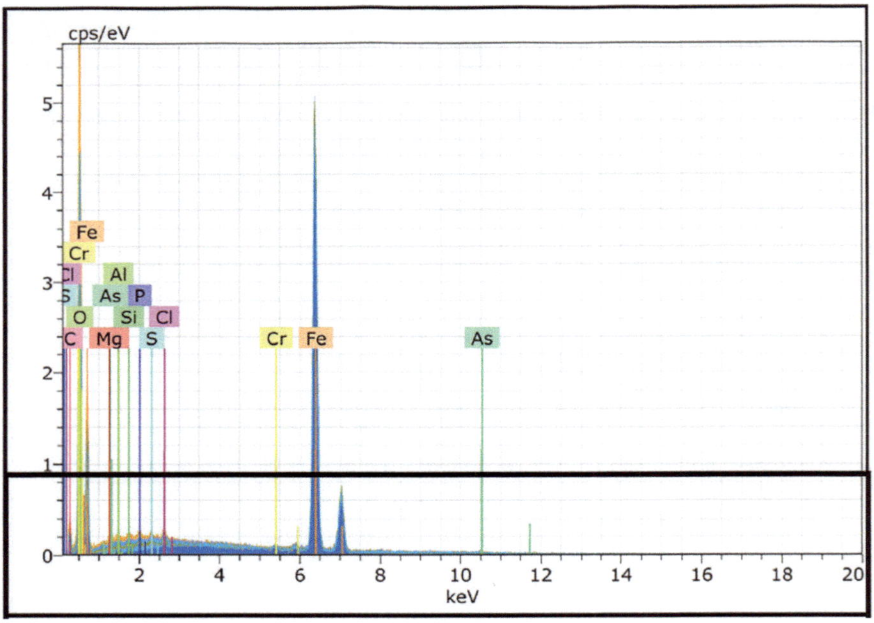

Fig. 10 EDX spectrum of green Fe_3O_4 NPs after adsorption

4.1.4 X-ray Diffraction (XRD)

XRD has confirmed that the synthesized green Fe_3O_4 NPs have a crystalline nature; the emerging patterns are displayed in Fig. 11. Table 3 shows the characteristic peaks and corresponding crystal planes of green Fe_3O_4 NPs.

Table 1 EDX analysis showing the elemental composition (mass%) of the green Fe_3O_4 NPs before the adsorption of *p*-ASA

Spectrum	C	O	Na	Mg	Al	Si	Cl	Ca	Cr	Fe	As
P 2	7.36	32.39	0.34	0.14	0.12	0.02	0.18	0.08	0.17	59.16	0
P 3	19.42	41.92	0.66	0	0.07	0.04	0.13	0.01	0	37.64	0.08
P 4	47.46	36.60	0.34	0	0.05	0.02	0.01	0	0	15.49	0
P 5	23.26	34.36	0.38	0.04	0.03	0.05	0.06	0.00	0.07	41.70	0
P 6	39.97	38.78	0.42	0	0	0.04	0.02	0	0.04	20.70	0

Table 2 EDX analysis showing the elemental composition (mass%) of the green Fe_3O_4 NPs after the adsorption of *p*-ASA

Spectrum	C	O	Na	Mg	Al	Si	Cl	Ca	Cr	Fe	As
P 13	5.31	33.85	0.06	0	0.09	0	0.08	0.08	0.15	59.34	1.00
P 14	3.53	24.83	0.03	0.12	0.08	0.12	0	0.03	0.25	69.87	1.10
P 15	3.74	30.91	0	0	0.08	0.00	0.02	0.12	0.11	64.02	0.95
P 16	6.27	40.62	0.15	0.11	0.06	0.10	0	0.15	0	52.29	0.21
P 17	2.68	8.78	0	0.21	0	0	0	0.17	0.47	86.02	1.65

The XRD pattern confirms that cubic spinel Fe_3O_4 was formed without impurities, such as Fe_2O_3. The characteristic planes for the spinel structure of green Fe_3O_4 are the planes (220), (311), (400), (422), (511), and (440). This phase purity indicates effective green synthesis with proper control of reaction parameters. The crystallite size was calculated using the Debye–Scherrer equation:

$$D = \frac{k\lambda}{\beta \cos\theta}$$

where,

D is the crystallite size (nm),
k is a Sherrer constant (0.9),
λ is the X-ray wavelength ($\lambda = 1.541$ Å),
β is the full width at half maxima (FWHM),
θ is the Braggs diffraction angle corresponding to the lattice plane.

The crystallite sizes range between (~ 66 nm to 120.5 nm), slightly larger than typical dimensions in the nanoscale. Larger sizes are more likely a result of some agglomeration of the particles or the development of crystalline domains in some manner during the synthesis procedure.

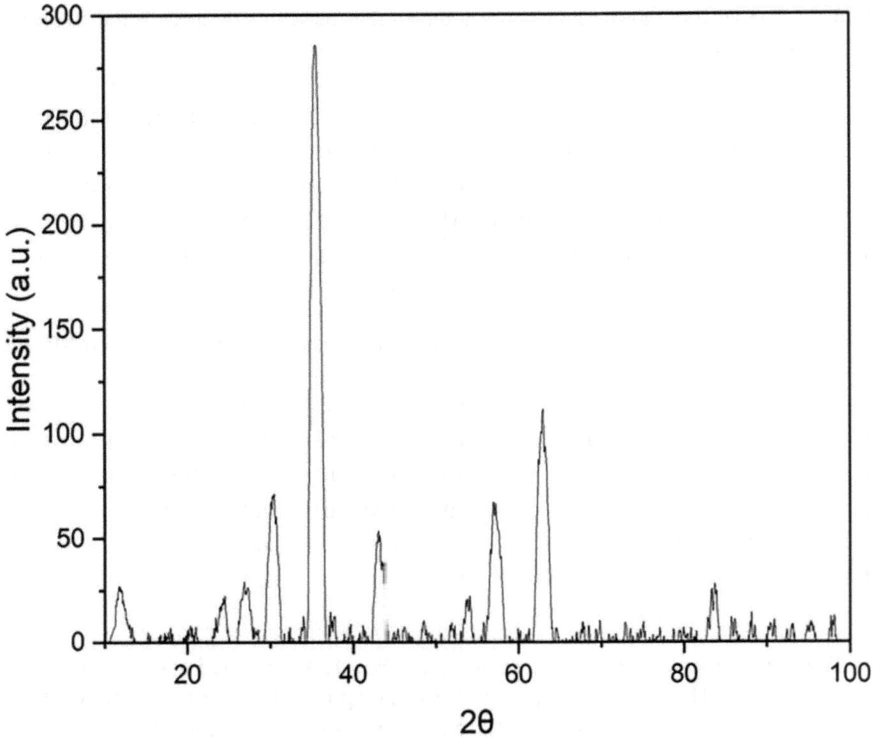

Fig. 11 XRD patterns of green Fe_3O_4 NPs

Table 3 Characteristics peaks and corresponding crystal planes of green Fe_3O_4 NPs

2θ values	Crystal plane
30.30	220
35.52	311
43.13	400
53.88	422
57.20	511
62.90	440

4.2 Removal of p-ASA Using Green Fe_3O_4 NPs

4.2.1 Effect of Adsorbent Dose and pH

Green Fe_3O_4 NPs at two different dosages, i.e., 10 mg and 50 mg in 10 ppm 50 ml *p*-ASA solution, were applied to study the effect of the adsorbent dose. An increase in dosage resulted in an increase in the removal of *p*-ASA. The major influencing

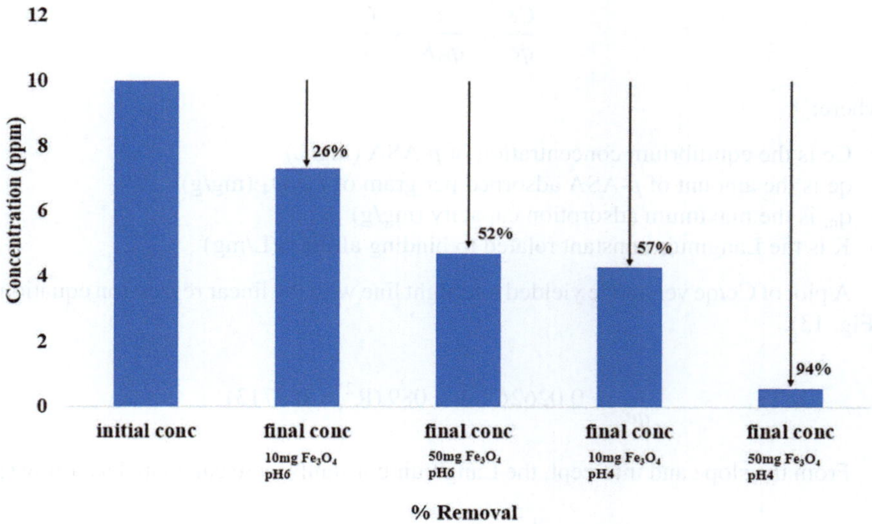

Fig. 12 Effect of dose and pH variations on the removal of *p*-ASA by green Fe_3O_4 NPs

factor for the removal of *p*-ASA was the pH of the solution; p-ASA removal was observed at acidic pH. The iron NPs were utilized to remove *p*-ASA in the pH range of 4–6, and a high adsorption efficiency of 94% was observed at pH 4. In an acidic pH, the presence of H^+ ions accelerates iron corrosion. This increases the rate of iron corrosion, which, in turn, promotes the removal of *p*-ASA by removing the oxide passive layer formed on the surface of the iron, maintaining its reactivity. The OH− ions compete for the adsorption sites on iron at higher pH levels; hence, the removal was not observed in alkaline pH. The removal of *p*-ASA reached 94% at acidic pH4 with a 50 mg dose of green Fe_3O_4, whereas at slightly acidic pH6 and a lower green Fe_3O_4 dose of 10 mg, the removal was just 26%, as shown in Fig. 12.

4.2.2 Adsorption Isotherm Modelling

To gain deeper insight into the adsorption behavior of *p*-ASA on green Fe_3O_4 nanoparticles, several isotherm models namely Langmuir, Freundlich, and Temkin were applied by varying *p*-ASA concentration 10–50 ppm under fixed adsorbent dosage (50 mg), pH (4) and contact time (120 min). Among these, the Langmuir model provided the best fit. This model describes adsorption as a monolayer process occurring on a surface with uniform active sites, assuming no interaction between the adsorbed molecules.

The Langmuir equation is expressed as:

Where:

$$\frac{Ce}{qe} = \frac{1}{q_m K} + \frac{Ce}{q_m}$$

where:

- Ce is the equilibrium concentration of *p*-ASA (mg/L)
- qe is the amount of *p*-ASA adsorbed per gram of Fe_3O_4 (mg/g)
- q_m is the maximum adsorption capacity (mg/g)
- K is the Langmuir constant related to binding affinity (L/mg)

A plot of Ce/qe versus Ce yielded a straight line with the linear regression equation (Fig. 13):

$$\frac{Ce}{qe} = 0.0262Ce + 0.089\,(\mathrm{R}^2 = 0.9713)$$

From the slope and intercept, the Langmuir constants were calculated as follows:

$$(q_m) = \frac{1}{Slope} = \frac{1}{0.0262} = 38.87mg/g$$

$$(K) = Intercept * q_m = 0.306L/mg$$

The high R^2 value (0.9713) indicates an excellent fit to the Langmuir model, confirming that p-ASA adsorption onto Fe_3O_4 nanoparticles follows a monolayer pattern on a surface with uniform energy sites. To further assess the adsorption behaviour, the experimental data were also analyzed using the Freundlich and Temkin

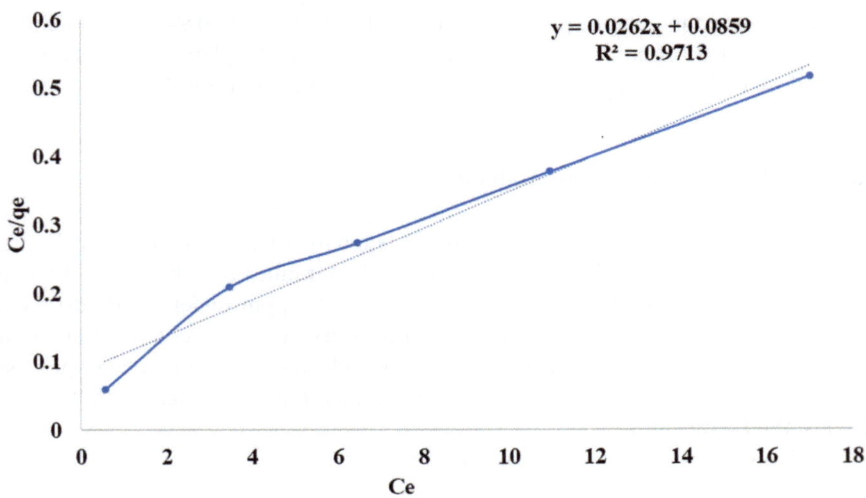

Fig. 13 Langmuir isotherm plot for *p*-ASA adsorption

isotherm models. However, both models produced lower correlation coefficients compared to the Langmuir model, reinforcing Langmuir's suitability for describing this system.

5 Degradation of *p*-ASA by Fenton's Reagent

The *p*-ASA degradation using Fenton's reagent was studied. The hydroxyl radicals generated by the Fenton reaction degraded the *p*-ASA into inorganic arsenic, which was qualitatively detected using a colorimetric method.

5.1 *Standard Curve of p-ASA*

A series of dilutions at concentrations of 2, 4, 6, 8, and 10 ppm was made using a 100 PPM stock solution of *p*-ASA and a standard curve was prepared to calculate the samples' residual concentration of *p*-ASA after the adsorption (Fig. 14).

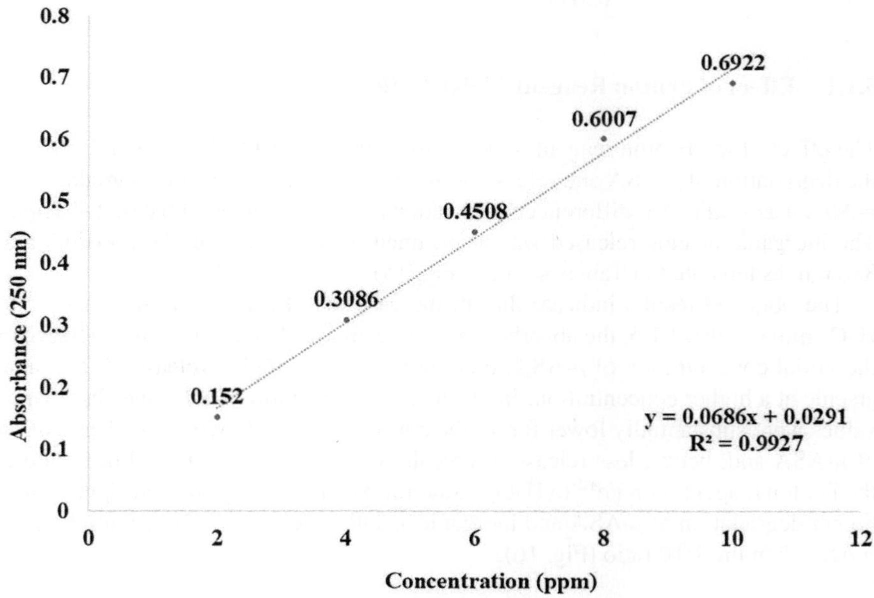

Fig. 14 Standard curve of *p*-ASA

Table 4 The colorimetric detection of inorganic arsenic released after 2 h of treatment with Fenton reagent at 1:5 and 1:10 Fe^{+2} to H_2O_2 ratios

Concentration of *p*-ASA (ppm)	Fenton reagent (Fe^{+2} to H_2O_2 ratio 1:5)	Fenton reagent (Fe^{+2} to H_2O_2 ratio 1:10)
2	0.1062	0.0246
4	0.2265	0.0211
6	0.4107	0.0185
8	0.5565	0.0264
10	0.585	0.0309

Table 5 The colorimetric detection of inorganic arsenic released after 4 h of treatment with Fenton reagent at 1:5 and 1:10 Fe^{+2} to H_2O_2 ratios

Concentration of *p*-ASA (ppm)	Fenton reagent (Fe^{+2} to H_2O_2 ratio 1:5)	Fenton reagent (Fe^{+2} to H_2O_2 ratio 1:10)
2	0.7474	0.1145
4	0.7729	0.1266
6	0.9046	0.1133
8	0.9904	0.1474
10	0.9973	0.1299

5.1.1 Effect of Fenton Reagent Molar Ratio

The effect of the Fenton reagent molar ratio of Fe^{2+} to H_2O_2, i.e., 1:5 and 1:10, on the degradation of *p*-ASA and release of inorganic arsenic during the degradation of *p*-ASA was studied for different concentrations of *p*-ASA in the range of 2–10 ppm. The inorganic arsenic released was determined with the help of UV absorption at 880 nm, as tabulated in Tables 4 and 5 (Fig. 15).

The obtained results indicate that in the case of a Fenton dose with a Fe^{+2} to H_2O_2 molar ratio of 1:5, the absorbance value increases linearly with the increase in the initial concentration of *p*-ASA, thereby reflecting a higher release of inorganic arsenic at a higher concentration. In the case of a dose ratio of 1:10, the absorbance values were substantially lower for all the concentrations, showing less degradation of *p*-ASA and, hence, less release of arsenic under these conditions. Furthermore, the Fenton reagent with Fe^{+2} to H_2O_2 molar ratio of 1:5 clearly showed significantly higher degradation of *p*-ASA and inorganic arsenic release for all concentrations of *p*-ASA than the 1:10 ratio (Fig. 16).

5.1.2 Degradation of *p*-ASA as a Function of Concentration and Time

Degradation of *p*-ASA in the 2–10 ppm concentration range was carried out with the Fenton reagent at 1:5 Fe^{+2}: H_2O_2 molar ratios. Experiments were conducted for

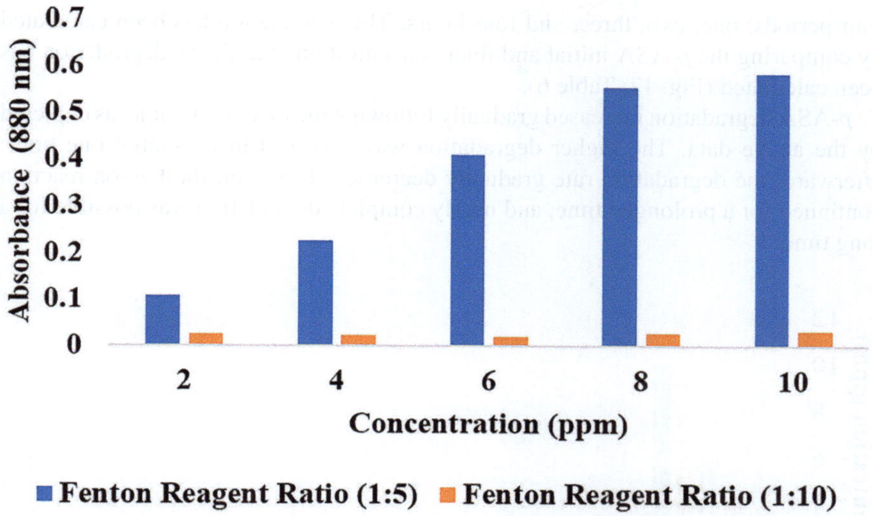

Fig. 15 Colorimetry detection of inorganic arsenic released in 2 h with Fenton reagent at 1:5 and 1:10 molar ratio of Fe^{+2} to H_2O_2

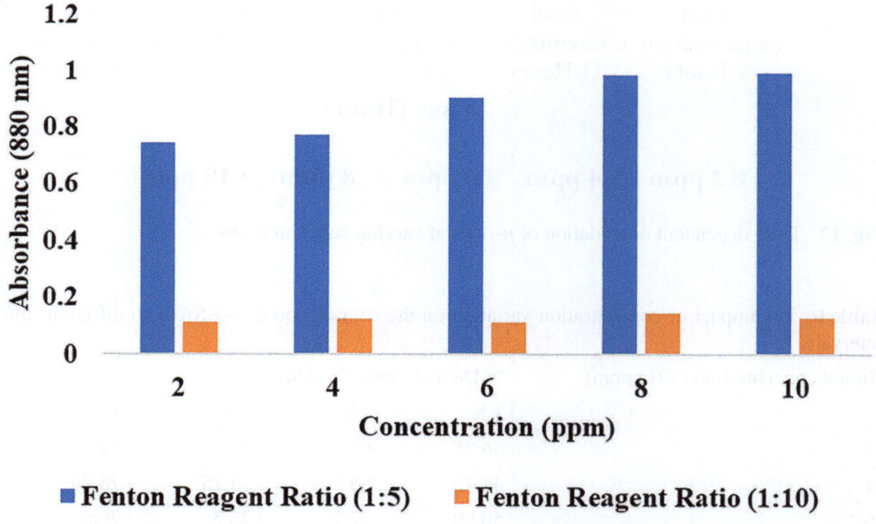

Fig. 16 Colorimetry detection of inorganic arsenic released in 4 h with Fenton reagent at 1:5 and 1:10 molar ratio of Fe^{+2} to H_2O_2

four periods: one, two, three, and four hours. The degradation has been calculated by comparing the *p*-ASA initial and final concentration, and the % degradation has been calculated (Fig. 17; Table 6).

p-ASA degradation increased gradually following the increase in time, as depicted by the above data. The higher degradation was observed in the initial one hour; afterward, the degradation rate gradually decreased. However, the Fenton reaction continued for a prolonged time, and nearly complete degradation was possible for a long time.

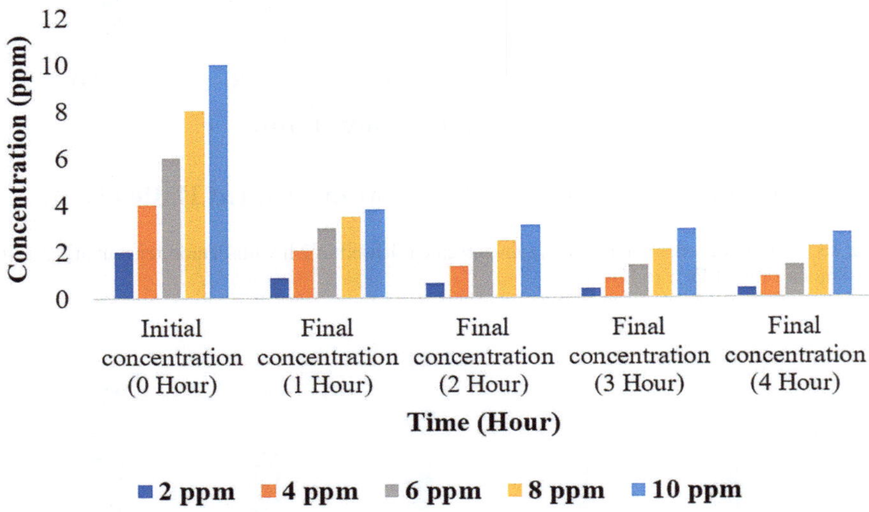

Fig. 17 Time-dependent degradation of *p*-ASA at varying concentrations

Table 6 The impact of concentration variation on the degradation of *p*-ASA with different time intervals

Initial concentration of (in ppm)	% Degradation of *p*-ASA			
	1 h	2 h	3 h	4 h
2	56.75	68.5	79	79.5
4	48.75	66.5	79.25	78.25
6	50.00	67.5	76.5	76.5
8	56.75	69.5	74.75	73
10	62.00	68.9	71	72.4

5.1.3 Kinetic Modelling and Rate Constant Determination

To understand how the analyte breaks down over time, the concentration–time data were analyzed using integrated rate laws for zero-order, first-order, and second-order reaction models. The results showed that the experimental data aligned best with the first-order kinetic model. This was supported by the strong linear relationship in the plot of ln C versus time (t), where *C* represents the concentration (mg/L) and *t* represents time (h).

The first-order integrated rate law is expressed as:

$$\ln C = \ln C_0 - kt$$

where:

- C_0 is the initial concentration,
- C is the final concentration,
- k is the first-order rate constant (h^{-1}),
- t is the reaction time (h).

Linear regression analysis of lnC versus time yielded a slope of − 0.0871, corresponding to a rate constant $k = 0.0871\ h^{-1}$. This kinetic behaviour indicates that the degradation rate is proportional to the concentration of the compound, consistent with typical first-order reaction mechanisms observed in similar environmental degradation studies. The calculated rate constant provides a fundamental parameter for understanding the temporal behaviour of the compound under the tested conditions.

5.1.4 Colorimetry Detection of Inorganic Arsenic

To determine the inorganic arsenic released by the degradation of *p*-ASA, samples of *p*-ASA were prepared, having different concentrations of *p*-ASA, i.e., 2–10 ppm, and each sample was taken in a volumetric flask. All the samples were treated with an identical dosage of the Fenton reagent at a fixed dosage rate of 1:5. After treatment at room temperature for 4 and 24 h, the UV absorbance measurements were made. The amount of inorganic arsenic released during *p*-ASA degradation was monitored by colorimetry at 880 nm and limit of detection for the colorimetric determination of inorganic arsenic using the molybdenum blue method was 10 μg/L. Recorded absorbance, which is proportional to the inorganic arsenic concentration, was noted after 4 h and 24 h. Results are shown in Fig. 18; Table 7.

From the above data, it can be seen that the absorbance values increased linearly with the increasing initial concentration of *p*-ASA, from 0.7474 (at 2 ppm) to 0.9973 (at 10 ppm). A positive correlation was obtained between the initial concentration and the absorbance, meaning that the higher the *p*-ASA concentration, the greater the release of inorganic arsenic within the first 4 h. The absorbance for the 24-h readings was greater than that for the 4-h readings and ranged from 0.9580 at 2 ppm to 1.1739

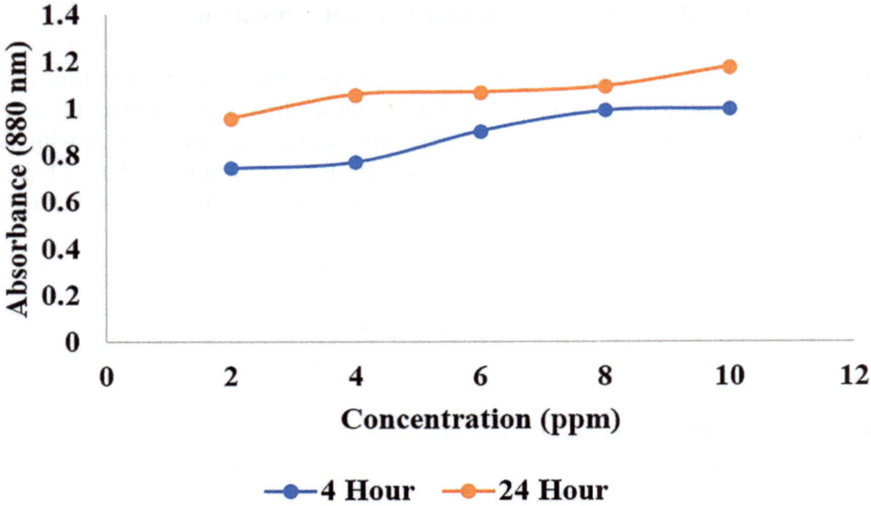

Fig. 18 Colourimetry detection of inorganic arsenic released in 4 and 24 h

Table 7 Shows the colorimetry detection of inorganic arsenic released in 4 and 24 h

Concentration of *p*-ASA (ppm)	Absorbance (880 nm)	
	4 h	24 h
2	0.7474	0.9580
4	0.7729	1.0593
6	0.9046	1.0687
8	0.9904	1.0911
10	0.9973	1.1739

at 10 ppm. In the same manner as the 4-h interval, there was a positive relationship between the initial *p*-ASA concentration and absorbance.

5.2 Reusability of Green Fe_3O_4 NPs

To evaluate their reusability, the Fe_3O_4 NPs were rigorously washed with 5 mL of distilled water after each cycle in a sonication bath, which facilitated the desorption of most of the adsorbed *p*-ASA. These washed NPs are subsequently reused in the next adsorption cycle. Over eight cycles, a total of 40 mL of desorbed *p*-ASA solution was collected. This *p*-ASA solution was then subjected to degradation using an optimized dose of Fenton reagent (1:5), which achieved a 94% degradation of *p*-ASA along with the release of inorganic arsenic.

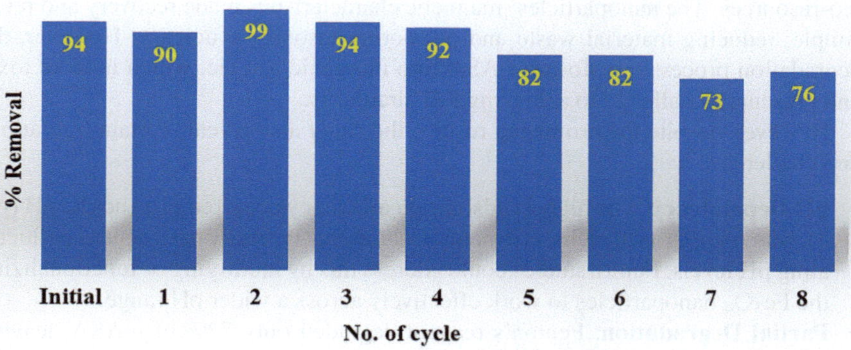

Fig. 19 Percentage removal showing the reusability of green Fe_3O_4 NPs

The percentage removal efficiency of *p*-ASA over eight cycles of adsorption–desorption is illustrated. During the first cycle, the green Fe_3O_4 nanoparticles exhibited the highest removal efficiency of 94%, demonstrating strong adsorption capacity. However, as the cycles progressed, the efficiency gradually declined. By the 5th cycle, the efficiency had decreased to approximately 85–88%. By the 8th cycle, the efficiency stabilized at around 70–75%, suggesting a moderate reduction due to possible saturation of adsorption sites or structural degradation of the nanoparticles (Fig. 19).

The green Fe_3O_4 NPs synthesized using *Calotropis procera* leaf extract proved highly effective for pollutant removal and remained functional over multiple uses. This data highlights the material's potential for sustainable wastewater treatment, offering reduced operational costs and minimal environmental impact.

6 Conclusion

Here, we report a green, efficient, and sustainable approach for the removal and degradation of the emerging organoarsenic contaminant p-arsanilic acid (p-ASA) by employing green-synthesized magnetite nanoparticles (Fe_3O_4 NPs) along with Fenton's reagent. The major outcome validates the fact that the green Fe_3O_4 NPs exhibited the optimal adsorption efficiency of 94% under acidic pH (pH 4) values, with high removal efficiency of 76% even after reuse for eight times, suggesting good reusability along with the feasibility of low-cost operation. Moreover, Fenton's reagent reduced up to 72% of the desorbed p-ASA within a contact time of 4 h, substantiating the effectiveness of the dual-mode approach of adsorption–degradation.

Furthermore, the circular economy's tenets are partially supported by this study. In addition to reducing dependency on hazardous chemicals, the use of *Calotropis procera* leaf extract for nanoparticle production emphasises the value of renewable bio-resources. The nanoparticles' magnetic characteristics made recovery and reuse simple, reducing material waste and promoting resource efficiency. However, the degradation process transforms *p*-ASA into inorganic arsenic, which is more toxic and presents a challenge to achieving full circularity.

However, despite the promising results, the study also reveals certain limitations and challenges:

- **pH Dependency**: The highest adsorption efficiency occurred at an acidic pH (pH 4), which limits its usefulness in natural waters that typically have neutral or fluctuating pH levels. Future studies could address this by modifying or functionalizing the Fe_3O_4 nanoparticles to work effectively across a wider pH range.
- **Partial Degradation**: Fenton's reagent degraded only 72% of *p*-ASA, leaving behind residual arsenic species particularly inorganic arsenate that still pose toxicity risks. A possible solution is to integrate advanced oxidation processes (AOPs), such as photo-Fenton, or combine the method with biofiltration systems to achieve complete mineralization.
- **Nanoparticle Agglomeration**: SEM imaging and zeta size analysis revealed some nanoparticle clumping, which can reduce surface area and reactivity. This could be addressed by adding biocompatible dispersants or coating agents to keep the nanoparticles stable and well-dispersed.
- **Lack of Real-World Testing**: Since the experiments were conducted under controlled lab conditions, they might not reflect the complexities of actual environmental systems. Conducting pilot-scale or field trials such as in wastewater treatment facilities or polluted agricultural runoff would help verify effectiveness under realistic conditions.

Combining green nanotechnology with oxidative degradation presents a practical and eco-friendly solution for treating water polluted with organoarsenic compounds. To move from the lab to large-scale application, it will be important to refine synthesis methods, implement real-time degradation monitoring, conduct life-cycle assessments, and evaluate cost-effectiveness. Moreover, collaboration across disciplines spanning environmental chemistry, toxicology, and engineering will be key to addressing existing challenges and improving overall system efficiency.

Abbreviations

p-ASA	*p*-Arsanilic acid
Fe_3O_4 NPs	Magnetite nanoparticles
SEM-EDX	Scanning eectron microscopy-energy dispersive X-ray spectroscopy
XRD	X-ray diffraction
FTIR	Fourier-transform Infrared spectroscopy
AOPs	Advance oxidation processes

γ-Fe_2O_3	Maghemite
α-Fe_2O_3	Hematite
$FeCl_3.6H_2O$	Ferric chloride hexahydrate
$FeCl_2.H_2O$	Ferrous chloride hydrate
NaOH	Sodium hydroxide
HCl	Hydrochloric acid
PDI	Polydispersity index
UV–VIS-NIR Spectrophotometer	Ultraviolet visible near infrared spectrophotometer
C_0	Initial concentration
C_e	Final concentration
•OH	Hydroxyl radicle
H_2O_2	Hydrogen peroxide
Conc	Concentration

References

1. Li, S., Xu, J., Chen, W., Yu, Y., Liu, Z., Li, J., et al.: Multiple transformation pathways of *p*-arsanilic acid to inorganic arsenic species in water during UV disinfection. J. Environ. Sci. **47**, 39–48 (2016)
2. Chen, J., Garbinski, L.D., Rosen, B., Zhang, J., Xiang, P., Ma, L.Q.: Organoarsenical compounds: occurrence, toxicology and biotransformation. Crit. Rev. Environ. Sci. Technol. **50**, 217–243 (2020)
3. Argos, M., Ahsan, H., Graziano, J.H.: Arsenic and human health: epidemiologic progress and public health implications. Rev. Environ. Health **27**(4), 191–195 (2012)
4. Wu, S., Yang, D., Zhou, Y., Zhou, H., Ai, S., Yang, Y., et al.: Simultaneous degradation of *p*-arsanilic acid and inorganic arsenic removal using M-rGO/PS Fenton-like system under neutral conditions. J. Hazard. Mater. **15**(399), 123032 (2020)
5. Sharma, V.K., Sohn, M.: Aquatic arsenic: toxicity, speciation, transformations, and remediation. Environ. Int. **35**(4), 743–759 (2009)
6. Lourens, A., Falch, A., Malgas-Enus, R.: Magnetite immobilized metal nanoparticles in the treatment and removal of pollutants from wastewater: a review. J. Mater. Sci. **58**(7), 2951–2970 (2023)
7. Yang, T., An, L., Zeng, G., Jiang, M., Li, J., Liu, C., et al.: Efficient removal of *p*-arsanilic acid and arsenite by Fe(II)/peracetic acid (Fe(II)/PAA) and PAA processes. Water Res. **1**(241), 120091 (2023)
8. Wu, Q., Ye, X., Lv, Y., Pei, R., Wu, M., Liu, M.: Lignin-based magnetic activated carbon for *p*-arsanilic acid removal: applications and absorption mechanisms. Chemosphere **1**(258), 127276 (2020)
9. Vega, M.P.B., Hinojosa-Reyes, M., Hernández-Ramírez, A., Mar, J.L.G., Rodríguez-González, V., Hinojosa-Reyes, L.: Visible light photocatalytic activity of sol–gel Ni-doped TiO2 on p-arsanilic acid degradation. J. Sol-Gel Sci. Technol. **85**(3), 723–731 (2018)
10. Biswas, P., Wu, C.Y.: Nanoparticles and the environment. J. Air Waste Manag. Assoc. **55**(6), 708–746 (2005)
11. Bhoj, Y., Pandey, G., Bhoj, A., Tharmavaram, M., Rawtani, D.: Recent advancements in practices related to desalination by means of nanotechnology. Chem. Phys. Impact. **1**(2), 100025 (2021)

12. Pandey, G., Tharmavaram, M., Rawtani, D.: Halloysite nanotubes: an 'Aluminosilicate Nanosupport' for energy and environmental applications. In: Ledwani, L., Sangwai, J.S. (eds.) Nanotechnology for Energy and Environmental Engineering [Internet], pp. 125–44. Springer International Publishing, Cham (2020) [cited 2022 Jan 14]. (Green Energy and Technology). Available from: https://doi.org/10.1007/978-3-030-33774-2_5
13. Li, C., He, S., Mo, H., Xu, X., Yang, P., Liu, M.: Recent developments of iron oxide-based photocatalysts in water treatment technology: a review. Environ. Sci. Water Res. Technol. **11**(6), 1369–1385 (2025)
14. Siddiqa, A., Khatun Most, H., Mostafa, M.D.G.: Green-synthesized magnetite-maghemite nanocomposites for the removal of lead and cadmium from water: competitive adsorption behavior and mechanism studies. Environ. Sci. Pollut. Res. **32**(29), 17439–65 (2025)
15. Iconaru, S.L., Guégan, R., Popa, C.L., Motelica-Heino, M., Ciobanu, C.S., Predoi, D.: Magnetite (Fe3O4) nanoparticles as adsorbents for as and Cu removal. Appl. Clay Sci. **1**(134), 128–135 (2016)
16. Kundu, S., Sarkar, T., Al-Ahmadi, A.A., Ali, E., Bhattacharjee, A.: Thermally synthesized hematite (α-Fe2O3) nanoparticles as efficient photocatalyst for visible light dye degradation. RSC Adv. **14**(39), 28944–28955 (2024)
17. Panda, S.K., Aggarwal, I., Kumar, H., Prasad, L., Kumar, A., Sharma, A., et al.: Magnetite nanoparticles as sorbents for dye removal: a review. Environ. Chem. Lett. **19**(3), 2487–2525 (2021)
18. Nguyen, M.D., Tran, H.V., Xu, S., Lee, T.R.: Fe3O4 Nanoparticles: structures, synthesis, magnetic properties, surface functionalization, and emerging applications. Appl. Sci. (Basel) **11**(23), 11301 (2021)
19. Osman, A.I., Zhang, Y., Farghali, M., Rashwan, A.K., Eltaweil, A.S., Abd El-Monaem, E.M., et al.: Synthesis of green nanoparticles for energy, biomedical, environmental, agricultural, and food applications: a review. Environ. Chem. Lett. **22**(2), 841–887 (2024)
20. Jadoun, S., Arif, R., Jangid, N.K., Meena, R.K.: Green synthesis of nanoparticles using plant extracts: a review. Environ. Chem. Lett. **19**(1), 355–374 (2021)
21. Chaudhary, R.G., Mahmood, S., Jotania, R.B.: Green Nanomaterials for Clean and Sustainable Environment. http://www.eurekaselect.com [Internet]. 5 Nov 2023 [cited 22 Dec 22 2024]; Available from: https://www.eurekaselect.com/article/131706
22. Martinez-Boubeta, C., Simeonidis, K.: Chapter 20—magnetic nanoparticles for water purification. In: Thomas, S., Pasquini, D., Leu, S.Y., Gopakumar, D.A. (eds.) Nanoscale Materials in Water Purification [Internet], pp. 521–52. Elsevier (2019) [cited 2024 Dec 22]. (Micro and Nano Technologies). Available from: https://www.sciencedirect.com/science/article/pii/B9780128139264000264
23. Pignatello, J.J., Oliveros, E., MacKay, A.: Advanced oxidation processes for organic contaminant destruction based on the fenton reaction and related chemistry. Crit. Rev. Environ. Sci. Technol. **36**(1), 1–84 (2006)
24. Corwin, C.J., Summers, R.S.: Adsorption and desorption of trace organic contaminants from granular activated carbon adsorbers after intermittent loading and throughout backwash cycles. Water Res. **45**(2), 417–426 (2011)
25. Jain, A., Kumari, S., Agarwal, S., Khan, S.: Water purification via novel nano-adsorbents and their regeneration strategies. Process. Saf. Environ. Prot. **1**(152), 441–454 (2021)
26. Gawade, V.V., Gavade, N.L., Shinde, H.M., Babar, S.B., Kadam, A.N., Garadkar, K.M.: Green synthesis of ZnO nanoparticles by using Calotropis procera leaves for the photodegradation of methyl orange. J. Mater. Sci. Mater. Electron. **28**(18), 14033–14039 (2017)
27. Yew, Y.P., Shameli, K., Miyake, M., Kuwano, N., Bt Ahmad Khairudin, N.B., Bt Mohamad, S.E., et al.: Green synthesis of magnetite (Fe3O4) nanoparticles using seaweed (Kappaphycus alvarezii) extract. Nanoscale Res. Lett. **11**(1), 276 (2016)
28. Khan, S.A., Khan, S.B., Khan, L.U., Farooq, A., Akhtar, K., Asiri, A.M.: Fourier transform infrared spectroscopy: fundamentals and application in functional groups and nanomaterials characterization. In: Sharma, S.K. (ed.) Handbook of Materials Characterization [Internet], pp. 317–44. Springer International Publishing, Cham (2018) [cited 2024 Dec 22]. Available from: https://doi.org/10.1007/978-3-319-92955-2_9

29. Babuponnusami, A., Muthukumar, K.: A review on Fenton and improvements to the Fenton process for wastewater treatment. J. Environ. Chem. Eng. **2**(1), 557–572 (2014)
30. Xie, X., Hu, Y., Cheng, H.: Rapid degradation of p-arsanilic acid with simultaneous arsenic removal from aqueous solution using Fenton process. Water Res. **1**(89), 59–67 (2016)
31. Ablat, H., Nurmamat, X., Tian, J., Zhao, Z.: Progress of photocatalytic oxidation–adsorption synergistic removal of organic arsenic in water. Water Environ. Res. **96**(5), e11057 (2024)
32. Dhar, R.K., Zheng, Y., Rubenstone, J., van Geen, A.: A rapid colorimetric method for measuring arsenic concentrations in groundwater. Anal. Chim. Acta **526**(2), 203–209 (2004)
33. Adabavazeh, F., Nadernejad, N., Pourseyedi, S., Razavizadeh, R., Mozafari, H.: Synthesis of magnetic nanoparticles and their effects on growth and physiological parameters of Calotropis procera seedlings. Environ. Sci. Pollut. Res. **29**(39), 59027–59042 (2022)

Harnessing Nanomaterials for Sustainable Water Purification

Sanjay Singh Bisht, Seema Garg, Suresh Sagadeven, and Lalita Ledwani

Abstract Water pollution is a critical worldwide problem that significantly threatens human well-being and environmental sustainability. Conventional water treatment methods are effective for many contaminants, but they do not remove emerging pollutants or meet increasingly stringent quality standards. Nanotechnology could provide some answers thanks to the use of nanomaterials with special physico-chemical properties. In this chapter, various classes of nanomaterials used for water treatment are discussed in detail, including carbon-based, metal-based, metal oxide, composite materials, etc. This chapter discuss the mechanisms of contaminants removal with these nanomaterials including adsorption, catalytic decomposition, photocatalysis, disinfection, and membrane filtration. Moreover, the chapter elaborated on the specific uses of nanomaterials for various pollutants such as organic compounds, pathogens, heavy metals, nutrients from several sources of water such as industrial wastewater, groundwater, and saline water. Furthermore, the health and environmental consequences in water treatment are reviewed using nanomaterials, and approaches for reducing possible risks are discussed along with associated regulatory aspects. The chapter concludes by addressing recommendations regarding policy and practice to promote the responsible and effective adoption of nanomaterial-based water purification technologies, followed by a discussion on technical and operational challenges, scalability, economic feasibility, and future research directions in this fastmoving area.

Keywords Nanomaterials · Water purification · Wastewater treatment · Environmental remediation · Nanotechnology · Contaminant removal · Adsorption · Membrane filtration

S. S. Bisht · S. Garg (✉)
Department of Chemistry, Amity Institute of Applied Sciences, Amity University, Noida, Uttar Pradesh, India
e-mail: sgarg2@amity.edu

L. Ledwani
Nanotechnology and Catalysis Research Centre, University of Malaya, Kuala Lumpur, Malaysia

S. Sagadeven
Manipal University Jaipur, Dehmi Kalan, Jaipur, Rajasthan, India

L. Ledwani and S. Srivastava (eds.), *Experimental and Computational Advances in Materials*, Engineering Materials, https://doi.org/10.1007/978-981-95-4180-5_2

1 Introduction

1.1 Background and Significance of Water Pollution

As a fundamental resource for life on earth, water is increasingly threatened by pollution from a variety of human activities. Industrialization, coupled with the discharge of untreated industrial and municipal wastewater, is a leading contributor by introducing harmful contaminants into important water bodies [1]. Agriculture further exacerbates water pollution by using large amounts of pesticides and nitrogen fertilizers and producing organic farm wastes [2]. The rapid pace of urbanization and the underinvestment in basic water supply and sewage treatment infrastructure worsen this problem, resulting in a rise in infectious and parasitic diseases [3]. The response to this pervasive pollution is severe, poor water quality is related with 50% of all child deaths and 80% of diseases wide-reaching [1]. These health effects include common waterborne diseases, commonly induced by enteroviruses, and less common ones like skin diseases, malnutrition and even increased risk of cancer due to heavy metals and industrial chemicals exposure [4]. The gradual decline of the quality of water potentially threatens human health and quality of life, the basis of maintaining the integrity of ecosystems and sustainable social development [1]. This issue is particularly acute in developing countries, where industrial water pollution correlates positively with foreign direct investment, indicating a more complicated interrelation between economic development and environmental degradation. Additionally, the recent global health crises have highlighted the essential link between water quality in the spread of viral contaminants and human health, which also brings into sharp focus the need for effective and robust water treatment strategies [5].

1.2 Overview of Conventional Water Treatment Methods

Conventional water treatment methodologies typically employ a multistep process to purify water available for different applications [6]. This treatment involves a sequence of processes: **Primary treatment;** coagulation, flocculation, and sedimentation to aggregate and remove suspended solids (where agents are added to bind suspended particles into larger flocs. These heavier flocs then settle out of the water). This is followed by **Secondary treatment;** Filtration (usually through media such as sand or anthracite) removes any remaining finer particulates. Then pathogen inactivation is achieved using disinfection processes such as chlorination or UV irradiation [6, 7]. These traditional processes were mainly developed for the elimination of suspended solids, organic substances, and some like phosphorus and nitrogen [7]. Additionally, processes such as sludge dewatering and stabilization are crucial components of Conventional wastewater treatment [8]. Depending on regulatory quality standards, **advanced or tertiary treatments** may be applied, including nutrient removal, adsorption, or advanced oxidation processes. Advanced treatment

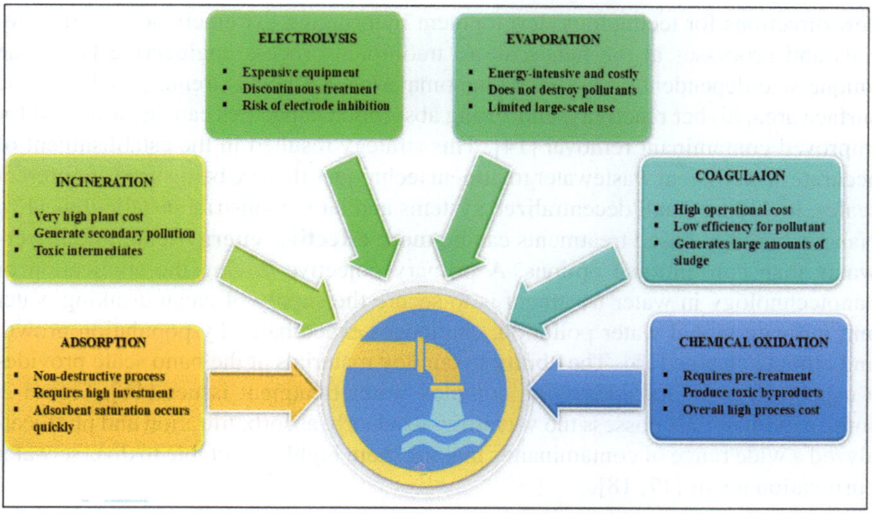

Fig. 1 Conventional wastewater treatment methods and their drawbacks

methods include granular activated carbon (GAC) for removing taste-and Odor-producing compounds, organic contaminants, and disinfection byproducts; aeration for volatile organic compounds (VOCs); and ion exchange for removing positively or negatively charged contaminants like heavy metals [8, 9].

Despite their widespread use, conventional water treatment methods have several constraints, as shown in Fig. 1, particularly in the addressing the increasingly nature of water pollution. Not only do they fail to sufficiently eliminate new threats, including pharmaceuticals, microplastics and heavy metals, that flow through the systems and examples of the dangers they pose to environmental and public health [10]. The inability to consistently meet increasingly stringent regulatory limit values for nutrients and emerging pollutants results in the costly upgrading of existing treatment plants [9]. In addition, most conventional approaches are energy demanding, oftentimes producing large quantities of sludge or undesirable by-products needing additional care [8]. Certain pollutants, including certain metal particles and dyes, can be extremely intractable to conventional treatment regimens [1, 11]. In many cases, these methods lead to ineffective remediation, requiring extensive infrastructure, technical know-how, and investment that is often inefficient and time-intensive, particularly for low concentrations of pollutants [12].

1.3 Emergence of Nanotechnology in Water Treatment

In response to these limitations, nanotechnology has emerged as a growing field as advanced solutions for water and wastewater treatment. Nanotechnology promises new directions for technology development by bringing extremely advanced materials and processes at the nanoscale to traditional process engineering [13]. The unique size-dependent properties of nanomaterials, such as extremely high specific surface area, higher reactivity, and strong absorption capacities can be harnessed for improved contaminant removal [14]. This strategy resulted in the establishment of accurate and efficient wastewater treatment techniques that are being used at different scales, including small decentralized systems and large industrial installations [13]. Nanotechnology-based treatments can be **more effective, energy-lean, and lower-waste** than conventional options. A primary objective driving the application of nanotechnology in water treatment is to secure the supply of clean drinking water and mitigate global water pollution, challenges exacerbated by population growth and climate change [15]. The ability to employ materials at the nano scale provides a capable frontier for addressing complex water treatment issues [16]. Figure 2, some nanomaterials possess the versatility to adsorb, absorb, filtration and photocatalyzed a wide range of contaminants, making them highly adaptable to diverse water purification needs [17, 18].

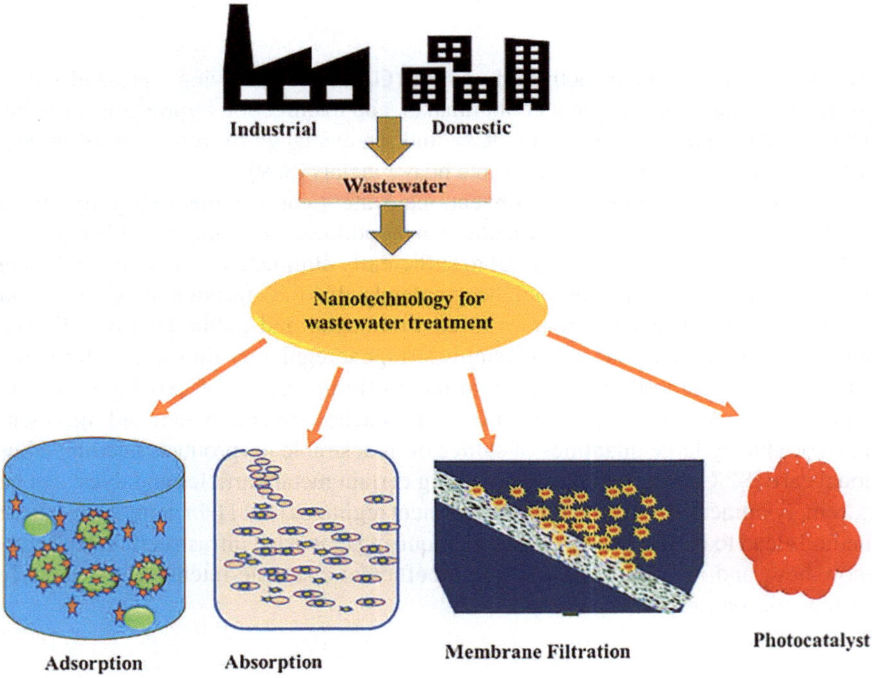

Fig. 2 Nanotechnological processes in the wastewater treatment [17]

1.4 *Objectives and Scope of the Chapter*

This chapter aims to provide an overview of the specific applications of nanomaterials in water treatment and purification. It will cover different categories of nanomaterials used in the field, and will be followed by a more detailed explanation on the mechanisms in which these materials remove contaminants from water. Then, environmental nanotechnology utilized in the treatment of organic pollutants, pathogen and heavy metals contaminants, in different water sources is discussed in this chapter. It will also critically assess the possible health and environmental effects of using nanomaterials for water treatment, discuss approaches to minimize these risks, as well as relevant regulatory aspects. Lastly, it will highlight the challenges with regards to the technical and operational aspects, scalability and economic viability, and future research needs within this rapidly evolving field, finishing on policy and practical recommendations on how to facilitate the responsible and effective implementation of nanomaterial-based water purification technologies.

2 Classification of Nano Materials Utilized in Water Treatment

The field of nanotechnology has introduced a wide range of nanomaterials with exclusive properties that are being harnessed for water treatment and purification. As seen in Fig. 3.

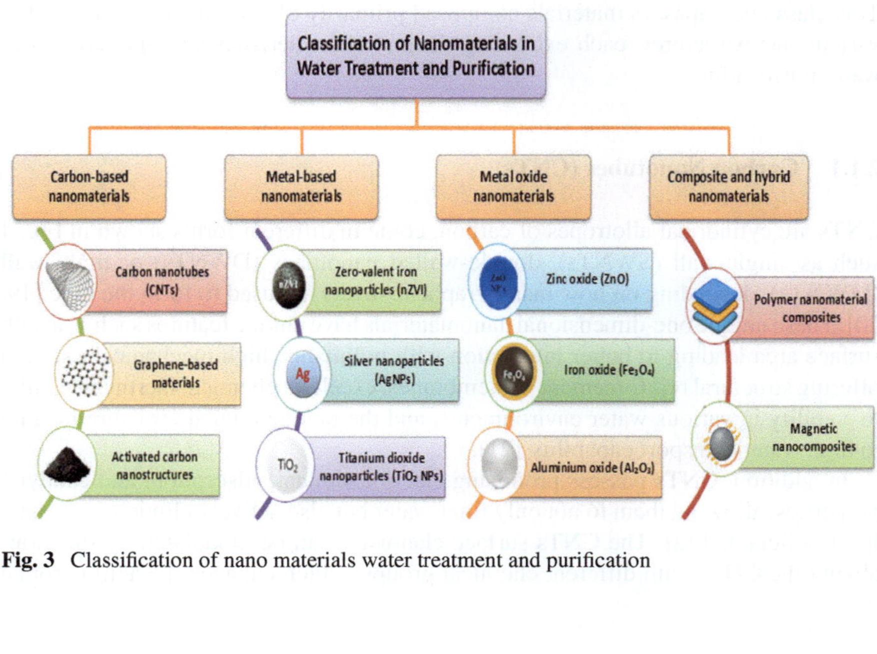

Fig. 3 Classification of nano materials water treatment and purification

Selection of Nanomaterials:
In this article, four classes of nanomaterials are identified that are highlighted in the review, namely: carbon-based, metal-based, metal oxide and composite/hybrid nanomaterials. These categories were chosen on purpose since they are the most studied (with the most practical significance) classes for water and wastewater treatment. Carbon-based nanomaterials (CNTs, graphene, activated carbon) possess large adsorption capacity and surface modifiability. Metal containing nanomaterials (e.g., silver, ZVI) reveal excellent redox and anti-microbiological features. The metal oxide nanomaterials (such as TiO_2, ZnO, Fe_3O_4, and Al_2O_3) are extensively used for photocatalysis and adsorption. Lastly, composite and hybrid nano-materials combine more than one functional feature for a synergistic operation. Though other nanomaterials are also available, overall, these four groups represent the research scope in nanomaterial applications and provide a kind of balanced way to think about how they are contributing to the field of water purification.

These nanomaterials are broadly classified based on their structure and dimensionality, including carbon-based, metal-based, metal oxide, composite and hybrid, and other emerging nanomaterials.

2.1 Carbon Based Nanomaterials

This class encompasses materials composed primarily of carbon atoms arranged in various nanostructures, each exhibiting distinct characteristics and applications in water purification.

2.1.1 Carbon Nanotubes (CNTs)

CNTs are cylindrical allotropes of carbon, come in different forms shown in Fig. 4 such as single-wall (SWNTs), double-walled nanotubes (DWNTs) or multi-wall (MWNTs), depending on how many graphene layers are used to form the tube [19, 20]. These unique one-dimensional nanomaterials have unique features such as a high surface area leading to better interaction with pollutants, high mechanical strength offering structural reinforcement in membranes, excellent chemical inertness leading to stability in various water environments, and the unique tubular structure leading to high water-transport capability [21].

In addition, CNTs possess promising electrochemical, adsorption, and catalytic properties, allowing them to not only filter water but also actively eliminate or transform pollutants [22]. The CNTs surface chemistry can be modulated by functionalizing the CNTs with different chemical groups, which enhance the antimicrobial

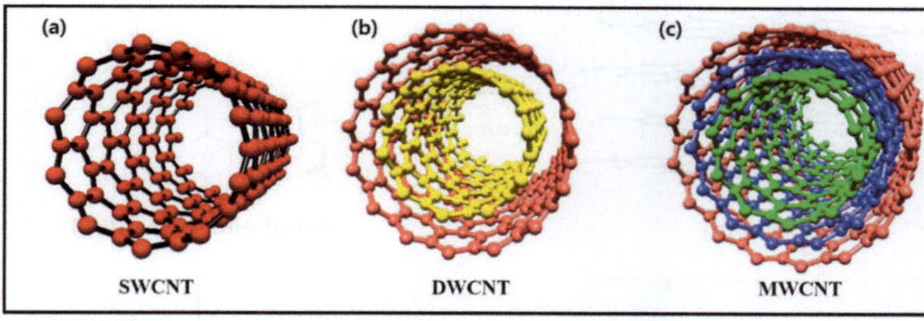

Fig. 4 Surface and internal view of, **a** SWNTs, **b** DWNTs, and **c** MWNTs [20]

ability of CNTs and also their binding to certain pollutants like heavy metals and organic compounds [23]. As a result, CNTs have been employed for the preparation of composite membranes for applications such as oil–water separation, removal of heavy metal ions, and seawater desalination, and emerging pollutants from water [24]. CNTs have also been reported to possess antimicrobial activity through stimulating the reactive oxygen species production causing oxidative stress in bacteria which damages their cell membranes [23]. They can be produced via numerous methods (for instance, laser ablation, chemical vapor deposition (CVD), and electric-arc discharge), with CVD being the widely favoured approach thanks to its relative ease and ideal for conversion of high-purity CNTs at high yields [24]. Due to their incredibly high dimensions and hollow structure, CNTs are especially conducive for adsorption-based removal, and their incorporation into membranes overcomes mechanical weakness to improve their robustness in water treatment applications.

2.1.2 Graphene-Based Materials

Graphene-derived materials represent another important class of carbon nanomaterials. Graphene is a two-dimensional substance composed of a single layer of sp2-bonded atoms of carbon organized into a honeycomb-shaped structure [25]. GO is usually synthesized by oxidizing graphite using techniques such as Hummers' process or its variations shown in the Fig. 5, then exfoliating the graphite to produce individual GO layers [26, 27].

GO is a key derivative, is produced through the oxidation of graphene, introducing oxygen-containing functional groups such as hydroxyl, epoxy, and carboxyl groups [18]. These functional groups impart GO with high dispersity in aqueous solutions and provide a large surface area for the adsorption of various contaminants [28].

Graphene based materials such as graphene oxide hybrids or graphene GM-hybrid matrix systems possess precise molecular sieving capabilities due to their controllable interlayer spacing or pore structure, along with high water permeability, excellent chemical stability, and outstanding mechanical strength, making them highly suitable for separation membranes [25]. As a result, they are extensively explored for a wide

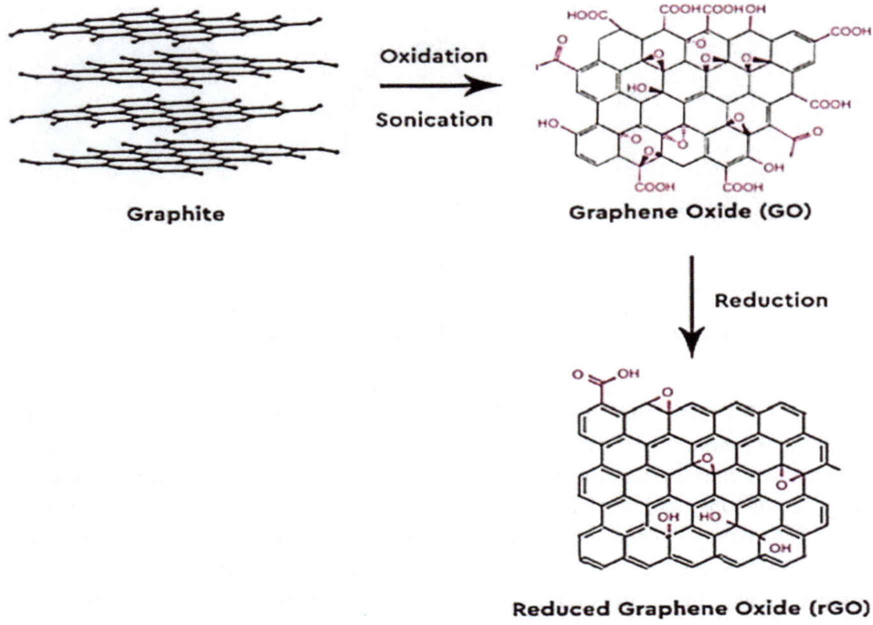

Fig. 5 Synthesis of graphene oxide (GO) and reduced G(rGO) from graphite powder [27]

array of water treatment applications, including desalination (via reverse osmosis and forward osmosis), nanofiltration, diafiltration, recovery of rare metal ions, pervaporation, and membrane distillation. **Graphene oxide–nanoparticle conjugates.** Beyond pristine GO/rGO, **GO–nanoparticle conjugates** (e.g., GO–AgNPs, GO–Fe_3O_4, GO–TiO_2) integrate complementary functions high-capacity adsorption from GO basal/edge sites with **antimicrobial** (Ag), **magnetic separability/Fenton chemistry** (Fe_3O_4), or **visible-to-UV photocatalysis** (TiO_2). Such hybrids enable **multi-mechanism removal** and **facile recovery** (magnetic capture), making them attractive for simultaneous removal of trace organics, heavy metals, and microbes. As seen in Fig. 6, research at the convergence of several disciplines, including ecological restoration, has been driven by the usage of graphene-based materials typical GM-hybrid systems towards diverse environmental fields [28, 29].

2.1.3 Activated Carbon Nanostructures

Activated carbon nanostructures, while a more established adsorbent material, also exist in nanostructured forms that offer enhanced properties. It is a highly porous material characterized by an exceptionally large surface area, making it an effective adsorbent for a wide range of contaminants [30–32]. It has proven effective eliminating harmful metal ions, pathogenic microorganisms, and a range of inorganic

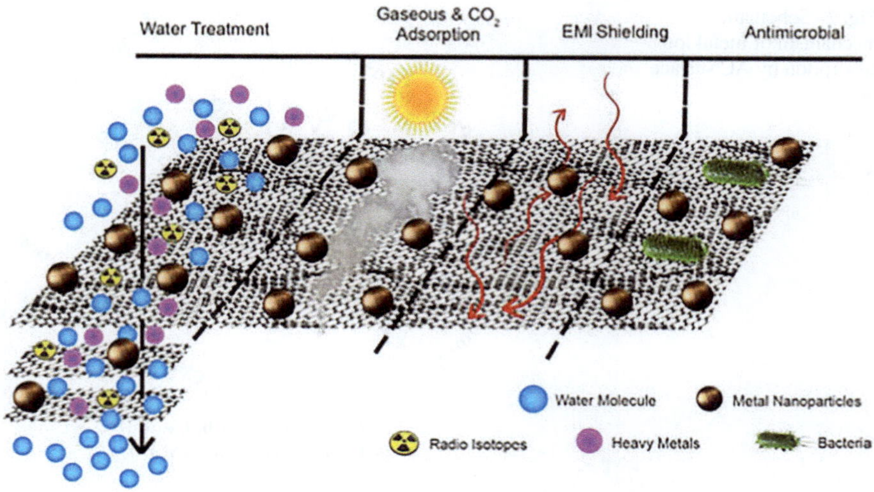

Fig. 6 Schematic illustration of graphene oxide hybrids for environmental applications, gaseous CO_2 adsorption, electromagnetic shielding, and antimicrobial elimination [29]

as well as organic compounds from water [31, 33]. Activated carbon nanostructures can be produced from diverse carbonaceous precursors, including agricultural wastes, through either chemical or physical activation processes, which enhance their porosity and surface area [31, 34]. For instance, the mechanism of metal ion adsorption on the surface of activated carbon is depicted in Fig. 7, While activated carbon remains a popular choice due to its broad-spectrum removal capabilities and cost-effectiveness, it tends to exhibit slower adsorption kinetics and can be more challenging to regenerate compared to more advanced carbon nanomaterials like CNTs [31, 35]. However, activated carbon fibres (ACFs), a form of activated carbon with a one-dimensional structure, offer higher adsorption kinetics due to the shorter diffusion distances for pollutants to reach the adsorption sites [36]. The extensive network of micropores in activated carbon provides a vast surface area for adsorption, and its production from diverse and inexpensive precursors contributes to its economic viability [30, 34].

Challenges: A significant challenge associated with CNTs and other carbon-based materials is their propensity to aggregate in aqueous solutions due to van der Waals forces and π–π interactions [37]. This aggregation drastically reduces their effective surface area, which is the primary driver of their high adsorption capacity, thereby negatively impacting their performance. Additionally, the inherent hydrophobicity of pristine carbon nanomaterials can lead to poor dispersion in water [37, 38].

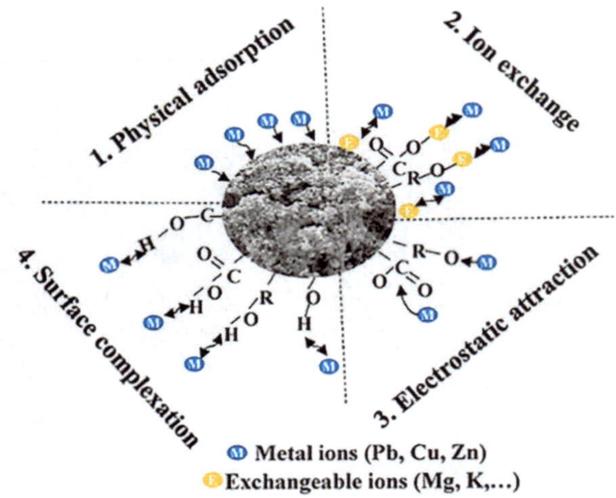

Fig. 7 Schematic mechanism of metal ion adsorption by AC surface [35]

2.2 *Metal-Based Nanomaterials*

This category includes nanomaterials composed primarily of metals, often in their elemental form, which exhibit unique reactivity and adsorption properties beneficial for water treatment.

2.2.1 Zero-Valent Iron Nanoparticles (nZVI)

One of the most represntative metal-based nanomaterials for water treatment is zero-valent iron nanoparticles (nZVI). These nanoparticles are composed of elemental (zero-valent) iron, which are extremely reactive because of the great reducing power and the large specific surface area [39]. Such a high reactivity of nZVI can decontaminate different classes of organic and inorganic pollutants via reductive transformation, adsorption, and catalytic processes to remove halogenated hydrocarbons, organic dyes, antibiotics, and heavy metal ions [39, 40]. nZVI is mainly applicable for in-situ groundwater remediation, treatment of industrial wastewater, and removal of particular toxic pollutants such as mercury and arsenic [41, 42]. The nZVI properties, including their reactivity plus mobility in aqueous as well as and particle size, purity, surface properties, the presence of modifying capping material, and the width of the oxide coat covering the iron surface and co-supports added to minimize aggregation [41]. These nanoparticles can be fabricated via either top-down techniques, where bulk iron is grinded into nanosized- particles, or bottom-up approaches, like iron salts chemical reduction in a solution using reducing agents like sodium borohydride, shown in Fig. 8 [43]. nZVI readily donates electrons due to its high reactivity to reduce contaminants to a less toxic state. Its nanoscale size offers a high surface area for these reactions to happen effectively.

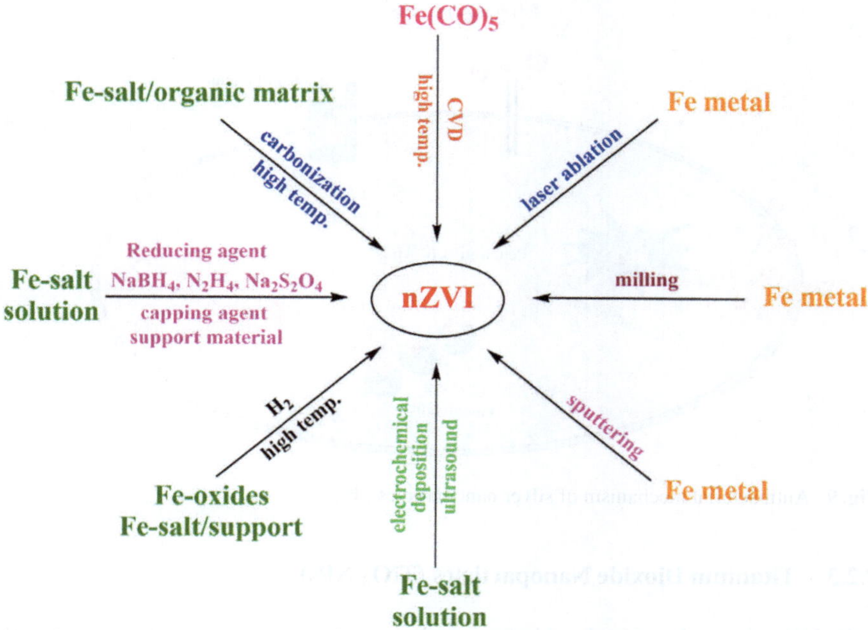

Fig. 8 Top-down and bottom-up synthetic methods for the production of nZVI [43]

2.2.2 Silver Nanoparticles (AgNPs)

Metal-based nanomaterials have also attracted a great deal of attention, with the silver nanoparticles (AgNPs) being a major class of nanomaterials owing to their unique physicochemical properties and extensive utilization in electronics, catalysis, biomedicine, and detection [44]. The antimicrobial properties of AgNPs are one of the main characteristics making them useful for water treatment as they have been evidenced against all microorganisms and afterwards (infectious) viruses such as bacteria, fungi and viruses [45, 46]. Various methods have been described for AgNPs synthesis, these are mainly divided into biological (green synthesis using plant extracts or microorganisms), chemical (where silver salts are reduced using substance such as sodium borohydride or sodium citrate where sol–gel or chemical vapor deposition is also used) or physical (also including techniques like sputtering, laser ablation and arc discharge) [44, 46, 47]. For instance, Antibacterial mechanism of silver nanoparticlesis shown in Fig. 9, due to their strong antimicrobial properties, AgNPs find a use in wastewater treatment processes, to coat surfaces to avoid the growth of bacteria, and as colorimetric sensors for detecting impurities in the water. [45, 47]. However, the potential release of silver ions (Ag^{+}) from these nanoparticles into the environment raises concerns regarding their broader ecological impact [45, 48].

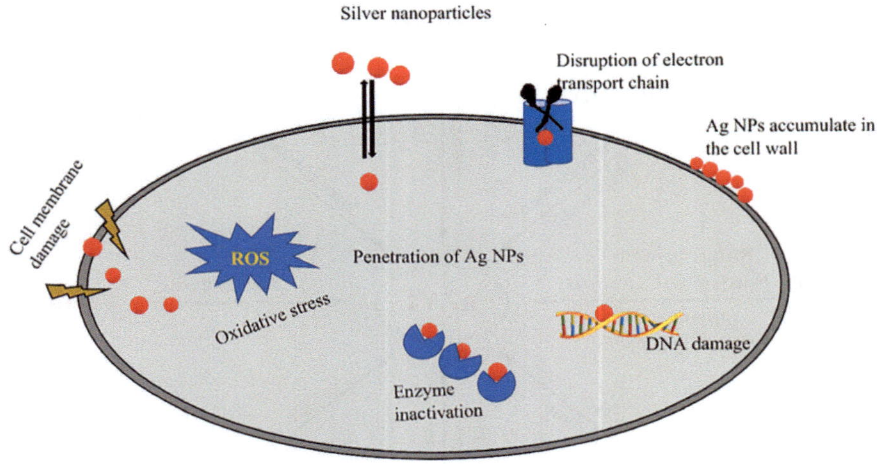

Fig. 9 Antibacterial mechanism of silver nanoparticles [49]

2.2.3 Titanium Dioxide Nanoparticles (TiO_2 NPs)

TiO_2 NPs are another metal oxide frequently included in metal-based nanomaterials studies owing to their widespread application and special characteristics. TiO_2 NPs is a promising candidate for environmental pollution control, as well as in water treatment [50–53] In photocatalysis, TiO_2 NPs are used to absorb light, usually from UV or near-UV sources, to promote oxidation and degradation processes capable of degrading a broad spectrum of organic pollutants like dyes, drugs, and other hazardous compounds dissolved in water and the resulting hydroxyl radical production is shown in Fig. 10 [54, 55]. These nanoparticles can be synthesized through several physical methods, chemical methods including sol–gel and hydrothermal methods, and environmental green methods of synthesis [50, 52]. In other words, the photocatalytic activity of TiO_2 NPs is controlled by the generation of electron–hole pairs as a function of irradiation, forming highly reactive radicals that oxidizes and decompose organic pollutants [56].

Challenges: Despite their potential, nZVI nanoparticles face significant challenges with aggregation and stability [57]. Their high reactivity makes them prone to rapid oxidation and agglomeration in water, which limits their mobility and long-term performance, particularly in applications like in-situ groundwater remediation [41]. A major drawback of AgNPs is the potential for environmental ecotoxicity. The release of silver ions from these nanoparticles into aquatic environments can be hazardous to non-target aquatic life, including algae and invertebrates, posing a significant ecological risk [45, 58].

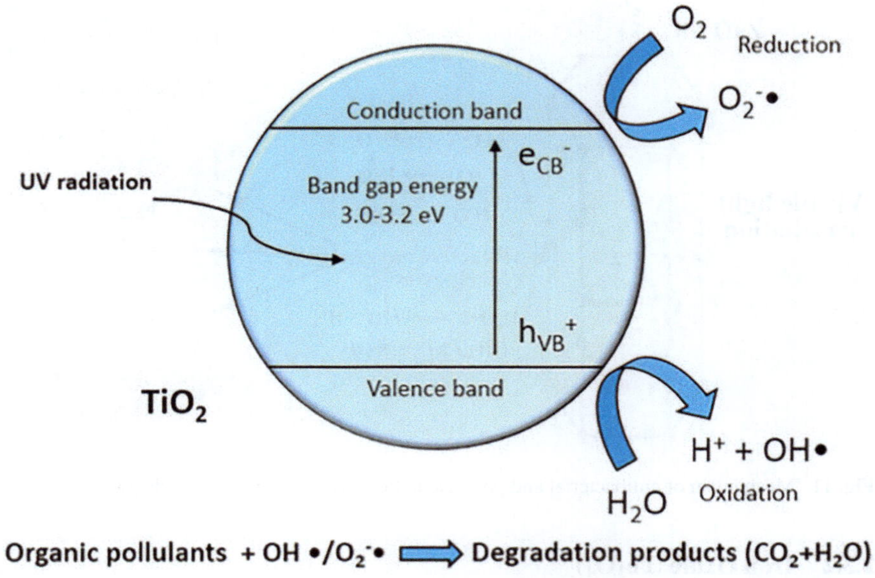

Fig. 10 Photocatalytic activation of TiO_2 and generation of reactive oxygen species for organic pollutant degradation [54]

2.3 Metal Oxide Nanomaterials

This class includes nanomaterials where metal atoms are bonded to oxygen atoms, forming structures that often exhibit excellent adsorption capacities and catalytic activities in water treatment.

2.3.1 Zinc Oxide (ZnO)

ZnO is a semiconductor metal oxide nanomaterial with notable qualities such as excellent chemical resistivity, a large electrochemical coupled value, a broad electromagnetic spectrum, and high photostability [59]. ZnO nanoparticles exhibit both antimicrobial and photocatalytic activities shown in Fig. 11, making them highly suitable for water decontamination and the degradation of various impurities [60]. These nanoparticles can be produced through several approaches, including ion exchange processes, hydrothermal reactions, co-precipitation techniques and green methods [61, 62]. ZnO has been shown to be works well against a variety of microorganisms, including both Gram-positive and Gram-negative bacteria., as well as yeast [60]. Similar to TiO_2, ZnO can act as a photocatalyst, but it can also be activated by visible light to some extent, offering advantages for water purification applications.

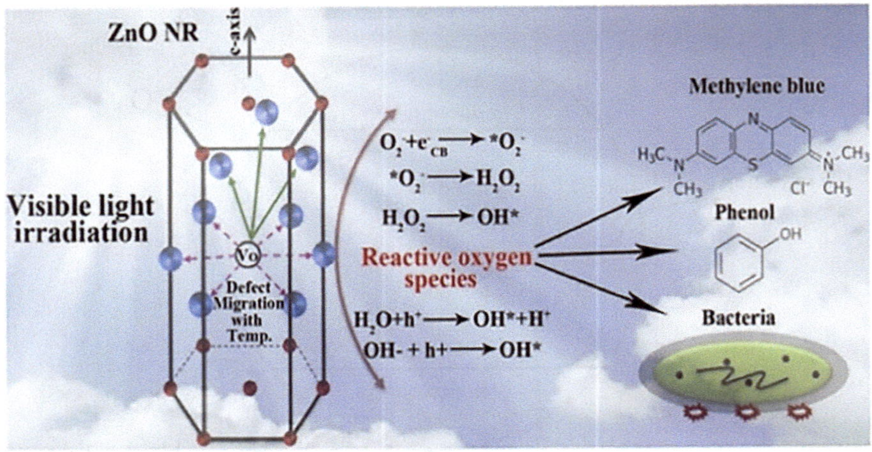

Fig. 11 Mechanism of antibacterial and photocatalytic activity of ZnO nanorods [60]

2.3.2 Iron Oxide (Fe_3O_4)

Another metal oxide nanomaterial is iron oxide (Fe_3O_4), or magnetite. It has specific magnetic properties, a high surface area and its surface chemistry can be tailored for different applications [63]. Fe_3O_4 nanoparticles have magnetic properties, which can thus be utilized for coagulation process for water treatment, and the particles can be easily controlled and separated from the liquid with an external magnetic field [64]. The metal nanoparticles had a high surface area and were useful for adsorbing a wide variety of toxins, such as biological contaminants and heavy metals. Furthermore, Fe_3O_4 can be utilized as a reagent in Fenton-type methods for the organic pollutant's degradation shown in Fig. 12 [65, 66].

Fe_3O_4 nanoparticles can be produced using different approaches, for example mechanochemical methods and various chemical methods as hydrothermal synthesis, sol–gel methods and co-precipitation [63]. Moreover, the superparamagnetic properties of Fe_3O_4 nanoparticles enable their effective separation from treated water, and their participation in Fenton reactions offers a powerful mechanism for the degradation of persistent organic contaminants.

2.3.3 Aluminium Oxide (Al_2O_3)

Al_2O_3 Nanoparticles possess several properties that make them advantageous for use in treatment of water, such as low cytotoxicity, high mechanical strength, robustness, and chemical stability [67]. These nanoparticles have been employed as adsorbents for the degradation of organic dyes and removal of heavy metals, including hexavalent chromium from aqueous solutions [68, 69]. Methods including sol–gel processing, microemulsion techniques, and chemical precipitation can be found in the literature

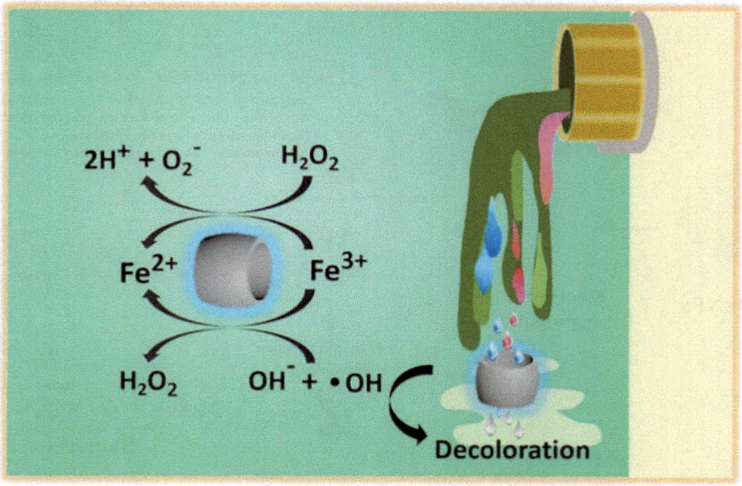

Fig. 12 Schematic of organic pollutant degradation by Fe_3O_4 nanoparticles via a Fenton mechanism [66]

for the synthesis of Al_2O_3 nanoparticles [70]. In addition, AlOxNPs are generally bacteriostatic due to an electrostatic interaction caused by the membrane of bacteria, and the behavior of aluminum cations, which contribute to the production of ROS and the oxidation of biopolymers as illustrated in Fig. 13 [69]. Al_2O_3 has a high surface area, amphoteric nature so both acidic and basic pollutants can be adsorbed, depending on the pH of the water, due to its inert nature with regard to impurities as well as its low toxicity [67, 70].

Challenges: A significant challenge for these materials is their activation energy; most photocatalytic applications rely on UV light, which constitutes only a small fraction of the solar spectrum, limiting their energy efficiency and large-scale applicability [58, 71, 72]. A thorough understanding of their potential long-term environmental and health impacts is essential, as their ecotoxicological risks to aquatic organisms and their potential to accumulate in aquatic systems are still being investigated [73, 74]. There is also evidence that these materials can facilitate horizontal gene transfer in bacteria, which could contribute to the development of antimicrobial resistance [58].

2.4 *Composite and Hybrid Nanomaterials*

This category encompasses nanomaterials that combine two or more different materials at the nanoscale, often resulting in improved performance and synergistic properties in water treatment.

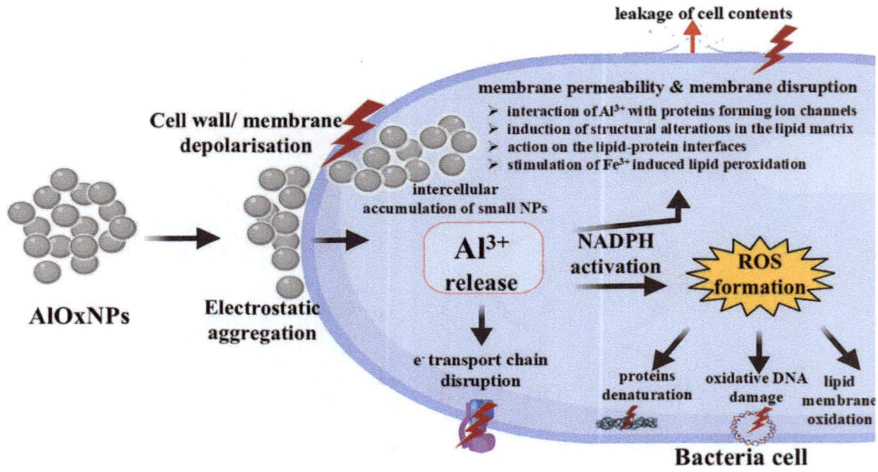

Fig. 13 Schematic of the main mechanisms of the antibacterial action of AlOxNPs [69]

2.4.1 Polymer-Nanomaterial Composites

Polymer-nanomaterial composites are engineered by incorporating nanomaterials into a polymer matrix to significantly enhance the performance of the subsequent composite in wastewater treatment [75]. The inclusion of nanomaterials can lead to improvements in membrane properties such as salt rejection, antifouling characteristics, antimicrobial activity, water permeability, and mechanical and thermal stability compared to pristine polymer membranes [76]. A wide range of nanomaterials, including graphene-based materials, metal and metal oxide nanoparticles, CNTs, and zeolites, can be utilized in these composites [77]. The synthesis of polymer-nanomaterial composites involves various methods shown in Fig. 14 such as blending, interfacial polymerization, grafting, coating techniques, and chemical crosslinking [76, 78]. These composite materials find applications in desalination processes, the removal of organic pollutants, oil–water separation, heavy metals, and disinfection [79]. The combination of polymers, providing mechanical strength and processability, with nanomaterials, offering specific functionalities, allows for the creation of highly specialized and effective water treatment solutions [80].

2.4.2 Magnetic Nanocomposites

Magnetic nanocomposites are designed by integrating magnetic nanoparticles, typically iron oxides like magnetite, with additional materials such as polymers, hydrogels, carbon, and silica shown in Fig. 15 [81]. The key advantage of these nanocomposites is the magnetic component, which allows for facile separation and recovery of the adsorbent loaded with pollutants from the aqueous solution using an external

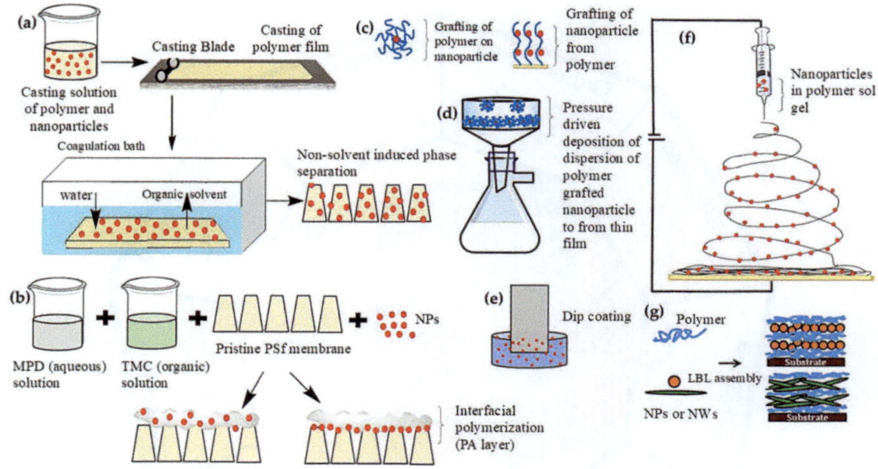

Fig. 14 Various methods for integrating nanomaterials into polymer membranes, including, **a** blending via phase inversion, **b** interfacial polymerization, **c** grafting, **d** pressure-assisted deposition of polymer-grafted nanoparticles as a thin film on a support, **e** dip-coating of a polymer membrane in a nanoparticle-containing dispersion, **f** electrospinning of a polymer sol–gel loaded with nanoparticles to form nanofibrous membranes, and **g** layer-by-layer (LBL) assembly of polymer and nanoparticle (or nanowire) layers [76]

magnetic field. This magnetic separability enables easy reusability and regeneration of the material for upcoming distillation cycles, making them economical and eco-friendly option [53]. Magnetic nanocomposites often exhibit enhanced adsorption capacities for a wide range of impurities, including organic dyes, oils, drugs, pesticides, and heavy metals, due to synergistic effects between the magnetic nanoparticles and the base material [82]. These composites can be formed via diverse methods, including filling, pulse-laser irradiation, ball milling, and electro spinning techniques [83, 84]. The incorporation of magnetic components provides a practical solution for the removal and recovery of nanomaterials used in water treatment, addressing a key challenge in preventing environmental contamination.

Challenges: The primary challenges associated with these materials are their complex and often costly fabrication methods [53, 80]. Furthermore, there is a risk of the nanomaterials leaching out of the polymer matrix over time, which not only degrades the material's performance but also poses a risk of secondary environmental contamination [53, 80, 85, 86]. Maintaining the long-term integrity and stability of the synergistic properties is a significant challenge, as is the complexity of their synthesis.16 Without proper chemical interactions like cross-linking or covalent bonds, these materials can experience phase separation and loss of function [85, 87, 88].

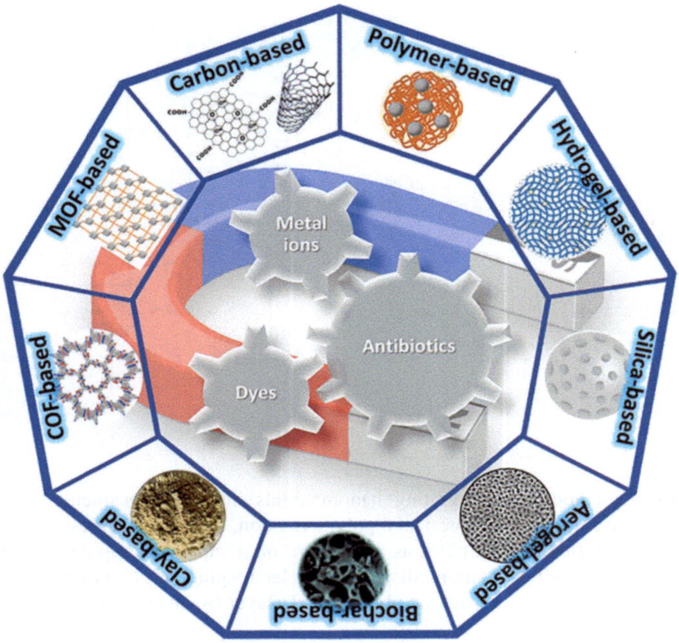

Fig. 15 Schematic representation of magnetic composite materials for the removal of organic and inorganic pollutants from contaminated water [81]

3 Mechanisms of Contaminant Removal by Nanomaterials

The effectiveness of nanomaterials in water treatment stems from their capability to interact with and remove contaminants through a variety of mechanisms shown in Fig. 16, including adsorption, catalytic degradation, photocatalysis, disinfection, and filtration/membrane separation.

3.1 Adsorption Mechanisms

Adsorption is a surface phenomenon in which contaminant molecules stick to the nanomaterial adsorbent surface. This process can be very weak as physisorption or by the establishment of chemical bonds between the contaminant and the nanomaterial surface (chemisorption) [89, 90]. The degree of adsorption depends on many aspects, such as the surface area and pore structure of the nanomaterials, the functional group type on the surface that can bind with the contaminant, the size and polarity of the contaminant molecules, their concentration in the water, the temperature of the system, and the pH of the solution [89, 91]. The **Adsorption** mechanism is presented

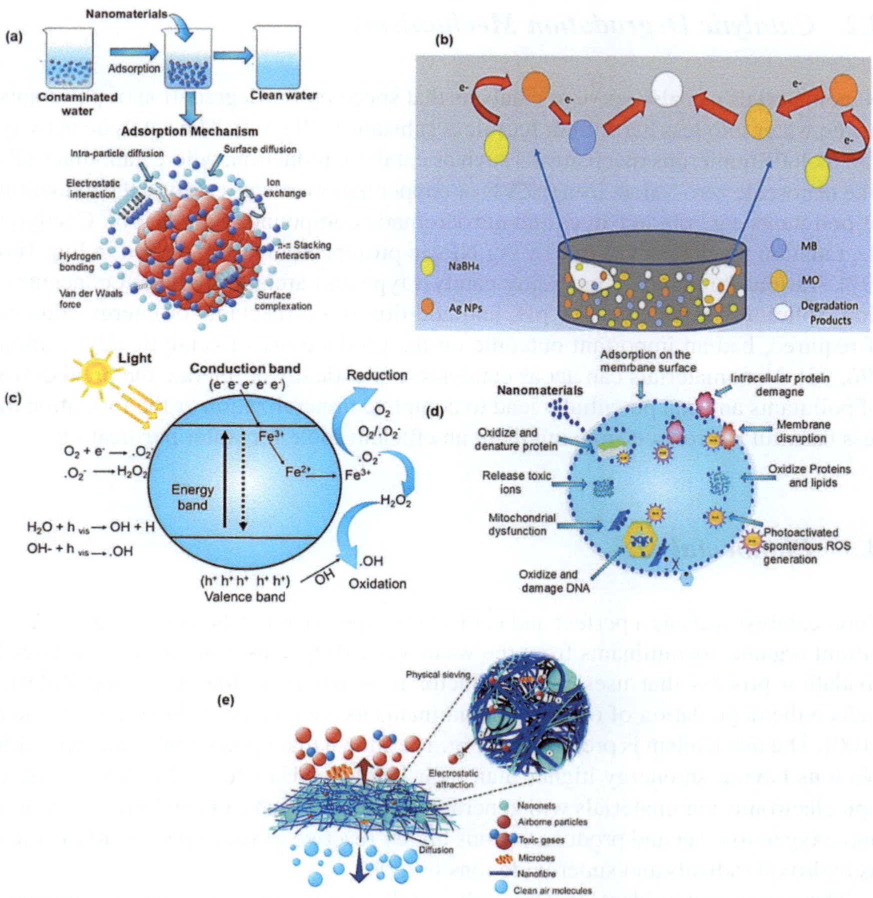

Fig. 16 Mechanisms of contaminant removal by nanomaterials including adsorption **a**, catalytic degradation **b**, photocatalysis **c**, disinfection **d**, and filtration/membrane separation **e** [92, 97]

in Fig. 16a [92]. To characterize and improve the adsorption process, the most widely used models include adsorption isotherms, such as the Langmuir, Freundlich, and Sips models, which describe how the amount of contaminant adsorbed onto the nanomaterial relates to the concentration of contaminant remaining in the water at equilibrium. This allows the maximum adsorption capacity of the nanomaterial to be determined, as well as to understand the nature of the adsorption process Endpoint and Review of the models [91].

3.2 Catalytic Degradation Mechanisms

Nanomaterials can also serve as catalysts that speed up the degradation of pollutants in the water into less harmful or harmless substances [93, 94]. The catalytic activity can include numerous mechanisms such as catalytic reduction, where nanomaterials like nanoscale zero-valent iron (nZVI) or copper nanoparticles catalyse the reduction of pollutants e.g., organic dyes and nitroaromatic compounds [90, 95, 96]. Catalytic degradation mechanism of dye by Ag NPs in presence of $NaBH_4$ shown in Fig. 16b [97]. Various parameters, including catalyst type and amount, pollutant concentration, solution temperature and pH, and addition of co-reactants or energy sources if required, had an important outcome on the productivity of catalytic degradation [96, 98]. Nanomaterials can act as catalysts to drastically accelerate the breakdown of pollutants and can potentially lead to complete mineralization or the formation of less harmful byproducts, making them an efficient route for water purification.

3.3 Photocatalysis

Photocatalysis acts as a perfect and ecofriendly approach for the degrading of recalcitrant organic contaminants from the wastewater [99]. This process is an advanced oxidation process that uses semiconductor nanomaterials, like TiO_2 and ZnO to induce the degradation of organic contaminants as well as the disinfection of water [100]. The mechanism is presented in Fig. 16c shown that upon being irradiated with photons having an energy higher than their bandgap (UVs to visible lights), these optoelectronic nanomaterials will generate e–h pairs, which can combine with water and oxygen to react and produce various highly reactive oxygen species (ROS) such as hydroxyl radicals and superoxide ions [92, 99].

They are potent oxidants that not only can decompose most organic contaminants into less toxic substances, but also can inactivate or kill microorganisms [45]. The effectiveness of photocatalytic treatment is affected by the intensity and wavelength of the light source, the physical/chemical properties of the photocatalyst (such as the band gap energy, surface area, and crystallinity of the material), the concentration of the pollutant, and the pH and temperature of the water [100]. The performance of Semiconductor photocatalyst can be improved further, by combining carbon-based nanomaterials, such as graphene, to improve charge separation and extend light absorption window [98].

3.4 Disinfection Mechanisms

Nanomaterials have been shown to be highly active for the disinfection of water by acting through different mechanisms resulting in inactivation or killing of pathogenic microorganisms (including microbes, germs, and fungi) [45]. The mechanism is presented in Fig. 16d. AgNPs, for example, can puncture bacterial cell membranes, prevent them from inducing an enzyme, or create reactive oxygen species that kill the cell [101]. TiO_2 and ZnO NPs can produce ROS when irradiated by light, leading to microbial cell injury. Carbon nanotubes (CNTs) tend to physically damage bacterial cell walls [45]. Nanomaterials, widely classified as Eco-friendly disinfectants, can react with a range of different microorganisms through various mechanisms.

3.5 *Filtration and Membrane Separation Mechanisms*

Nanomaterials, have greatly improved the performance of filtration membranes used for water purification in terms of durability, separation productivity, and water permeability, as well as fouling resistance. The incorporation of nanostructures into reversible osmosis (RO) and nanofiltration (NF) membranes has enhanced solute rejection performance and may reduce energy requirements in desalination [102]. The mechanism of filtration and membrane is presented in Fig. 16e. Nanomaterials have been reported to improve membrane separation through several different mechanisms; size exclusion, where contaminants that are physically larger than the membrane pores are retained; electrostatic repulsion, which can reject charged contaminants based on the charge of the surface membrane; and specific interactions between the membrane material often modified with nanomaterials—and contaminants, such as chemical binding or affinity [103].

By forming more uniform and smaller pore sizes, enhancing overall surface area for filtration or adsorption, changing the surface chemistry and wettability of the membrane material, and providing structural support to prevent compaction, nanomaterials can alter the structure of membranes [104]. To exemplify, graphene oxide (GO) membranes show their unique interlayer nano-structure, which serves as a reliable foundation of selective and effective molecular sieving, since they permit fast transport for water molecules and effectively exclude ions and bigger molecules [25]. The integration of CNTs into a matrix polymer as composite membranes can dramatically improve water flux thanks to the smooth, hydrophobic inner surfaces of CNTs and enhance ion selectivity according to their diameter and surface charge [105, 106].

4 Applications of Nanomaterials in Water Treatment

The nanomaterials exclusive properties have directed to their application in addressing a wide variety of water pollutants across various water sources as shown Fig. 17.

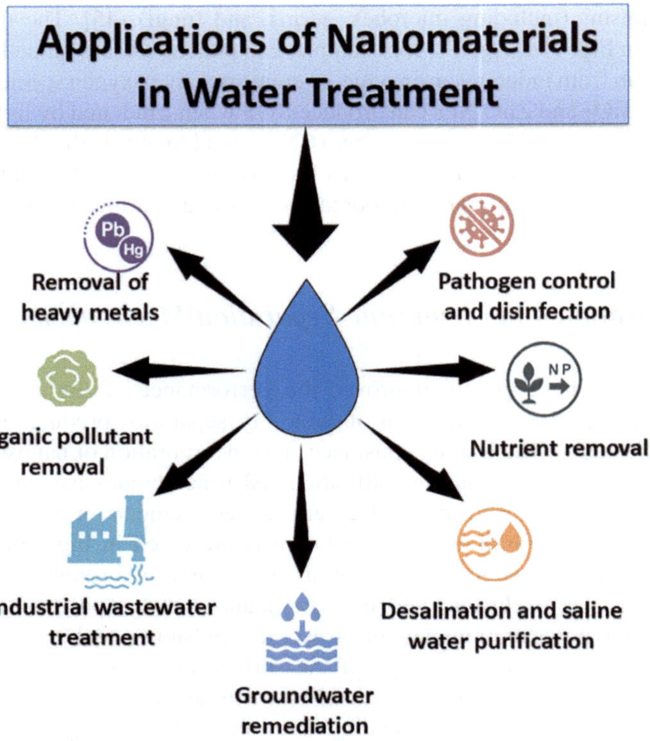

Fig.17 Applications of nanomaterials in advanced water treatment processes

4.1 Removal of Heavy Metals

Nanomaterials have emerged as highly effective adsorbents for the removal of toxic heavy metals, including lead, arsenic, mercury, and cadmium, from contaminated water. The mechanisms involved in heavy metal removal by nanomaterials include adsorption onto their high surface area, ion exchange with functional groups on the nanomaterial surface, precipitation of metal compounds, and complexation with specific ligands attached to the nanomaterial [57, 107–109]. Table 1 provides a comparative review of heavy metal removal mechanism, revealing a range of advantages, disadvantages applications.

Table 1 Nanomaterials for heavy-metal removal: representative materials, mechanisms, and target metals

Nanomaterial example	Havey metal	Removal mechanism	Advantages	Disadvantages	References
CNTs, Gr, GO, rGO	Pb (II)	Adsorption/ surface complexation	Very high surface area; tunable functional groups; fast kinetics	Fouling; regeneration cost; potential nanosheet release	[110, 111]
Fe_3O_4/α-Fe_2O_3/ γ-Fe_2O_3	As (III/ V)	Inner-sphere ligand exchange; adsorption	Strong as affinity; magnetic separability (Fe3O4)	Aggregation; capacity loss in complex matrices; Fe leaching control	[112, 113]
Activated carbon (functionalized)	Hg (II)	Adsorption (functional group binding)	Mature technology; broad availability; stable	Slower kinetics than CNT/GO; thermal/ chemical regeneration burden	[114]
Al2O_3, ZnO, MnO_2, TiO_2	Cd (II)	Adsorption/ precipitation	pH-tunable uptake; robust oxides	Particle recovery; selectivity in competitive ion mixtures	[115, 116]
MOFs	Pb (II), As(V)	Adsorption/ complexation	Very high capacities; tailored chemistry	Moisture/ chemical stability; synthesis cost/ scale	[111, 113]
Zeolite nanoparticles and composites	Pb (II), Cd (II)	Ion exchange + adsorption	Fast ion exchange; regenerable	Selectivity varies with competing cations; fouling	[107, 111]

4.2 Organic Pollutant Removal

Nanomaterials have found extensive applications in the exclusion of numerous organic contaminants from water, including pharmaceuticals, colorants, and pesticides. These pollutants are removed through a combination of catalytic degradation, photocatalysis and adsorption mechanisms [79, 117, 118]. Table 2, summarizes nanomaterials applied for organic-pollutant treatment such as dyes, pesticides, and pharmaceuticals. It contrasts adsorption and photocatalysis pathways while noting the advantages and limits of each system.

Table 2 Nanomaterials for organic-pollutant removal: adsorbents, photocatalysts, and nanocomposites

Nanomaterial example	Organic-pollutant	Mechanism	Advantages	Limitations	References
TiO2, ZnO	Dyes, pesticides, APIs	Photocatalysis (ROS oxidation)	Mineralization potential; reusable	UV bias (unless modified); e^-/h^+ recombination; catalyst recovery	[117–119]
CNTs, graphene	Dyes, APIs	Adsorption	High capacity; π–π interactions; easy blending into membranes	Fouling; difficult regeneration without loss of capacity	[117, 120]
Activated carbon nano-forms	Broad organics	Adsorption	Scalable; robust; widely available	Saturation and regeneration overhead	[121]
Polymer/magnetic nanocomposites	Dyes, APIs	Adsorption ± catalytic effects	Magnetic/easy recovery; multifunctionality	Long-term stability; possible leaching of active phase	[79] [122]
Zeolites/zeolite-carbon	Pharmaceuticals	Ion exchange/adsorption	Selective uptake of cationic drugs	Reduced performance in high-salinity matrices	[123]

4.3 Pathogen Control and Disinfection

Nanomaterials have emerged as powerful tools for controlling and disinfecting waterborne pathogens, including bacteria, viruses, and fungi. Table 3 shows nanomaterials for pathogen control and disinfection: mechanisms and deployment modes and their application.

Nanomaterials inactivate microbes through membrane disruption, ion release, and ROS generation. AgNPs are broad-spectrum antibacterial/antiviral agents; TiO2/ZnO generate ROS under irradiation; CNTs physically damage cell walls; chitosan nanoparticles provide electrostatic antimicrobial activity; composite systems can combine these effects.

4.4 Nutrient Removal

The removal of excess nutrients, such as phosphorus and nitrogen, from water is key to avert eutrophication, and nanomaterials are being investigated for their potential in this area [23].

Table 4 To prevent eutrophication, nanomaterials target phosphorus and nitrogen. Modified biochar supports nitrogen removal; lanthanum-modified sorbents and iron-oxide nanoparticles remove phosphate via adsorption/precipitation; ZrO_2-coated magnetic spheres and Al-based nanomaterials also adsorb phosphate; zeolites exchange NH_4^+ efficiently; nZVI and its composites are used for phosphate elimination.

4.5 Desalination and Saline Water Purification

Nanomaterial-enhanced membranes have emerged as a promising technology for improving the productivity of desalination and saline water purification processes [131, 132], Shown in Table 5. Nanomaterial-enhanced membranes (GO, CNTs, zeolites; and nanoporous single-layer graphene) improve water transport, salt rejection, and may lower energy demand relative to conventional polymeric membranes; FO membranes modified with nanomaterials are also being pursued for lower-energy desalination.

4.6 Groundwater Remediation

Nanomaterials have shown significant promise for the in-situ remediation of contaminated groundwater, offering cost-effective and efficient solutions for a variety of

Table 3 Nanomaterials for pathogen control and disinfection: mechanisms and deployment modes

Nanomaterial	Mechanism	Deployment	Target pathogens	Advantages	Challenges	References
AgNPs	Ion release, ROS, cell membrane damage	Coatings, powders, membranes	Bacteria (Gram + /Gram −), fungi, viruses	Broad-spectrum, low doses effective	Ag^+ release monitoring; resistance/ ecotoxicity concerns	[45, 124]
TiO_2, ZnO	Photo-ROS (•OH, O2•−)	Light-activated systems	Bacteria, algae, protozoa, viruses	Non-selective microbial inactivation; low chemical footprint	Light requirement; photo corrosion; recovery	[45]
CNTs	Physical disruption	Often embedded in membranes	Bacteria (E. coli, S. aureus, etc.)	Mechanical inactivation; anti-biofouling	Safety/regulatory scrutiny; dispersion stability	[45, 124]
Chitosan NPs	Electrostatic interactions	Suspensions, coatings	Bacteria, fungi, some viruses	Biopolymer; good antimicrobial range	pH-dependent efficacy; biodegradation	[124, 125]
Nanocomposites	Combined mechanisms	Coatings, membranes	Bacteria, fungi, viruses (depending on embedded NP)	Synergy (adsorption+disinfection)	Complexity; leaching control	[79]

Table 4 Nanomaterials for nutrient removal: nitrogen and phosphorus sorbents and exchange media

Nanomaterial	Nutrient	Mechanism	Advantages	Limitations	References
Modified biochar	N species	Adsorption/ microbial support	Low-cost; valorizes waste biomass	Variable quality; saturation in high-N loads	[126]
La-modified sorbents	Phosphate (PO4^3 −)	Adsorption/ precipitation	High P affinity; selective	La cost; regeneration chemistry	[127]
Fe-oxide NPs	Phosphate	Adsorption/ precipitation	Proven P uptake; robust	Aggregation; retrieval from water	[127]
ZrO_2-coated magnetic spheres	Phosphate	Adsorption	Magnetic recovery; good stability	Coating durability; cost	[128]
Zeolites	Ammonia (NH3/NH4+)	Ion exchange	Fast, regenerable	Competes with other cations; brine handling	[129]
nZVI and composites	Phosphate	Precipitation/ adsorption	Multi-pathway removal; possible in-situ use	Passivation; transport in porous media	[130]
Al-based nanomaterials	Phosphate	Adsorption	Additional P sorbents; available	Al release control; pH sensitivity	[128]

pollutants. Table 6, for in-situ cleanup, nZVI reduces or immobilizes chlorinated solvents, metals, and pesticides; bimetallic Fe–Pd accelerates degradation (e.g., TCE); EZVI improves DNAPL contact; CNTs adsorb metals/organics; and stabilized nanocomposites (e.g., PEI-coated nZVI or sulfide-modified nZVI on biochar) increase efficiency and longevity in the subsurface.

4.7 *Industrial Wastewater Treatment*

Industrial wastewater often holds a multifaceted combination of contaminants, including dyes, pharmaceuticals, heavy metals, and other organic and inorganic compounds, and nanomaterials have shown great potential in addressing these challenges [93]. Industrial effluents often contain mixed contaminant loads (dyes, APIs, heavy metals). Carbon adsorbents (activated carbon, CNTs, graphene) provide broad removal; ZnO/TiO_2 photocatalyzed dye/organic degradation; magnetic nanoparticles aid adsorption with easy magnetic separation; nanofiber membranes filter metals/dyes; MOFs are explored for selective metal removal Table 7.

Table 5 Nanomaterial-enhanced desalination and saline-water purification membranes

Membrane/ material	Reported advantage	Key advantages	Challenges	References
GO membranes	Tunable interlayer spacing; ion sieving	High flux/ selectivity; tunable channels	Swelling control; fouling; long-term stability	[131, 132]
CNT-based composites	Smooth hydrophobic channels; high flux	Enhanced permeability; mechanical strength	Fabrication cost; alignment/control of CNTs	[105]
Single-layer graphene (nanoporous)	Near-perfect salt rejection; very high transport	Atomically thin; precise pores	Scalable pore fabrication; defect management	[133]
Zeolite membranes	Ion-exchange assisted separations	Chemically robust; selective	Brittleness; integration with modules	[104]
FO membranes (nano-modified)	Lower energy potential versus RO	Reduced pressure/ energy; tunable layers	Draw-solution challenges; reverse solute flux	[25]

5 Environmental and Health Impacts of Nanomaterials in Water Treatment

While nanomaterials offer noteworthy returns in water treatment, it is decisive to reflect their potential conservational and health impacts to ensure their responsible and sustainable application.

5.1 Potential Risks and Ecotoxicity Concerns

In water treatment applications, the ecotoxicological impacts of nanomaterials on aquatic life are a major issue. Through various processes, including oxidative stress, genotoxicity, and membrane damage, nanomaterials may have deleterious impacts on marine life, including fish, invertebrates, and algae [73, 142]. Even at low concentrations, nanoparticles' high surface area can have serious harmful impacts [74]. At high quantities, metal-based nanoparticles with antibacterial qualities, such as CuNPs and AgNPs, can also be hazardous to aquatic life's [142, 143]. Similarly, because of their large surface area and capacity for bioaccumulation, carbon-based nanomaterials such as CNTs and GO may have negative impacts on aquatic creatures. Once within an organism, nanoparticles can build up inside cells, which might result in intracellular alterations [73, 143].

Table 6 Nanomaterials for groundwater remediation: in-situ technologies and their mechanisms

Nanomaterial	Pollutants	Mechanism	Advantages	Disadvantages	References
nZVI	Chlorinated solvents, metals, pesticides	Reductive transformation; immobilization	Field-tested; rapid dichlorination	Passivation; delivery and distribution	[134, 135]
Fe–Pd BNPs	TCE and similar	Catalyzed reductive degradation	Faster kinetics; lower activation energy	Pd cost; potential metal release	[136]
EZVI	DNAPLs	Emulsified ZVI improves contact	Better DNAPL contact; stabilized	Emulsion stability; injection logistics	[136]
CNTs	Metals, organics	Adsorption	High capacity; complements ZVI	Potential mobility/ toxicity concerns	[137]
Stabilized nZVI (PEI-coated; sulfide-mod; biochar-supported)	Mixed contaminants	Stabilization+reduction/ adsorption	Improved longevity; transport	Additive cost; site-specific tuning	[136]

Table 7 Nanomaterials in industrial wastewater treatment: broad contaminant classes and treatment routes

Nanomaterial	Target	Mechanism	Advantages	Challenges	References
Activated carbon, CNTs, graphene	Dyes, APIs, metals	Adsorption	Broad-spectrum; scalable; retrofit-friendly	Saturation/ regeneration; fouling	[138, 139]
ZnO, TiO_2	Dyes/ organics	Photocatalysis	Effective on recalcitrant dyes; reusable	UV bias; recombination; catalyst handling	[72, 140]
Magnetic nanoparticles	Dyes, metals	Adsorption+magnetic recovery	Easy separation/ reuse; fast kinetics	Stability; potential leaching	[72]
Nanofiber membranes	Metals, dyes	Size/charge sieving	High surface area; tailored pores	Mechanical durability; fouling	[141]
MOFs	Metals	Adsorption/ complexation	Tunable selectivity/ capacity	Hydrolytic stability; scale-up cost	[111]

5.2 Transport and Fate of Nanomaterials in Aquatic Environments

Nanoparticles used in water treatment can enter aquatic systems through various pathways, including direct manufacturing discharges, disposal of wastewater treatment wastes, and indirect surface excess from soils [144]. Once in these environments, nanomaterials undergo various transformations, such as agglomeration, aggregation, dissolution, and sulfidation, which are influenced by their physicochemical characteristics and the properties of the receiving water, including concentration, pH, and temperature, of natural organic matter (NOM), ionic strength, salinity, and water rigidity [145]. These transformations can lead to the formation of a "corona" on the surface of nanomaterials due to the adsorption of organic ligands, metals, and naturally occurring colloids, which can further modify their fate and bioavailability. There is also evidence that nanomaterials can be bioavailable to plants, potentially leading to trophic transfer and impacting other organisms through biomagnification within aquatic food webs [146]. Notably, nanoparticles can even be transported from aquatic environments back to terrestrial ecosystems via emerging aquatic insects, potentially decoupling aquatic and terrestrial food webs [147].

5.3 Strategies for Minimizing Environmental Risks

To minimize the potential environmental risks of nanomaterials in water treatment, numerous strategies are being actively investigated [148]. That is, to develop intrinsically safer nanomaterials with lower toxicity, and better stability in aquatic environments. Nanomaterials can also modify their surfaces to control their aggregation state and improve their dispersibility, thus modulating their interactions with biological systems. The most effective removal and recovery of the nanomaterial following the water treatment step, such as magnetic separation in the case of magnetic nanoparticles, is important so as not to release them into the environment in large amounts [57, 117]. In addition, full life cycle assessments of various nanomaterial-based water treatment technologies can provide insight into the environmental impact of all phases of the technology from production to disposal and guide sustainability efforts in this space [117, 149].

5.4 Regulatory Considerations and Guidelines

The regulatory landscape for nanomaterials, including their use in water treatment, is still evolving globally, many countries acknowledge the potential risks and benefits of nanomaterials, but regulations are still lacking in many jurisdictions. In Europe, nanomaterials are controlled under the REACH (Registration, Evaluation, Authorization and Restriction of Chemicals) framework, as they fall under the definition of a chemical "substance", with specific stipulations for nanomaterials are still being refined [150]. In the United States, Under the Toxic Substances Control Act (TSCA), the Environmental Protection Agency (EPA) has issued administrative decisions allowing the restricted production of novel nanoscale chemical compounds, with ongoing development of specific rules for nanomaterials [151]. Despite the evolving nature of regulations, regulatory bodies issued guidelines and assessments [16], some regions have also established voluntary or obligatory registers for nanomaterials to gather information on their production, distribution, and potential risks [152]. Developing clear regulatory frameworks and standards is considered crucial for ensuring the safety and environmental sustainability of nanomaterial applications in water treatment. Achieving harmonization of these regulations across different regions and continents remains a significant challenge.

6 Limitations, Challenges, and Future Perspectives

The application of nanomaterials in water treatment, while exhibiting significant potential, is also associated with several challenges, limitations, and areas requiring further research and development.

6.1 Technical and Operational Challenges

Aggregated nanoparticles are one of the major technical challenges facing these systems; the reduction of effective surface area has been associated with the loss of activity for many processes which negatively impacts the candidate to remove contaminants [153, 154]. Another critical operational challenge is the recovery of nanomaterials from treated water. Without an efficient and low-cost method for separation, nanomaterials could be released into the environment, causing secondary contamination [155]. While magnetic separation offers a solution for certain materials, it is not universally applicable [156]. Furthermore, maintaining the long-term stability and performance of nanomaterials, particularly in complex water matrices with varying pH, salinity, and temperature, remains a significant hurdle [155].

6.2 Scalability and Economic Feasibility

High production costs of some nanomaterials inhibit the wide development of nanomaterial-based water treatment technologies in large scale applications. For these technologies to be affordable for widespread use, there is an acute need for developing cheaper and more sustainable synthesis approaches for nanomaterials [79]. While addressing these challenges requires a thorough investigation of the life-cycle of nanomaterials, the potential for recycling and reusing nanomaterials post water treatment application may be an avenue to further decrease costs and improve the economic feasibility of these technologies. Additionally, nanotechnology-based decentralized and point-of-use water treatment systems may provide economically viable alternatives in remote or neglected communities [117]. In this context, life cycle assessments and cost–benefit analyses can be conducted to assess the economic feasibility and sustainability of nanoscale solutions for water treatment [57].

6.3 Synergistic and Multi-Pollutant Strategies

Many waters carry **co-occurring organics, metals, and pathogens**, making single-function steps inefficient. **Multifunctional hybrids** co-locate mechanisms to shorten trains and improve resilience:

- Carbon-supported photocatalysts (e.g., GO–TiO_2/GO–ZnO) in which Gr/GO can pre-concentrate aromatics and improve charge separation at TiO_2/ZnO interfaces, accelerating ROS-driven degradation under light [98], while the photocatalytic pathway is standard in your mechanism section [99, 100]. This pairing targets organics and can assist microbial inactivation in the same stage.

- GO–Fe_3O_4 or Fe_3O_4–polymer composites, Fe3O4/Zinc have high-capacity adsorption with magnetic recovery; Fe_3O_4 can contribute Fenton-like oxidation for polishing [63, 66, 112].
- Ag-decorated carbon membranes: size/charge sieving plus broad antimicrobial action to control biofouling and pathogens [44, 48, 97, 101].
- nZVI–carbon composites-based hybrid provides reductive transformation of halogenated organics alongside adsorption of metals/organics on carbon [29, 87].

To demonstrate **true synergy**, test **mixed feeds** (e.g., dye + Pb^{2+} + E. coli), track **TOC mineralization** and $\geq$ **3-log** microbial reduction, quantify **nanoparticle release**, and verify $\geq$ **80–90%** performance retention over **5–10** regeneration cycles with **easy retrieval** (magnetic or embedded). These designs align with the removal mechanisms in Fig. 16 (adsorption, catalytic/photocatalytic oxidation, disinfection, membrane separation) and with your **composite/magnetic** sections, providing a direct bridge from mechanisms to compact treatment modules.

6.4 Future Research Directions and Technological Advancements

Future research in this field should focus on several key areas to overcome current limitations and realize the full potential of nanotechnology in water treatment.

6.4.1 Visible-light Photocatalysts and Z-schemes

Future work should prioritize nanomaterials that can be readily activated by visible light, rather than requiring short wavelengths in the ultraviolet (UV) range, because utilizing visible light, which is the most abundant part of the sunlight, would allow for more sustainable and energy-saving techniques for photocatalytic water purification [157]. For example, doped TiO_2 or heterojunctions with ZnO and narrow-band semiconductors while preserving strong redox potentials and stability over repeated cycles. Mechanistic and materials choices already discussed in the chapter (TiO_2, ZnO, band-gap tuning, and green synthetic routes) provide the basis for such designs [54, 55, 60].

6.4.2 Multifunctional Nanomaterials Materials

Future directions in this area should aim at designing multifunctional nanomaterials capable of simultaneous removal of several pollutants simultaneously and disinfection to help in making the water purification more efficient and less complex [158]. Practical directions include carbon-supported photocatalysts (e.g., graphene/TiO_2) to enhance charge separation and broaden light use [98], magnetic cores for easy retrieval and optional Fenton-like polishing (Fe_3O_4) [65, 66], and composites that retain performance after regeneration. Report durability and leaching alongside removal.

6.4.3 Sensing-integrated Membranes and Adaptive Control

Advanced monitoring and sensing techniques based on nanomaterials will be required to screen water quality in real-time and to optimize the performance of nanomaterial-based treatment systems [28, 159]. Embedding pressure/UV254/ROS sensors into nanomaterial-enabled membranes can sharpen backwash/UV dosing and reduce fouling, complementing the membrane performance gains summarized earlier (flux, rejection, antifouling) [24, 103, 104].

6.4.4 Hybridization with Conventional Systems and Long-Term Impact Studies

The effect of nanotechnology-based solutions on existing conventional water treatment infrastructure needs additional studies to investigate their potential for incorporating nanotechnology-based solutions to enable sustainable processes to be hybridized with existing plants, thus capitalizing on the strengths of both methods [117]. Further investigation and additional research into the long-term ecological and human health effects of nanomaterials released from water treatment plants is necessary to ensure the safe and sustainable utilization of nanomaterials [160].

6.4.5 Green Synthesis and safe-by-design

A novel bio-inspired and green synthesis approaches in the manufacturing of nanomaterials will also help build more sustainable and environmentally friendly water treatment technologies [79]. We review that green methods reported for TiO_2 and ZnO; a practical next step is to scale these approaches to kilogram–tonne levels with tight control over particle size distribution and band-gap-relevant surface states. Priorities include immobilization (polymer/ceramic matrices) or magnetic cores to simplify recovery and limit leaching, and reporting of manufacturing metrics alongside performance (E-factor, solvent use, kWh kg^{-1}, batch reproducibility). Because cost is a barrier, pair green routes with life-cycle assessment and cost-per-m^3 to quantify benefits relative to conventional syntheses and to 6.2's economic benchmarks. Finally, integrate release monitoring (e.g., Ag^+, Fe) and end-of-life plans (magnetic retrieval, low-temperature regeneration) to align synthesis with environmental safety and regulatory expectations [15, 17, 161]. The major challenge with green synthesis is the lack of standardization and control. The phytochemical composition of a plant extract can vary based on the plant species, growth conditions, and harvesting time, which can lead to inconsistencies in the size, shape, and stability of the resulting nanoparticles. Therefore, future research must focus on developing standardized protocols and quality control measures to ensure the consistency and reliability of green-synthesized nanomaterials for industrial applications.

7 Conclusion

Nanomaterials have been attracting growing interest as a promising and versatile family of materials with significant utility for the treatment and purification of water. Their exclusive characteristics at the nanoscale allow them to promote multiple contaminant elimination techniques such as highly efficient adsorption, catalytic and photocatalytic degradation, effective elimination of bacteria and microorganisms, and considerable improvements to membrane separation technologies. They enable the elimination of pollutants from a broad range, including heavy metals, organic compounds, pathogenic microorganisms, excess nutrients, and advanced applications of separation such as desalination and groundwater remediation. Nanomaterials have the potential to foster access to fresh and harmless water to an increasingly large international population by providing solutions that can be more efficient, selective, sustainable, and cost-effective than traditional methods of water purification. Nonetheless, the wide-ranging applications of nanomaterials in water treatment will necessitate further investigations of their long-term effects on human healthiness and the environment. And, emerging strong monitoring agendas and affordable yet scalable technologies are also important next steps in making the promise of nanotechnology a reality, not just in the laboratory but in providing practical solutions to the problems of the purifying water and protecting this resource critical to all life.

References

1. Lin, L., Yang, H., Xu, X.: Effects of water pollution on human health and disease heterogeneity: a review. Front Environ. Sci. **10** (2022) https://doi.org/10.3389/fenvs.2022.880246
2. Mancuso, G., Bencresciuto, G.F., Lavrnić, S., Toscano, A.: Diffuse water pollution from agriculture: a review of nature-based solutions for nitrogen removal and recovery. MDPI AG. (2021). https://doi.org/10.3390/w13141893
3. Saquib, S., Gupta, A., Joshi, A.: Chapter 21—emerging water crisis: impact of urbanization on water resources and constructed wetlands as a nature-based solution (NbS). In: Srivastav, A.L., Madhav, S., Bhardwaj, A.K., Valsami-Jones, E. (eds.) Current Directions in Water Scarcity Research, vol. 6, pp. 447–468. Elsevier (2022). https://doi.org/10.1016/B978-0-323-91838-1.00021-X
4. Mustafa, B., Hassan, N.: Water contamination and its effects on human health: a review. J. Geogr. Environ. Earth Sci. Int. **28**:38–49 (2024). https://doi.org/10.9734/JGEESI/2024/v28i1743
5. Adelodun, B., et al.: Assessment of socioeconomic inequality based on virus-contaminated water usage in developing countries: a review. Environ. Res. **192** (2021). https://doi.org/10.1016/j.envres.2020.110309

6. Angreni, E.: Review on optimization of conventional drinking water treatment plant. World Appl. Sci. J. **7**(9), 1144–1151 (2009)
7. Wang, S.: From coagulation to disinfection: key steps and challenges in wastewater treatment. In: E3S Web of Conferences. EDP Sciences (2025). https://doi.org/10.1051/e3sconf/202560605002
8. Nasir, O., et al.: Conventional and advanced wastewater treatment techniques; development of sustainable environment and production of biofertilizers. Int. J. Res. Publ. Rev. **5**(1), 5491–5499 (2024). https://doi.org/10.55248/gengpi.5.0124.0364
9. Shamshad, J., Ur Rehman, R.: Innovative approaches to sustainable wastewater treatment: a comprehensive exploration of conventional and emerging technologies. R. Soc. Chem. (2024). https://doi.org/10.1039/d4va00136b
10. Liu, Z., Gan, Y., Luo, J., Luo, X., Ding, C., Cui, Y.: Current status of emerging contaminant models and their applications concerning the aquatic environment: a review. Multidiscip. Digit. Publ. Inst. (MDPI) (2025). https://doi.org/10.3390/w17010085
11. Crini, G., Lichtfouse, E.: Wastewater Treatment: An Overview. pp. 1–21 (2018). https://doi.org/10.1007/978-3-319-92111-2_1
12. Matesun, J., Petrik, L., Musvoto, E., Ayinde, W., Ikumi, D.: Limitations of wastewater treatment plants in removing trace anthropogenic biomarkers and future directions: a review. Ecotoxicol. Environ. Saf. **281**, 116610 (2024). https://doi.org/10.1016/j.ecoenv.2024.116610
13. Gehrke, I., Geiser, A., Somborn-Schulz, A.: Innovations in Nanotechnology for Water Treatment. Dove Medical Press Ltd. (2015) https://doi.org/10.2147/NSA.S43773
14. Jain, K., Patel, A.S., Pardhi, V.P., Flora, S.J.S.: Nanotechnology in wastewater management: a new paradigm towards wastewater treatment. MDPI AG (2021). https://doi.org/10.3390/molecules26061797
15. Nishu, Kumar, S.: Smart and innovative nanotechnology applications for water purification. Hybrid Adv. **3**, 100044 (2023). https://doi.org/10.1016/j.hybadv.2023.100044
16. Dada, M.A., Oliha, J.S., Majemite, M.T., Obaigbena, A., Nwokediegwu, Z.Q.S., Daraojimba, O.H.: Review of nanotechnology in water treatment: adoption in the USA and prospects for Africa. World J. Adv. Res. Rev. **21**(1), 1412–1421 (2024). https://doi.org/10.30574/wjarr.2024.21.1.0170
17. Kumar, S., Chauhan, J.S.: Application of Plant-Based Nanoparticles in Wastewater Decontamination. pp. 89–113 (2022). https://doi.org/10.1007/978-3-031-00812-2_4
18. Assad, H., Lone, I.A., Kumar, A., Kumar, A.: Unveiling the contemporary progress of graphene-based nanomaterials with a particular focus on the removal of contaminants from water: a comprehensive review. Frontiers Media SA (2024). https://doi.org/10.3389/fchem.2024.1347129
19. Rathinavel, S., Priyadharshini, K., Panda, D.: A review on carbon nanotube: an overview of synthesis, properties, functionalization, characterization, and the application. Mater. Sci. Eng. B **268**, 115095 (2021). https://doi.org/10.1016/j.mseb.2021.115095
20. Patil, T.V., Patel, D.K., Dutta, S.D., Ganguly, K., Randhawa, A., Lim, K.T.: Carbon nanotubes-based hydrogels for bacterial eradiation and wound-healing applications. Appl. Sci. **11**, 9550 (2021). https://doi.org/10.3390/APP11209550
21. Arora, B., Attri, P.: Carbon nanotubes (CNTs): a potential nanomaterial for water purification. MDPI AG (2020). https://doi.org/10.3390/jcs4030135
22. Mishra, S., Sundaram, B.: Efficacy and challenges of carbon nanotube in wastewater and water treatment. Environ. Nanotechnol. Monit. Manag. **19**, 100764 (2023). https://doi.org/10.1016/j.enmm.2022.100764
23. Marghade, D., Karunanidhi, D.: 10—Properties and adsorption mechanism of biological contaminants by carbon nanotubes. In: Hadi Dehghani, M., Karri, R.R., Mubarak, N.M. (eds.) Water Treatment Using Engineered Carbon Nanotubes, pp. 271–305. Elsevier (2024) https://doi.org/10.1016/B978-0-443-18524-3.00012-X
24. Ma, L., et al.: Fabrication and water treatment application of carbon nanotubes (CNTs)-based composite membranes: a review. Membranes **7**(1), 16 (2017). https://doi.org/10.3390/MEMBRANES7010016

25. Kang, J., et al.: Graphene membrane for water-related environmental application: a comprehensive review and perspectives. ACS Environmental Au. **5**(1), 35–60 (2025). https://doi.org/10.1021/acsenvironau.4c00088
26. Hou, Y., Lv, S., Liu, L., Liu, X.: High-quality preparation of graphene oxide via the Hummers' method: understanding the roles of the intercalator, oxidant, and graphite particle size. Ceram. Int. **46**(2), 2392–2402 (2020). https://doi.org/10.1016/j.ceramint.2019.09.231
27. Plastiras, O.E., Deliyanni, E., Samanidou, V.: Applications of graphene-based nanomaterials in environmental analysis. Appl. Sci. **11**(7), 3028 (2021). https://doi.org/10.3390/APP11073028
28. Khalajiolyaie, A., Jian, C.: Advances in graphene-based materials for metal ion sensing and wastewater treatment: a review. Multidiscip. Digit. Publ. Inst. (MDPI) (2025). https://doi.org/10.3390/environments12020043
29. Joel, E.F., Lujanienė, G.: Progress in graphene oxide hybrids for environmental applications. Environments **9**(12), 153 (2022). https://doi.org/10.3390/ENVIRONMENTS9120153
30. Ganjoo, R., Sharma, S., Kumar, A., Daouda, M.M.A.: Activated carbon: fundamentals, classification, and properties. In: Activated Carbon, The Royal Society of Chemistry, pp. 1–22 (2023). https://doi.org/10.1039/BK9781839169861-00001
31. Soonmin, H., Kabbashi, N.: Review on activated carbon: synthesis, properties and applications. Int. J. Eng. Trends Technol. **69**, 124–139 (2021). https://doi.org/10.14445/22315381/IJETT-V69I9P216
32. Tarikuzzaman, M.: A review on activated carbon: synthesis, properties, and applications. Eur. J. Adv. Eng. 114–123 (2023)
33. Shahrokhi-Shahraki, R., Benally, C., El-Din, M.G., Park, J.: High efficiency removal of heavy metals using tire-derived activated carbon vs commercial activated carbon: insights into the adsorption mechanisms. Chemosphere **264**, 128455 (2021). https://doi.org/10.1016/j.chemosphere.2020.128455
34. Muhammad, S., Yahya, E.B., Abdul Khalil, H.P.S., Marwan, M., Albadn, Y.M.: Recent advances in carbon and activated carbon nanostructured aerogels prepared from agricultural wastes for wastewater treatment applications. MDPI (2023). https://doi.org/10.3390/agriculture13010208
35. Lazarenko, N.S., Golovakhin, V.V., Shestakov, A.A., Lapekin, N.I., Bannov, A.G.: Recent advances on membranes for water purification based on carbon nanomaterials. Membranes (Basel) **12**(10), 915 (2022). https://doi.org/10.3390/membranes12100915
36. Zhu, L., Yao, Y., Chen, D., Lan, P.: The effective removal of Pb2+ by activated carbon fibers modified by l-cysteine: exploration of kinetics, thermodynamics and mechanism. RSC Adv. **12**(31), 20062–20073 (2022). https://doi.org/10.1039/D2RA01521H
37. Dubey, R., Dutta, D., Sarkar, A., Chattopadhyay, P.: Functionalized carbon nanotubes: synthesis, properties and applications in water purification, drug delivery, and material and biomedical sciences. Nanoscale Adv. **3**(20), 5722–5744 (2021). https://doi.org/10.1039/D1NA00293G
38. Saifuddin, N., Raziah, A.Z., Junizah, A.R.: Carbon nanotubes: a review on structure and their interaction with proteins. J. Chem. **2013**(1), 676815 (2013). https://doi.org/10.1155/2013/676815
39. Liu, M., Chen, G., Xu, L., He, Z., Ye, Y.: Environmental remediation approaches by nanoscale zero valent iron (nZVI) based on its reductivity: a review. RSC Adv. **14**(29), 21118–21138 (2024). https://doi.org/10.1039/D4RA02789B
40. Al-Zuhairy, M.S., et al.: Efficient pollutant removal in wastewater using nanosized zero-valent iron: mechanisms and applications. Mater. Sci. Eng. B **305**, 117432 (2024). https://doi.org/10.1016/j.mseb.2024.117432
41. Ken, D.S., Sinha, A.: Recent developments in surface modification of nano zero-valent iron (nZVI): remediation, toxicity and environmental impacts. Environ. Nanotechnol. Monit. Manag. **14**, 100344 (2020). https://doi.org/10.1016/j.enmm.2020.100344
42. Dan-Iya, B.I., Khan, A., Shukor, M.Y.A., Sabullah, M., Masdor, N.A.: Zero-valent iron nanoparticles for environmental Hg (II) removal: a review. Peer J. Mater. Sci. **5**, e29 (2023). https://doi.org/10.7717/peerj-matsci.29

43. Pasinszki, T., Krebsz, M.: Synthesis and application of zero-valent iron nanoparticles in water treatment, environmental remediation, catalysis, and their biological effects. Nanomaterials **10**(5):917 (2020). https://doi.org/10.3390/NANO10050917
44. Abbas, R., et al.: Silver nanoparticles: synthesis, structure, properties and applications. Multidisciplinary Digital Publishing Institute (MDPI) (2024). https://doi.org/10.3390/nano14171425
45. Olawade, D.B., Wada, O.Z., Fapohunda, O., Egbewole, B.I., Ajisafe, O., Ige, A.O.: Nanoparticles for microbial control in water: mechanisms, applications, and ecological implications. Front. Nanotechnol. **6** (2024). https://doi.org/10.3389/fnano.2024.1427843
46. Asif, M., Yasmin, R., Asif, R., Ambreen, A., Mustafa, M., Umbreen, S.: Green synthesis of silver nanoparticles (AgNPs), structural characterization, and their antibacterial potential. Dose-Response **20**(2) (2022). https://doi.org/10.1177/15593258221088709
47. Nkosi, N.C., Basson, A.K., Ntombela, Z.G., Dlamini, N.G., Pullabhotla, R.V.S.R.: Green synthesis, characterization and application of silver nanoparticles using bioflocculant: a review. Multidiscip. Digit. Publ. Inst. (MDPI) (2024). https://doi.org/10.3390/bioengineering11050492
48. López-Herrera, A., Gómez-Merino, F.C., Zavaleta-Mancera, H.A., Avalos-Borja, M., García-Nava, J.R., Trejo-Téllez, L.I.: Effect of silver nanoparticles (AgNPs) on aquatic and wetland plants. Multidiscip. Digit. Publ. Inst. (MDPI) (2024). https://doi.org/10.3390/environments11120297.
49. Sameena, V.P., Thoppil, J.E.: Green synthesis of silver nanoparticles from Euphorbia and its biological activities. Nanotechnol. Environ. Eng. **7**(2), 377–392 (2022). https://doi.org/10.1007/S41204-022-00232-6
50. Lusvardi, G., Barani, C., Giubertoni, F., Paganelli, G.: Synthesis and characterization of TiO2 nanoparticles for the reduction of water pollutants. Materials 10(10) (2017). https://doi.org/10.3390/ma10101208
51. Al-Ani, F.H., Alsalhy, Q.F., Raheem, R.S., Rashid, K.T., Figoli, A.: Experimental investigation of the effect of implanting tio2-nps on pvc for long-term uf membrane performance to treat refinery wastewater. Membranes (Basel) **10**(4) (2020). https://doi.org/10.3390/membranes10040077
52. Tetteh, E.K., Rathilal, S., Asante-Sackey, D., Chollom, M.N.: Prospects of synthesized magnetic TiO2-based membranes for wastewater treatment: a review. Materials **14**(13), 3524 (2021). https://doi.org/10.3390/ma14133524
53. Mehmood, A., et al.: Magnetic nanocomposites for sustainable water purification—a comprehensive review. Environ. Sci. Pollut. Res. **28**(16), 19563–19588 (2021). https://doi.org/10.1007/s11356-021-12589-3
54. Rosa, D., Abbasova, N., Di Palma, L.: Titanium dioxide nanoparticles doped with iron for water treatment via photocatalysis: a review. Nanomaterials **14**(3), 293 (2024). https://doi.org/10.3390/NANO14030293
55. Nam, Y., Lim, J.H., Ko, K.C., Lee, J.Y.: Photocatalytic activity of TiO2 nanoparticles: a theoretical aspect. J. Mater. Chem. A Mater. **7**(23), 13833–13859 (2019). https://doi.org/10.1039/C9TA03385H
56. Nam, Y., Lim, J.H., Ko, K.C., Lee, J.Y.: Photocatalytic activity of TiO2 nanoparticles: a theoretical aspect. J. Mater. Chem. A **7**(23), 13833–13859 (2019). https://doi.org/10.1039/C9TA03385H
57. Karnwal, A., Malik, T.: Nano-revolution in heavy metal removal: engineered nanomaterials for cleaner water. Frontiers Media SA (2024). https://doi.org/10.3389/fenvs.2024.1393694
58. Ngoepe, M.P., Schoeman, S., Roux, S.: Challenges associated with the use of metal and metal oxide nanoparticles as antimicrobial agents: a review of resistance mechanisms and environmental implications. Biotechnol. J. **20**(7), e70066 (2025). https://doi.org/10.1002/BIOT.70066
59. Abou Zeid, S., Leprince-Wang, Y.: Advancements in ZnO-based photocatalysts for water treatment: a comprehensive review. Multidiscip. Digit. Publ. Inst. (MDPI) (2024). https://doi.org/10.3390/cryst14070611

60. Aftab, S. et al.: Highly efficient visible light active doped ZnO photocatalysts for the treatment of wastewater contaminated with dyes and pathogens of emerging concern. MDPI (2022). https://doi.org/10.3390/nano12030486
61. Rajeswari, S., Venckatesh, R.: Chapter 24—ZnO-based nanoparticles for wastewater treatment: a review. In: Abd-Elsalam, K.A. (ed.) Zinc-Based Nanostructures for Environmental and Agricultural Applications, pp. 485–507 Elsevier (2021). https://doi.org/10.1016/B978-0-12-822836-4.00022-7
62. Spoială, A., et al.: Zinc oxide nanoparticles forwater purification. Materials **14**(16) (2021). https://doi.org/10.3390/ma14164747
63. Nguyen, M.D., Tran, H.V., Xu, S., Lee, T.R.: Fe3O4 nanoparticles: structures, synthesis, magnetic properties, surface functionalization, and emerging applications. MDPI (2021). https://doi.org/10.3390/app112311301
64. Arrisujaya, D., Yulizar, Y., Saefumillah, A.: Accelerated Water Purification With Magnetite (Iron(II, III) Oxide) Nanoparticles: Coagulation Applications. Sami Publishing Company (2025). https://doi.org/10.48309/jcr.2025.476858.1372
65. Aragaw, T.A., Bogale, F.M., Aragaw, B.A.: Iron-based nanoparticles in wastewater treatment: a review on synthesis methods, applications, and removal mechanisms. J. Saudi Chem. Soc. **25**(8), 101280 (2021). https://doi.org/10.1016/j.jscs.2021.101280
66. Li, X., Li, J., Shi, W., Bao, J., Yang, X.: A fenton-like nanocatalyst based on easily separated magnetic nanorings for oxidation and degradation of dye pollutant. Materials **13**(2), 332 (2020). https://doi.org/10.3390/MA13020332
67. Gharbi, A.H., et al.: Eco-friendly synthesis of Al2O3 nanoparticles: comprehensive characterization properties, mechanics, and photocatalytic dye adsorption study. Coatings **14**(7) (2024). https://doi.org/10.3390/coatings14070848
68. Dhawale, V.P., Khobragade, V., Kulkarni, S.D.: Synthesis and characterization of aluminium oxide (Al2O3) nanoparticles and its application in azodye decolourisation. Int. J. Environ. Chem. **2**(1), 10–17 (2018). https://doi.org/10.11648/j.ijec.20180201.13
69. Gudkov, S.V., Burmistrov, D.E., Smirnova, V.V., Semenova, A.A., Lisitsyn, A.B.: A mini review of antibacterial properties of Al2O3 nanoparticles. MDPI (2022). https://doi.org/10.3390/nano12152635
70. Rajendran, S., et al.: A review of synthesis and applications of Al2O3 for organic dye degradation/adsorption. Multidiscip. Digit. Publ. Inst. (MDPI) (2023). https://doi.org/10.3390/molecules28237922
71. Olawade, D.B., et al.: Metal and metal oxide nanomaterials for heavy metal remediation: novel approaches for selective, regenerative, and scalable water treatment. Frontiers Media SA (2024). https://doi.org/10.3389/fnano.2024.1466721
72. Senthil Rathi, B., et al.: Recent trends and advancement in metal oxide nanoparticles for the degradation of dyes: synthesis, mechanism, types and its application. Nanotoxicology **18**(3):272–298 (2024). https://doi.org/10.1080/17435390.2024.2349304
73. Gambardella, C., Pinsino, A.: Nanomaterial ecotoxicology in the terrestrial and aquatic environment: a systematic review. MDPI (2022). https://doi.org/10.3390/toxics10070393
74. Ray, P.C., Yu, H., Fu, P.P.: Toxicity and environmental risks of nanomaterials: challenges and future needs. (2009). https://doi.org/10.1080/10590500802708267
75. Agboola, O., et al.: A review on polymer nanocomposites and their effective applications in membranes and adsorbents for water treatment and gas separation. MDPI AG (2021). https://doi.org/10.3390/membranes11020139
76. Sahu, A., Dosi, R., Kwiatkowski, C., Schmal, S., Poler, J.C.: Advanced polymeric nanocomposite membranes for water and wastewater treatment: a comprehensive review. Polymers 15(3), 540 (2023). https://doi.org/10.3390/POLYM15030540
77. Adeola, A.O., Nomngongo, P.N.: Advanced polymeric nanocomposites for water treatment applications: a holistic perspective. MDPI (2022). https://doi.org/10.3390/polym14122462
78. Pandey, M.: Advances in nanoparticles and nanocomposites for water and wastewater treatment: a review. (2024). https://doi.org/10.3390/w16111481

79. Tripathy, J., et al.: Advances in nanoparticles and nanocomposites for water and wastewater treatment: a review. Multidiscip. Digit. Publ. Inst. (MDPI) (2024). https://doi.org/10.3390/w16111481
80. Darwish, M.S.A., Mostafa, M.H., Al-Harbi, L.M.: Polymeric nanocomposites for environmental and industrial applications. MDPI (2022). https://doi.org/10.3390/ijms23031023
81. Niculescu, A.-G., Mihaiescu, B., Mihaiescu, D., Hadibarata, T., Grumezescu, A.: An updated overview of magnetic composites for water decontamination. Polymers (Basel) **16**(5), 709 (2024). https://doi.org/10.3390/polym16050709
82. Govan, J.: Recent advances in magnetic nanoparticles and nanocomposites for the remediation of water resources. MDPI AG (2020). https://doi.org/10.3390/MAGNETOCHEMISTRY6040049
83. Zhao, Y., et al.: Preparation and application of magnetic composites using controllable assembly for use in water treatment: a review. Multidiscip. Digit. Publ. Inst. (MDPI) (2023). https://doi.org/10.3390/molecules28155799
84. Sukoviene, A., Ali, S., Jagminas, A., Ramanavicius, S.: Magnetic cobalt and other types of ferrite nanoparticles: synthesis aspects and novel strategies for application in wastewater treatment (Review). Appl. Sci. **15**(2), 857 (2025). https://doi.org/10.3390/app15020857
85. Chkirida, S., Zari, N., el kacem Qaiss, A., Bouhfid, R.: Future Challenges and Perspectives in Water Purification by Hybrid Materials, pp 317–330 (2022). https://doi.org/10.1007/978-981-19-2332-6_13
86. Ursino, C., Castro-Muñoz, R., Drioli, E., Gzara, L., Albeirutty, M.H., Figoli, A.: Progress of nanocomposite membranes for water treatment. MDPI AG (2018). https://doi.org/10.3390/membranes8020018
87. Singh, A., Verma, N., Kumar, K.: Hybrid composites: a revolutionary trend in biomedical engineering. Mater. Biomed. Eng. Bioact. Mater. Prop. Appl. 33–46 (2019). https://doi.org/10.1016/B978-0-12-818431-8.00002-7
88. Kickelbick, G.: Hybrid materials—past, present and future. Hybrid Mater. **1**(1) (2014). https://doi.org/10.2478/HYMA-2014-0001
89. Satyam, S., Patra, S.: Innovations and challenges in adsorption-based wastewater remediation: a comprehensive review. Heliyon **10**(9) (2024). https://doi.org/10.1016/j.heliyon.2024.e29573
90. Liu, W.X., Song, S., Ye, M.L., Zhu, Y., Zhao, Y.G., Lu, Y.: Nanomaterials with excellent adsorption characteristics for sample pretreatment: a review. MDPI (2022). https://doi.org/10.3390/nano12111845
91. Ibrahim, Q., Creedon, L., Gharbia, S.: A literature review of modelling and experimental studies of water treatment by adsorption processes on nanomaterials. MDPI (2022). https://doi.org/10.3390/membranes12040360
92. Asghar, N., et al.: Advancement in nanomaterials for environmental pollutants remediation: a systematic review on bibliometrics analysis, material types, synthesis pathways, and related mechanisms. J. Nanobiotechnology **22**(1), 1–28, (2024). https://doi.org/10.1186/S12951-023-02151-3
93. Ben Mbarek, W., et al.: Nanomaterials as a sustainable choice for treating wastewater: a review. MDPI (2022). https://doi.org/10.3390/ma15238576
94. Yaqoob, A.A., Parveen, T., Umar, K., Ibrahim, M.N.M.: Role of nanomaterials in the treatment of wastewater: a review. MDPI AG (2020). https://doi.org/10.3390/w12020495
95. Saxena, M., Jain, K., Saxena, R.: Role of nanomaterials in catalytic reduction of organic pollutants (2022)
96. Pasinszki, T., Krebsz, M.: Synthesis and application of zero-valent iron nanoparticles in water treatment, environmental remediation, catalysis, and their biological effects. MDPI AG (2020). https://doi.org/10.3390/nano10050917.
97. Chandra Paul, S., et al. (2020) Silver nanoparticles synthesis in a green approach: size dependent catalytic degradation of cationic and anionic dyes. Orient. J. Chem. **36**(03), 353–360 (2020). https://doi.org/10.13005/OJC/360301

98. Mohapatra, L., Cheon, D., Yoo, S.H.: Carbon-based nanomaterials for catalytic wastewater treatment: a review. MDPI (2023). https://doi.org/10.3390/molecules28041805
99. Mishra, S., Sundaram, B.: A review of the photocatalysis process used for wastewater treatment. Mater Today Proc **102**, 393–409 (2024). https://doi.org/10.1016/j.matpr.2023.07.147
100. Mushtaq, A.: Wastewater Treatment by Photocatalysis: Approaches, Mechanisms, Applications, and Challenges, vol. 24, pp. 2023 (2023)
101. More, P.R., Pandit, S., De Filippis, A., Franci, G., Mijakovic, I., Galdiero, M.: Silver nanoparticles: bactericidal and mechanistic approach against drug resistant pathogens. MDPI (2023). https://doi.org/10.3390/microorganisms11020369
102. Aydin, D., Gübbük, I.H., Ersöz, M.: Recent advances and applications of nanostructured membranes in water purification. TUBITAK (2024). https://doi.org/10.55730/1300-0527.3635
103. Nain, A., Sangili, A., Hu, S.-R., Chen, C.-H., Chen, Y.-L., Chang, H.-T.: iScience recent progress in nanomaterial-functionalized membranes for removal of pollutants. iScience **25**, 104616 (2022). https://doi.org/10.1016/j.isci
104. Khraisheh, M., Elhenawy, S., Almomani, F., Al-Ghouti, M., Hassan, M.K., Hameed, B.H.: Recent progress on nanomaterial-based membranes for water treatment. MDPI (2021). https://doi.org/10.3390/membranes11120995
105. Wang, R., Chen, D., Wang, Q., Ying, Y., Gao, W., Xie, L.: Recent advances in applications of carbon nanotubes for desalination: a review. MDPI AG (2020). https://doi.org/10.3390/nano10061203
106. Ma, L., et al.: Fabrication and water treatment application of carbon nanotubes (CNTs)-based composite membranes: a review. MDPI AG (2017). https://doi.org/10.3390/membranes7010016
107. Adabi, S., Yazdanbakhsh, A., Shahsavani, A., Sheikhmohammadi, A., Hadi, M.: Removal of heavy metals from the aqueous solution by nanomaterials: a review with analysing and categorizing the studies. J. Environ. Health Sci. Eng. (2023). https://doi.org/10.1007/s40201-023-00863-0
108. De Silva, M., Cao, G., Tam, K.C.: Nanomaterials for the removal and detection of heavy metals: a review. R. Soc. Chem. (2025). https://doi.org/10.1039/d4en01041h
109. Tahoon, M.A., Siddeeg, S.M., Alsaiari, N.S., Mnif, W., Ben Rebah, F.: Effective heavy metals removal fromwater using nanomaterials: a review. MDPI AG (2020). https://doi.org/10.3390/PR8060645
110. Tonu, N., et al.: Removal of Lead and Mercury from Water Using Nanoparticles: A Concise Review on Adsorption, pp. 11–42 (2024). https://doi.org/10.22034/jchr.2023.1996328.1825
111. Yang, J., et al.: Nanomaterials for the removal of heavy metals from wastewater. MDPI AG (2019). https://doi.org/10.3390/nano9030424
112. Huang, X., et al.: Removal of arsenic from wastewater by using nano Fe3O4/Zinc organic frameworks. Int. J. Environ. Res. Public Health **19**(17) (2022). https://doi.org/10.3390/ijerph191710897
113. Jain, R.: Recent advances of magnetite nanomaterials to remove arsenic from water. R. Soc. Chem. (2022). https://doi.org/10.1039/d2ra05832d
114. Al-Yaari, M., Saleh, T.A.: Mercury removal from water using a novel composite of polyacrylate-modified carbon. ACS Omega **7**(17), 14820–14831 (2022). https://doi.org/10.1021/acsomega.2c00274
115. Kumar, R., Chawla, J.: Removal of cadmium ion fromwater/wastewater by nano-metal oxides: a review. Water Qual. Expo. Health **5** (2014). https://doi.org/10.1007/s12403-013-0100-8
116. Jawor, A., Hoek, E.: Removing cadmium ions from water via nanoparticle-enhanced ultrafiltration. Environ. Sci. Technol. **44**, 2570–2576 (2010). https://doi.org/10.1021/es902310e
117. Elhenawy, S., Khraisheh, M., AlMomani, F., Al-Ghouti, M., Selvaraj, R., Al-Muhtaseb, A.: Emerging nanomaterials for drinking water purification: a new era of water treatment technology. Multidiscip. Digit. Publ. Inst. (MDPI) (2024). https://doi.org/10.3390/nano14211707

118. Alguacil, F.J., Alonso, M., Robla, J.I.: Removal of hazardous organic dyes from liquid wastes using advanced nanomaterials. Multidiscip. Digit. Publ. Inst. (MDPI) (2024). https://doi.org/10.3390/ijms25179671
119. Singh, S. et al.: Competence of Nanoparticles for Removal of Pesticides from Wastewater: An Overview, pp. 253–266 (2022). https://doi.org/10.1016/B978-0-323-90893-1.00012
120. González-Poggini, S., Rosenkranz, A., Colet-Lagrille, M.: Two-dimensional nanomaterials for the removal of pharmaceuticals from wastewater: a critical review. MDPI (2021). https://doi.org/10.3390/pr9122160
121. Homaeigohar, S.: The nanosized dye adsorbents for water treatment. MDPI AG (2020). https://doi.org/10.3390/nano10020295
122. Freyria, F.S., Geobaldo, F., Bonelli, B.: Nanomaterials for the abatement of pharmaceuticals and personal care products from wastewater. MDPI AG (2018). https://doi.org/10.3390/app8020170
123. Bajda, T. et al.: Using zeolite materials to remove pharmaceuticals from water. Materials **17**(15) (2024). https://doi.org/10.3390/ma17153848
124. Ojha, A.: Nanomaterials for removal of waterborne pathogens. In: Waterborne Pathogens: Detection and Treatment, pp. 385–432. Elsevier (2020). https://doi.org/10.1016/B978-0-12-818783-8.00019-0
125. Puri, S., Divakar, S., Pramoda, K., Praveen, B.M., Padaki, M.: A review on bio-inspired nanoparticles and their impact on membrane applications. RSC Sustain. **3**(3), 1212–1233 (2025). https://doi.org/10.1039/D4SU00460D
126. Huang, W.-H., Lee, D.-J., Huang, C.: Modification on biochars for applications: a research update. Bioresour. Technol. **319**, 124100 (2021). https://doi.org/10.1016/j.biortech.2020.124100
127. Lu, Y.-Y., Wang, H., Chang, R.-R., Li, W.-B., Huang, B.-C., Jin, R.-C.: A literature review on phosphate removal from wastewater by lanthanum-based adsorbents: from mechanisms to applications. J. Water Process. Eng. **61**, 105282 (2024). https://doi.org/10.1016/j.jwpe.2024.105282
128. Zhang, L., et al.: Low-cost efficient magnetic adsorbent for phosphorus removal from water. ACS Omega **5**(39), 25326–25333 (2020). https://doi.org/10.1021/acsomega.0c03657
129. Savage, N., Diallo, M.S.: Nanomaterials and water purification: opportunities and challenges. J. Nanoparticle Res. 331–342 (2005). https://doi.org/10.1007/s11051-005-7523-5
130. Suazo-Hernández, J. et al.: nZVI-based nanomaterials used for phosphate removal from aquatic systems. MDPI (2023). https://doi.org/10.3390/nano13030399
131. Ahmed, M.A., Mahmoud, S.A., Mohamed, A.A.: Nanomaterials-modified reverse osmosis membranes: a comprehensive review. R. Soc. Chem. **12** (2024) https://doi.org/10.1039/d4ra01796j
132. Amari, A., Ali, M.H., Jaber, M.M., Spalevic, V., Novicevic, R.: Study of membranes with nanotubes to enhance osmosis desalination efficiency by using machine learning towards sustainable water management. Membranes (Basel) **13**(1) (2023). https://doi.org/10.3390/membranes13010031
133. Surwade, S.P., Smirnov, S.N., Vlassiouk, I.V., Unocic, R.R., Dai, S., Mahurin, S.M. Water desalination using nanoporous single-layer graphene with tunable pore size. Nat. Nanotechnol. **10**(5), 459–464
134. Chen, H., Qian, L.: Performance of field demonstration nanoscale zero-valent iron in groundwater remediation: a review. Sci. Total. Environ. **912**, 169268 (2024). https://doi.org/10.1016/j.scitotenv.2023.169268
135. Galdames, A., Ruiz-Rubio, L., Orueta, M., Sánchez-Arzalluz, M., Vilas-Vilela, J.L.: Zero-valent iron nanoparticles for soil and groundwater remediation. Int. J. Environ. Res. Public Health **17**(16), 1–23 (2020). https://doi.org/10.3390/ijerph17165817
136. Bhardwaj, C.K., Chandra, U., Rafi, S., Dubey, O.P.: An analytical survey on the role of nanotechnology in groundwater remediation. Water Supply **22**(11), 8092–8103 (2022). https://doi.org/10.2166/ws.2022.367

137. Del Prado-Audelo, M.L., García Kerdan, I., Escutia-Guadarrama, L., Reyna-González, J.M., Magaña, J.J., Leyva-Gómez, G.: Nanoremediation: nanomaterials and nanotechnologies for environmental cleanup. Front. Environ. Sci. **9** (2021). https://doi.org/10.3389/fenvs.2021.793765
138. Rao, V.P., Gunasekhar, T., Sarada, B.A.: Carbon-based sustainable nanomaterials for waste water treatment: a review. J. Res. Chem. **5**(2), 08–14 (2024). https://doi.org/10.22271/reschem.2024.v5.i2a.139
139. de Carvalho Lima, E.N., Souza, L.H.M., Aguiar, E.M., Octaviano, A.L.M., Justo, J.F., Piqueira, J.R.C.: Can bacteria and carbon-based nanomaterials revolutionize nanoremediation strategies for industrial effluents? Frontiers Media SA (2024). https://doi.org/10.3389/fnano.2024.1389107
140. Zahra, Z., Habib, Z., Chung, S., Badshah, M.A.: Exposure route of TiO2 nps from industrial applications to wastewater treatment and their impacts on the agro-environment. MDPI AG (2020). https://doi.org/10.3390/nano10081469
141. Li, L., Guo, W., Zhang, S., Guo, R., Zhang, L.: Electrospun nanofiber membrane: an efficient and environmentally friendly material for the removal of metals and dyes. MDPI (2023). https://doi.org/10.3390/molecules28083288
142. Andreani, T., et al.: Ecotoxicity to freshwater organisms and cytotoxicity of nanomaterials: are we generating sufficient data for their risk assessment? Nanomaterials **11**(1), 1–25 (2021). https://doi.org/10.3390/nano11010066
143. Corsi, I., et al.: Eco-interactions of engineered nanomaterials in the marine environment: towards an eco-design framework. MDPI AG (2021). https://doi.org/10.3390/nano11081903
144. Ugwuanyi, E., Nwokediegwu, Z., Dada, M., Majemite, M., Obaigbena, A.: The impact of nanomaterials in enhancing wastewater treatment processes: a review. Magna Sci. Adv. Res. Rev. **10**, 273–285 (2024). https://doi.org/10.30574/msarr.2024.10.1.0030
145. Selck, H., Handy, R.D., Fernandes, T.F., Klaine, S.J., Petersen, E.J.: Nanomaterials in the aquatic environment: a european union-united states perspective on the status of ecotoxicity testing, research priorities, and challenges ahead. Environ. Toxicol. Chem. **35**(5), 1055–1067 (2016). https://doi.org/10.1002/etc.3385
146. Singh, S., Prasad, S.M., Bashri, G.: Fate and toxicity of nanoparticles in aquatic systems. Acta. Geochimica. **42**(1), 63–76 (2023). https://doi.org/10.1007/s11631-022-00572-9
147. Bundschuh, M., et al.: Nanoparticles transported from aquatic to terrestrial ecosystems via emerging aquatic insects compromise subsidy quality. Sci. Rep. **9**(1) (2019). https://doi.org/10.1038/s41598-019-52096-7
148. Jasawa, R.: Environmental transformations of nanomaterials. (2024). https://doi.org/10.37421/2380-2391.2024.11.375
149. Saleem, H., Zaidi, S.J.: Developments in the application of nanomaterials for water treatment and their impact on the environment. MDPI AG (2020). https://doi.org/10.3390/nano10091764
150. Allan, J., et al.: Regulatory landscape of nanotechnology and nanoplastics from a global perspective. Regul. Toxicol. Pharmacol. **122**, 104885 (2021). https://doi.org/10.1016/j.yrtph.2021.104885
151. Rodríguez-Gómez, F.D., Monferrer, D., Penon, O., Rivera-Gil, P.: Regulatory pathways and guidelines for nanotechnology-enabled health products: a comparative review of EU and US frameworks. Front. Med. (Lausanne) **12** (2025). https://doi.org/10.3389/fmed.2025.1544393
152. Chávez-Hernández, J.A., Velarde-Salcedo, A.J., Navarro-Tovar, G., Gonzalez, C.: Safe nanomaterials: from their use, application, and disposal to regulations. Nanoscale Adv. **6**(6), 1583–1610 (2024). https://doi.org/10.1039/D3NA01097J
153. Zhou, X., Huang, B., Zhou, T., Liu, Y., Shi, H.: Aggregation behavior of engineered nanoparticles and their impact on activated sludge in wastewater treatment. Chemosphere **119**, 568–576 (2015). https://doi.org/10.1016/j.chemosphere.2014.07.037
154. Bathi, J.R., et al.: Dispersion and aggregation fate of individual and co-existing metal nanoparticles under environmental aqueous suspension conditions. Materials **15**(19), 6733 (2022). https://doi.org/10.3390/ma15196733

155. Ajith, M.P., Rajamani, P.: Nanotechnology for water purification—current trends and challenges. J. Nanotechnol. Nanomater. **2**(Issue 2), 88–91 (2021). https://doi.org/10.33696/NANOTECHNOL.2.025
156. Saud, A., Gupta, S., Allal, A., Preud'homme, H., Shomar, B., Zaidi, S.J.: Progress in the sustainable development of biobased (Nano)materials for application in water treatment technologies. ACS Omega **9**(27), 29088–29113 (2024). https://doi.org/10.1021/ACSOMEGA.3C08883/ASSET/IMAGES/LARGE/AO3C08883_0008.JPEG
157. Mousa, H.M., et al.: Engineering of multifunctional nanocomposite membranes for wastewater treatment: oil/water separation and dye degradation. Membranes (Basel) **13**(10), 810 (2023). https://doi.org/10.3390/membranes13100810
158. Kamath, S.V., Mruthunjayappa, M.H., Mondal, D., Sanna Kotrappanavar, N.: Nanocomposite-based high-performance adsorptive water filters: recent advances, limitations, nanotoxicity and environmental implications. Environ. Sci. Nano. **9**(7), 2320–2341 (2022). https://doi.org/10.1039/D2EN00155A
159. Nasrollahzadeh, M., Sajjadi, M., Iravani, S., Varma, R.S.: Carbon-Based Sustainable Nanomaterials for Water Treatment: State-of-Art and Future Perspectives. Elsevier Ltd. (2021) https://doi.org/10.1016/j.chemosphere.2020.128005
160. Choi, W.S., Lee, H.-J.: Nanostructured materials for water purification: adsorption of heavy metal ions and organic dyes. Polymers (Basel) **14**(11), 2183 (2022). https://doi.org/10.3390/polym14112183
161. Yitagesu, G.B., Leku, D.T., Workneh, G.A.: Green synthesis of TiO2 using impatiens rothii hook.f. Leaf extract for efficient removal of methylene blue dye. ACS Omega **8**(46), 43999–44012 (2023). https://doi.org/10.1021/ACSOMEGA.3C06142/ASSET/IMAGES/LARGE/AO3C06142_0013.JPEG

Biochar-Enhanced Biofertilization: Boosting Soil Microbial Activity, Disease Resistance, and Plant Growth

Rajnandani Rathore, Lalita Ledwani, and Rakesh Kumar Sharma

Abstract Biochar is a multifunctional biomaterial with many important uses in agriculture, particularly as a biofertilizer. This chapter discusses the unique features of biochar that have made it a valuable biofertilizer, highlighting its potential to support a number of microbial communities in the soil, suppress plant diseases, and contribute to sustainable crop growth. The impact of biochar on plant growth is being investigated, with an emphasis upon how it may encourage the formation of roots, retention of water, and nutrient retention – all of which in turn lead to higher agricultural yields. Biochar's dual purpose as a microbial inoculant carrier increases its value as a biofertilizer through offering a substrate which boosts the longevity as well as efficiency of advantageous microorganisms in the environment. The chapter compares modern pyrolysis techniques to traditional biochar production techniques such as pit and mound burning. Conventional approaches are readily available and inexpensive, but they are less effective and produce greater amounts of pollutants. With higher temperature and combustion rate control achievable by means of advanced pyrolysis, biochar with higher permeability, pore size, and nutritional value can be produced, improving its uniformity and sustainability as a biofertilizer. A comparative investigation illustrates the long - term benefits of organic fertilizers, such as biochar, over chemical alternatives, highlighting their ability to enhance agricultural resilience and their lesser environmental impact. The section covers the physical and chemical properties of the biochar of the biochar, including its surface structure, porosity, and amount of nutrients, which maximize its interactions with soil. In addition to promoting the use of biochar in agricultural systems to improve soil health, reduce disease pressure, and advance sustainable agriculture, this chapter ends with research directions for the future.

R. Rathore
Department of Biosciences, Manipal University Jaipur, Jaipur, India

L. Ledwani (✉)
Department of Chemistry, Manipal University Jaipur, Jaipur, India
e-mail: lalita.ledwani@jaipur.manipal.edu

R. K. Sharma
Department of Life Sciences, Chhatrapati Shahu Ji Maharaj University, Kanpur, India

L. Ledwani and S. Srivastava (eds.), *Experimental and Computational Advances in Materials*, Engineering Materials, https://doi.org/10.1007/978-981-95-4180-5_3

Keywords Biochar · Biofertilizer · Soil microbial activity · Plant disease suppression · Sustainable agriculture · Crop growth

1 Introduction

Biomaterials, that may be natural or synthetic, have reshaped industries by offering functionally different and environmentally sustainable remedies. Materials like these are crucial in industries such as agriculture, ecological restoration, and medicine since they have been engineered for interaction with biological processes. Biomaterials have become increasingly known for its ability to sustain or boost performance while advancing sustainability, particularly within applications whose resource efficiency and environmental impact are significant. One such promising biomaterial is biochar, a carbon-rich byproduct of organic biomass pyrolyzed in an oxygen-limited environment. Though biochar has been historically known for its role in agriculture as a soil conditioner and biofertilizer, its various other applications are gaining attention. It is ideal for strengthening soil health by providing nutrient retention, stimulating microbial diversity, and preventing plant diseases by virtue of its porous structure, extensive surface area, and functionalized chemical composition. The application of biochar in composite materials enhances its flexibility beyond agriculture. Biochar—polymer composites, for instance are taking advantage of the synergy among the rigidity of biochar and the flexibility of polymers, bringing up opportunities in sensors, proficient environmental filters, as well as other cutting—edge technologies. Biochar's ability as an essential component in the production of sustainable, versatile materials has been shown by this incorporation, thereby improving its usefulness. Biochar's transformational potential can be seen by its ability to work in collaboration with other materials. Biochar is an agricultural additive that improves soil fertility and promotes ecologically sound farming methods. In novel uses, it serves as a material that addresses environmental issues along with offering answers that integrate industrial innovation and ecological treatment.

2 Overview of Biochar

Global socioeconomic conditions underwent a dramatic change as a result of the Industrial Revolution. The quality of life has increased along with the speed of life due to scientific and technical advancements. The effects of this change and evolution have now affected all facets of human existence. Global warming, soil pollution, water scarcity and pollution, resource depletion, and food shortages are unfortunate outcomes [30]. Organic pollutants, which are mostly the products and byproducts of anthropogenic processes, include polychlorinated biphenyls, poly—and perfluoroalkyl substances, pesticides, petroleum-based hydrocarbons, polybrominated diphenyl ethers, dibenzofurans, dioxins, pharmaceuticals, and personal care products. The widespread use of pesticides, organic solvents, pharmaceuticals, and other chemicals has led to the distribution of organic pollutants throughout global ecosystems [62]. Biochar may hold the key to creating a sustainable future while incorporating valuable items into the circular economy model, addressing current and impending issues. By using biochar as a soil amendment, researchers hope to find a way to enhance soil quality and increase agricultural sustainability [2].

The solid byproduct of heating biomass under circumstances of zero or extremely low oxygen is called biochar. It is a black, porous, carbon-rich substance that has significant potential advantages for soil and plant growth in addition to storing carbon in a stable form for hundreds of years. When biomass is cooked without oxygen, its chemical composition changes, losing oxygen, nitrogen, and hydrogen in proportion to carbon. Microorganisms find it extremely difficult to break down biochar because of the molecular structure formed by the carbon atoms becoming firmly bonded to one another [7]. The addition of biochar enhances the physicochemical and biological characteristics of soil. It improves soil structure, changes soil microbial communities, raises soil pH and cation exchange capacity (CEC), and promotes nutrient retention. Despite its stubborn nature, biochar is a beneficial soil investment because of its capacity to interact with soil qualities. Applying biochar over an extended period of time alters the soil's physiochemical characteristics. As a result, the bacterial community in the soil changes [10].

Nowadays, research and commercial development are being conducted all over the world to produce biochar from the incomplete combustion of woody biomass. The main objective is to generate renewable energy, such as electricity and thermal energy, with the secondary goal of producing biochar as a marketable byproduct. The potential economic value generated by producing biochar as an additional marketable product can be the decisive factor in otherwise economically marginal wood-to-energy operations, as two recent proprietary studies in Louisiana and California have shown. Small to medium-sized producers have an entrepreneurial opportunity to diversify and increase their market coverage because charcoal and biochar are produced under the identical conditions, with the only difference being the final product's particle size. Natural charcoal producers can take advantage of price and demand by turning charcoal into biochar through a size reduction process (grinding) or "producing" biochar from their fines. Nevertheless, natural charcoal producers will have little control over the properties of the final biochar to the degree that the feedstock they use varies. Biomass energy producers also face difficulties in producing biochar since they typically employ a consistent feedstock that is relatively small (chips or pellets, for example) to ensure combustion efficiency, hence excluding the manufacture of a charcoal product [16].

3 History of Biochar

A key component of gunpowder, biochar was accidentally created by alchemists in ninth-century China as they were trying to figure out the secrets of life. From the military upheaval brought about by gunpowder weaponry in the fifteenth and sixteenth centuries to its use as blasting powder in mining in the early seventeenth century, gunpowder should have had a significant impact on history. In China, a significant quantity of charcoal-infused black pottery was discovered among the artifacts recovered from the Hemudu site about 7000 years ago. Ancestors in Hemudu purposefully incorporated charcoal into the clay to decrease the cohesiveness of the clay and boost the production of finished goods. Charcoal use from the Shang and Zhou dynasties is also documented, indicating that China transitioned from an agrarian culture into the Bronze Age and ultimately the Iron Age. Charcoal was also discovered in 1971 at the Han dynasty's Mawangdui Tomb archeological site. The owner of the tomb was buried over 2100 years ago, but when she was discovered, her body was in good condition. Her remains was probably preserved in the tomb because of the more than 5000 kg of special charcoal that was found inside. Bai Juyi, a poet from the Tang Dynasty, highlighted the use of charcoal for energy in ancient China. In addition to the traditional industrial use of carbon in electrodes, carbon black, and dynamo brushes, as well as in carbon fiber, pyrolytic graphite, and other new industrial carbon products, the household application of carbon materials, such as charcoal, coal, ink, dyes, and preservatives, has expanded since modern civilization.

However, the most valuable reductant in the metallurgical business as of right now is biochar. In addition, it has long been a popular fuel for cooking. Around 1500 BC, the ancient Egyptians were already using biochar's adsorptive qualities. They used it to absorb foul odors from putrefying sores. Carl W. Scheele, a Swedish scientist, was the first to conduct a scientific investigation into the adsorptive properties of biochar in Europe in the late 1700s. His work was rarely carried out in the 19th century due to the lack of interest in adsorptive materials outside of a few specialized fields like sugar refining. Charles F. Hartt presented Terra Preta to the scientific community in the late 1800s. A committed soil scientist named Wim Sombroek released "Amazon Soils: A reconnaissance of the soils of the Brazilian Amazon region" in 1966, about 90 years later. It marked the start of Terra Preta's scientific investigation, which included archeology as well as soil sciences. Sombroek's goal was to develop a new black, productive soil called Terra Preta Nova that would enable sustainable land usage on soils that couldn't sustain ongoing farming without a lot of fertilizer [14, 39].

4 Properties of Biochar

Each variety of biochar has distinct physio-chemical properties, including surface area, pH, and the concentration of various elements and nutrients (such as carbon (C), oxygen (O), nitrogen (N), phosphorous (P), potassium (K), and calcium (Ca)). The physicochemical properties of BC are influenced by the feedstock type and pretreatment, the pyrolysis parameters (temperature, time, heating rate, and sweeping gas type and flow rate), and the post-pyrolysis treatment. Biochars from seaweeds, manures, and agricultural wastes generally have higher pH values, less stable carbon, and higher nutrient contents than lignocellulosic-rich feedstocks like wood.

In contrast to slow pyrolysis, fast pyrolysis tends to increase the surface area of biochar while having little influence on the C:O ratio [21]. Higher temperature-generated BC is typically more alkaline, has higher concentrations of ash and fixed carbon, EC, WHC, and specific surface area (SSA), as well as higher aromaticity, fewer surface functional groups, and lower CEC values than that produced at a lower temperature. Additionally, using BC as an amendment helps mitigate climate change by increasing soil carbon absorption and lowering greenhouse gas (GHG) emissions [1]. The diverse nature of the biochar composition results in both stable and unstable components. The primary ingredients of biochar for characterization are its surface area, carbon, and mineral matter (ash). Biochar has two intrinsic properties that allow it to hold carbon in the soil for long periods: (1) Stability, encompassing the soil's ability to tolerate deterioration from living and non-living sources, and (2) the fact that biochar has more carbon than biomass [54] (Fig. 1).

4.1 Physical Properties

The surface area of biochar, which is controlled by its porosity, has a significant impact on its capacity to adsorb chemicals. By changing the geometry of the

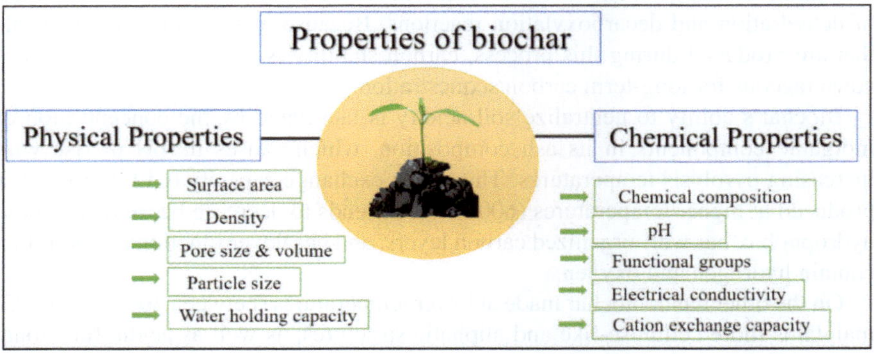

Fig. 1 Properties of biochar [34, 46]

pores, variations in the pyrolysis temperature can significantly affect the porosity and specific surface area. Electron microscopy reveals different pore topologies at different pretreatment temperatures. Biochar contains three different types of holes: micropores (less than 2 nm), mesopores (between 2 and 50 nm), and macropores (more than 50 nm). Zhang and colleagues' 2021 [65] study found that higher pyrolysis temperatures increase the adsorption capacity of biochar by producing more complex pore structures and more sumptuous surface surfaces. The bulk density (BD) represents the arrangement of soil particles. Biochar improves soil aggregation, lowers soil BD and particle density, and increases total porosity through its porosity (70–90%). While macropores facilitate drainage, micropores improve water retention. By increasing soil porosity (5–25 μm), biochar additions enhance the transport of heat, water, and gases. The type of feedstock has a big influence on porosity patterns [41, 42]. Depending on the production conditions, such as temperature, water content, soil texture, and application techniques, the hydrophobic microstructure of biochar can cause soils to become water-repellent. Low-temperature and fresh biochar tend to repel more water. Biochar may reduce hydraulic conductivity and infiltration in sandy soils by obstructing macropores, which would lessen leaching and preserve nutrients. However, in fine-textured soils, biochar can enhance water flow, which will improve water uptake and storage. To maximize water retention and nutrient availability, biochar slows water transport in sandy soils and increases it in clayey soils [8].

4.2 Chemical Properties

The temperature at which pyrolysis occurs and the type of feedstock utilized have a major impact on the chemical characteristics of biochar. It is mostly made up of carbon, with different proportions of hydrogen, oxygen, and inorganic minerals that affect its stability and availability of nutrients. The hydrogen-to-carbon (H/C) and oxygen-to-carbon (O/C) ratios drop with increasing pyrolysis temperatures, indicating a change from labile to more resistant carbon structures as a result of dehydration and decarboxylation reactions. Because of the aromatic chemicals that are produced during this process, carbon stability is improved and biochar is advantageous for long-term carbon sequestration [57].

Biochar's ability to neutralize soil acidity is facilitated by the concentration of inorganic components in its ash composition, which causes its pH to rise with increasing pyrolysis temperatures. The cation exchange capacity (CEC) of biochar produced at higher temperatures (600–700 °C) tends to decrease because it is more hydrophobic, has well-organized carbon layers, and has fewer functional groups that contain hydrogen and oxygen.

On the other hand, biochar made at lower temperatures (between 300 and 400 °C) maintains more cellulose-like and aliphatic structures, as well as acidic functional groups like phenolic and carboxyl acids, which support ion exchange and negative surface charge [29].

Biochar's nutrient content varies greatly; potassium (K) is usually found in higher quantities than nitrogen (N) and phosphorus (P), whereas sodium (Na) is frequently found in higher concentrations than calcium (Ca) and magnesium (Mg). With significant variances in total N (1.8–56.4 g/kg), P (2.7–480 g/kg), and K (1.0–58 g/kg), the carbon content might vary from 172 to 905 g/kg. Lower temperatures (< 500 °C) help maintain phosphorus, potassium, and sulfur in accessible forms, while higher temperatures (~ 700 °C) can cause more nitrogen loss and decreased nutrient availability, according to the research. Depending on the feedstock, the ash level may contain undesired pollutants and advantageous chemicals like calcium carbonate ($CaCO_3$), which have liming qualities. The pH, nutritional composition, cation exchange capacity, adsorption ability, and organic feedstock type of biochar are all crucial quality characteristics. [50].

4.3 Thermal Properties of Biochar

The thermal behaviour of lignocellulosic biomass is a decisive factor in governing its decomposition, conversion into biochar, and the ultimate performance of the derived material in environmental and catalytic applications [60]. Hemicellulose, being amorphous and highly branched, is the least thermally stable fraction, decomposing rapidly within 200–350 °C and releasing CO, CO_2, acetic acid, and other light volatiles, which contributes to pore formation but results in a low char yield. Cellulose, a semi-crystalline polymer, shows sharper degradation between 315 and 400 °C, mainly producing levoglucosan and volatile anhydrosugars, and its degree of crystallinity plays a key role in determining the thermal stability and porosity of the resulting char [55]. In contrast, lignin displays the broadest and most gradual thermal decomposition, spanning 250–500 °C, due to its complex aromatic network, and contributes significantly to the residual char yield, aromaticity, and fixed carbon content, thereby enhancing the overall thermal stability of biochar [51]. Compared to raw biomass, biochar demonstrates considerably higher thermal stability, with the extent largely dependent on feedstock composition and pyrolysis temperature [57]. Low-temperature biochars (< 400 °C) retain labile organic fractions and are less stable in soil, whereas high-temperature biochars (> 600 °C) are enriched in condensed aromatic structures, exhibit lower oxygen and hydrogen content, and display superior resistance to microbial and chemical degradation [9]. Thermogravimetric analysis (TGA/DTG) has been widely employed to characterize these transitions, revealing distinct multi-stage weight loss patterns corresponding to hemicellulose, cellulose, and lignin decomposition, followed by the stabilization of carbonaceous residues [20]. Such thermal characteristics are of immense importance, as they not only dictate the long-term persistence of biochar in soil for carbon sequestration but also influence surface area, pore structure, and sorption capacity, which are critical parameters in wastewater remediation and catalytic applications.

5 Biochar Production

The production of biochar shifted the traditional way of production which was based on the use of earth pits, mound kilns and covered soil furnaces towards modern production with higher rates of efficiency and safety to environment. Simple and cheap, traditional techniques only were widely practiced, and sometimes, an incomplete carbonization process (partially carried out), was a consequence with the yield being inconsistent. The pyrolysis, torrefaction and gasification systems used in the modern system offer a steady control of temperature and oxygen levels, minimized emissions and are used to produce biochar of relative quality. This change indicates how agriculture, carbon sequestration, renewable energy has become viable with biochar as a renewable resource when moving on beyond indigenous knowledge-based practices to the use of scientific inventions.

5.1 Traditional Methods

In the ancient times, many communities of the world came up with simple but efficient techniques in which they could produce charcoal and biochar through the use of biomass that was available locally. Traditional methods were often associated with low technology and low-cost kilns or pits, in which wood and crop remains or plants were burnt in oxygen limited environment. The desired product was frequently the production of fuel to the households, blacksmithing, pottery, or metallurgy, though, biochar as a side effect, led to greater soil health and long-term carbon sequestration. There are some of the historical ways of biochar production all over the world, which were incorporated into the rural economies, agricultural systems, and environmental management in regions. Such ancient techniques of biochar production were typical in different parts of the world, and they constituted an element of local food production, rural life, and sustainable environment.

- **Catalonia, Spain**

  In the Catalonia region, "foreigner" (anthill) kilns were traditionally used to dispose of agricultural and forestry biomass. These kilns involved burning dry, woody plant waste under a 10–20 cm thick soil cover, characterized by slow and incomplete combustion. This method produced biochar with an optimal nutrient profile, enriching the soil with essential nutrients like phosphorus, potassium, and nitrogen [33].

- **Yogyakarta, Indonesia**

  Traditional charcoal manufacturing in Yogyakarta relies on several methods, including earth mound and pit kilns, transportable steel kilns, oil drum kilns, brick kilns, concrete kilns, and fired clay kilns. The yield from these methods varies between 22.88% and 35.98%, depending on factors such as the type of kiln, raw materials used, processing time, and carbonization conditions [38].

- **Hungary (GemerRegion)**

  In Hungary, particularly in the Gemer region, traditional pit kilns known as "boksas" were commonly used for charcoal production. These shallow, plate-shaped pits, measuring approximately 3 × 3 meters, were leveled and cleared of vegetation before stacking wood inside. The setup was covered with leaf litter and a thin 2–3 cm layer of dry soil. This method resulted in an efficient charcoal yield of about 25% [32].

- **Sweden**

  Charcoal production in Sweden dates back to the early Iron Age and was widely practiced during the 18th to early 20th centuries. Historic charcoal kiln platforms in boreonemoral forests have left a lasting ecological legacy, modifying soil properties at these sites and transforming them into distinct environmental niches within the forest landscape [18].

- **Tuscany, Italy**

  In central Italy's Mediterranean forest ecosystems, historical charcoal production practices, particularly in Tuscany, have led to significant ecological changes. The soil properties at former production sites have been modified, creating distinctive microhabitats within the broader forest landscape, illustrating the long-term impact of these practices [11].

- **Wallonia, Belgium**

  The long-term effects of places enriched with charcoal residues from more than 150 years ago on soil carbon dynamics have been investigated in Wallonia, Belgium. Research emphasizes biochar's historical and geographical significance in agricultural operations and its ability to improve soil health and sequester carbon [64].

- **India (Earth Kilns, Pit Kilns and Bhatti Systems)**

  Traditional biochar production in India has long been associated with rural households, agriculture, and small-scale industries, where earthen kilns, pit kilns, and "bhatti" systems were widely used [5]. In the pit kiln method, woody biomass such as crop residues, twigs, or firewood was stacked in shallow pits and covered with soil or leaves to limit oxygen supply, resulting in slow pyrolysis and partial carbonization[59].These kilns were inexpensive, but the yields were generally low (15–25%) due to incomplete combustion and loss of volatiles. In many regions, traditional earth mound kilns were constructed by stacking biomass in conical heaps, plastering with a thick layer of clay or soil, and allowing controlled smoldering over several days [17]. The "bhatti" system, still used in parts of northern and central India for making charcoal from wood, bamboo, or coconut shells, involves brick-lined chambers where carbonization occurs with limited air supply [13]. Although simple and widely accessible, these methods suffered from poor efficiency, non-uniform heating, and higher greenhouse gas emissions. Nonetheless, they played a crucial role in rural energy security and soil amendment practices, especially in subsistence farming systems, highlighting the deep cultural and agricultural connection of biochar production in India.

5.2 *Carbonization Process*

Carbonization, the earliest known thermochemical conversion process, has been integral to human civilization since the Stone Age, enabling the transformation of biomass into energy-dense and functional carbon materials [53]. The process typically involves heating biomass at elevated temperatures (above 400 °C) under limited or no oxygen conditions, leading to thermal decomposition of complex carbonaceous materials such as wood, crop residues, and agricultural by-products [36]. Depending on the process conditions, carbonization produces a range of products including charcoal for fuel, biochar for soil fertility enhancement, biocoke for metallurgical applications, and activated carbon for adsorption and purification [48]. Over time, several controlled carbonization approaches have evolved to improve efficiency and product selectivity. Among these, pyrolysis (slow and fast), flash carbonization, microwave pyrolysis, and torrefaction are the most widely studied techniques, each characterized by distinct operational parameters such as heating rate, residence time, and pressure. These modern variations of carbonization not only increase yield and quality but also diversify the applications of biochar and related carbon materials in energy, agriculture, and environmental remediation (Fig. 2).

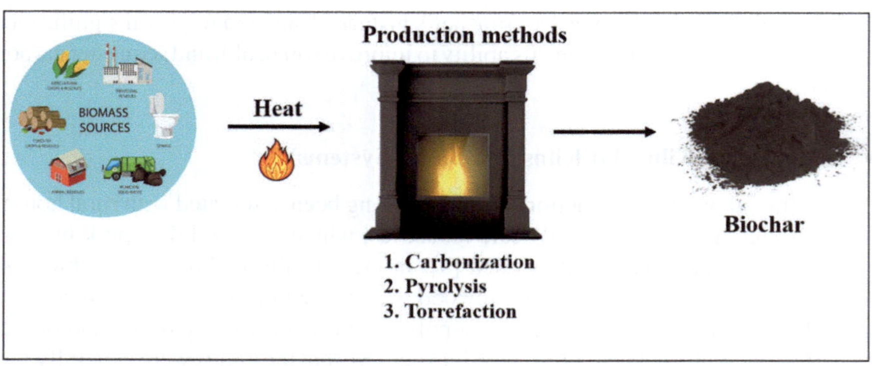

Fig. 2 Biochar preparation techniques [4, 12]

5.2.1 Pyrolysis

Pyrolysis is the process of a fuel breaking down thermochemically at high temperatures without the need for outside oxygen. The process starts with the drying of biomass. The particle is further heated, releasing volatile matter from the solid. The volatile compounds can form permanent gases (such as CO_2, CO, CH_4, and H_2) or condensable organic compounds (e.g., methanol and acetic acid) [62]. A wide range of pyrolysis methods (slow pyrolysis, fast pyrolysis, gasification, and torrefaction) yield different proportions of biochar/char (solid) and co-products (gases and bio-oil), depending on the pyrolysis temperature, heating rate, and reaction time, have been developed [3]. The key differences between slow and fast pyrolysis are the maximum reaction temperature and heating rate. Slow pyrolysis is a robust and energy-efficient process. Biochar production is found to be the maximum compared to the gaseous and liquid products during slow pyrolysis. This method is widely used for farm-based or small-scale biochar production. The temperature range for slow pyrolysis is 300–700 °C, with long residence time (minutes to hours), and a low heating rate of 0.1–10 °C min-1, which removes the vapors during the heating process. Fast pyrolysis differs from slow pyrolysis in terms of its operating conditions, while different reaction conditions could produce different targeted products. In fast pyrolysis, the biomass is heated rapidly at 600–1000 °C under oxygen-free conditions to generate pyrolysis vapor and biochar. It employs fast heating rates (10–10,000 °C min-1) and short residence time (0.5–5 s) to maximize the bio-oil production [56].

5.2.2 Flash Carbonization

Flash carbonation is a process in which biomass is heated by periodically introducing a flash fire at high pressures (greater than 1 MPa). The pyrolysis chamber is filled with biomass, and a flash fire is introduced from the bottom in an upward direction while maintaining airflow in a downward direction. Under these conditions, the biomass is converted into gaseous fuel, leaving behind carbon content as biochar (BC) [28].

5.2.3 Microwave Pyrolysis

The manufacturing of biochar has advanced significantly with the advent of microwave pyrolysis, a faster and more material-selective heating method than the traditional one. It provides a more effective method that achieves comparable outcomes to conventional pyrolysis while lowering the temperature requirement by 200°. A group investigated experimental conditions for maximum biochar and hydrogen production and reported 450 °C temperature, 400 W microwave power, and 4–6 min resident time as the best conditions for microwave pyrolysis for biochar production. However, maximum hydrogen production was obtained at 700 °C temperature, 400 W microwave power, and 4–6 min retention time. The biochar produced by microwave pyrolysis not only has the potential for CO_2 adsorption but also inspires hope for its diverse applications in the renewable energy sector [19].

5.2.4 Torrefaction

Torrefaction involves heating biomass in a controlled nitrogen environment within a muffle furnace. First, biomass samples are placed in a stainless-steel reactor, and nitrogen gas is used to purge the reactor, eliminating any oxygen. The reactor is then heated to specific temperatures ranging from 200 to 300 °C for set durations of either 30 or 60 min, with a thermocouple ensuring precise temperature control throughout the process. After heating, the furnace is rapidly cooled under nitrogen to prevent oxidation. This procedure results in the production of torrefied biomass, which is then collected for further analysis [31].

6 Applications of Biochar

(See Fig. 3).

6.1 Biochar in Lithium and Sodium-ion Batteries

Materials based on biochar have demonstrated great promise for lithium-ion (Li-ion) and sodium-ion batteries because of their cost-effectiveness, sustainability, and advantageous electrochemical characteristics. Biochar can improve electrochemical processes in Li-ion batteries because of its moderate surface area, facilitating effective ion diffusion during charging and discharging cycles. Additionally, biochar is a source of amorphous carbon for anode materials, offering higher specific capacity than conventional graphite.

Advanced biochar-based materials, such as highly graphitic carbon nanosheets (HGCNS), can be synthesized via hydrothermal and graphitization processes, forming two-dimensional (2D) nanostructures that enable efficient Li-ion storage and rapid ion and electron transport. These properties make biochar a promising candidate for energy storage applications in sustainable battery technologies [49].

6.2 Biochar for Water Purification and Filtration

Biochar is emerging as a powerful and cost-effective tool for water filtration and purification, capable of removing heavy metals like Cd^{2+}, Pb^{2+}, Cu^{2+}, and Hg^{2+}, as well as organic contaminants from industrial wastewater. Its versatility extends to

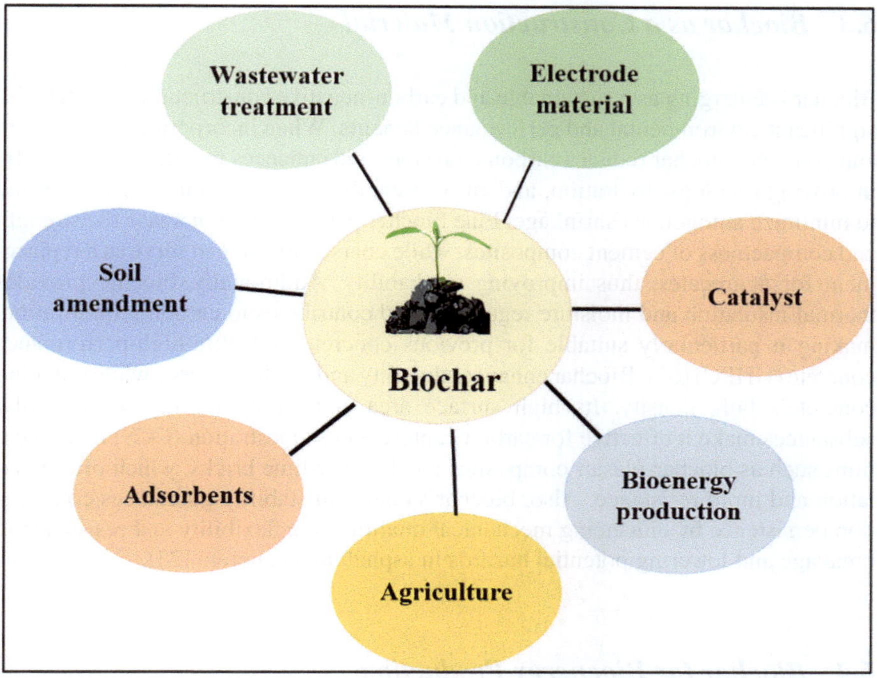

Fig. 3 Applications of biochar [58]

decentralized and household-level water treatment systems, particularly in rural and vulnerable communities, where it can be produced from locally available biowastes through pyrolysis [25]. Biochar has been successfully integrated into bioretention filters to manage stormwater runoff, and studies show that adding 30% wood chip biochar to sand filters can achieve up to 95% removal of *Escherichia coli* [27]. Modified biochar, such as zero-valent iron (ZVI)-biochar composites, has demonstrated further effectiveness by reducing *E. coli* growth by nearly 50% and, when combined with silver ions, forming a nanocomposite that completely inhibited bacterial growth. Pilot systems such as N-E-W TechTM demonstrate the potential scalability of biochar by improving pollutant removal and enabling the repurposing of adsorbed minerals as fertilizers, both of which support sustainable practices [61]. With the capacity to rival commercial activated carbon in performance, biochar-based systems such as biochar-sand filters and multi-layer adsorption reactors offer practical, sustainable solutions to address water quality challenges, though continued research is necessary to fully explore biochar's catalytic potential in advanced oxidation processes (AOPs).

6.3 *Biochar as a Construction Material*

Biochar is emerging as a sustainable and carbon-negative construction material with significant environmental and performance benefits. When incorporated into cement and concrete, biochar reduces carbon emissions and enhances material properties by improving rheology, hydration, and early strength while promoting internal curing to minimize autogenous shrinkage. Fine biochar particles help increase the strength and compactness of cement composites, while coarse biochar can serve as a replacement for aggregates, thus improving workability. Additionally, biochar provides thermal insulation and moisture regulation and contributes to carbon sequestration, making it particularly suitable for previous concrete and ultra-high-performance concrete (UHPC) [65]. Biochar enhances ductility and fracture energy while reducing concrete's bulk density. Its high surface area and strong affinity for nonpolar substances make it effective for carbon capture and sequestration (CCS) in applications such as biochar-plaster composites and biochar-lime bricks, which offer insulation and mold resistance. Also, biochar's chemical stability guarantees construction persistence by enhancing mechanical qualities like flexibility and resistance to breakage and lowering potential hazards in asphalt and concrete [23].

6.4 *Biochar for Bioenergy Production*

Biochar is becoming a vital instrument for bioenergy production, directly replacing coal in co-firing facilities to generate power in a renewable manner and lowering carbon dioxide emissions worldwide. In addition to direct combustion, biochar plays a role in microbial fuel cells (MFCs), where it acts as an electrode to facilitate redox reactions, generating electricity without heat consumption and aiding wastewater treatment by reducing sludge and chemical oxygen demand [43]. Biochar's abundant availability, low-cost production, and carbon sequestration potential further enhance its utility in energy storage applications such as lithium-ion batteries and supercapacitors. Its high surface area, microporous structure, and heteroatom doping contribute to improved energy density and capacitance, with studies reporting capacitance retention of 88% after 5000 cycles and specific capacitance values up to 428 Fg^{-1} [45]. As an alternative to traditional materials, biochar's adaptability and affordability are demonstrated by the development of high-performance supercapacitors made from it. According to [26], biochar is produced from a variety of biomass sources, such as marine biomass waste and chicken manure. This diverse use of biochar addresses issues with waste management and pollution while highlighting its promise in sustainable bioenergy solutions.

6.5 Adsorption of Greenhouse Gases

Biochar demonstrates significant potential for greenhouse gas (GHG) adsorption, particularly for CO_2, with adsorption capacities reaching up to 2 mmol/g, comparable to commercial adsorbents [47]. Due to its absorption ability, soil investigations have shown lower CO_2 emissions, suggesting that biochar both directly absorbs CO_2 from the gas phase and prevents microbial carbon mineralization. High porosity, sp^2 hybridization, and wide surface area are surface properties that improve biochar's sorption efficacy and impact its capacity to reduce greenhouse gas emissions. Biochar is essential for reducing methane (CH_4) emissions because it improves soil aeration and enhances the activity of methanotrophic bacteria. When combined with soil amendments, biochar can reduce methane emissions by up to 91%. Biochar enhances soil structure and fosters microbial activity, which reduces methane production, particularly in well-aerated soils.

Biochar's high porosity and the presence of numerous active functional groups contribute to the stabilization of organic fertilizers, which in turn helps to reduce nitrous oxide (N_2O) emissions. However, research indicates that the effectiveness of biochar in mitigating N_2O emissions is modest, with a reduction rate of 24.64% [40]. This dual mechanism of adsorption and microbial interaction positions biochar as a sustainable solution for mitigating CO_2, CH_4, and N_2O emissions in agroecosystems [52].

6.6 Biochar in Plant Disease Suppression

Biochar has emerged as a promising tool for suppressing plant diseases, particularly soil-borne pathogens while promoting soil health and maintaining environmental sustainability. It reduces the incidence of diseases such as *Rhizoctonia solani* in cucumber, *Ralstonia solanacearum* in tomato, and *Phytophthora aphanidermatum* by enhancing the population of beneficial microbes in the soil and inducing disease resistance in host plants through the production of defense compounds and alterations in the soil microbiome [6]. Cell wall-degrading enzymes are efficiently adsorbed by biochar, slowing the course of illness. A healthier growth environment is produced by its sorptive capacity, which also increases microbial activity in the rhizosphere, decreases nutrient leaching, immobilizes dangerous contaminants, and improves soil fertility [15]. Furthermore, biochar promotes the development and colonization of arbuscular mycorrhizal fungus, which interferes with microbial chemical signaling networks and lessens the impacts of phytotoxic and fungitoxic substances. Biochar is a useful addition for long-term plant disease control because of these combined benefits, which also enhance plant resilience and effectively suppress disease.

6.7 Biochar in Soil Remediation

Biochar has shown considerable potential in addressing soil contamination by stabilizing toxic elements and reducing their bioavailability [22]. It immobilizes heavy metals by adsorbing metal ions onto its surface and transforming them into less soluble forms, such as hydroxides, carbonates, and phosphates. This process helps to reduce ecotoxicity and restore soil functionality for safe crop production. The mechanisms for stabilizing heavy metals include cation exchange with elements like Ca^{2+}, Mg^{2+}, K^{+}, and Na^{+}, inner-sphere complexation, and precipitation with mineral oxides and humic matter [42]. Biochar's alkalinity and its ability to increase soil pH are crucial for promoting metal precipitation, which reduces the bioavailability of harmful metals.

Additionally, the presence of functional groups on biochar—such as carboxyl, hydroxyl, and phenolic groups—enhances mechanisms like electrostatic attraction, ion exchange, and surface complexation, which contribute to metal stabilization. Moreover, biochar enhances the physicochemical characteristics of soil, which helps crops absorb and move heavy metals less easily. In addition to helping with soil reclamation, its porous structure encourages microbial activity [24]. Although biochar cannot completely remove heavy metals, its ability to encapsulate pollutants provides a long-term solution to lessen their negative impacts.

7 Conclusion

Biochar's unique characteristics make it a versatile biomaterial with great promise for environmental sustainability and agriculture. These include specialized surface functional groups, high porosity, and rich nutrient content, all of which improve soil health, maintain microbial communities, and aid in the suppression of plant diseases. When paired with advanced pyrolysis methods, biochar's ability to enhance nutrient and water retention makes it a viable substitute for chemical fertilizers with applications beyond agriculture. In addition to soil and environmental benefits, biochar production also enhances the energy potential of biomass, as reflected by improvements in calorific value. Comparative studies have reported that raw lignocellulosic materials such as crop residues (CV $\approx$ 14–18 MJ/kg), wood biomass (CV $\approx$ 17–19 MJ/kg), and agro-wastes (CV $\approx$ 12–16 MJ/kg) display significantly lower heating values than the biochar derived from them, which can reach 25–30 MJ/kg depending on pyrolysis temperature and feedstock type [37, 44]. This demonstrates that biochar production not only stabilizes carbon but also improves the energy density of biomass, thereby contributing to renewable energy strategies.

Future studies should concentrate on improving the qualities of biochar by tailoring its manufacturing procedures to specific applications. Investigating the potential of biochar to mitigate climate change and sequester carbon remains essential. Equally important is exploring how biochar can be integrated with advanced technologies such as microbial inoculants and nanotechnology. To verify and promote its use, comprehensive economic analyses, supportive legislative frameworks, and extensive field trials are needed. Furthermore, analyzing biochar's function in circular economy models—such as waste management and renewable energy generation—as well as conducting long-term environmental impact studies will improve its incorporation into sustainable practices. Such a strategy directly addresses key global challenges related to food security, soil health, energy security, and climate resilience.

Acknowledgement Authors acknowledge the financial support received by Department of Science and Technology, Govt of India under PURSE scheme [SR/PURSE/2022/142 (G)].

References

1. Alkharabsheh, H.M., Seleiman, M.F., Battaglia, M.L., Shami, A., Jalal, R.S., Alhammad, B.A., Almutairi, K.F., Al-Saif, A.M.: Biochar and its broad impacts in soil quality and fertility, nutrient leaching and crop productivity: a review. Agronomy **11**(5) (2021). https://doi.org/10.3390/agronomy11050993
2. Allohverdi, T., Mohanty, A.K., Roy, P., Misra, M.: A review on current status of biochar uses in agriculture. Molecules **26**(18) (2021). https://doi.org/10.3390/molecules26185584
3. Al-Wabel, M.I., Hussain, Q., Usman, A.R.A., Ahmad, M., Abduljabbar, A., Sallam, A.S., Ok, Y.S.: Impact of biochar properties on soil conditions and agricultural sustainability: a review. Land Degrad. Dev. **29**(7), 2124–2161 (2018). https://doi.org/10.1002/ldr.2829
4. Amalina, F., Razak, A.S.A., Krishnan, S., Zularisam, A.W., Nasrullah, M.: A comprehensive assessment of the method for producing biochar, its characterization, stability, and potential applications in regenerative economic sustainability—a review. Clean. Mater. **3** (2022). https://doi.org/10.1016/j.clema.2022.100045
5. Banerjee, T., Srivastava, R.K.: Plastics waste management and resource recovery in India. Int. J. Environ. Waste Manage. **10**(1), 90–111 (2012). https://doi.org/10.1504/IJEWM.2012.048153
6. Bhatt, B., Gupta, S.K., Mukherjee, S., Kumar, R.: A comprehensive review on biochar against plant pathogens: Current state-of-the-art and future research perspectives. Heliyon **10**(17) (2024). https://doi.org/10.1016/j.heliyon.2024.e37204
7. Biochar in European Soils and Agriculture (n.d.): [Reference incomplete—no publisher, journal, or provided].
8. Blanco-Canqui, H.: Biochar and soil physical properties. Soil Sci. Soc. Am. J. **81**(4), 687–711 (2017). https://doi.org/10.2136/sssaj2017.01.0017
9. Bolan, N., Hoang, S.A., Beiyuan, J., Gupta, S., Hou, D., Karakoti, A., Joseph, S., Jung, S., Kim, K.H., Kirkham, M.B., Kua, H.W., Kumar, M., Kwon, E.E., Ok, Y.S., Perera, V., Rinklebe, J., Shaheen, S.M., Sarkar, B., Sarmah, A.K., Van Zwieten, L.: Multifunctional applications of biochar beyond carbon storage. Int. Mater. Rev. **67**(2), 150–200 (2022). https://doi.org/10.1080/09506608.2021.1922047
10. Brtnicky, M., Dokulilova, T., Holatko, J., Pecina, V., Kintl, A., Latal, O., Vyhnanek, T., Prichystalova, J., Datta, R.: Long-term effects of biochar-based organic amendments on soil microbial parameters. Agronomy **9**(11) (2019). https://doi.org/10.3390/agronomy9110747
11. Carrari, E., Ampoorter, E., Verheyen, K., Coppi, A., Selvi, F.: Former charcoal kiln platforms as microhabitats affecting understorey vegetation in Mediterranean forests. Appl. Veg. Sci. **19**(3), 486–497 (2016). https://doi.org/10.1111/avsc.12238
12. Cha, J.S., Park, S.H., Jung, S.C., Ryu, C., Jeon, J.K., Shin, M.C., Park, Y.K.: Production and utilization of biochar: a review. J. Ind. Eng. Chem. **40**, 1–15 (2016). https://doi.org/10.1016/j.jiec.2016.06.002

13. Chakraborty, A., Pal, A., Saha, B.B.: A critical review of the removal of radionuclides from wastewater employing activated carbon as an adsorbent. Materials **15**(24), 8818 (2022). https://doi.org/10.3390/MA15248818
14. Chen, W., Meng, J., Han, X., Lan, Y., Zhang, W.: Past, present, and future of biochar. Biochar **1**(1), 75–87 (2019). https://doi.org/10.1007/s42773-019-00008-3
15. Dorjee, L., Nishmitha, K., Pattanayak, S., Wangmu, T., Meshram, S., Chongtham, S., Gogoi, R.: Biochar: a comprehensive review on a natural approach to plant disease management. J. Pure Appl. Microbiol. **18**(1), 29–45 (2024). https://doi.org/10.22207/JPAM.18.1.58
16. dovetailbiochar0316. (n.d.). [Reference incomplete – no publisher, journal, or provided].
17. ELDeeb, I.S., Bakry, A., Nasr, A.M., Showaib, E.A.: Sustainable clay brick manufacturing: a review of mechanical properties, firing technologies, emissions, and control strategies. J. Eng. Res. **8**(5) (2024). https://doi.org/10.70259/engJER.2024.851886
18. Eriksson, O., Glav Lundin, L.: Legacies of historic charcoal production affect the forest flora in a Swedish mining district. Nord. J. Bot. **39**(11) (2021). https://doi.org/10.1111/njb.03312
19. Gabhane, J.W., Bhange, V.P., Patil, P.D., Bankar, S.T., Kumar, S.: Recent trends in biochar production methods and its application as a soil health conditioner: a review. SN Appl. Sci. **2**(7) (2020). https://doi.org/10.1007/s42452-020-3121-5
20. Gerassimidou, S., Velis, C.A., Williams, P.T., Komilis, D.: Characterisation and composition identification of waste-derived fuels obtained from municipal solid waste using thermogravimetry: a review. Waste Manage. Res. **38**(9), 942–965 (2020). https://doi.org/10.1177/0734242X20941085
21. Gul, S., Whalen, J.K., Thomas, B.W., Sachdeva, V., Deng, H.: Physico-chemical properties and microbial responses in biochar-amended soils: mechanisms and future directions. Agric. Ecosyst. Environ. **206**, 46–59 (2015). https://doi.org/10.1016/j.agee.2015.03.015
22. Guo, M., Song, W., Tian, J.: Biochar-facilitated soil remediation: mechanisms and efficacy variations. Front. Environ. Sci. **8** (2020). https://doi.org/10.3389/fenvs.2020.521512
23. Gupta, S., Kua, H.W.: Factors determining the potential of biochar as a carbon capturing and sequestering construction material: Critical review. J. Mater. Civ. Eng. **29**(9) (2017). https://doi.org/10.1061/(asce)mt.1943-5533.0001924
24. Gu, Z.R., Hui, X.Y., Zeng, H.J., Qu, Q.C., Deji, Q., Xie, P.C., Huang, R.Q., Li, W., Xiong, J.: Research progress on the application of biochar in the remediation of heavy metals in tailings soil. Pol. J. Environ. Stud. **33**(3), 3123–3138 (2024). https://doi.org/10.15244/pjoes/176043
25. Gwenzi, W., Chaukura, N., Noubactep, C., Mukome, F.N.D.: Biochar-based water treatment systems as a potential low-cost and sustainable technology for clean water provision. J. Environ. Manage. **197**, 732–749 (2017). https://doi.org/10.1016/j.jenvman.2017.03.087
26. Ibitoye, S.E., Loha, C., Mahamood, R.M., Jen, T.C., Alam, M., Sarkar, I., Das, P., Akinlabi, E.T.: An overview of biochar production techniques and application in iron and steel industries. Bioresour. Bioprocess. **11**(1) (2024). https://doi.org/10.1186/s40643-024-00779-z
27. Inyang, M., Dickenson, E.: The potential role of biochar in the removal of organic and microbial contaminants from potable and reuse water: a review. Chemosphere **134**, 232–240 (2015). https://doi.org/10.1016/j.chemosphere.2015.03.072
28. Jeyasubramanian, K., Thangagiri, B., Sakthivel, A., Dhaveethu Raja, J., Seenivasan, S., Vallinayagam, P., Madhavan, D., Malathi Devi, S., Rathika, B.: A complete review on biochar: production, property, multifaceted applications, interaction mechanism and computational approach. Fuel **292** (2021). https://doi.org/10.1016/j.fuel.2021.120243
29. Jindo, K., Mizumoto, H., Sawada, Y., Sanchez-Monedero, M.A., Sonoki, T.: Physical and chemical characterization of biochars derived from different agricultural residues. Biogeosciences **11**(23), 6613–6621) (2014). https://doi.org/10.5194/bg-11-6613-2014
30. Kamarudin, N.S., Dahalan, F.A., Hasan, M., An, O.S., Parmin, N.A., Ibrahim, N., Hamdzah, M., Zain, N.A.M., Muda, K., Wikurendra, E.A.: Biochar: A review of its history, characteristics, factors that influence its yield, methods of production, application in wastewater treatment and recent development. Biointerface Res. Appl. Chem. **12**(6), 7914–7926 (2022). https://doi.org/10.33263/BRIAC126.79147926

31. Khairy, M., Amer, M., Ibrahim, M., Ookawara, S., Sekiguchi, H., & Elwardany, A.: The influence of torrefaction on the biochar characteristics produced from sesame stalks and bean husk. Biomass Convers. Biorefinery **14**(15), 17127–17148 (2024). https://doi.org/10.1007/s13399-023-03822-9
32. Kocsis, T., Biró, B., Ulmer, Á., Szántó, M., Kotroczó, Z.: Time-lapse effect of ancient plant coal biochar on some soil agrochemical parameters and soil characteristics. Environ. Sci. Pollut. Res. **25**(2), 990–999 (2018). https://doi.org/10.1007/s11356-017-8707-0
33. Köves, M., Madár, V., Ringer, M., Kocsis, T.: Overview of traditional and contemporary industrial production technologies for biochar along with quality standardization methods. Land **13**(9), 1388 (2024). https://doi.org/10.3390/land13091388
34. Kumari, K., Kumar, R., Bordoloi, N., Minkina, T., Keswani, C., Bauddh, K.: Unravelling the recent developments in the production technology and efficient applications of biochar for agro-ecosystems. Agric. (Switz.) **13**(3) (2023). https://doi.org/10.3390/agriculture13030512
35. Kuo, P.C., Illathukandy, B., Özdemir, F., Woudstra, T., Aravind, P.V.: Thermodynamic analysis of a novel integrated biomass pyrolysis-solid oxide fuel cells-combined heat and power system for co-generation of biochar and power. Front. Energy Res. **10** (2022). https://doi.org/10.3389/fenrg.2022.731191
36. Libra, J. A., Ro, K.S., Kammann, C., Funke, A., Berge, N.D., Neubauer, Y., Titirici, M.M., Fühner, C., Bens, O., Kern, J., Emmerich, K.H.: Hydrothermal carbonization of biomass residuals: a comparative review of the chemistry, processes and applications of wet and dry pyrolysis. Biofuels **2**(1), 71–106 (2011). https://doi.org/10.4155/BFS.10.81
37. Maggiotto, G., Colangelo, G., Milanese, M., de Risi, A.: Thermochemical technologies for the optimization of olive wood biomass energy exploitation: a review. Energies **16**(19), 6772 (2023). https://doi.org/10.3390/EN16196772
38. Mencarelli, A., Cavalli, R., Greco, R., Grigolato, S.: Comparison of technical and operational conditions of traditional and modern charcoal kilns: a case study in Italy. Energies **16**(23), 7757 (2023). https://doi.org/10.3390/EN16237757
39. Meyer, D.: Biochar—A survey. ENER-1346 Special Assignment in Energy and Process Engineering. [Institutional report—no available] (2009)
40. Mosa, A., Mansour, M. M., Soliman, E., El-Ghamry, A., El Alfy, M., El Kenawy, A.M. (2023). Biochar as a soil amendment for restraining greenhouse gases emission and improving soil carbon sink: current situation and ways forward. Sustain. (Switz.) **15**(2). https://doi.org/10.3390/su15021206
41. Murtaza, G., Ahmed, Z., Eldin, S. M., Ali, B., Bawazeer, S., Usman, M., Iqbal, R., Neupane, D., Ullah, A., Khan, A., Hassan, M.U., Ali, I., Tariq, A.: Biochar–soil–plant interactions: a cross talk for sustainable agriculture under changing climate. Front. Environ. Sci. **11** (2023). https://doi.org/10.3389/fenvs.2023.1059449
42. Murtaza, G., Ahmed, Z., Eldin, S.M., Ali, I., Usman, M., Iqbal, R., Rizwan, M., Abdel-Hameed, U.K., Haider, A.A., Tariq, A.: Biochar as a green sorbent for remediation of polluted soils and associated toxicity risks: a critical review. Separations **10**(3) (2023). https://doi.org/10.3390/separations10030197
43. Osman, A.I., Lai, Z.Y., Farghali, M., Yiin, C.L., Elgarahy, A.M., Hammad, A., Ihara, I., Al-Fatesh, A.S., Rooney, D.W., Yap, P.S.: Optimizing biomass pathways to bioenergy and biochar application in electricity generation, biodiesel production, and biohydrogen production. Environ. Chem. Lett. **21**(5), 2639–2705 (2023). https://doi.org/10.1007/s10311-023-01613-2
44. Perez, A., Luna, M.Y., Morales, R., Saldaña-Trinidad, E., Blanco, R., De La Cruz-Arreola, L., Pérez-Sariñana, S., Robles-Ocampo, B.Y., Perea-Moreno, A.-J., Díaz-González, A., Pérez-Sariñana, B.Y., Robles-Ocampo, J.B.: Assessment of the pretreatments and bioconversion of lignocellulosic biomass recovered from the husk of the cocoa pod. Energies **15**(10), 3544 (2022). https://doi.org/10.3390/EN15103544
45. Prabakar, P., Mustafa Mert, K., Muruganandam, L., Sivagami, K.: A comprehensive review on biochar for electrochemical energy storage applications: an emerging sustainable technology. Front. Energy Res. **12** (2024). https://doi.org/10.3389/fenrg.2024.1448520

46. Ravindiran, G., Rajamanickam, S., Janardhan, G., Hayder, G., Alagumalai, A., Mahian, O., Lam, S.S., Sonne, C.: Production and modifications of biochar to engineered materials and its application for environmental sustainability: a review. Biochar 6(1) (2024). https://doi.org/10.1007/s42773-024-00350-1
47. Ringsby, A.J., Ross, C.M., Maher, K.: Sorption of soil carbon dioxide by biochar and engineered porous carbons. Environ. Sci. Technol. 58(19), 8313–8325 (2024). ACS. https://doi.org/10.1021/acs.est.4c02015
48. Sajdak, M., Muzyka, R., Gałko, G., Ksepko, E., Zajemska, M., Sobek, S., Tercki, D.: Actual trends in the usability of biochar as a high-value product of biomass obtained through pyrolysis. Energies **16**(1), 355 (2022). https://doi.org/10.3390/EN16010355
49. Sakhiya, A.K., Anand, A., Kaushal, P.: Production, activation, and applications of biochar in recent times. Biochar **2**(3), 253–285 (2020). https://doi.org/10.1007/s42773-020-00047-1
50. Sekaran, C., Elangovan, R., Rangasami, S., Murugaragavan, R.: Characteristics of biochar: a review. Pharma Innov. J. **11**(12), 243–246 (2022). [No DOI, publisher = TPI]. www.thepharmajournal.com
51. Sen, S., Patil, S., Argyropoulos, D.S.: Thermal properties of lignin in copolymers, blends, and composites: a review. Green Chem. **17**(11), 4862–4887 (2015). https://doi.org/10.1039/C5GC01066G
52. Shrestha, R.K., Jacinthe, P.A., Lal, R., Lorenz, K., Singh, M.P., Demyan, S. M., Ren, W., Lindsey, L.E.: Biochar as a negative emission technology: a synthesis of field research on greenhouse gas emissions. J. Environ. Qual. **52**(4), 769–798 (2023). https://doi.org/10.1002/jeq2.20475
53. Singh, A., Bin Abu Sofian, A.D.A., Chan, Y.J., Chakrabarty, A., Selvarajoo, A., Abakr, Y.A., Show, P.L.: Hydrothermal carbonization: Sustainable pathways for waste-to-energy conversion and biocoal production. GCB Bioenergy **16**(6), e13150 (2024). https://doi.org/10.1111/GCBB.13150
54. Singh Yadav, S.P., Bhandari, S., Bhatta, D., Poudel, A., Bhattarai, S., Yadav, P., Ghimire, N., Paudel, P., Paudel, P., Shrestha, J., Oli, B.: Biochar application: a sustainable approach to improve soil health. J. Agric. Food Res. **11** (2023). https://doi.org/10.1016/j.jafr.2023.100498
55. Syguła, E., Ciolkosz, D., Białowiec, A.: The significance of structural components of lignocellulosic biomass on volatile organic compounds presence on biochar: a review. Wood Sci. Technol. 58(3), 859–886 (2024). https://doi.org/10.1007/S00226-024-01557-Y
56. Tan, H., Lee, C.T., Ong, P.Y., Wong, K.Y., Bong, C.P.C., Li, C., Gao, Y.: A review on the comparison between slow pyrolysis and fast pyrolysis on the quality of lignocellulosic and lignin-based biochar. In: IOP Conference Series: Materials Science and Engineering, vol. 1051, no. 1, p. 012075. IOP Publishing (2021). https://doi.org/10.1088/1757-899x/1051/1/012075
57. Tomczyk, A., Sokołowska, Z., Boguta, P.: Biochar physicochemical properties: Pyrolysis temperature and feedstock kind effects. Rev. Environ. Sci. Biotechnol. **19**(1), 191–215 (2020). https://doi.org/10.1007/s11157-020-09523-3
58. Varkolu, M., Gundekari, S., Omvesh, Palla, V. C. S., Kumar, P., Bhattacharjee, S., Vinodkumar, T.: Recent advances in biochar production, characterization, and environmental applications. Catalysts, **15**(3), 243 (2025). https://doi.org/10.3390/CATAL15030243
59. Wang, K., Remón, J., Jiang, Z., Ding, W.: Recent advances in the preparation and application of biochar derived from lignocellulosic biomass: a mini review. Polymers **16**(6), 851 (2024). https://doi.org/10.3390/POLYM16060851
60. Wang, X., Guo, Z., Hu, Z., Zhang, J.: Recent advances in biochar application for water and wastewater treatment: a review. PeerJ **8** (2020). PeerJ Inc. https://doi.org/10.7717/peerj.9164
61. Weber, K., Quicker, P.: Properties of biochar. Fuel **217**, 240–261 (2018). https://doi.org/10.1016/j.fuel.2017.12.054
62. Xiang, L., Harindintwali, J. D., Wang, F., Redmile-Gordon, M., Chang, S. X., Fu, Y., He, C., Muhoza, B., Brahushi, F., Bolan, N., Jiang, X., Ok, Y.S., Rinklebe, J., Schaeffer, A., Zhu, Y. G., Tiedje, J.M., Xing, B.: Integrating biochar, bacteria, and plants for sustainable remediation of soils contaminated with organic pollutants. Environ. Sci. Technol. **56**(23), 16546–16566 (2022). https://doi.org/10.1021/acs.est.2c02976

63. Zanutel, M., Garré, S., Bielders, C.L.: Long-term effect of biochar on physical properties of agricultural soils with different textures at pre-industrial charcoal kiln sites in Wallonia (Belgium). Eur. J. Soil Sci. **73**(1) (2022). https://doi.org/10.1111/ejss.13157
64. Zhang, Y., He, M., Wang, L., Yan, J., Ma, B., Zhu, X., Ok, Y.S., Mechtcherine, V., Tsang, D.C.W.: Biochar as construction materials for achieving carbon neutrality. Biochar **4**(1) (2022). https://doi.org/10.1007/s42773-022-00182-x
65. Zhang, Y., Wang, J., Feng, Y.: The effects of biochar addition on soil physicochemical properties: a review. Catena **202** (2021). https://doi.org/10.1016/j.catena.2021.105284

Green Remediation: Exploring Biosurfactants and Nanobiosurfactant for Chemical Pollutant Removal

Tanisha Singh, Indu Tomar, and Mousumi Debnath

Abstract Given the current environmental alterations and population growth, there is an urgent need to explore cost-effective and efficient bioremediation strategies. Currently, nanobiosurfactants have received recognition as an eco-friendly material for remediation technology, with better physicochemical properties. The mechanism of soil washing using biosurfactants and nanobiosurfactants, as well as their efficiency as detergents, is discussed in this review. Biosurfactants function by minimizing interfacial surface tension, consequently maximizing surface area to solubilize the least soluble compounds. They are armoured with both water and lipid-loving components, enabling them to form micelles. Micelles expedite the solubilization of substances that are the least soluble compounds. When in proximity with heavy metals, micelles form circular structures with a hydrophobic tail and metal particles at the core and a hydrophilic head towards the outside. The categorization of microorganisms can broadly be divided into glycolipids and lipopeptides. *Pseudomonas aeruginosa* has been identified as a source of biosurfactant that synthesizes rhamnolipids, while others, like *Bacillus paramycoides*, are responsible for synthesizing lipopeptide biosurfactant types, surfactin and iturin. A comparative analysis of the microorganisms for soil washing using rhamnolipids & surfactin biosurfactant in contrast to their nanobiosurfactant counterparts for soil washing results in distinct efficiency and varied environmental impact. The role of nanobiosurfactant in enhanced performance for the bioremediation of pollutants in soil can help in reducing the pollutant level and make it appropriate for agricultural use again.

Keywords Biosurfactant · Nanobiosurfactant · Soil washing · Bioremediation · Clean soil · Good health

---

T. Singh · I. Tomar · M. Debnath (✉)
Department of Biosciences, Manipal University Jaipur, Jaipur 303007, Rajasthan, India
e-mail: mousumi.debnath@jaipur.manipal.edu

L. Ledwani and S. Srivastava (eds.), *Experimental and Computational Advances in Materials*, Engineering Materials, https://doi.org/10.1007/978-981-95-4180-5_4

1 Introduction

The surge in industrial production and emissions has led to a significant increase in pollutant levels, posing environmental and health risks [1]. The contamination of the soil is primarily due to various industrial sectors, fertilizers, gasoline, improper disposal of toxic waste, pesticides, textile waste, coal combustion, etc. [2]. This may result in the accumulation of chemical contaminants like dyes, polycyclic aromatic compounds, or heavy metals, thereby existing as micropollutants and nanopollutants. The heavy metals often form residues and settle in the atmosphere through evaporation or as sediment in the [1]. Bioremediation of these pollutants can be followed using living organisms. Unlike plants, which require extended growth periods and are more challenging to engineer, microbes offer a more efficient and expedient option for genetic modification and application. The faster growth rate and relative ease of manipulation make microbes an attractive and worthwhile choice for various biotechnological applications. Hence, working with microbes is easier and worthwhile [3].

This method of using living microorganisms was first executed in 1989 for cleaning massive oil spills from a tanker ship named Exxon Valdez in Alaska [4]. Hydrocarbons are transformed into more environmentally friendly compounds by microorganisms, plants, or other microbial species that transform them into CO_2, water, and other inorganic components (Fig. 1). The type of pollutant, the surrounding environment, and the availability of phosphorus and nitrogen are among the factors that influence the progress of microorganisms and their ability to clean up [5] (Fig. 1).

Heavy metals are commonly found in contaminated soil, and often include metals present above the optimum concentrations [6]. There are multiple implications of this contamination, as it affects the soil respiration. These metals severely disturb the normal respiration and hamper the metabolism of microorganisms [1]. These metals (copper, lead, cadmium, chromium, mercury, nickel, manganese, zinc, and iron) present in the soil are often captured by plants and thus present in the diet consumed, which results in ill health in human beings and animals. This could cause physiological dysfunction, such as vascular damage, nerve disorders, congenital abnormalities, kidney and gastrointestinal system dysfunction, immune dysregulation, and skin lesions [7]. Therefore, bioremediation using microbes for the removal of metals is a crucial strategy for achieving sustainability, as it offers an eco-friendly and efficient approach to mitigate environmental pollution. This method can restore ecosystems, reduce toxicity, and promote a healthier environment.

Multiple fused aromatic rings composed of hydrogen and carbon atoms make up complex chemical molecules known as polycyclic aromatic hydrocarbons (PAHs). Their carcinogenic and mutagenic qualities make them known environmental contaminants that are frequently produced by the incomplete burning of fossil fuels, industrial processes, and natural sources. They are also linked to possible health hazards [8]. Many of these PAH are cancer-inducing molecules [9]. PAH present in soil, air, and water, as environmental pollutants, have a plethora of concerns, such as carcinogenic, mutagenic, and teratogenic chemicals [10]. The soil easily absorbs

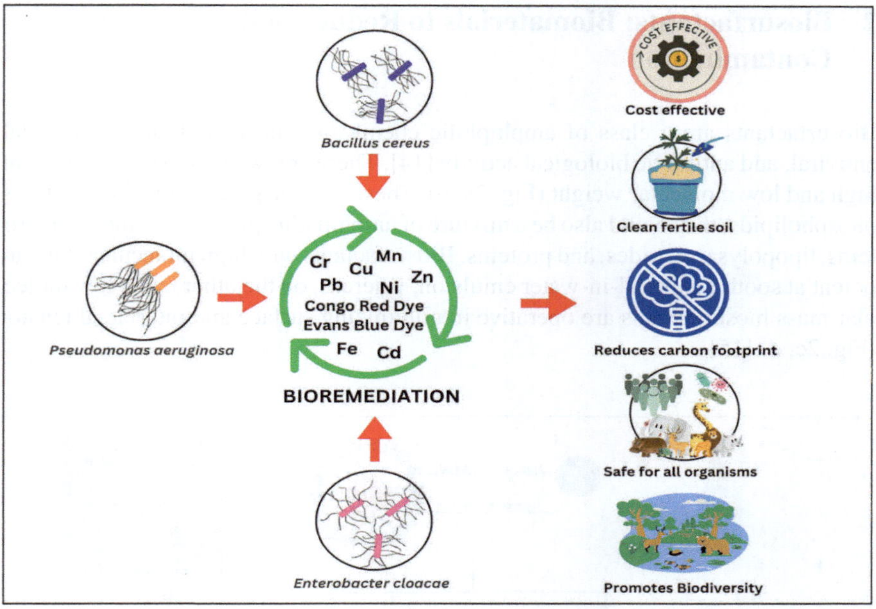

Fig. 1 Bioremediation of different pollutants with the help of microorganisms such as *Pseudomonas aeruginosa, Enterobacter cloacae,* and *Bacillus cereus*

it due to it hydrophobicity and lipophilicity [11]. This organic pollutant, PHA, is produced due to incomplete combustion and is resistant to biodegradation. They are toxic and carcinogenic, have low solubility, and mutagenic effects. Furthermore, since they are highly hydrophobic and are easily adsorbed in the soil, they cause disturbances and the sustainability of the environment [12].

The production of dyes worldwide has increased to over one million tons per year, making them one of the primary contaminants in industrial textile effluents. The textile industry is a significant consumer of dyes, generating substantial amounts of wastewater during various production processes. Wastewater from the textile industry, contains synthetic dyes, acid colorants, sulphur colorants, dyestuffs, vat dyes, mordant colouring agents, solvent colorants, gum thickener (guar), ammoniacal nitrogen, and heavy metals [13]. Given its overall environmental impact, this is a global concern.

2 Biosurfactants: Biomaterials to Reduce Soil Contamination

Biosurfactants are a class of amphiphilic chemicals with hemolytic, insecticidal, antiviral, and antibiotic biological activity [14]. There are two types of biosurfactant, high and low molecular weight (Fig. 2a, b). These include glycolipids, lipopeptides, phospholipids; they could also be a mixture of amphipathic polysaccharides, lipoproteins, lipopolysaccharides, and proteins. Biosurfactants with high molecular mass are potent at soothing the oil-in-water emulsion, whereas, on the other hand, low molecular mass biosurfactants are operative in minimizing surface and interfacial tension (Fig. 2c, d) [15].

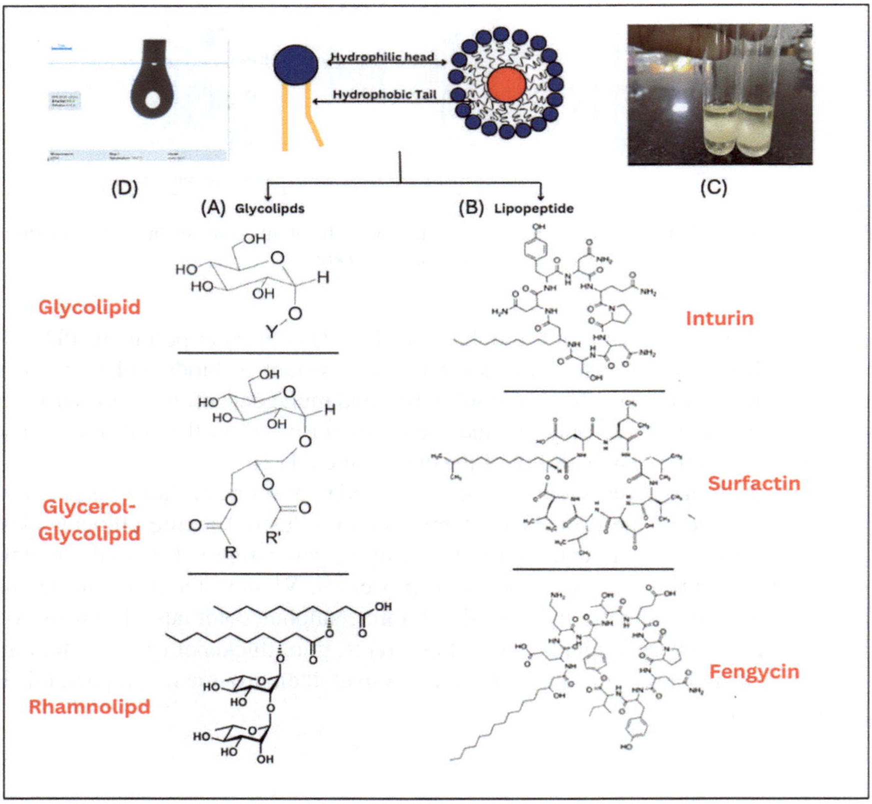

Fig. 2 Common types of biosurfactants used for soil washing: **a** glycolipid, **b** lipopeptide; screening tests for biosurfactants: **c** emulsification activity, and **d** the reduced surface tension between two liquids by biosurfactants

Polymeric microbial surfactant is a combination of various types of chemicals that are produced by microorganisms under specific conditions. Their physicochemical properties are governed by the fatty acid residues they contain, which can vary depending upon the bacterial strain, as well as the growth conditions and the type of nutrient present. *Bacillus* species, *Bacillus subtilis*, is responsible for synthesizing lipopeptides like surfactin and iturin, containing a cyclic peptide and fatty acid residue. *Pseudomonas aeruginosa* is responsible for synthesizing glycolipids like rhamnolipids. Rhamnolipid is one such example that is highly sought for its surfactancy in the detergent industry [16]. Biosurfactants are also categorized as polymeric and particulate surfactants, which belong to the low-mass surfactant category and comprise glycolipids, lipopeptides, and phospholipids [17]. Biosurfactants are produced under aerobic conditions, diverse in chemical nature, and can be lipid-containing molecules [18, 19].

When compared against synthetic alternatives, biosurfactants remain cost-competitive. Chemical surfactants are less environmentally friendly than biosurfactants derived from microorganisms. In comparison to chemical surfactants, these are relatively harmless and biodegradable materials, including excellent foaming ability, high selectivity, and the ability to function in harsh environmental pH, temperature, and salinity conditions [20]. Biosurfactants populate at the interfaces between two different phases and enhance aqueous phase solubility of nonaqueous phase liquids. With the reduction of the surface tension, there is an increase in aqueous surfactant concentration, thereby forming micelles. The critical micelle concentration (CMC) is the minimum concentration of surfactant above which micelles spontaneously form. There is a correlation between the level of concentration and the point at which the surfactant indicates the least surface tension. A multitude of properties is used to denote surfactants, relying on the CMC, including oil solubilization, foaming, detergency, surface tension, and emulsion formation. These characteristics are employed to assess the appropriateness of biosurfactants for soil washing and environmental restoration [16].

Rhamnolipids consist of a water-loving (polar) head with rhamnose units and include a lipid-loving (non-polar) tail incorporating hydrocarbon chains (Fig. 2a). The positively charged metals can be eliminated by rhamnolipids combined with the soil and sediments. This takes place due to the anionic properties of rhamnolipids obtained by their carboxylic moiety. It has been studied that rhamnolipids reduce toxicity and strengthen *Azotobacter and Rhizobium* microbial activity, and improve the quality of soil [21].

3 Biosurfactant in Soil Washing

The concept that both organic and inorganic pollutants tend to attach to particles like silt, clay, and organic particles is the foundation of the soil washing mechanism (Fig. 3). The first stage is mechanical screening, which removes any coarse particles

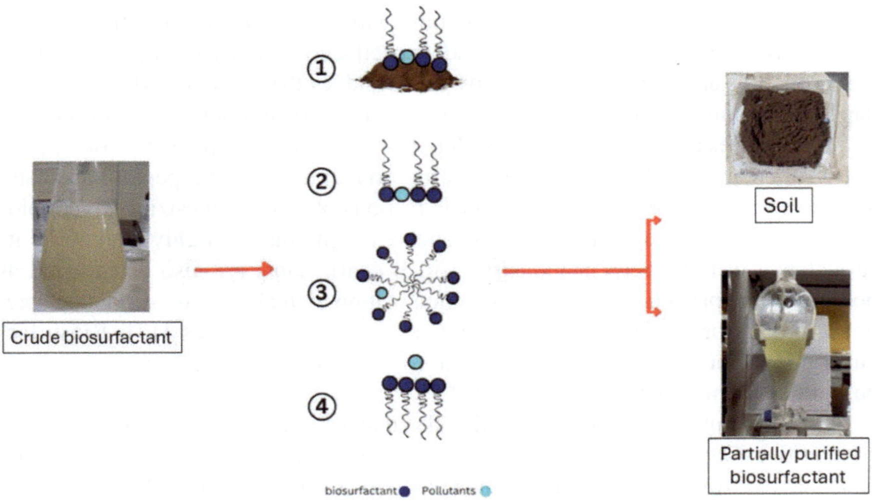

Fig. 3 Mechanism of soil washing transpiring in a laboratory using biosurfactant

and produces clean coarse soil devoid of materials such as rocks, wood, plastic, and stones.

The clean soil is then sent into a washing plant, where it is cleaned with a washing solution to guarantee that all leftover waste is eliminated. Gravity aids in the collection of sludge at the bottom of the unit. After passing through a treatment plant, the soil is bioremediated using biosurfactant, resulting in clean soil devoid of contaminants (Fig. 4).

4 Biosurfactant of *Pseudomonas aeruginosa* for Soil Washing

The gram-negative isolate of bacteria, *Pseudomonas aeruginosa,* is known to synthesize the biosurfactant rhamnolipids, which constitute an attractive and non-toxic alternative for bioremediation. Rhamnolipids can remove heavy metals to 80% which cause potential hazardous contamination, such as mercury, lead, and arsenic [22]. Table 1 presents the potential of this bacterium to remove various contaminants during soil washing. Rhamnolipids extracted from *P. aeruginosa* can bioremediate heavy metals, hydrocarbons, and PAH. The rhamnolipid biosurfactant produced by *P. aeruginosa* strain PF2 was explored for desorbing pollutants like anthracene, pyrene, and phenanthrene, different types of PAH, followed by breaking down of the desorbed material using electrokinetic oxidation, Utilizing a specific type of nanoparticle for enhancing the oxidation, the magnetite nanoparticles modified graphite (MNMG)

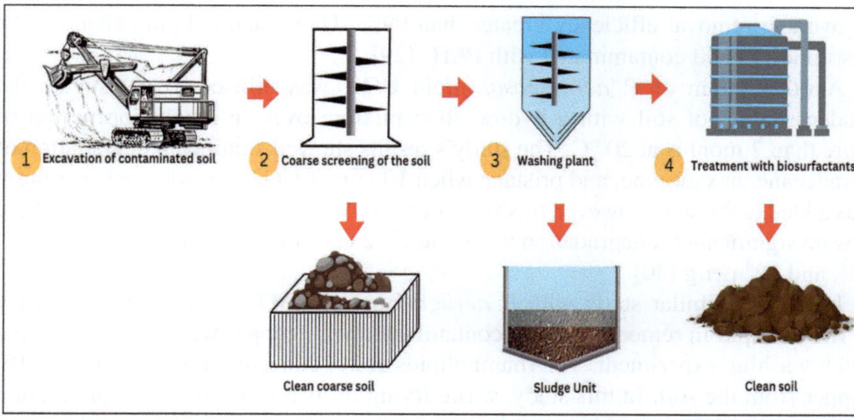

Fig. 4 Soil washing mechanism: once excavation of soil is done, it is put through a mechanical screening, which removes any coarse materials. The clean course soil is washed to ensure all the remaining pollutants are removed. The effect of gravity helps the sludge to collect at the bottom. Finally, a treatment plant bioremediated with biosurfactant, which results in contaminant-free soil

was combined with the rhamnolipid to undertake the innovative approach for remediation by biosurfactant aided soil washing. PF2 strain showed a CMC of 60 mg/L, could reduce surface tension to 35 mN/m and showed an emulsification value ranging from 55 to 60% for different solvents.

The rhamnolipid produced by this bacterium showed a critical micelle concentration (CMC) of 60 mg/L. For the maximum desorption of the different types of PAHs, it was experimentally found that CMC, contact hours, and pH were the noteworthy factors. The maximum desorption of PAH occurred at a CMC of 3, pH of 6, with a contact time of 34 h. With a modified graphite electrode, PAH degradation was enhanced exponentially. The optimal results were achieved at an electrolyte concentration of 25 mg/L, pH 5, voltage of 3 V, and a contact time of 6 h, resulting in

Table 1 Strains of *Pseudomonas aeruginosa* and types of pollutant degraded

Microorganism	Types of biosurfactant	Type of pollutant degraded	Reference
Pseudomonas aeruginosa	Lipopeptide	Heavy metal	[23]
Pseudomonas sp. IITISM 19	Lipopeptide	Total petroleum hydrocarbon	[24]
Pseudomonas sp. SA3	Lipopeptide	Petroleum	[25]
Pseudomonas aeruginosa C1501	Rhamnolipid	Biodegrade crude oil	[26]
Pseudomonas aeruginosa (DKB1)	Rhamnolipid	Rhamnolipid	[27]
Pseudomonas aeruginosa	Rhamnolipid	Remediation of oil-contaminated saline soil	[28]

an average removal efficiency greater than 99%. Thus, rhamnolipid enhanced the desorption of soil contaminated with PAHs [29].

Another strain of *P. aeruginosa,* strain UG2, was utilised to investigate the biodegradation of soil with a hydrocarbon mixture over an incubation period of more than 2 months at 20 °C. The study's results showed significant degradation of tetradecane, hexadecane, and pristane when 100 μg of UG2 biosurfactants per gram was added to the soil. However, this was not effective on 2-methylnaphthalene. There was no significant biodegradation when the UG2 cell was added at a density of 10^6, 10^7, and 10^8 per g [30].

In another similar study with *P. aeruginosa* strain MTCC 2297, the efficiency of rhamnolipid in remediating soil contaminated with copper was explored using a batch washing experiment. The rhamnolipids at 2% concentration remediated 71% copper from the soil. In this study, waste frying oil was utilized for the production of rhamnolipid that could bioremediate copper-contaminated soil. Hence, this work resulted in an eco-friendly and budget-effective alternative to many other popular physical and chemical methods of remediation [31].

A study on microbial-assisted phytodegradation using alfalfa and sunflower, along with *P. aeruginosa* and *Aspergillus oryzae* was used to improve the condition of pyrene-contaminated soil. To evaluate and assess pyrene removal from soil, three remediation strategies were applied. The first one consisted of bioremediation considering the bacterial strains, followed by phytoremediation with plants, subsequently, microbial-assisted phytoremediation was done. The result of this study indicated that *P. aeruginosa* enhanced the tolerance and growth of the plants and reduced the concentration of the pollutant, pyrene, present in the soil as compared to the plants in pyrene-contaminated soil without inoculation. *P. aeruginosa* inoculated alfalfa exhibited the maximum (91%) pyrene removal, followed by *A. oryzae* inoculated alfalfa (83.96%). The findings of the study led to the understanding that the rhizosphere relationship of plants and microorganisms is advantageous for the bioremediation of pyrene. This combinatorial approach using *P. aeruginosa* to degrade the pollutant, in conjunction with phytodegradation, leverages plants' capacity to absorb and break down the contaminants. This synergistic method can enhance the remediation of pyrene-contaminated soil and offer an effective solution rather than using either approach alone [32].

A study in 2023 investigated the cellular toxicity and damage caused during PAH degradation by *P. aeruginosa* G24. The incubation for 21 days resulted in 43.8% removal of phenanthrene and 36.8%. removal of pyrene, respectively. It was also shown that the *P. aeruginosa* G24 can degrade PAH, as there was an increased zeta potential and cell permeability, due to which the bacterial surface properties were altered. Probably due to this, the exopolysaccharide content decreased, and there was injury in the cell membrane, finally leading to cell apoptosis. The cellular studies on the bacteria show that such damage may limit the bacteria from continuously degrading the pollutant PAH, especially in the later stages of incubation [33].

5 Biosurfactant of *Bacillus cereus* for Soil Washing

Bacillus cereus is a gram-positive bacterium that is commonly present in soil, marine sponges and food. *Bacillus cereus* can be highly toxic and can lead to gastrointestinal illness in humans. *Bacillus cereus* NWUAB01 was isolated from the niche habitat of a gold mining site in South Africa. Thus demonstrating a high resistance to grow in an environment where there is a likelihood of finding other metals also apart from gold. *B.cereus* strain NWUAB01 produced biosurfactant of surfactin type and showcased its effectiveness for the removal of heavy metals. The production of biosurfactant was assessed for the soil washing of contaminated soil. The results show that the strain can tolerate cadmium and also has its removal efficiency of 54%; Similarly, strain NWUAB01 can grow in chromium and lead-containing soil also has the metal removal efficacy too [34].

The biosurfactant extracted from *Bacillus cereus* isolated from saline soil, was utilized to test its ability to extract crude oil by soil washing. To test the washing ability of this bacterial crude biosurfactant, two conditions were applied: contaminated soil containing oil was washed with crude biosurfactant, synthetic surfactant polyoxyethylene (20) sorbitan monolaurate (polysorb 20), and water as a control. The extracted biosurfactant showed its potential to reduce the surface tension value of 41.4 ± 1.3 mN/ m using crude oil in the culture media. The emulsification index obtained was 78.4%. Utilizing the crude biosurfactant, around 60% of oil was removed from the contaminated soil, which was 8% more than that achieved with the synthetic surfactant (polysorb 20). Despite having a few parameters modified in the culture medium, *Bacillus cereus* was able to synthesize biosurfactants and use them to remove hydrocarbons [35].

In another interesting study, Assam's Lakwa oilfield was used to acquire a soil sample for studying the remediation of total petroleum hydrocarbon (TPH) in the contaminated paddy soil. The biosurfactant-producing bacterial isolates utilized in this study were a mixed microbial culture (consortium) of 2 *Bacillus* strains viz. *Bacillus cereus* R2 and *Bacillus pumilus* KS2. Following a month of incubation in an earthen pot with the consortia, the Gas chromatography–mass spectrometry analysis of the soil sample confirmed that the microbial activity led to the degradation of TPH. The properties that were examined in addition to TPH degradation were pH, total organic content (TOC), conductivity, water holding capacity, and changes in total nitrogen concentration. With each passing month, there was an upward trend in the TPH degradation. After the 6-month mark, the sample that had the lowest initial TPH concentration also had more than 90% degradation. The bacterial consortium was noted to degrade a broad spectrum of aliphatic hydrocarbons that was found in soil samples. Different low- and high-molecular-weight PAHs present in the soil were fully mineralized or drastically reduced. With such a decrease in TPH levels, an observable improvement of the soil's physicochemical qualities was also noted [36, 37]. Table 2 shows the use of *Bacillus cereus* for the removal of different types of pollutants by soil washing.

Table 2 Strains of *Bacillus cereus* and types of pollutants degraded

Microorganism	Type of biosurfactant	Type of pollutant degraded	References
Bacillus malacitensi	Glycolipid	PAH and soil washing	[38]
Bacillus safensis J2	Glycolipid	Oil recovery	[17]
Bacillus pumilus 1529 bacteria	Glycolipid	Biodegradation, phytotoxicity	[39]
Bacillus mycoides	Glycolipid	Bioremediation of textile effluents, phytotoxicity	[40]
Bacillus amyloliquefaciens	Glycolipid	Phytotoxicity, oil contamination	[41]
Bacillus cereus	Rhamnolipid	Crude oil	[42, 35]
Bacillus amyloliquefaciens C11	Surfactin	Pesticides	[43]
Bacillus cereus KH1	Rhamnolipid	Dye decolorization	[44]
Bacillus subtilis	Surfactin	Phenanthrene removal	[45]
Bacillus pumilus NITDID1	Lipopeptide	Lead, cadmium, chromium, oil degradation	[46]
Bacillus subtilis MTCC 2423	Surfactin	Zinc	[47]
Bacillus albus (VK8)	Glycolipid	Degradation of thiocyanate	[48]
Bacillus alveayuensis	Lipopeptides	Oil removal	[49]
Bacillus subtilis	Surfactin and fengycin	Hydrocarbons and metals	[50]
Bacillus sp. Z-13	Lipopeptide	Petroleum hydrocarbons	[51]

6 Biosurfactant of *Enterobacter Sp.* For Soil Washing

The soil contamination by raw petroleum is a relentless issue with negative consequences due to its mutagenic, unmanageable, and toxic properties. Soil contaminated with hydrocarbons was collected, and different bacterial isolates were isolated in pure form. Findings uncovered that 40% of bacterial isolates showed the potential to biodegrade hydrocarbons, solubilize phosphorus, and produce exopolysaccharides. An interesting experiment examined the synthesis of biosurfactant using bacterial strains isolated from the marine sponge *Clathria* sp. collected from an oil-soiled coastal area of V.O. Chidambaranar Harbor, Thoothukudi, Tamil Nadu, India. They also assessed the biosurfactant production of six morphologically distinct biosurfactant-producing bacterial strains (SJ1–SJ6) that were cultured on oil agar plates.

The isolate SJ2 was selected as the highest possible biosurfactant-producing strain based on the screening results, which included a surface tension decrease of 31.6 mN/m, an oil displacement zone of 26 mm, and an emulsification index of less than 70%. It was hinted that this bacterial strain, SJ2, later identified using 16S rRNA sequencing, was *Enterobacter cloacae* with potential bioremediation activity. Analytical examinations using the chemical fingerprinting technique of FTIR and MALDI-TOF-MS examination of the biosurfactant confirmed it to be rhamnolipid. Against a variety of bacterial microorganisms, this rhamnolipid biosurfactant demonstrated antibacterial activity [52]. Table 3 shows the activity of *E. cloacae* for the reduction/removal of diverse forms of pollutants by soil washing.

Table 3 Strains of *Enterobacter cloacae* and types of pollutants degraded

Microorganism	Type of biosurfactant	Type of pollutant degraded	References
Enterobacter cloacae SD4–1	Rhamnolipid	Evans blue dye	[53]
Enterobacter cloacae	Rhamnolipid	Copper, zinc, iron, and manganese	[54]
Enterobacter cloacae	Rhamnolipid	Hydrogen sulphide, endotoxins	[55]
Enterobacter cloacae	Rhamnolipid	Congo red dye	[56]
Enterobacter cloacae (WAG45)	Rhamnolipid	Organophosphate herbicide (Glycophosate)	[57]
Enterobacter cloacae OM180597 SCUF0000312	Rhamnolipid	Zn^{+2}, Fe^{+2}, Pb^{+2}, Co^{+2}, Mn^{+2}, Ni^{+2}, and Cd^{+2}	[58]
Enterobacter cloacae AR-IASST(01)	Rhamnolipid	Heavy metal	[37]
Enterobacter cloacae	Rhamnolipid	Lead	[59]
Enterobacter cloacae UT25	Rhamnolipid	Cr (VI), Pb (II), Cu (II), Co (II), Ni (II), Hg (II), and Zn (II)	[60]
Enterobacter cloacae	Rhamnolipid	Triclosan (TCS)	[61]
Enterobacter cloacae	Rhamnolipid	Sulfate	[62]
Enterobacter cloacae	Rhamnolipid	Petroleum oil &hydrocarbon	[63]
Enterobacter cloacae	Rhamnolipid	Phosphate	[64]
Enterobacter cloacae	Rhamnolipid	Chromium and lithium	[59]
Enterobacter cloacae strain Cu6	Rhamnolipid	Cadmium	[65]
Enterobacter cloacea	Rhamnolipid	Copper and lead	[66]

7 Nanobiosurfactant and Its Properties

There is a recent rise in the scientific community for nanoscience and nanotechnology. Nanomaterials typically range from 1 to 100 nm (nm). They can contribute significantly to the challenges in the environment, water treatment, and environmental bioremediation [67, 68]. Various methods, like biological, chemical, and physical, are used to prepare nanoparticles. Nanoparticles are particularly advantageous as they are stabilizing and reducing agents, making them a better option to use. There are a few types of nanomaterials, however, silver nitrate is typically used for the degradation of polyaromatic hydrocarbons (PAHs). Silver nitrate is considered non-toxic for the environment, antimicrobial, stable, and cost-effective. The affinity of the reductase can reduce Ag^{2+} ions into Ag^0 at high temperatures. Hence, at high temperature with mid-pressure, the thiol group is responsible for minimizing silver [69].

Metal oxide nanoparticles are promising but they are not stable and often agglomerate [70]. Biosurfactants prevent agglomeration of nanoparticles by reducing the surface tension and offer stability over time by forming a stable layer on the surface of nanoparticles.

An advancement in the green synthesis of nanoparticle formation is biosurfactant-mediated nanoparticles. Biosurfactants, bearing their stabilization and reducing properties, help to stabilize the structure of the nanoparticle. Biosurfactant can reduce cluster formation through the attraction due to electrostatic forces, which facilitates a uniform nanoparticle morphology [71]. Biosurfactant must be encapsulated with nanoparticles to support the surface of the molecules. Biosurfactants maintain stability by micro-emulsion formation or a reverse micelle system [1].

Capping is a method of coating or stabilizing nanoparticles with a surfactant molecule. It is used to achieve stabilization of nanoparticles and facilitate bioremediation of heavy metals; in addition, it also controls the size, dispersibility, and surface properties. To form a nanobiosurfactant, the biosurfactant molecules are introduced to a solution containing nanoparticles. This leads to the formation of clusters with the hydrophobic tail; on the other hand, the hydrophilic head retains the original orientation, which is facing the aqueous environment. The following key step is encapsulation, which takes place when the hydrophobic surface of the nanoparticle interacts with the hydrophobic tails of the surfactant. Hence, a stable and protective layer is created [72, 73] as shown in Fig. 5.

Capping agents are stabilizers that hinder the extra nanoparticles and intercept their aggregation and/or coagulation in colloidal synthesis [72]. During capping, a type of biosurfactant, sophorolipids produced by *Starmerella bombicola*, is capped with nanoparticles. Rhamnolipids, synthesised by *Pseudomonas aeruginosa*, can also be used as capping agents for specific nanoparticles that use zinc sulphate. The capping results in significantly smaller nanoparticles. These miniature nanoparticles facilitate the rapid decolorization and degradation of dyes used in the textile industry.

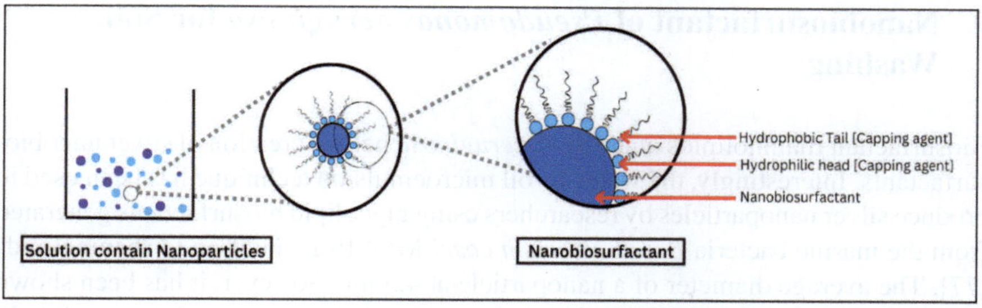

Fig. 5 Formation of nanobiosurfactant: the addition of the biosurfactant molecules to a solution containing nanoparticles results in cluster formation. Encapsulation, the next crucial stage, occurs when the hydrophobic surface of the nanoparticle interacts with the hydrophobic surfactant tails. Thus, a solid shielding layer is created that helps to stabilise the newly formed nanobiosurfactant

Zinc sulphate nanoparticles synthesised by Rhamnolipids also act as a photocatalyst, which enables it to disintegrate any organic compound in the presence of light [19, 73, 74].

8 Nanobiosurfactant in Soil Washing

The heavy metals are often noted as pollutants found in the soil and water, which are unsafe for humans as well as other living organisms. In general, metals are not biodegradable, although the mobility and toxicity can be altered when there is a change in their chemical state (used redox processes or alkylation) [75]. The soil quality of agricultural land can be improved with the utilisation of biosurfactants. The bioremediation of the soil matrix is determined by factors such as the pH, particle size of the material, metal exchange capacity, contamination time, and metal persistence, which is erratic. It is predominantly based on their capacity to form complexes with metals. The anionic nanobiosurfactant constructs complexes through ionic bonds with a cationic metallic form. These bonds are considerably stronger than those bonds formed between the metal and the soil. By reducing the interfacial tension, metal nanobiosurfactant complexes effectively competes with the soil surface for binding to heavy metal ions and the formation of micelles allows the solubilization and elimination of these metal ions from the soil [76].

9 Nanobiosurfactant of *Pseudomonas aeruginosa* for Soil Washing

Biosurfactant rhamnolipids made by *P. aeruginosa* help in creation of silver nanobiosurfactants. Interestingly, the water-in-oil microemulsion technique has been used to produce silver nanoparticles by researchers using glycolipid biosurfactants generated from the marine bacteria *Brevibacterium casei* MSA19 in addition to rhamnolipids [77]. The average diameter of a nanoparticle at 4.5 nm, however, it has been shown that *P. aeruginosa* MM1011 produces a nanobiosurfactant of 631.4 nm [78]. In another case where the diameter of the nanoparticles varies, the synthesis of silver nanoparticles with a diameter of 1.13 nm was also produced for *P. aeruginosa*. These rhamnolipids were made using vegetable oil refinery residue as well as corn steep liquor, 2.5%, which was biodegradable and showed no toxicity [79].

There have been studies indicating microbial remediation using biosurfactant [80] and their use in synthesis of nanoparticles for various applications including their potential to be a viable option as well as efficient in removing or degrading heavy metals from soil, water, sediment, etc [81–86]. These biosurfactant help in stabilizing the nanoparticle during its formation [81, 83]. *P. aeruginosa* is found frequently in contaminated areas. They are known to be resistant to a large spectrum of antibiotics, including carbapenems, alongside being resistant to detergents and organic solvents. Therefore, it could be said that *P. aeruginosa* is highly versatile and highly resilient to cadmium isolated from various environmental regimes. Thus, *P. aeruginosa*, can also be utilized as a potent bioinoculant and for nano biosurfactant production.

10 Conclusion

With the advancement of industrialization, there is a staggering abundance of pollutants in the soil, which will harm the environment and life on earth, including the crops grown in that soil and ultimately consumed by us. Nanobiosurfactant offers a greener and safer approach for soil washing of pollutants. As during its production, the major two components work synergistically. The biosurfactants help the metallic nanoparticle to be stable and prevent them from their inherent nature to aggregate. They reduce the surface tension and also form a stable layer on the surface of the nanoparticle and help them to maintain their integrity for longer duration. Hence, it can be said that nanobiosurfactant can be in the active state for longer duration and play an active role in the bioremediation of pollutants.

References

1. Xie, Y., Ye, R., Liu, H.: Synthesis of silver nanoparticles in reverse micelles stabilized by natural biosurfactant. Colloids Surf A Physicochem Eng Asp **279**, 175–178 (2006). https://doi.org/10.1016/j.colsurfa.2005.12.056
2. Wuana, R.A., Okieimen, F.E.: Heavy metals in contaminated soils: a review of sources, chemistry, risks and best available strategies for remediation. ISRN Ecol. **2011**, 1–20 (2011). https://doi.org/10.5402/2011/402647
3. Hussain, S., Khan, M., Sheikh, T.M.M., et al.: Zinc essentiality, toxicity, and its bacterial bioremediation: a comprehensive insight. Front Microbiol. **13**, 900740 (2022). https://doi.org/10.3389/fmicb.2022.900740
4. Lindeberg, M.R., Maselko, J., Heintz, R.A., et al.: Conditions of persistent oil on beaches in Prince William sound 26 years after the Exxon Valdez spill. Deep Sea Res. Part II: Top. Stud. Ocean. **147**, 9–19 (2018). https://doi.org/10.1016/j.dsr2.2017.07.011
5. Sales da Silva, I.G., Gomes de Almeida, F.C., Padilha da Rocha e Silva, N.M., et al.: Soil bioremediation: overview of technologies and trends. Energ. (Basel) **13**, 4664 (2020). https://doi.org/10.3390/en13184664
6. Priya, A.K., Jalil, A.A., Vadivel, S., et al.: Heavy metal remediation from wastewater using microalgae: recent advances and future trends. Chemosphere **305**, 135375 (2022). https://doi.org/10.1016/j.chemosphere.2022.135375
7. Balali-Mood, M., Naseri, K., Tahergorabi, Z., et al.: Toxic mechanisms of five heavy metals: mercury, lead, chromium, cadmium, and arsenic. Front. Pharmacol. **12**, 643972 (2021). https://doi.org/10.3389/fphar.2021.643972
8. Gu, Y.-G., Ke, C.-L., Liu, Q., Lin, Q.: Polycyclic aromatic hydrocarbons (PAHs) in sediments of Zhelin Bay, the largest mariculture base on the eastern Guangdong coast, South China: characterization and risk implications. Mar. Pollut. Bull. **110**, 603–608 (2016). https://doi.org/10.1016/j.marpolbul.2016.06.025
9. Wang, D., Ma, J., Li, H., Zhang, X.: Concentration and potential ecological risk of PAHs in different layers of soil in the petroleum-contaminated areas of the loess plateau, China. Int. J. Environ. Res. Public Health **15**, 1785 (2018). https://doi.org/10.3390/ijerph15081785
10. Cao, W., Yin, L., Zhang, D., et al.: Contamination, sources, and health risks associated with soil PAHs in rebuilt land from a coking plant, Beijing, China. Int. J. Environ. Res. Public Health **16**, 670 (2019). https://doi.org/10.3390/ijerph16040670
11. Cao, Y., Yang, B., Song, Z., et al.: Wheat straw biochar amendments on the removal of polycyclic aromatic hydrocarbons (PAHs) in contaminated soil. Ecotoxicol. Environ. Saf. **130**, 248–255 (2016). https://doi.org/10.1016/j.ecoenv.2016.04.033
12. Haritash, A.K., Kaushik, C.P.: Biodegradation aspects of polycyclic aromatic hydrocarbons (PAHs): a review. J. Hazard. Mater. **169**, 1–15 (2009). https://doi.org/10.1016/j.jhazmat.2009.03.137
13. Kumari, S., Khan, A.A., Chowdhury, A., et al.: Efficient and highly selective adsorption of cationic dyes and removal of ciprofloxacin antibiotic by surface modified nickel sulfide nanomaterials: Kinetics, isotherm and adsorption mechanism. Colloids Surf A Physicochem Eng Asp **586**, 124264 (2020). https://doi.org/10.1016/j.colsurfa.2019.124264
14. Xi, W., Ping, Y., Alikhani, M.A.: A review on biosurfactant applications in the petroleum industry. Int. J. Chem. Eng. **2021**, 1–10 (2021). https://doi.org/10.1155/2021/5477185
15. Pacwa-Płociniczak, M., Płaza, G.A., Piotrowska-Seget, Z., Cameotra, S.S.: Environmental applications of biosurfactants: recent advances. Int. J. Mol. Sci. **12**, 633–654 (2011). https://doi.org/10.3390/ijms12010633
16. Mulligan, C.N. Sustainable remediation of contaminated soil using biosurfactants. Front. Bioeng. Biotechnol. **9**, 635196 (2021). https://doi.org/10.3389/fbioe.2021.635196
17. Das, A.J., Kumar, R.: Production of biosurfactant from agro-industrial waste by Bacillus safensis J2 and exploring its oil recovery efficiency and role in restoration of diesel contaminated soil. Environ. Technol. Innov. **16**, 100450 (2019). https://doi.org/10.1016/j.eti.2019.100450

18. Kosaric, N.: Biosurfactants and their application for soil bioremediation. Food Technol. Biotechnol. **39**(4), 295–304 (2001)
19. Sarubbo, L.A., Silva M da, G.C., Durval, I.J.B., et al.: Biosurfactants: production, properties, applications, trends, and general perspectives. Biochem. Eng. J. **181**, 108377 (2022). https://doi.org/10.1016/j.bej.2022.108377
20. Costa, J.A.V., Treichel, H., Santos, L.O., Martins, V.G.: Solid-state fermentation for the production of biosurfactants and their applications. In: Current Developments in Biotechnology and Bioengineering, pp. 357–372. Elsevier (2018)
21. Ron, E.Z., Rosenberg, E.: Biosurfactants and oil bioremediation. Curr. Opin. Biotechnol. **13**, 249–252 (2002). https://doi.org/10.1016/S0958-1669(02)00316-6
22. Abalos-Rodríguez, A., Rodríguez-Gámez, O., Martin, Y.B.-S.: Rhamnolipids from Pseudomonas aeruginosa in the cleaning of polluted environments. In: Multifunctional Microbial Biosurfactants, pp. 231–249. Springer Nature Switzerland, Cham (2023)
23. Gaur, L., Chandola, I., Negi, N., Rawat, P.: Microbial profile and antibiotic resistance pattern of water supply in a tertiary care hospital of Uttarakhand. Iranlan J. Microbiol. **15**(2), 225 (2023)
24. Ray, M., Kumar, V., Banerjee, C.: Kinetic modelling, production optimization, functional characterization and phytotoxicity evaluation of biosurfactant derived from crude oil biodegrading Pseudomonas sp. IITISM 19. J. Environ. Chem. Eng. **10**(2), 107190 (2022). https://doi.org/10.1016/j.jece.2022.107190
25. Ambust, S., Das, A.J., Kumar, R.: Bioremediation of petroleum contaminated soil through biosurfactant and Pseudomonas sp. SA3 amended design treatments. Curr. Res. Microb. Sci. **2**, 100031 (2021). https://doi.org/10.1016/j.crmicr.2021.100031
26. Adetunji, C.O., Oloke, J.K., Osemwegie, O.O.: Environmental fate and effects of granular pesta formulation from strains of Pseudomonas aeruginosa C1501 and Lasiodiplodia pseudotheobromae C1136 on soil activity and weeds. Chemosphere **195**, 98–107 (2018). https://doi.org/10.1016/j.chemosphere.2017.12.056
27. Deivakumari, M., Sanjivkumar, M., Suganya, A.M., et al.: Studies on reclamation of crude oil polluted soil by biosurfactant producing Pseudomonas aeruginosa (DKB1). Biocatal. Agric. Biotechnol. **29**, 101773 (2020). https://doi.org/10.1016/j.bcab.2020.101773
28. Ebadi, A., Khoshkholgh Sima, N.A., Olamaee, M., et al.: Effective bioremediation of a petroleum-polluted saline soil by a surfactant-producing Pseudomonas aeruginosa consortium. J. Adv. Res. **8**, 627–633 (2017). https://doi.org/10.1016/j.jare.2017.06.008
29. Pourfadakari, S., Ahmadi, M., Jaafarzadeh, N., et al.: Remediation of PAHs contaminated soil using a sequence of soil washing with biosurfactant produced by Pseudomonas aeruginosa strain PF2 and electrokinetic oxidation of desorbed solution, effect of electrode modification with Fe_3O_4 nanoparticles. J. Hazard. Mater. **379**, 120839 (2019). https://doi.org/10.1016/j.jhazmat.2019.120839
30. Jain, D.K., Lee, H., Trevors, J.T.: Effect of addition of Pseudomonas aeruginosa UG2 inocula or biosurfactants on biodegradation of selected hydrocarbons in soil. J. Ind. Microbiol. **10**, 87–93 (1992). https://doi.org/10.1007/BF01583840
31. Venkatesh, N.M., Vedaraman, N.: Remediation of soil contaminated with copper using rhamnolipids produced from Pseudomonas aeruginosa MTCC 2297 using waste frying rice bran oil. Ann Microbiol **62**(1), 85–91 (2012)
32. Panwar, R., Mathur, J.: Microbial-assisted phytodegradation for the amelioration of pyrene-contaminated soil using Pseudomonas aeruginosa and Aspergillus oryzae with alfalfa and sunflower. 3 Biotech **13**(7), 251 (2023)
33. Gao, H., Wu, M., Liu, H., et al.: Cell toxic damages during polycyclic aromatic hydrocarbons biodegradation by Pseudomonas aeruginosa G24. J. Water Process. Eng. **54**, 103992 (2023). https://doi.org/10.1016/j.jwpe.2023.103992
34. Ayangbenro, A.S., Babalola, O.O.: Genomic analysis of Bacillus cereus NWUAB01 and its heavy metal removal from polluted soil. Sci. Rep. **10**(1), 19660 (2020). https://doi.org/10.1038/s41598-020-75170-x
35. Gracida, J., Corona, A., Amaro, A., et al.: Washing soil contaminated with crude oil using biosurfactant of halotolerant Bacillus cereus, grown under different conditions. Biologia (Bratisl) (2023)

36. Patowary, K., Bhuyan, T., Patowary, R., et al.: Soil treatment using a biosurfactant producing bacterial consortium in rice fields contaminated with oily sludge— a sustainable approach. Environ. Res. **220**, 115092 (2023). https://doi.org/10.1016/j.envres.2022.115092
37. Patowary, R., Rajbongshi, B., Devi, A., et al.: Biosurfactant-Enhanced Bioremediation of Petroleum Sludge: Unveiling the Potential of Enterobacter Cloacae Ar-Iasst. SSRN (2023)
38. Christopher, J.M., Sridharan, R., Somasundaram, S., Ganesan, S.: Bioremediation of aromatic hydrocarbons contaminated soil from industrial site using surface modified amino acid enhanced biosurfactant. Environ. Pollut. **289**, 117917 (2021). https://doi.org/10.1016/j.envpol.2021.117917
39. Khajavi-Shojaei, S., Moezzi, A., Enayatizamir, N., Mokhtari, B.: Biodegradation and phytotoxicity assessment of phenanthrene by biosurfactant-producing *Bacillus pumilus* 1529 bacteria. Chem. Ecol. **36**, 396–409 (2020). https://doi.org/10.1080/02757540.2020.1754807
40. Mahmood, R., Sharif, F., Ali, S., Hayyat, M.U.: Enhancing the decolorizing and degradation ability of bacterial consortium isolated from textile effluent affected area and its application on seed germination. Sci. World J. **2015**, 1–9 (2015). https://doi.org/10.1155/2015/628195
41. Nedoroda, V., Trokhymenko, G., Magas, N.: Bioremediation possibilities of oil-contaminated soil by biosurfactant based on <i>Bacillus</i> strain. J. Ecol. Eng. **23**, 49–55 (2022). https://doi.org/10.12911/22998993/150672
42. Gracida, J., Corona, A., Amaro, A., et al.: Correction to: washing soil contaminated with crude oil using biosurfactant of halotolerant Bacillus cereus, grown under different conditions. Biologia (Bratisl.) **78**, 2203–2203 (2023). https://doi.org/10.1007/s11756-023-01365-y
43. Schalchli, H., Lamilla, C., Rubilar, O., et al.: Production and characterization of a biosurfactant produced by Bacillus amyloliquefaciens C11 for enhancing the solubility of pesticides. J. Environ. Chem. Eng. **11**, 111572 (2023). https://doi.org/10.1016/j.jece.2023.111572
44. Nor, F.H.M., Abdullah, S., Ibrahim, Z., et al.: Role of extremophilic Bacillus cereus KH1 and its lipopeptide in treatment of organic pollutant in wastewater. Bioprocess Biosyst. Eng. **46**, 381–391 (2023). https://doi.org/10.1007/s00449-022-02749-1
45. Lyu C, Zhu Y, Zhang G, Li H (2023) Phenanthrene removal from soil washing eluent by *Bacillus subtilis* embedded in alginate-carboxymethyl cellulose-diatomite beads. Environ. Technol. 1–11. https://doi.org/10.1080/09593330.2023.2246106
46. Dasgupta, A., Saha, S., Ganguli, P., et al.: Characterization of pumilacidin, a lipopeptide biosurfactant produced from Bacillus pumilus NITDID1 and its prospect in bioremediation of hazardous pollutants. Arch. Microbiol. **205**, 274 (2023). https://doi.org/10.1007/s00203-023-03619-4
47. Jayalatha, N.A., Prabhu, A., Adi, V.K.: Assessing the potential of biosurfactant production by Bacillus subtilis MTCC 2423 to remediate the zinc-contaminated soil: a process optimization approach. Arab. J. Sci. Eng. **49**, 43–57 (2024). https://doi.org/10.1007/s13369-023-07938-y
48. Argentin, M.N., Martins, L.F., Sousa, M.P., Bossolan, N.R.S.: Biosurfactant from a thermohalophilic strain of Bacillus alveayuensis isolated from a Brazilian oil reservoir: production, chemical characterization, antimicrobial activity, and efficiency in wettability reversal and oil removal from oil-soaked sand. Geoenergy Sci. Eng. **231**, 212324 (2023). https://doi.org/10.1016/j.geoen.2023.212324
49. Bae, S.S., Lee, J.H., Kim, S.J.: Bacillus alveayuensis sp. nov., a thermophilic bacterium isolated from deep-sea sediments of the Ayu Trough. Int. J. Syst. Evol. Microbiol. **55**(3), 1211–1215 (2005)
50. Krucoń, T., Ruszkowska, Z., Pilecka, W., et al.: Bioprospecting of the Antarctic Bacillus subtilis strain for potential application in leaching hydrocarbons and trace elements from contaminated environments based on functional and genomic analysis. Environ. Res. **227**, 115785 (2023). https://doi.org/10.1016/j.envres.2023.115785
51. Zhang, S., Zhang, M., Han, F., et al.: Enhanced degradation of petroleum in saline soil by nitrogen stimulation and halophilic emulsifying bacteria Bacillus sp. Z-13. J. Hazard. Mater. **459**, 132102 (2023). https://doi.org/10.1016/j.jhazmat.2023.132102

52. Harikrishnan, S., Sudarshan, S., Alsalhi, M.S., Parivallal, M., Devanesan, S., SenthilBalan, S., Moovendhan, M., Rajasekar, A., Jayalakshmi, S.: Production and characterization of biosurfactant from Enterobacter cloacae SJ2 isolated from marine sponge Clathria sp. Biomass Convers. Biorefinery **14**(21), 27525–27536 (2024)
53. Ravi, A., Krishnan, R., Ravuri, M. et al.: Sustainable approach for the degradation of contrast dye Evans blue by Enterobacter cloacae strain SD4-1. J. Taiwan Inst. Chem. Eng. **166**, 105323 (2025). https://doi.org/10.1016/j.jtice.2023.105323
54. Borodina, E., Asyakina, L., Proskuryakova, L., et al.: The potential of using plant-growth-stimulating bacteria in phytoremediation of coal dumps. BIO Web Conf **82**, 06011 (2024). https://doi.org/10.1051/bioconf/20248206011
55. Daae, H.L., Graff, P., Foss, O.A.H., et al.: A cross-sectional study on occupational exposure to microorganisms, endotoxin, hydrogen sulfide, and dust during work at drilling waste treatment plants. Ann Work Expo Health **68**, 58–77 (2024). https://doi.org/10.1093/annweh/wxad069
56. El-Taweel, G., Hemdan, B.: Eco-friendly biological degradation and detoxification of congo red dye indigoes monospecies and mixed culture of bacterial strains isolated from textile industrial effluents. Egypt J. Chem. **0**:0–0 (2023). https://doi.org/10.21608/ejchem.2023.213881.8035
57. Mohy-Ud-Din, W., Javed Akhtar, M., Bashir, S., et al.: Isolated bacterial strains efficiently degrade glyphosate under different environmental conditions. Pak. J. Bot. **56**(2), 765–74 (2024). https://doi.org/10.30848/PJB2024-2(28)
58. Kelany, M.S., El-sawy, M.A., El-Gendy, A.R., Beltagy, E.A.: Bioremediation of industrial wastewater heavy metals using solo and consortium Enterobacter spp. Environ. Monit. Assess. **195**, 1357 (2023). https://doi.org/10.1007/s10661-023-11951-x
59. Tahir, Z., Hayyat, M.U., Khan, Q.F., et al.: Phyto- and bio-management of metal (loid)-contaminated soil by inoculating resistant bacteria: evaluating tolerance of treated rice plant and soil with its efficiency. Environ. Sci. Pollut. Res. **30**, 122524–122536 (2023). https://doi.org/10.1007/s11356-023-30769-1
60. Kalsoom, A., Jamil, N., ul Hassan, S.M., et al.: Chromate removal by Enterobacter cloacae strain UT25 from tannery effluent and its potential role in Cr (VI) remediation. Curr Microbiol **80**, 99 (2023). https://doi.org/10.1007/s00284-023-03194-3
61. Ghafouri, M., Pourjafar, F., Ghobadi Nejad, Z., Yaghmaei, S.: Biological treatment of triclosan using a novel strain of Enterobacter cloacae and introducing naphthalene dioxygenase as an effective enzyme. J. Hazard. Mater. **459**, 131833 (2023). https://doi.org/10.1016/j.jhazmat.2023.131833
62. Abduallah, E.R., Saad, E.M., Aboseidah, A.A., Rasmey, A.-H.M.: Efficient bioreduction of sulfate from industrial wastewater effluents using Enterobacter cloacae emr69. J. Adv. Vet. Res. **13**(10), 1949–1953 (2023)
63. Fahim Muhsin, K., Fadhil AboKsour, M., Hadi, S.: Bioremediation by bacteria isolated from water contaminated with hydrocarbons. Bionatura **8**, 1–10 (2023). https://doi.org/10.21931/RB/2023.08.03.94
64. Fathy, F. E., Mahmoud, S. M., Ahmed, A. I. et al.: Biological removal of phosphate from synthetic wastewater using Enterobacter cloacae. AUJES, **4** (4), 303–314 (2023). https://doi.org/10.21608/aujes.2023.222742.1174
65. Xia, Y., Xu, Y., Zhou, Y., et al.: Comparative genome analyses uncovered the cadmium resistance mechanism of enterobacter cloacae. Int. Microbiol. **26**, 99–108 (2022). https://doi.org/10.1007/s10123-022-00276-3
66. Sadullaeva, M.S, Turgunova L.S., Kadirova, G.Kh., et al.: Effect of bacteria on growth and development of wheat in soils contaminated with Cu and Pb cations. Sci. Innov. Int. Sci. J. (2023)
67. Horváthová, H., Lászlová, K., Dercová, K.: Bioremediation vs. nanoremediation: degradation of polychlorinated biphenyls (PCBS) using integrated remediation approaches. Water Air Soil Pollut. **230**(8), 204 (2019)
68. Hidangmayum, A., Debnath, A., Guru, A., et al.: Mechanistic and recent updates in nano-bioremediation for developing green technology to alleviate agricultural contaminants. Int. J. Environ. Sci. Technol. **20**, 11693–11718 (2023)

69. Nehal, N., Singh, P.: Role of nanotechnology for improving properties of biosurfactant from newly isolated bacterial strains from Rajasthan. Mater. Today Proc. **50**, 2555–2561 (2022). https://doi.org/10.1016/j.matpr.2021.05.682
70. Dhanapal, A. R., Thiruvengadam, M., Vairavanathan, J. et al.: Nanotechnology approaches for the remediation of agricultural polluted soils. ACS omega, **9**(12), 13522–13533 (2024). https://doi.org/10.1021/acsomega.3c09776
71. Kumar Sur, U., Ankamwar, B., Karmakar, S., et al.: Green synthesis of Silver nanoparticles using the plant extract of Shikakai and Reetha. Mater Today Proc. **5**, 2321–2329 (2018). https://doi.org/10.1016/j.matpr.2017.09.236
72. Javed, R., Zia, M., Naz, S., et al.: Role of capping agents in the application of nanoparticles in biomedicine and environmental remediation: recent trends and future prospects. J. Nanobiotechnology **18**(1), 172(2020)
73. Mathur, S., Singh, D., Ranjan, R.: Remediation of heavy metal (loid) contaminated soil through green nanotechnology. Front. Sustain. Food Syst. **6**, 932424 (2022). https://doi.org/10.3389/fsufs.2022.932424
74. Yogalakshmi, K.N., Das, A., Rani, G., et al.: Nano-bioremediation: a new age technology for the treatment of dyes in textile effluents. In: Bioremediation of Industrial Waste for Environmental Safety (2019)
75. Eras-Muñoz, E., Farré, A., Sánchez, A., et al.: Microbial biosurfactants: a review of recent environmental applications. Bioengineered **13**, 12365–12391 (2022). https://doi.org/10.1080/21655979.2022.2074621
76. Priyashantha, A.K.H., Mahendranathan, C.: Biosurfactants: an alternative approach to synthetic surfactants. Int. J. Res. Rev. (ijrrjournal.com) **8**, 550 (2021)
77. Kiran, G.S., Ninawe, A.S., Lipton, A.N., et al.: Rhamnolipid biosurfactants: evolutionary implications, applications and future prospects from untapped marine resource. Crit. Rev. Biotechnol. 1–17 (2015). https://doi.org/10.3109/07388551.2014.979758
78. Saeedi, L.H., Assadi, M.M., Heydarian, S.M., Jahangiri, M.: The Production and Evaluation of a Nano-biosurfactant. Pet. Sci. Technol. **32**, 125–132 (2014). https://doi.org/10.1080/10916466.2011.574176
79. Farias, C.B.B., Silva, A.F., Rufino, R.D., et al.: Synthesis of silver nanoparticles using a biosurfactant produced in low-cost medium as stabilizing agent. Electron. J. Biotechnol. **17**, 122–125 (2014). https://doi.org/10.1016/j.ejbt.2014.04.003
80. Araújo, H.W.C., Andrade, R.F.S., Montero-Rodríguez, D., et al.: Sustainable biosurfactant produced by Serratia marcescens UCP 1549 and its suitability for agricultural and marine bioremediation applications. Microb. Cell Fact. **18**, 2 (2019). https://doi.org/10.1186/s12934-018-1046-0
81. Durval, I.J.B., Meira, H.M., de Veras, B.O., et al.: Green synthesis of silver nanoparticles using a biosurfactant from Bacillus cereus UCP 1615 as stabilizing agent and its application as an antifungal agent. Fermentation **7**, 233 (2021). https://doi.org/10.3390/fermentation7040233
82. Liu, G., Zhong, H., Yang, X., et al.: Advances in applications of rhamnolipids biosurfactant in environmental remediation: a review. Biotechnol. Bioeng. **115**, 796–814 (2018). https://doi.org/10.1002/bit.26517
83. Malhotra, S.P.K., Alghuthaymi, M.A.: Biomolecule-assisted biogenic synthesis of metallic nanoparticles. In: Agri-Waste and Microbes for Production of Sustainable Nanomaterials, pp. 139–163. Elsevier (2022)
84. Priya, A.K., Muruganandam, M., Ali, S.S., Kornaros, M.: Clean-up of heavy metals from contaminated soil by phytoremediation: a multidisciplinary and eco-friendly approach. Toxics **11**, 422 (2023). https://doi.org/10.3390/toxics11050422
85. Kuppusamy, S., Maddela, N.R., Megharaj, M., Venkateswarlu, K.: Approaches for Remediation of Sites Contaminated with Total Petroleum Hydrocarbons (2019)
86. Wang, Y., O'Connor, D., Shen, Z., et al.: Green synthesis of nanoparticles for the remediation of contaminated waters and soils: constituents, synthesizing methods, and influencing factors. J. Clean. Prod. **226**, 540–549 (2019). https://doi.org/10.1016/j.jclepro.2019.04.128

Nanotechnology-Based Approaches for Degradation and Removal of Organoarsenicals: Review of Recent Advancements

Akanksha Pandey, Shruti Patel, and Hemen Dave

Abstract 'Arsenic' is an omnipresent environmental metalloid that emerged as a global threat to public health. Broadly, arsenic in the environment is found in inorganic and organic forms. The aliphatic organoarsenic compounds found in the environment are biogenic in origin, and aromatic, primarily anthropogenic in origin. Aromatic organoarsenic compounds (AOCs) are increasingly recognized as significant contributors to arsenic exposure due to their industrial usage and environmental mobility, in addition to geogenic sources of contamination. This article initially summarized the use of AOCs in industries such as swine and poultry and their environmental fate. Various treatment approaches have been explored for the degradation/removal of AOCs, such as microbial degradation, adsorption, and advanced oxidation. These conventional methods can be significantly enhanced through nanotechnology-based approaches. A nanotechnology-based approach for the degradation and removal of AOCs from water/wastewater and other environmental matrices is reviewed explicitly in this article. Various nanomaterials such as nanoparticles, nanocomposites, nanotubes, nanocubes, nanocrystals, nanofibrous membranes, nanosheets, nanoplates, and nanobiochar, etc. have been developed and utilized for the degradation and removal of AOCs. Nanotechnology can potentially be used for catalytic degradation of the AOCs and simultaneous adsorption of degradation products/by-products. Also, nanomaterials such as nanobiochar, nanofibrous membranes, and nanocrystals have been studied for the adsorptive removal of AOCs from water/wastewater. Comprehensively, this article provides treasured information about how nanotechnology-based approaches can be utilized unaided or in combination with other methods for the degradation/removal of AOCs.

Graphical Abstract

A. Pandey · S. Patel
School of Doctoral Studies and Research, National Forensic Sciences University, Gandhinagar, Gujarat, India

H. Dave (✉)
School of Pharmacy, National Forensic Sciences University, Gandhinagar, Gujarat, India
e-mail: hemen.dave@nfsu.ac.in

L. Ledwani and S. Srivastava (eds.), *Experimental and Computational Advances in Materials*, Engineering Materials, https://doi.org/10.1007/978-981-95-4180-5_5

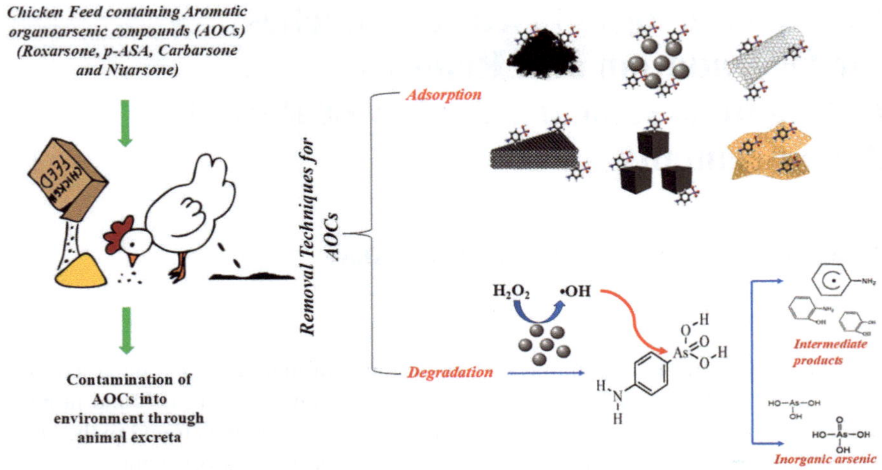

Keywords Aromatic organoarsenic compounds · AOC removal · Nanotechnology · Roxarsone · Nanomaterial

1 Introduction

Arsenic is a metalloid found in the Earth's crust at an average concentration of 1–1.8 mg/kg and in the soil at an average concentration of 2–20 mg/kg [1, 2]. It is ubiquitous in almost every environmental sphere, whether hydrosphere, pedosphere, or biosphere [3]. However, arsenic disrupts the ecosystem's integrity when it enters the environment at elevated concentrations through anthropogenic and natural sources. It is renowned as a poison because of its venomous effects on the health of humans, animals, and plants [4]. Natural sources of this metalloid include arsenic-containing minerals such as carbonates, sulfide, silicates, and oxide minerals, volcanoes, and some biogenic sources such as freshwater flora/fauna (green algae, diatoms/fishes, mussel, frogs) as well as marine flora/fauna (plankton, algae, seaweeds/fishes, bivalves, snails) [5]. Anthropogenic sources of arsenic are burning coal, agricultural use of arsenic compounds as pesticides and insecticides, arsenic compounds used as preservatives in wood industries, mining, and non-ferrous smelting, etc. [3]. The toxicity of arsenic varies depending on its elemental speciation; generally, inorganic arsenic is more toxic than the organic form. Arsenite (As(III)) and arsenate (As(V)), the two inorganic forms of arsenic, are primarily found in reducing and oxygenated environments, respectively [6]. Arsenic in these forms contaminates the drinking water, depletes soil fertility, and affects crop production when arsenic-contaminated water is used for irrigation [7]. As per the WHO drinking water standards, the permissible limit for total inorganic arsenic (As (III) + As (V)) in drinking water is 10 μg L^{-1}. However, many parts of the world (almost 108 countries), specifically Southeast

Asian countries (32 countries) such as Bangladesh and India, have arsenic in drinking water beyond this limit due to geogenic contamination in groundwater [4, 8].

Arsenic in drinking water is considered a class-1 carcinogen according to the International Agency for Research on Cancer (IARC) [9] and systemic toxicant, which affects almost all the organs of the body and has the potential to cause cancers of the skin, liver, kidney, lung, urinary bladder, gastrointestinal, and brain [10]. Like organochlorine pesticides and methyl mercury, arsenic also biotransforms and bioaccumulates; its concentrations increase with each successive trophic level [8, 11]. Exposure to inorganic arsenic due to the use of arsenic-contaminated groundwater (in 90% of cases) is a well-recognized threat [12]; it is reported that globally, 230 million people, of which 180 million people are from Asia, are at risk [8].

Due to the extreme toxicity of inorganic arsenic [13–16], most research has focused on monitoring [17, 18] and reduce the inorganic arsenic exposure from the environment [19, 20] and removal of arsenic from water [21–29].

At the same time, less research is focused on the organic form of arsenic [30]. However, environmental exposure to organoarsenic compounds contributes as much to human health threats as inorganic arsenic pollution [31–33]. Organoarsenic compounds are naturally and artificially found on Earth [30]. Organoarsenic compounds found in the environment can be broadly classified as aliphatic (mainly of biogenic origin) and aromatic (primarily derived from anthropogenic activities) [30, 32]. In the aquatic environment, organoarsenic compounds such as arsenosugar, arsenobetaine, arsenocholine, and arsenolipid with incorporated inorganic arsenic species (As(III) or As(V)) are synthesized naturally by the microorganisms [30]. These organoarsenic compounds are metabolized or mineralized by bacteria into inorganic or methyl arsenicals [34]. Aquatic organisms accumulate or transform the organoarsenic compounds inside the body and become a significant source of arsenic exposure for human beings [35, 36].

Furthermore, synthetic arsenic compounds such as inorganic arsenic, methyl arsenicals, and AOCs have been used as pesticides, herbicides, and feed additives in animal husbandry [37]. Inorganic arsenic compounds such as calcium arsenite and sodium arsenite have been used as pesticides, while dimethyl arsenicals have been applied as herbicides to control weeds in cotton cultivation. Apart from this, mono- and disodium methyl arsenate and dimethylarsinic acid were used as defoliants before cotton was harvested. The occurrence of these compounds was examined in the soil, surface water, and groundwater in nearby cotton production areas. According to the preliminary findings, surface water samples from cotton-producing regions contain higher amounts of methyl arsenic species (>10 μg of As/L) than background regions (<1 μg of As/L) [38].

AOCs such as Roxarsone (4-hydroxy-3-nitrophenylarsonic acid), p-arsanilic acid (4-aminophenylarsonic acid), carbarsone ([4-(carbamoylamino) phenyl] arsonic acid) and nitarsone ((4-nitrophenyl) arsenic acid) have been extensively used in poultry and swine industries as a feed additive to control infectious diseases in chicken and swine, to promote growth rate in animals, improve meat pigmentation, and increase feed efficiency [39–41]. AOCs do not metabolize in the animal's body, and $\geq$ 90% of these compounds are excreted without any chemical modification through urine and faeces, while the rest of these compounds get deposited in the animal's body parts [39, 42]. Previous studies on animals that consumed organoarsenic reported elevated arsenic concentrations in feathers, tissues, eggs, and liver [43].

Globally, thousands of metric tons of poultry litter/waste are generated annually and dumped into the soil as an effective fertilizer to increase soil fertility or for disposal. The poultry litter contaminated with AOCs causes the AOCs to be introduced into the environment, elevating the soil arsenic concentration [40, 44, 45]. As a result, there is a rise in the amount of arsenic in the soil, sediment, groundwater, and surface water [46]. Even crops grown in that soil have been found to contain trace amounts of arsenic [40]. According to a previous study, rice plants accumulate arsenic in their various parts with the increasing concentration of AOCs added to the soil [47]. Rice shoots, pepper, and bush beans quickly accumulate the organic form of arsenic compared to inorganic arsenic species [47].

Apart from this, recent literature has also shown that AOCs get converted or transformed into more toxic species such as As(III), dimethylarsinic acid (DMA), As(V), and monomethyl arsenic acid (MMA) [48]. These are more mobile than organoarsenic, causing damage to groundwater, surface water, the quality of soil, and food security by damaging the crops grown in that soil (J. Chen et al., 2020). The AOCs have emerged as a significant source of arsenic in soil and water apart from the geogenic sources [49]. Therefore, researchers have a substantial interest in AOCs' existence and behavior in the environment and their removal measures [40, 48].

Scientists have tried several approaches such as adsorption [50], photocatalytic degradation [51], hydrothermal treatment [52], microbial degradation [53], and chemical oxidation [54], including homogeneous/heterogeneous Fenton [55], electrofenton [56], and heterogeneous catalytic peroxymonosulfate (PMS) processes [57] for degradation/removal of AOCs [40]. Still, a synergistic combination of degradation-adsorption approach is needed to remove AOCs completely (Hlongwane et al., 2019).

Nanotechnology is a potential approach that has emerged as an efficient technology for removing AOCs. It offers excellent potential for expanding next-generation water and wastewater treatment and could overcome the limitations of conventional technologies [58]. Nanotechnology includes using various nanomaterials such as nanoparticles, nanolayers, nanotubes, nanocomposites, etc. [59]. The main differences among these nanostructures involve the method of their fabrication and the arrangement of their atoms and molecules [60].

2 Bibliometric Analysis

2.1 *Growing Interest in Aromatic Organoarsenic Compounds: Annual Patterns and Worldwide Distribution of Research Publications*

The data used in this study were obtained from the SCOPUS online database. The search was done in (ARTICLE TITLE, ABSTRACT, KEYWORDS) by searching ((Aromatic AND organoarsenic AND compound) OR (roxarsone) OR (p-arsenic AND acid) AND (degradation) OR (adsorption) OR (nanomaterials)). The data extracted from SCOPUS was screened according to year and top countries. A total of 153 articles were extracted, and only documents limited to the period from 2010 to 2024 in the subject area of environmental science encompassing research on the AOCs, their detection, toxicological effects, and degradation or removal of AOCs were included. Year-wise, the number of documents is given in Fig. 1, and 2022 was the year with the highest number of papers (22 papers). Concerning the countries. AOCs articles, AOCs research publishing country, China has the highest number of published articles (123 out of 153), followed by the United States (24). Country-wise, the number of publications is given in Fig. 2.

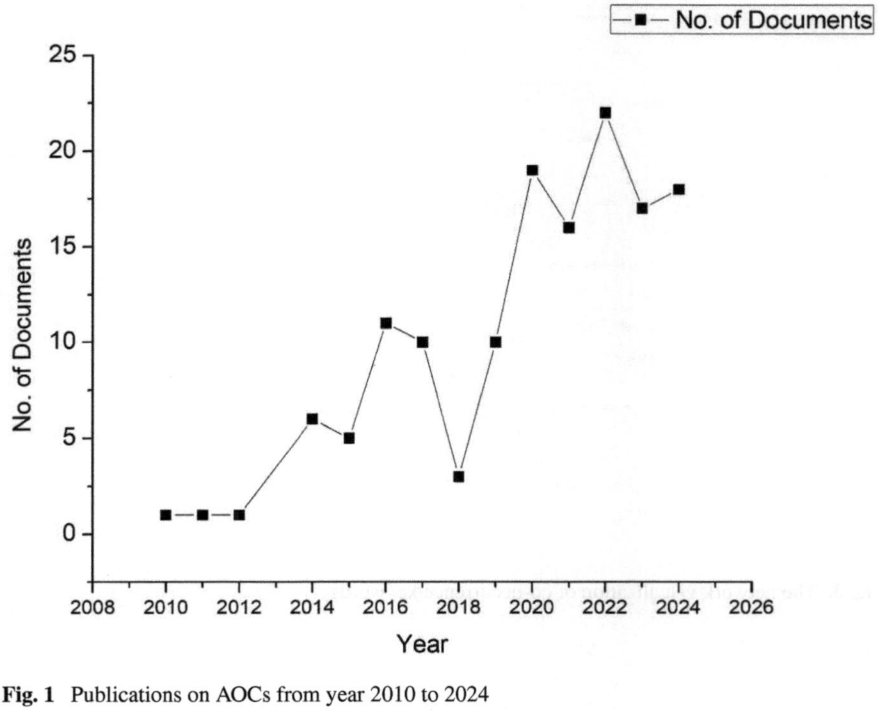

Fig. 1 Publications on AOCs from year 2010 to 2024

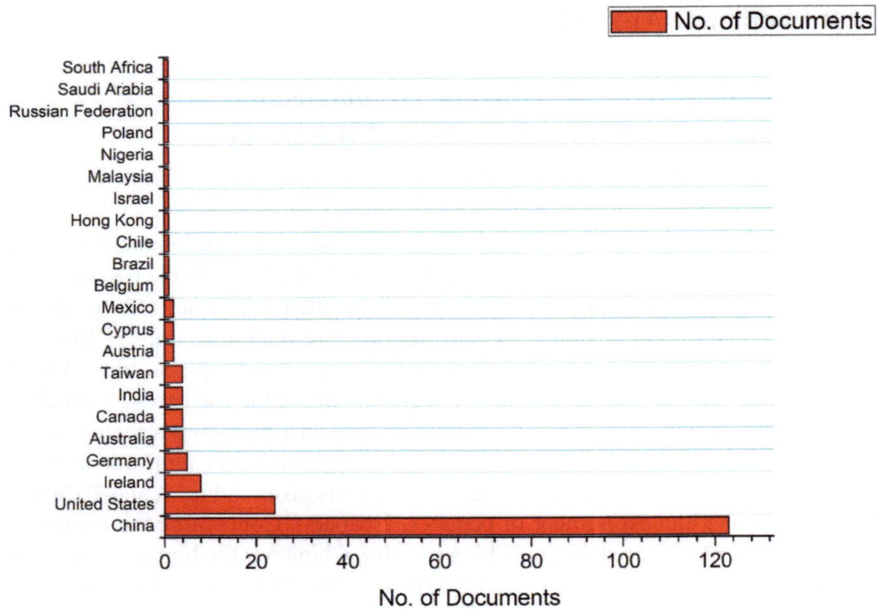

Fig. 2 Publications on AOCs from various countries (2010 to 2024)

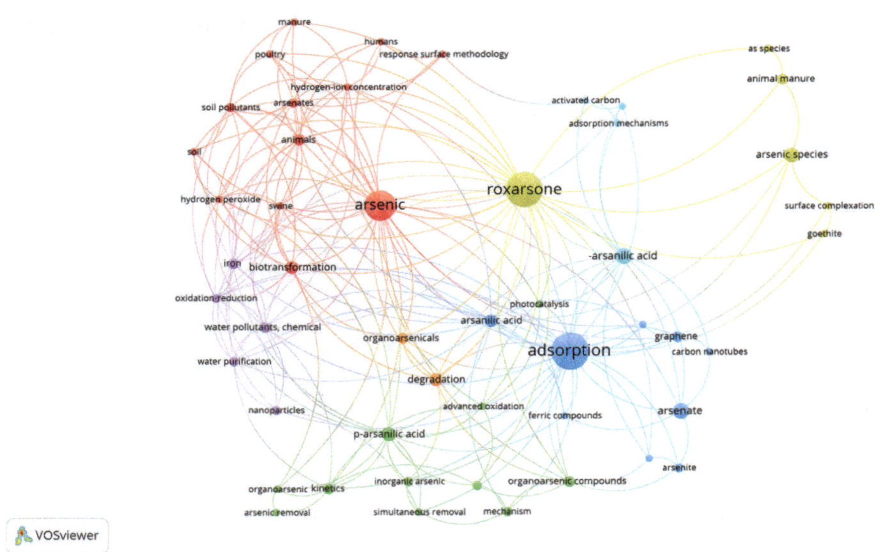

Fig. 3 The network visualization of co-occurrence keywords

Therefore, this analysis highlights the growing interest in AOCs, their detection, and their degradation methods, especially through advanced nanotechnology approaches. The high dominance of China reflects significant investment and progress in environmental science and nanomaterials. The study offers a fascinating look at research trends, geographic contributions, and publication patterns informing this crucial area and will shape its future research and collaboration.

2.2 VOS Viewer Analysis

The network was generated in VOSviewer, version 1.6.19, and visualizes the co-occurrence of keywords by title and abstracts containing the topic of interest in aromatic organoarsenic compounds. The major clusters are the most highly related terms that repeatedly occur in the same publication; these clusters provide information about research trends and interrelationships among the topics. The network's major nodes—arsenic, roxarsone, and adsorption—highlight their importance in studies about aromatic organoarsenic compounds as a source of arsenic exposure and research work related to the removal of AOCs. These keywords are linked to most of the terms studied from one end to another to show the interdisciplinary nature of the topic under discussion. Cluster Red (Arsenic) outlines contaminations in the soil by this kind of arsenic, accumulating in animals and humans alike.

Yellow Cluster (Roxarsone and Arsenic Species) highlights the role of roxarsone in generating arsenic species. Links to terms from this cluster include animal manure and surface complexation, and point to research directed to agricultural runoff and resultant arsenic mobilization. Blue Cluster (Adsorption Mechanisms) emphasizes adsorption as one of the chief methods of removal of such contaminants. Keywords such as graphene, carbon nanotubes, and photocatalysis indicate that advanced nanomaterials effectively remove AOCs/arsenic contamination. The transformation and elimination of organoarsenic compounds are the main topics of the Green Cluster (Organoarsenic Kinetics and Removal). Terms like p-arsanilic acid, simultaneous removal, and arsenite removal are examples of this cluster, where removal methods have been the main emphasis. The keywords defining the Purple Cluster (Chemical Processes) include oxidation–reduction, iron, and hydrogen peroxide; hence, this cluster also represents interest in redox reactions and chemical treatments of arsenic and detoxification. The close associations between arsenic and contaminants found in soil, poultry, and water highlight how arsenic affects ecosystems. Associations between roxarsone and arsenic species imply the importance of speciation in arsenic and the probable risks associated with using the compound. Adsorption is related to some advanced nanomaterials, such as graphene and carbon nanotubes, showing a rise of interest in the usage of nanotechnology in removal processes for arsenic (Fig. 3).

3 Aromatic Organoarsenic Compounds: Usage and Environmental Fate

3.1 Industrial Uses of AOCs

In the 1920s, modern farming practices replaced traditional farming, i.e., concentrated animal feeding operations (CAFOs), to maximize food production and profit [61]. It is an industrial, agricultural facility that proficiently raises a significant number of animals, around 200,000 birds, by a small team of farmers in a confined area to consume poultry products such as eggs, milk, or meat. This modern farming practice is economical because it raises animals in close vicinity [62]. However, because of proximity, there is a great threat of various parasitic infections and diseases in animals. Therefore, to control such infections, AOCs, Roxarsone, carbarsone, nitarsone, and *p*-arsanilic acid-containing functional groups (hydroxy, carbamoylamino, nitro, amino) are used as a feed supplement in the swine and poultry sectors. The functional groups of AOCs interact with thiol-containing enzymes in microorganisms, and the arsenic moiety interferes with vital metabolic processes. By increasing membrane permeability, the aromatic ring makes it easier for the molecule to penetrate microbial cells. Furthermore, functional groups like –OH, $-NO_2$, $-NH_2$, and $-CH_3$ impact the compound's toxicity, selectivity, and overall antimicrobial activity by modifying its lipophilicity, electron distribution, and binding affinity [32].

These compounds also accelerate growth and increase feed utilization efficiency [63]. In the United States, these arsenical compounds were permitted as animal feed additives in 1958; in China, they were approved in 1996 [64]. Roxarsone and p-arsenic acid are the most commonly used compounds among these four AOCs.

Roxarsone, 3-NHPAA, 3-Nitro-4-hydroxyphenyl arsonic acid, having 28.5% arsenic, is the most extensively used arsenical drug approved by the FDA since 1944, in combination with antibiotics to treat coccidiosis, an intestinal tract parasitic infection caused by coccidian protozoa in chickens [39]. It also improves feed efficiency and pigmentation and stimulates speedy growth. The recommended dosage of Roxarsone in poultry feeds is 25–50 ppm. In the turkey and swine industry, Roxarsone improves the productivity and prevention of enteric diseases at 25–37.5 ppm [65].

Arsanilic acid, p-ASA, and 4-Aminophenyl arsonic acid, having 34.6% arsenic, are the second most commonly used arsenic drugs in livestock industries. It is also used to enhance weight gain efficiency and prevent animal sickness. The recommended dosage of p-arsanilic acid in poultry feed is around 100 mg/kg [39].

Carbarsone, p-UPAA, 4-Carbamoyl aminophenylarsonic acid, and Nitarsone or Histostat-50, p-NPAA, 4-Nitrophenyl arsonic acid have 28.8 and 30.4% arsenic, respectively. Both drugs are antiprotozoal and regulate histomoniasis, blackhead disease, and other diseases, such as amebiasis and nematode infections in chickens and turkeys [66].

3.2 Fate of AOCs in the Environment

AOCs are excreted by more than 90% of the initial feed by poultry animals via urine and faeces without any chemical alteration in their structure. Through the usage/ disposal of poultry litter, it enters the environment and acts as a significant source of arsenic exposure apart from the geogenic one. The rest of the compound amount gets deposited in the bodies of animals [40]. A recent study reported that AOCs accumulate in the chicken body. There was a significant build-up of arsenic in the stomach and chest flesh, but not as much in the heart or legs. The research team also calculated the total amount of arsenic ingested through the intake of chicken. A person may consume 0.186–0.372 μg of arsenic daily if one consumes 60.0 g of chicken flesh (leg, breast, muscles, and stomach) daily [67]. This study unequivocally demonstrates that humans' exposure to arsenic comes from poultry animals, and the waste generated from animal husbandry industries also becomes a significant source of environmental contamination with arsenic. Figure 4 illustrates the fate, transport, and impact of AOCs on the environment and humans.

It has also been observed that the poultry industry's feathers are converted into food products used as fodder for other animals. According to some studies, Roxarsone, utilized as a feed supplement in the poultry industry, accumulates in the keratinous material of poultry feathers. Most accumulated arsenic is inorganic [68].

In the US, around 8.6 billion chickens are fed with AOCs, especially Roxarsone, and it is estimated that 900 metric tons of Roxarsone are released into the environment yearly [44]. The concentration of Roxarsone in waste generated through poultry industries has been reported to be 14–76 mg kg^{-1} [69]. In China, 21.6 mg/kg of arsenic was reported in chicken manure because of roxarsone consumption [70].

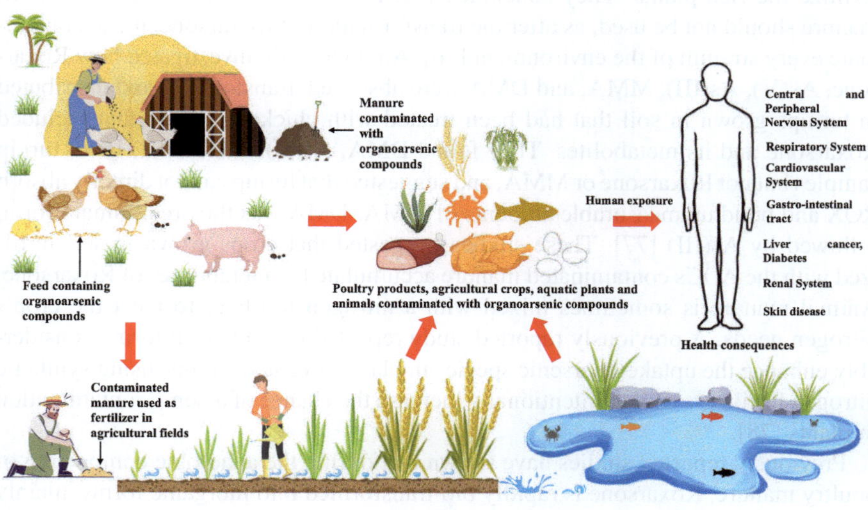

Fig. 4 Fate and transport of aromatic organoarsenic compounds

Because of potential dangers to humans and the environment, Roxarsone was banned as a feed additive in Europe, Canada, and the United States in 1998, 2011, and 2013, respectively [71]. Recently, China officially banned AOCs in 2019 and joined the worldwide phase-out initiative. However, numerous nations such as India, Brazil, Mexico, Argentina, Chile, and Vietnam still permit their use [72].

India is one of the world's largest countries, producing 74 billion eggs and 3.8 million tonnes of broiler meat annually, with total feed production around 22 million tons [73]. Only a few studies have been done on organoarsenic compounds in India, one of which was conducted by the West Bengal University of Animal and Fishery Sciences (WBUFAS), Kolkata. The study revealed that samples of chicken, mutton, and eggs contained levels of arsenic that were five times higher than permissible. The senior researcher of WBUFAS said they found 0.77 and 0.38 ppm of arsenic in broiler chickens and eggs, which are significantly higher than the permissible levels. It was reported that Roxarsone severely affects those who consume chicken meat and indirectly affects those who do not [74].

The contamination of poultry litter with AOCs is a serious concern. The disposal of excreted waste or poultry litter and its transportation to other locations is pricey, so farmers deposit the waste or poultry litter in rows known as windrows and use them as fertilizer. As reported earlier, using manure/fertilizers obtained from poultry litter/waste leads to an elevated concentration of arsenic in soil and crops grown [75]. A study showed that the surface soil of concentrated animal feeding operations (CAFOs) has higher arsenic levels (16.0 mg/kg) due to the extensive use of AOCs compared to nearby agricultural fields (12.2 mg/kg) of Pearl River Delta, China [62]. Also, as reported in a previous study, rice is more likely to accumulate arsenic than other. Researchers found the DMA and arsenite in rice grains, DMA, MMA, As(V), and As(III) in rice hull when roxarsone-contaminated chicken manure was used to fertilize the rice plants. They concluded that cultivating rice AOCs contaminated manure should not be used, as after the transformation of Roxarsone, it can contaminate every stratum of the environment [76]. Another study investigated how Roxarsone, As(V), As(III), MMA, and DMA were absorbed, translocated, and distributed in turnips grown in soil that had been treated with chicken manure that included Roxarsone and its metabolites. They found DMA, As(V), and As(III) in all turnip samples but not Roxarsone or MMA, and suggested that turnip cannot directly absorb ROX and build up measurable amounts of MMA. DMA was the predominant form, followed by As(III) [77]. These studies suggested that crops grown in soil fertilized with the AOCs contaminated manure accumulate the metabolites of Roxarsone. Animal manure is sometimes mixed with a nitrogen fertilizer to meet the crop's nitrogen needs. A previously reported study reported that nitrogen forms considerably enhance the uptake of arsenic species in plants. Because of this, using synthetic nitrogen fertilizers may unintentionally increase the chance of arsenic contamination in plants [78].

Previously reported studies have shown that during the anaerobic composting of poultry manure, Roxarsone is rapidly bio-transformed into inorganic forms, mainly arsenate, after a month, capable of horizontal transport [44]. *Clostridium* species were reported to rapidly transform Roxarsone in a chicken litter into inorganic species and 3-amino-4-hydroxyphenylarsonic acid [79]. It is evident that when roxarsone-containing manure is used in fields, it is transported to the soil and crops, contaminating the surface and groundwater [63]. Unlike anaerobic degradation, the abiotic transformation of AOCs in the environment results in partial or total breaking of the C-As bond and concomitant release of inorganic arsenic. In water, degradation of Roxarsone and *p*-arsanilic acid induced by UV radiation resulted in the formation of inorganic arsenic and thus elevated exposure to arsenic. *p*-arsanilic acid is also quickly photodegraded when exposed to sunlight, yielding inorganic arsenic [80, 81]. Plenteous research on the presence, fate, and toxicity of inorganic arsenic and AOCs in groundwater, surface water, soil, sediments, crops, and humans has been published; nonetheless, investigations on organoarsenic compounds are still in their infancy. Even though four types of AOCs exist, namely Roxarsone, p-arsanilic acid, carbarsone, and nitarsone, which are widely used in poultry and swine industries, the major research focus is on Roxarsone and p-arsanilic acid. The research gap still exists for the remaining AOCs, which need to be filled with newer scientific contributions.

4 Degradation and Removal of AOCs Through Various Nanomaterials

Since its discovery, interest in nanomaterials has grown among researchers. Potential applications for nanomaterials have been investigated in almost every field of study, including electronics, medicine, energy, food, transportation, the environment, etc. [58]. Treatment of water and wastewater using various nanomaterials such as nanoparticles, for instance, metal (Iron, zinc, silver) or metal oxide nanoparticles (zinc oxide, iron oxide, titanium oxide), nanocomposites, nanotubes (carbon nanotubes), nanocubes, nanocrystals, nanoplates, nanosheets, nano-biochar, and nanofibrous membrane have been successfully reported [82]. The removal/treatment of AOCs by adsorption or advanced oxidation using various nanomaterials has gained much attention in the last decade. Various nanomaterials explored for the degradation and removal of AOCs are illustrated in Fig. 5. The synthesis methods of various nanomaterials and their application in the degradation and removal of AOCs are presented in Table 1.

4.1 Nanoparticles

The most fundamental nanotechnology unit is the nanoparticle, a cluster of tens to thousands of atoms with an aggregate diameter between 1 and 100 nm. Nanoparticles are manufactured via two approaches: the bottom-up and top-down approaches. The bottom-up approach is the most productive and allowable approach for the fabrication

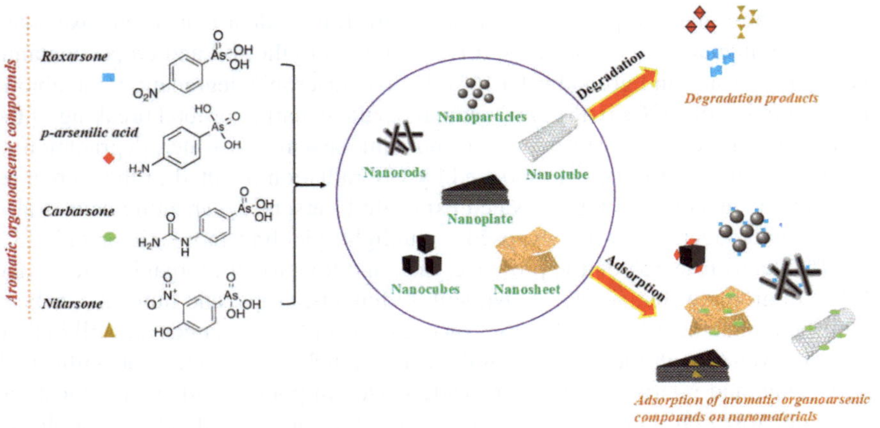

Fig. 5 Degradation and removal of aromatic organoarsenic compounds through various nanomaterials

of nanoparticles. In the bottom-up approach, nanoparticles are created atom by atom, so a particle's size, and often its shape, is controlled by experimental conditions [111]. In the top-down approach, nanoparticles are synthesized by breaking up bulk material in the 1–100 nm nano range. Because of the unique characteristics of nanoparticles, such as the vast surface area to volume ratio, size-dependent features, biocompatibility, high reactivity, and degree of functionalization, they are increasingly used in the purification of wastewater and its treatment [112]. Several research and review articles are available on removing inorganic arsenic using nanoparticles [28, 113, 114]. However, less research on AOC degradation/ removal using nanoparticles has been reported than on inorganic arsenic.

Nanoparticles, such as bimetallic oxide and zero-valent, have been reported for the adsorption or degradation of AOCs. In a study, manganese ferrite bimetal oxide nanoparticles were synthesized via the co-precipitation method for the adsorption of DMA, As(V), Roxarsone, and p-arsanilic acid. They found that the synthesized adsorbent showed maximum adsorption towards As(V) (68.25 mg g^{-1}), followed by *p*-arsanilic acid (59.45 mg g^{-1}), Roxarsone (51.49 mg g^{-1}), and DMA (35.77 mg g^{-1}). The adsorption of AOCs and arsenic on the nanoparticles is due to the linkages of As-O-Fe or As-O-Mn [37]. Likewise, bifunctional magnetite CeO_2–Fe_3O_4 & CuO-Fe_3O_4 nanoparticles were fabricated in other studies. The nanoparticles demonstrate photocatalytic degradation of *p*-arsanilic acid under UV and visible light irradiation. During the degradation process, hydroxyl radicals were formed, which cleaved the As-C bond, and *p*-arsanilic acid was converted into As(V), and for complete conversion, it took 28 min in UV and 36 min in visible light irradiation. At the same time, around 98% and 95% of As (V) were adsorbed onto the surface of CeO_2-Fe_3O_4 and CuO-Fe_3O_4, respectively [84, 85]. In another study, lignin-based magnetic-activated carbon nanoparticles removed 96% of *p*-arsanilic acid at pH 4 [86].

Table 1 Summary of previous studies on degradation and removal of aromatic organoarsenic compounds through various nanomaterials

S. No.	Nanomaterial	Synthesis method of nanomaterials synthesis	Detection and characterization techniques	Size of nanomaterial	Degraded arsenic compound
1.	*Nanoparticles*				
	Mesoporous bimetal oxide magnetic manganese ferrite nanoparticles	Co-precipitation	XRD, TEM, ATR-FTIR, XPS, ZETASIZER, UV–Vis, ICP-AES	30 nm	As(V)
					p-ASA
					ROX
					DMA
	Silicon dioxide-coated nano zero-valent iron		UPLC, GC–MS, ICP-MS, HPLC, XRD, XPS		p-ASA
	Bi-functional magnetic CeO_2-Fe_3O_4 nanoparticles		ICP-AES, HPLC, TOC-VCPH, XRD, EDS, HRTEM, UV–Vis	10–20 nm	p-ASA
	Bi-functional magnetic CuO-Fe_3O_4 nanoparticles	Co-precipitation	XRD, HR-TEM, BET, VSM, UV–Vis, FTIR, XPS, ICP-OES, HPLC, TOC-VCPH	9.47 nm,11.72 nm	p-ASA
	Lignin-based magnetic activated carbon nanoparticles		SEM, EDS, XRD, FTIR, VSM, XPS, UV–Vis	30–50 nm	p-ASA
	Bifunctional TiO_2/ ferrihydrite hybrid	Hydrothermal	FE-SEM, TEM, XRD, DRS, XPS, HPLC, AFS	20 nm	ROX
	Magnetic copper ferrite with peroxymonosulfate		XRD, TEM, FTIR, XPS, ZETASIZER, HPLC, UV–Vis, TOC, ICP-MS, HG-AFS		p-ASA
2.	*Nanocomposite*				
	Fe_3O_4@ graphene	Hydrothermal method	FTIR, RAMAN, XRD, XPS, VSM, HPLC, TEM	10 nm	Arsanilic acid

(continued)

Table 1 (continued)

S. No.	Nanomaterial	Synthesis method of nanomaterials synthesis	Detection and characterization techniques	Size of nanomaterial	Degraded arsenic compound
	Fe-Ti-Mn composite oxide	Oxidation and co-precipitation	SEM, HETEM-ICP-MS, FTIR, XPS, ZETASIZER	10–30 nm	p-ASA
	Nano-hydrated zirconium oxide encapsulated inside a hyper-cross-linked anion exchanger		AFS, ICP-OES, XRD, XPS, SEM, EDS	10–25 nm	p-ASA
	Biochar-supported $MnFe_2O_4$ magnetic nanocomposite	Co-precipitation	HPLC-ICP-MS, TOC, XRD, VSM, FESEM, HETEM-ICP-MS, FTIR, ZETASIZER, XPS	5–20 nm	p-ASA As(V)
	Fe_3O_4@3D graphene nanocomposite	In-situ compositing method	XRD, BET, HRTEM, SEM, XPS, UV–Vis, HPLC-ICP-MS, FTIR	20–50 nm	ROX
	Composite of reduced graphene oxide and carbon nanotubes impregnated with goethite		FE-SEM, TEM, XRD, FTIR, XPS, LC, TOF–SIMS, AFS		ROX
	Magnetic material-based reduced graphene oxide with persulfate	Co-precipitation	TEM, SEM, XRD, XPS, FTIR, UV–Vis, ZETASIZER, BET, BJH, LC/Q-TOF/MS, ICP-OES		*p*-ASA
	Cellulose-goethite composites		TGA, FTIR, RAMAN, BET, XRD, TEM		ROX

(continued)

Table 1 (continued)

S. No.	Nanomaterial	Synthesis method of nanomaterials synthesis	Detection and characterization techniques	Size of nanomaterial	Degraded arsenic compound
	Monoclinic zirconium dioxide anchored on reduced graphene oxide	Hydrothermal	HPLC, XRD, SEM, TEM, XPS		3-NHPAA
	Mix phases of zirconium dioxide anchored on reduced graphene oxide				
	Tetragonal zirconium dioxide anchored on reduced graphene oxide				
	Sludge-based biochar supported zerovalent iron nanocomposite Resin-based hydrated iron oxide		HPLC, AFS, UV–Vis, SEM, FTIR, XPS, EDS, UPLC-QTOF-MS, XRD, BET, TEM XPS, BET, XRD, ZETASIZER, SEM- EDS, FTIR, HPLC	50 nm	ROX p-ASA As(V)
	$MnFe_2O_4$/graphene	Solvothermal	XPS, SEM, TEM, FTIR, XRD, BET,	200–400 nm	p-ASA ROX
3.	*Nanotubes*				
	Multi-walled carbon nanotubes with electrolytic manganese residue	Ball milling	XRD, XPS, SEM, TEM	2 μm–200 nm	ROX
	Iron (hydr)oxide-modified multiwalled carbon nanotubes		HPLC, FTIR, XPS	10–20 nm	ROX

(continued)

Table 1 (continued)

S. No.	Nanomaterial	Synthesis method of nanomaterials synthesis	Detection and characterization techniques	Size of nanomaterial	Degraded arsenic compound
	Iron encapsulated B/N-doped carbon nanotubes		SEM, HETEM, XRD, TGA, AFS, TOC, LC–MS/MS	10–100 nm	p-ASA
	Goethite impregnated graphene oxide carbon nanotubes aerogel	Self-assembly method	XRD, FESEM, TEM, BET, BJH, FTIR, XPS, ICP-AES	2 μm–500 nm	As(V)
					DMA
					p-ASA
4.	*Nanocubes*				
	Mn – Fe mixed oxide with peroxymonosulfate		XRD, SEM, TEM, ICP-AES, SEM–EDX, ZETASIZER, XPS, FTIR, UPLC, AFS, TOC	~300 nm	p-ASA
5.	*Nanocrystals*				
	Hematite		XRD, RAMAN, SEM, TEM, XPS, HPLC, ATR-FTIR	30.3, 79.3 nm	PhAs
6.	*Nanofibrous membrane*				
	Amino-modified electrospun nanofibrous cellulose membrane		HPLC, FESEM, FTIR, XRD, BET, XPS	462 nm	ROX
					p-ASA
					PAA
7.	*Nanoplate*				
	Zinc oxide	Wet chemical method	XRD, FTIR, UV–Vis., BET, SEM, TEM, HPLC-GC–MS, EPR	1 μm	ROX

(continued)

Table 1 (continued)

S. No.	Nanomaterial	Synthesis method of nanomaterials synthesis	Detection and characterization techniques	Size of nanomaterial	Degraded arsenic compound
8.	*Nanosheet*				
	Magnetic greigite	Solvothermal	HPLC, XRD, SEM, VSM	25 nm	ROX
9.	*Nano-biochar*				
	Nano-sized goethite modified biochar	Coprecipitation	SEM–EDS, BET, FTIR, XRD, VSM, XPS	2–400 nm	Cd(II)
					As(III)
					ROX
					Phosphate

Degradation and removal approach	Best fitted model and order of reaction	Optimum pH	Degradation/removal capacity	References
Nanoparticles				
Adsorption	Langmuir and Pseudo-second-order	2	68.25 mg g^{-1}	[37]
			59.45 mg g^{-1}	
			51.49 mg g^{-1}	
			35.77 mg g^{-1}	
Oxidation and Adsorption	NA	3	99.60%	[83]
Photocatalysis and Adsorption	Freundlich and First-order	4–7	98%	[84]
Photocatalysis and Adsorption	First-order	4–7	95%	[85]
Adsorption	Langmuir and Pseudo-first-order	4	96%	[86]
Photocatalysis (254 nm) and adsorption	Pseudo-first order	7	96%	[87]
Oxidation and Adsorption	Pseudo-first-order	3–11	85%	[88]
Nanocomposite				
Adsorption	Langmuir and Pseudo-first-order	3–6	92.3 and 93.5%	[89]

(continued)

Table 1 (continued)

Degradation and removal approach	Best fitted model and order of reaction	Optimum pH	Degradation/removal capacity	References
Oxidation and Adsorption	Freundlich and Pseudo-second order	3–10	45.6 mg/g	[90]
Adsorption	Langmuir	3	> 23.00 $mg \cdot g^{-1}$	[91]
Adsorption	Langmuir and Pseudo-first and second order	3–7	105 mg/g	[92]
			90 mg/g	
Adsorption	Dubinin-Ashtakhov, Langmuir, and Pseudo-second-order non-linear	5	98.60%	[93]
Catalytic oxidation and adsorption	Langmuir and First-order	2.98	99.39%	[94]
Decomposition and adsorption		7	98.80%	[95]
Adsorption	Langmuir, Freundlich, and Pseudo-second–order	4–8	~97%	[96]
Adsorption	Langmuir and Pseudo-second order	3.8–8.8	207.2 mg g^{-1}	[97]
		3.8–7.1	164.6 mg g^{-1}	
		3.8–5.9	92.5 mg g^{-1}	
Adsorption	Langmuir and pseudo-second-Order Langmuir, Freundlich, Sips, and pseudo-second-order	3 5–9	975 mg g^{-1} 18 mg g^{-1} 55 mg As g^{-1}	[98] [99]
Adsorption	Langmuir	3–5	22.75 mg g^{-1} 30.59 mg g^{-1}	[100]
Nanotubes				
Oxidation and adsorption	Pseudo-first-order	3.04	90.96%	[101]
Adsorption	Freundlich, Polanyi–Manes and Pseudo-second-order	2	4.48 mg g^{-1}	[102]
Oxidation and adsorption		3	95 and 86%	[103]

(continued)

Table 1 (continued)

Degradation and removal approach	Best fitted model and order of reaction	Optimum pH	Degradation/removal capacity	References
Adsorption	Langmuir, Freundlich, and pseudo-second-order	3–9	56.43 mg g^{-1}	[104]
		3–8	24.43 mg g^{-1}	
		5–9	102.11 mg g^{-1}	
Nanocubes				
Oxidation and adsorption		4	90%	[105]
Nanocrystals				
Adsorption	Langmuir and Pseudo-second-order	3	4.44 As/nm^2	[106]
Nanofibrous membrane				
Adsorption	Langmuir and pseudo-second-order model with intraparticle diffusion	8	186.22 mg g^{-1}	[107]
		7	69.15 mg g^{-1}	
		9	62.77 mg g^{-1}	
Nanoplate				
Photodegradation with UV-A light (365 nm)		7	70%	[108]
Nanosheet				
Reduction and Adsorption	Pseudo-second-order	3.6–8.6	> 90%	[109]
Nano-biochar				
Adsorption	Langmuir and Pseudo-first and second order	2–9	92%	[110]
			98%	
			98%	
			95%	

In a recent study, another bifunctional nanomaterial, TiO_2/ferrihydrite hybrid, was prepared hydrothermally to remove Roxarsone altogether. For the simultaneous photocatalytic degradation of Roxarsone and adsorption of As(V), they used a 20W UV low-pressure lamp. The wavelength of the emitted monochromatic light was 254 nm. TiO_2/ ferrihydrite hybrid was irradiated by UV light, and •OH radicals were generated, which were responsible for the arsenate release after the attack on the As-C bond of Roxarsone. They observed that 96% of Roxarsone was degraded after 12 h of irradiation by a 1:1 TiO_2/ferrihydrite hybrid with the simultaneous removal of As(V) [115].

Apart from this, zero-valent nanoparticles have gained much attention in treating AOCs. A recent study used the Fenton process catalyzed by synthesized zero-valent iron nanoparticles coated with silica dioxide to treat *p*-arsanilic acid from natural water. Around 99.6% *p*-arsanilic acid was completely mineralized within 60 min into arsenate, ammonium ion, and *p*-hydroquinone. The maximum degradation rate was observed at pH 3 [83].

4.2 Nanocomposites

Composites are solid materials manufactured when two or more fundamental materials, each with distinct attributes, are linked to create an innovative compound with superior qualities to the original components [116]. The primary purpose of designing these composites is to provide strength and stiffness, resistance to deterioration, simple repair of damaged structures, ease of manufacturing, and a low coefficient of expansion to the material [117]. Nanocomposites are composites in which at least one phase (such as nanoparticles, nanotubes, lamellar nanostructures, etc.) possesses nanoscale morphology. The nanocomposite uses building blocks with dimensions in the nanometer range to design and create new materials with extraordinary flexibility and improvement in their numerous properties, for instance, mechanical, thermal, electrical conductivity, and surface appearance [117]. Nanocomposites generally are prepared using two types of materials: one is polymer-based (polymer/ceramic, inorganic/organic polymer, inorganic/organic hybrid, polymer/layered silicate), while the other is non-polymer-based or inorganic (ceramic, ceramic-ceramic, metal, metal–metal) [116].

Several studies are available on removing inorganic arsenic using nanocomposites [118, 119], and a few studies have investigated the removal of AOCs. In a study, magnetite@graphene nanocomposite was prepared via a hydrothermal method to synergistically eliminate arsanilic acid from pig urine and water collected near the pig farm. According to this research, the anionic nature of arsanilic acid and the positively charged surface of the developed nanocomposite exhibited an electrostatic attraction to each other—the nanocomposite exhibited increased adsorption in the pH range of 3 to 6. The lowest adsorption occurred at a high pH value due to the electrostatic repulsion between the anionic nature of arsanilic acid and the negatively charged surface of the nanocomposite. The pHPZC value of the synthesized nanocomposite determined the surface charge of the nanocomposite. When the pH of the solution was below the pHPZC value of the nanocomposite, the surface charge of the nanocomposite was positive, and vice versa. The maximum adsorption (313.7 mg g^{-1}) of arsanilic acid was obtained in the first 10 min of the contact [89].

4.3 Nanotubes

Nanotubes are nanometer-sized hollow tubular structures and have multiple applications. Carbon nanotubes (CNTs), which are the most common, are synthesized and available in a variety of diameters and sizes and can be single-walled (SWCNT) or multi-walled (MWCNT) [120]. CNTs can potentially adsorb various contaminants from water and wastewater due to their structural properties and excellent surface area. Several studies were reported for the removal of inorganic arsenic using nanotubes such as single-wall carbon nanotube coated with iron oxide [121], multi-walled carbon nanotubes functionalized with magnetic nanoparticles [122], titanium dioxide-coated carbon nanotube [123], graphene-carbon nanotube-iron oxide nanostructures [124], Fe_3O_4 nanoparticle-coated boron nitride nanotubes [125], single-walled aluminosilicate nanotube-iron oxide composite [126], titanate nanotubes [127], and many such studies reported in scientific literature.

For AOCs removal, MWCNTs were utilized to adsorb Roxarsone from an aqueous solution. Researchers reported that iron-modified MWCNT (10.04 mg/g) and unmodified CNT (9.864 mg/g) showed excellent Roxarsone adsorption. Low pH was found to be ideal for the maximum roxarsone adsorption on modified MWCNT due to the electrostatic interaction between the positively charged MWCNT and the protonated Roxarsone. Increased pH was responsible for the dissociation of Roxarsone; consequently, the dissociated species obstructed the hydrogen bond formation with MWCNT, resulting, less adsorption of Roxarsone [102]. Likewise, Roxarsone degradation via a sulfate radical-based advanced oxidation process was reported in another study. For that, a catalyst was prepared using a ball-milling method manufactured by MWCNT, which was combined with electrolytic manganese residue. Within 60 min, a 0.1 g dosage of the manufactured nanomaterial demonstrated remarkable catalytic activity with a 90.96% elimination rate for the Roxarsone at pH 3.04. When the pH was raised from 3.04 to 10.05, the degradation rate of roxarsone decreased. They investigated the formation of arsenite after the cleavage of the As-C bond. The produced As(III) was oxidized into As(V) by reactive oxygen species and then adsorbed by the formed nanomaterial [101].

Other AOCs, such as *p*-arsanilic acid, were also removed via CNTs. According to a previous study, AOCs *p*-arsanilic acid and inorganic arsenic species such as DMA and As(V) were adsorbed by the graphene oxide CNT aerogel, incorporated with goethite (a-FeOOH@GCA). The synthesized nanomaterial had a pH_{PZC} value of 9.5. When the pH was lower than 9.5, the surface charge of the synthesized nanomaterial was positive; therefore, negatively charged anions formed a bond with the nanomaterial. Hence, maximum adsorption of around 81% to 86% of As (V) occurred at a pH range of 3–9 and decreased when the pH increased from 9 to 12. The pH range of 5 to 9 had the highest adsorption rate for *p*-arsanilic acid, around 98% [104].

4.4 Nanocubes

The fabrication of nanocubes through aggregation and self-assembly of nanoparticles has proven to be one of the most promising approaches [128]. Using nanocubes to eliminate inorganic arsenic and AOCs has received significant attention from researchers. To eliminate inorganic arsenic, Fe_2O_3, nanocubes-impregnated graphene aerogel [129] was reported, whereas bimetallic Mn–Fe mixed oxide nanocubes were synthesized and utilized for AOCs removal.

Bimetallic Mn–Fe mixed oxide nanocubes were synthesized to remove *p*-arsanilic acid from the water sample. The prepared nanocubes and peroxymonosulfate (PMS) oxidized *p*-arsanilic acid and removed total arsenic. They found that combining nanocubes with PMS oxidation increased the oxidation rate of *p*-arsanilic acid and eliminated total arsenic compared to using only PMS or nanocubes. With an optimized dose of nanocubes (0.4 gL^{-1}) and PMS (325 μM) concentration, the maximum adsorption of around 90% was achieved. The role of manganese in the nanocubes was to oxidize the *p*-arsanilic acid, while Iron was used to adsorb the total arsenic. Pre-adsorption of *p*-arsanilic acid by nanocubes formed a ligand—oxide interface, improving the oxidation of *p*-arsanilic acid and removing total arsenic [105].

4.5 Nanocrystals

A nanocrystal is a crystalline particle composed of atoms in either a single- or poly-crystalline arrangement with at least one dimension less than 100 nm [130]. Inorganic arsenic removal from nanocrystals has been the subject of extensive investigation, whereas the removal of AOCs has received less attention. For instance, magnetite nanocrystals [131], $MnFe_2O_4$ nanocrystals [132], and many such nanocrystals have been reported for the removal of inorganic arsenic.

Only hematite nanocrystals were reported to remove AOCs [106]. Two facets of hematite nanocrystals, hematite nanocubes (012) and hematite nanoplates (001), were synthesized and explored for AOC removal. The study used the open circuit potential during adsorption to quantify the interface's potential change and detect the surface's interaction. It is reported that hematite nanocubes had a more considerable open circuit potential change. Consequently, more electrons were transported from hematite nanocubes to the adsorbed phenylarsonic acid. The hematite nanocubes showed a more robust adsorption capacity than hematite nanoplates in acidic conditions [106].

4.6 Nanofibrous Membranes

Membrane separation has been used in various fields, including water and wastewater treatment, separation, and purification of substances in industrial sectors such as food, chemicals, medicine, and biotechnology [133–135]. Compared to reverse osmosis and ultrafiltration, nanofiltration provides the advantages of high-solute rejection and low-energy consumption with a membrane pore size of 0.5–2.0 nm and operation pressure ranging from 4 to 30 bar [133, 136]. Using nanofibrous membranes to filter out different contaminants, including contaminants of emerging concern from water/wastewater, has gained significant interest from researchers globally. Several studies reported on the removal of inorganic arsenic by different nanofibrous membranes such as chitosan-based electrospun nanofiber membranes [137], chitosan/PVA/zero-valent iron biopolymeric nanofibers mat [138], modified electrospun nanofiber membranes with α-Fe_2O_3 nanoparticles [139], chitosan-functionalized sodium alginate-based electrospun nanofiber membrane [140] and many more, however, for removal of AOCs only a study reported in which AOCs were eliminated from the sample using the electrospun nanofibrous cellulose membrane that had undergone amino modification.

First, to synthesize an amino-modified electrospun nanofibrous cellulose membrane, the cellulose acetate membrane (CAM) was prepared using cellulose acetate and a mixture of acetone and N, N-dimethylacetamide. Hydroformylated CAM was synthesized from the synthesized CAM. In the end, the hydroformylated CAM was modified by polyvinylamine and formic acid to obtain an amino-modified electrospun nanofibrous cellulose membrane. The amino-modified nanofibrous membrane showed excellent adsorption towards Roxarsone (186.22 mg g^{-1}), followed by *p*-ASA (69.15 mg g^{-1}) and phenylarsonic acid (62.77 mg g^{-1}), compared to the unmodified membrane. The adsorption mechanism was hydrogen bonding and electrostatic interactions [107].

4.7 Nanosheets

Nanosheets are two-dimensional nanostructures commonly fabricated through the bottom-up approach, with a thickness range of 1 to 100 nm [141]. Several nanosheets were fabricated for the adsorption of inorganic arsenic. For instance, iron nanosheets [142], Fe_3O_4-functionalized boron nitride nanosheets [143], hexagonal boron nitride and graphene nanosheets [144], and many such nanosheets are prepared for the As(III) adsorption. For As(V) removal, magnetite-coated boron nitride nanosheets [145] and Mg_2Al-NO_3 layered double hydroxide nanosheets [146] were reported. γ-Fe_2O_3 nanosheets were used for the adsorption of total inorganic arsenic [147]. However, for AOCs, only a study was reported in which prepared magnetic greigite/iron sulfide (Fe_3S_4) nanosheets showed excellent adsorption of Roxarsone. The Fe_3S_4 nanosheets were synthesized with the help of the solvothermal method by using

$FeCl_3 \cdot 6H_2O$ (ferric chloride hexahydrate) and thiourea. The removal of Roxarsone with the synthesized nanosheets occurred in the pH 3.6 to 8.6 range, and the pH range of 3.6 to 4.6 was found to have the highest adsorption levels. However, the roxarsone removal rate was significantly reduced when the pH was increased to 8. During the removal process, intermediates such as 3-amino-4-hydroxyphenyl-arsonic acid, As(III), and As(V) were produced, and these intermediates were detected with the help of HPLC. It is reported that the formed 3-amino-4-hydroxyphenyl-arsonic acid revealed that Fe_3S_4 reduced the -NO_2 moiety of Roxarsone to -NH_2 [109].

4.8 Nanoplates

Nanoplates are defined as having only one dimension on a nanometer scale. Silicate clay minerals (montmorillonite) and graphite, both layered materials, are the most extensively used nanoplates [148]. Many scientific studies investigated the use of various nanoplates to remove inorganic arsenic. For instance, As(V) was adsorbed by nanoplates such as hematite nanoplates [149], Mg–Fe layered double hydroxide nanoplates [150], and iron(III) hydroxide nanoplates [151], MgO nanoplates [152]. Magnetite [153] and silver [154] nanoplates were used for the adsorption of As(III) and total inorganic arsenic, respectively. Several nanoplates were reported for the removal of inorganic arsenic; however, only zinc oxide nanoplate was reported for the removal of AOCs.

Zinc oxide (ZnO) nanoplates were manufactured using a wet chemical technique to photodegrade Roxarsone in an aqueous solution. The study used a UV-A light (365 nm) source for the photodegradation of Roxarsone at neutral pH. The combination of UV-A light with ZnO nanoplates showed excellent degradation of Roxarsone compared to the experiments without ZnO nanoplates. The degradation experiment was also performed under a nitrogen atmosphere, but only negligible degradation was observed. •OH radicals produced by ZnO nanoplates play a primary role in Roxarsone's degradation by cleaving the As-C bond of Roxarsone. As (V) was the dominant product produced after the Roxarsone's degradation [155].

4.9 Nano-biochar

Macro-biochar, colloidal biochar, and nano-biochar are biochars with particle sizes of 1 μm or more extensive, 1–100 nm, and less than 100 nm, respectively [156]. Nano-biochar has a significantly larger surface area-to-mass ratio than macro-biochar and has myriad applications compared to other biochars [157]. The most common way of producing nano-biochar is through mechanical grinding, and approximately more than 98% of nano-biochar was fabricated through this method [158]. Studies related to removing contaminants such as heavy metals from nano-biochar have been discussed extensively by Jiang and team [159]. According to a study, the modified nano-biochar removed 95% of dissolved arsenic [160]. Inorganic arsenic is produced after the degradation of AOCs; thus, AOCs and inorganic arsenic removal using nano-biochar can potentially prevent the release of non-geogenic arsenic in the environment.

A recent study examined the effectiveness of modified nano-biochar for the adsorption of Roxarsone, Cd(II), As(III), and phosphorus. The nano-sized biochar was modified by goethite through the co-precipitation method. The modified biochar showed a relatively higher adsorption capacity for all four contaminants than the unmodified. It was also found that pH greatly influenced the adsorption rate for these contaminants. Low pH was suitable for higher adsorption of As(III), Roxarsone, and phosphate, and high pH was suitable for cadmium [110].

5 Other Removal Materials for the Removal of AOCs

5.1 Adsorbents

Various adsorbents materials, such as activated carbon derived from corncob [161], montmorillonite modified by Fe/La [162], iron-modified sorghum straw biochar [163], Fe–Mn framework [164], magnetic microporous biochar derived from hazelnut shell [165], iron oxides [166], iron humate [167], hematite [168] are reported for removal of AOCs. Roxarsone was removed efficiently through different types of adsorbents, such as activated carbon derived from corncob [161], montmorillonite modified by Fe/La [162], and iron-modified sorghum straw biochar [163]. The adsorption of Roxarsone is facilitated by activated carbon derived from corncob in a lowered pH environment, and hydrogen bonding serves as the primary mechanism for roxarsone elimination [161]. Likewise, functional groups (Fe–O, OH, C = O, C–O, and C = C) in iron-modified sorghum straw biochar performed crucial roles in efficiently removing arsenic from the solution. The Fe–Mn framework was reported for the adsorption of *p*-ASA. The amino group of *p*-ASA, Mn (IV) release from the Fe–Mn framework was responsible for the adsorption of *p*-ASA [164]. Magnetic microporous biochar derived from hazelnut shell was used to remove the *p*-ASA, in which the prepared material contained functional groups (C–O and C = O), and aromatic units of the targeted compound played dominant roles in the adsorption process [165]. Iron oxides [166] and iron humate [167] were also reported for the adsorption of *p*-ASA, in which Fe ions and carboxyl groups were majorly involved during adsorption.

The adsorbent materials such as ferric and manganese binary oxides [169], Iron and aluminum oxides [170], metal oxide and hydroxide, and clay minerals [171] were reported for the removal of both Roxarsone and *p*-ASA. Ferric and manganese binary oxide effectively adsorbed the *p*-arsanilic acid and Roxarsone at pH 7 with maximum adsorption capacities of 0.52 and 0.25 mmol g^{-1}. Due to the combined effect of Mn

(IV) and Fe (III) oxides, the adsorbent had a strong affinity for *p*-arsanilic acid and Roxarsone in this binary oxide. FTIR and XPS spectra show significant oxidation of *p*-arsanilic acid, and Roxarsone occurred due to Mn release at low pH levels [169]. The study of the adsorption of Roxarsone and arsanilic acid through iron and aluminum oxides was regulated by the As (V) functional group of the compounds [170]. In the adsorption of Roxarsone and arsanilic acid through iron hydroxide and manganese oxide, the main mechanism was the development of inner-sphere complexes through the arsonic acid group.

On the other hand, adsorption via kaolinite and aluminum oxide, outer-sphere complexation, was significant [171]. Though various adsorbents and materials were used to remove the AOCs, due to drawbacks such as reduced surface area, poorly defined pore structure, and fewer active sites, conventional adsorbents are not as efficient as nanomaterials. The adsorption capacity is likewise lower compared to the nano-adsorbent; also, in comparison to nanomaterials, conventional adsorbents and materials have poor reusability. As reported in a comparative study, Roxarsone was adsorbed using both biochar and nano-biochar; however, nano-biochar adsorbs Roxarsone (66.5 mg/g) [110] more effectively than biochar (12.4 mg/g) [163].

5.2 Metal–Organic Frameworks (MOFs)

Unique sorts of very porous crystalline materials are MOFs, three-dimensional frameworks made of metal templates and organic linkers [172]. Various MOFs have recently been used to remove AOCs. Roxarsone was efficiently removed by the Zr-based MOF(UiO-66) [173], and *p*-ASA was removed by the amine-modified MOF (UiO-67-NH_2) [174], zeolitic imidazolate framework-8 [175], Cu(I)-tpp@ZIF-8 [176], and amino-functionalized indium-based MOF (NH_2-MIL-68(In)) [177]. Some other MOFs have been reported that can remove multiple AOCs. These were MIL-88A(Fe)/cotton fiber composites [178], amino-modified MOF (UiO-66-D-NH2) [179], MIL-101(Fe) [180], MIL-101$(OH)_3$ [181], iron benzene tri carboxylate (MIL-100-Fe) [182], Zr-MOFs (DUT-67, UiO-66, UiO-67, MOF-808, MOF-808F, NU-1000, NU-1000B) [183].

Functional groups present in MOFs, such as amino (-NH_2) or carboxyl (–COOH), and their porous structures are significant for MOFs' removal abilities by adsorption through electrostatic interactions, π-π stacking, and hydrogen bonding [184]. Transition metals like Fe, Cu, or Zr facilitate catalytic degradation, which encourages breaking C-As bonds in AOCs and transforming them into hazardous byproducts like inorganic arsenic for later elimination. MOFs have drawbacks despite their benefits, including poor stability in aqueous environments, expensive synthesis, limited selectivity for certain AOCs, and structural deterioration that reduces reusability [185]. To overcome these constraints, nanotechnology provides revolutionary solutions. Hybrid materials that combine MOFs with nanoparticles have improved stability, catalytic activity, and adsorptive capacity. Modified diameters of pores and functionalized active sites in nanostructured MOFs increase removal efficiency, while green synthesis methods that use nanotechnology lower costs and have less environmental impact. AOC removal can be revolutionized by incorporating nanotechnology into MOF design because it increases efficiency, adapts to settings with many contaminants, and promotes sustainability by using reusable materials that produce less waste.

5.3 Oxidizing Agents and Advanced Oxidation

Various oxidizing agents were used to treat AOCs. Ferrate, a green oxidizing agent, was used for the treatment of Roxarsone [186], *p*-ASA [187, 188], and both [189]. Ferrate with sulfur [190] and aluminum chloride [191] was also used to degrade AOCs. Arsenic in the form of arsenite (As(III)) or arsenate (As(V)) is released along with intermediate degradation products when the C-As bond in AOCs is broken by hydroxyl radicals produced by advanced oxidation approaches such as Fenton-based procedures, which include homogenous [192–194], heterogeneous [195] and plasma-Fenton [196]. Hence, it releases inorganic arsenic and breakdown products, which are the most important concern when AOCs are degraded by oxidizing agents or advanced oxidation. Inorganic arsenic and product degradation must be eliminated to avoid contaminating the environment. For this, nanomaterials have become effective solutions. Released arsenic species can be reduced by adsorption or catalytic degradation by nanoparticles such as Fe_3O_4, TiO_2, and ZnO. For instance, Fe_3O_4 nanoparticles have a high binding affinity for arsenite and arsenate [197].

5.4 Photocatalysts

The most widespread technique for transforming, degrading, and removing organic molecules is the photocatalysis of pollutants under UV or visible light. Various studies have reported the degradation of Roxarsone. UV/Fe^{3+} [198], solar/hematite/oxalate [199], BiOCl0.875Br0.125/PDA functionalized PVDF membranes/visible light [200]. For *p*-ASA Fe(III)-oxalate/near-visible UV irradiation [201], colloidal ferric hydroxide/black light lamps [202], UV/H_2O_2 [203], UV/O_3/H_2O_2 [204]. Roxarsone and *p*-arsenilic acid were both degraded with the help of UV/TiO_2 [205], and MMA, DMA, and As(III) were removed through TiO_2 suspension in the presence of sunlight [206].

Under UV or visible light, catalysts such as Fe(III)-oxalate complexes generate reactive oxygen species (ROS), which aid in breaking compounds like *p*-ASA and roxarsone. Nonetheless, concerns like catalyst stability and recovery under various circumstances continue to exist. Heterogeneous catalysis gives benefits in terms of stability and catalyst reuse by using solid catalysts such as TiO_2, ZnO nanoparticles, and haematite suspended in liquid. Systems like UV/TiO_2 are known to cleave C-As bonds, which efficiently degrade AOCs like *p*-ASA and roxarsone.

These materials also offer active sites for the adsorption and degradation of contaminants. Nanostructured catalysts provide increased active sites and better degradation efficiency because of their large surface area, greater light absorption, and tunability. The photocatalytic activity may be adjusted by doping and functionalization, which target certain contaminants and enable a longer light response. The photocatalytic effectiveness is increased by hybrid nanostructures, such as TiO_2 and graphene [207]. Green synthesis techniques for photocatalysts are also made possible by nanotechnology, which improves the process's environmental friendliness. By enabling systems to target a variety of pollutants and providing reusability and lower operating costs, nanotechnology in photocatalysis offers enormous potential for efficient and sustainable pollutant degradation [208].

5.5 Biodegradation

One of the natural mechanisms of removing chemical substances from the environment is microbial transformation. In previously reported studies, AOCs biotransformed under aerobic or anaerobic conditions into inorganic arsenic and another organic form of arsenic. The bacteria consortium was reported to degrade the Roxarsone under aerobic conditions [209]. The exoelectrogenic bacteria *Shewanella putrefaciens* were also reported for Roxarsone's degradation. *Shewanella putrefaciens* was also reported to transform methyl arsenate to methyl arsenite and methyl arsenite to arsenite. It also reduced nitarsone into p-aminophenyl arsenate [210]. Bioreduction of nitarsone was also reported by *Shewanella oneidensis*. Inorganic arsenic, aniline, and arsanilic acid were formed after the reduction of nitarsone [211]. Apart from this, *Tetrahymena thermophila*, a freshwater protozoan, was reported to degrade Roxarsone [212].

However, despite this promising capability, microbial transformation has several drawbacks concerning AOC degradation, such as incomplete degradation. Microbial processes often degrade AOCs into intermediates; some intermediates, like inorganic arsenic, are toxic and need further treatment. Efficiency in microbial degradation depends upon specific microbial strains, which may not always be present in contaminated environments. Such factors as temperature, pH, and oxygen levels can strongly impact microbial activity and, correspondingly, the efficiency of the processes. Generally, microbial processes are slower than chemical or nanotechnology-based methods, making them less suitable for large-scale or urgent remediation efforts. Nanotechnology can address some of the limitations of microbial transformation and offer synergistic benefits. Nanoparticles, including nZVI or TiO_2, could stimulate microbial activity by providing reactive surfaces or improving electron transfer. Nanomaterials can adsorb or degrade toxic intermediate products, such as inorganic arsenic, reducing environmental risk. For example, combining nanotechnology with microbial approaches may accelerate degradation rates and extend the scope of treatable contaminants. Nanotechnology can speed up microbial transformation time by improving reaction kinetics. Hybrid systems can handle complicated mixtures of contaminants, especially AOCs and co-existing pollutants. These hybrid systems contribute more to sustainability and efficiency in environmental remediation by incorporating natural microbial processes with nanomaterials [213].

6 Potential Challenges for the Applications of Nanotechnology

Although nanotechnology has great promise for removing and degrading AOCs, some obstacles prevent its widespread use. First, the widespread usage of engineered nanomaterials is hampered by their high manufacturing costs and scaling problems. Secondly, issues about the nanomaterials' stability, possible toxicity, and environmental fate create safety concerns that are not fully addressed. In many cases, insufficient regeneration and reuse capacity lead to reduced economic viability over multiple treatment cycles. Furthermore, the performance of nanomaterials can be adversely affected by complicated realworld matrices (such as the presence of cocontaminants, natural organic matter, or fluctuating pH). Finally, the lack of comprehensive regulatory guidelines for using and disposing nanomaterials in environmental systems remains a significant bottleneck. Addressing these challenges will be critical for translating nanotechnology-based AOC remediation strategies from laboratory-scale studies to practical field applications.

7 Conclusion and Future Implications

AOCs are utilized in the poultry and swine industries as preventative antibiotics and growth promoters. Inorganic arsenic's fate, transport, and toxicity have all been extensively studied, but far less attention has been paid to AOCs, particularly nitarsone and carbarsone, which remain underexplored. Research on the fate, transportation, and removal from environmental matrices of AOCs in the environment is still required. The current review described AOCs' presence, fate, and elimination through different nanomaterials. The majority of prior research has been on Roxarsone and arsanilic acid. Nitarsone and carbarsone, the other two AOCs, were also used as feed additives and are still used in some nations. This review contemplates the emerging threat of arsenic exposure apart from geogenic arsenic due to the usage of AOCs in industries such as swine and poultry, and their transportation/fate in the environment. Also, several potential treatment methods have been studied for the degradation/removal of AOCs. Various nanomaterials such as nanoparticles, nanocomposites, nanotubes, nanocubes, nanocrystals, nanofibrous membranes, nanosheets, nanoplates, and nanobiochar, etc. have been developed and utilized for the degradation and removal of AOCs. Out of these nanomaterials, nanobiochar, nanofibrous membranes, and nanocrystals have been studied for the adsorptive removal of AOCs from water/wastewater. The other nanomaterials were studied simultaneously for catalytic degradation of the AOCs and the simultaneous adsorption of degradation products/by-products. Inclusive, this article provides treasured information about how nanotechnology-based approaches can be utilized unaided or combined with other techniques for the degradation/removal of AOCs.

However, significant research gaps remain—especially in understanding long-term environmental behavior, degradation pathways, and by-product toxicity. Future research should focus on developing cost-effective, multifunctional, and regenerable nanomaterials, as well as real-world application studies under environmentally relevant conditions. Emphasis should also be placed on underrepresented compounds like nitarsone and carbarsone to assess their environmental impact fully.

Abbreviations

AOCs	Aromatic organoarsenic compounds
As(III)	Arsenite
As(V)	Arsenate
p-ASA	*p*-Arsanilic acid
ROX	Roxarsone
DMA	Dimethyl arsenate
XRD	X-ray diffraction
TEM	Transmission electron microscopy
ATR-FTIR	Attenuated total internal reflectance Fourier transform infrared spectroscopy
XPS	X-ray photoelectron spectroscopy
UV-Vis	Ultraviolet–visible spectroscopy
ICP-AES	Inductively coupled plasma atomic emission spectrometer
UPLC	Ultra-performance liquid chromatography
GC-MS	Gas chromatography–mass spectrometry
ICP-MS	Inductively coupled plasma mass spectrometry
HPLC	High performance liquid chromatography
TOC	Total organic carbon analyzer
EDS	Energy dispersive spectroscopy
HRTEM	High resolution transmission electron microscopy
BET	Brunauer-emmett-teller
VSM	Vibrating sample magnetometer
ICP-OES	Inductively coupled plasma optical emission spectrometry
FE-SEM	Field emission scanning electron microscope
DRS	Diffuse reflectance spectroscopy
AFS	Atomic fluorescence spectrometry
HG-AFS	Hydride generation atomic fluorescence spectrometry
RAMAN	Raman spectroscopy
LC	Liquid chromatography

TOF–SIMS,	Time-of-flight secondary ion mass spectrometry
BJH	Barrett-joyner-halenda
LC/Q-TOF/MS	Liquid chromatography/quadrupole time-of-flight mass spectrometry
TGA	Thermogravimetric analysis
3-NHPAA	3-Nitro-4-hydroxy-phenylarsonic acid
UPLC-QTOF-MS	Ultraperformance liquid chromatography quadrupole time-of-flight mass spectrometry
LC–MS/MS	Liquid chromatography–mass spectrometry/ mass spectrometry
PhAs and PAA	Phenylarsonic acid

References

1. Masuda, H.: Arsenic cycling in the Earth's crust and hydrosphere: interaction between naturally occurring arsenic and human activities. Prog. Earth Planet Sci. **5**(1), 68 (2018)
2. Handali, S., Rezaei, M.: Arsenic and weight loss: at a crossroad between lipogenesis and lipolysis. J. Trace Elem. Med. Biol. **68**, 126836 (2021)
3. Bhattacharya, P., Jacks, G., Frisbie, S., Smith, E., Naidu, R., Sarkar, B.: Arsenic in the Environment. Heavy Met Environ. [Internet]. Available from: https://www.academia.edu/3102558/Arsenic_in_the_environment_a_global_perspective (2002). cited 8 Jun 2022
4. Rajakovic, L., Rajaković-Ognjanović, V.: Arsenic in Water: Determination and Removal (2018)
5. Garelick, H., Jones, H., Dybowska, A., Valsami-Jones, E.: Arsenic pollution sources. Rev. Environ. Contam. Toxicol. **197**, 17–60 (2008)
6. Ganie, S.Y., Javaid, D., Hajam, Y.A., Reshi, M.S.: Arsenic toxicity: sources, pathophysiology and mechanism. Toxicol. Res. **13**(1), tfad111 (2023)
7. Bag, A., Khanam, R., Chatterjee, N., Mukherjee, A., Sahu, G., Plal, B., et al.: Effect of Arsenic on Soil, Human and Plant Growth (2020)
8. Shaji, E., Santosh, M., Sarath, K.V., Prakash, P., Deepchand, V., Divya, B.V.: Arsenic contamination of groundwater: a global synopsis with focus on the Indian Peninsula. Geosci. Front. **12**(3), 101079 (2021)
9. Kumar, M., Yadav, A., Ramanathan, A.L.: Arsenic contamination in environment, ecotoxicological and health effects, and bioremediation strategies for its detoxification. In: Saxena, G., Bharagava, R.N. (eds.) Bioremediation of Industrial Waste for Environmental Safety [Internet]. Singapore: Springer Singapore, p. 245–64. Available from: http://link.springer.com/https://doi.org/10.1007/978-981-13-1891-7_12 (2020). [cited 23 Aug 2021]
10. Guha Mazumder, D.N.: 6-Health effects chronic arsenic toxicity. In: Flora, S.J.S. (eds.) Handbook of Arsenic Toxicology [Internet], pp. 137–177. Academic Press, Oxford. Available from: https://www.sciencedirect.com/science/article/pii/B978012418688000006X (2020). [cited 23 Aug 2021]
11. SK, V., Raman, R.K., Talukder, A., Mahanty, A., Sarkar, D.J., Das, B.K., et al.: Arsenic bioaccumulation and identification of low-arsenic-accumulating food fishes for aquaculture in arsenic-contaminated ponds and associated aquatic ecosystems. Biol. Trace. Elem. Res. **200**(6), 2923–2936 (2022)
12. Genchi, G., Lauria, G., Catalano, A., Carocci, A., Sinicropi, M.S.: Arsenic: a review on a great health issue worldwide. Appl. Sci. **12**(12), 6184 (2022)

13. Ganie, S.Y., Javaid, D., Hajam, Y.A., Reshi, M.S.: Arsenic toxicity: sources, pathophysiology and mechanism. Toxicol. Res. **13**(1), tfad111 (2024)
14. Verma, S.K., Chaurasia, S.: Implicating the effects of consuming water with a high level of arsenic content: highlighting the cause and consequences of arsenic contamination in drinking water. Water Pract. Technol. **19**(4), 1071–1083 (2024)
15. Fatoki, J.O., Badmus, J.A.: Arsenic as an environmental and human health antagonist: a review of its toxicity and disease initiation. J. Haz. Mater. Adv. **5**, 100052 (2022)
16. Khosravi-Darani, K., Rehman, Y., Katsoyiannis, I.A., Kokkinos, E., Zouboulis, A.I.: Arsenic exposure via contaminated water and food sources. Water **14**(12), 1884 (2022)
17. Uppal, J.S., Zheng, Q., Le, X.C.: Arsenic in drinking water—recent examples and updates from Southeast Asia. Curr. Opin. Environ. Sci. Health **7**, 126–135 (2019)
18. Ramsay, L., Petersen, M.M., Hansen, B., Schullehner, J., van der Wens, P., Voutchkova, D., et al.: Drinking water criteria for arsenic in high-income, low-dose countries: the effect of legislation on public health. Environ Sci. Technol. **55**(6), 3483–3493 (2021)
19. Chaudhary, M.M., Hussain, S., Du, C., Conway, B.R., Ghori, M.U.: Arsenic in water: understanding the chemistry, health implications, quantification and removal strategies. Chem. Eng. **8**(4), 78 (2024)
20. Rahaman, M.D.S., Rahman, Md.M., Mise, N., Sikder, Md.T., Ichihara, G., Uddin, Md.K., et al.: Environmental arsenic exposure and its contribution to human diseases, toxicity mechanism and management. Environ. Pollut. **289**, 117940 (2021)
21. Rahidul, H.H.: A review on different arsenic removal techniques used for decontamination of drinking water. Environ. Pollut. Bioavailab. **35**(1), 2165964 (2023)
22. Alka, S., Shahir, S., Ibrahim, N., Ndejiko, M.J., Vo, D.V.N., Manan, F.A.: Arsenic removal technologies and future trends: a mini review. J. Clean. Prod. **278**, 123805 (2021)
23. Mahamallik, P., Swain, R.: A mini-review on arsenic remediation techniques from water and future trends. Water Sci. Technol. **87**(12), 3108–3123 (2023)
24. Hamid, N.H.A., Rushdan, A.I., Nordin, A.H., Faiz Norrrahim, M.N., Muhamad, S.N.H., Tahir, M.I.H.M., et al.: A review: the state-of-the-art of arsenic removal in wastewater. Water Reuse. **14**(3), 279–311 (2024)
25. Mojiri, A., Razmi, E., KarimiDermani, B., Rezania, S., Kasmuri, N., Vakili, M., et al.: Adsorption methods for arsenic removal in water bodies: a critical evaluation of effectiveness and limitations. Front. Water [Internet] (2024)
26. Dilpazeer, F., Munir, M., Baloch, M.Y.J., Shafiq, I., Iqbal, J., Saeed, M., et al.: A comprehensive review of the latest advancements in controlling arsenic contaminants in groundwater. Water **15**(3), 478 (2023)
27. Nguyen, T.H., Nguyen, T.V., Vigneswaran, S., Ha, N.T.H., Ratnaweera, H.: A review of theoretical knowledge and practical applications of iron-based adsorbents for removing arsenic from water. Minerals **13**(6), 741 (2023)
28. Habuda-Stanić, M., Nujić, M.: Arsenic removal by nanoparticles: a review. Environ. Sci. Pollut. Res. **22**(11), 8094–8123 (2015)
29. Jain, N., Singh, P., Bhatnagar, A., Maiti, A.: Arsenite oxidation and adsorptive arsenic removal from contaminated water: a review. Environ. Sci. Pollut. Res. **31**(30), 42574–42592 (2024)
30. Xue, X.M., Xiong, C., Yoshinaga, M., Rosen, B., Zhu, Y.G.: The enigma of environmental organoarsenicals: insights and implications. Crit. Rev. Environ. Sci. Technol. 1–28 (2021)
31. Muzaffar, S., Khan, J., Srivastava, R., Gorbatyuk, M.S., Athar, M.: Mechanistic understanding of the toxic effects of arsenic and warfare arsenicals on human health and environment. Cell Biol. Toxicol. **39**(1), 85–110 (2023)
32. Chen, J., Garbinski, L.D., Rosen, B., Zhang, J., Xiang, P., Ma, L.Q.: Organoarsenical compounds: occurrence, toxicology and biotransformation. Crit. Rev. Environ. Sci. Technol. **50**(3), 217–243 (2020)
33. Reimer, K.J., Koch, I., Cullen, W.R.: Organoarsenicals: distribution and transformation in the environment. Met. Ions Life Sci. **7**, 165–229 (2010)
34. Azizur Rahman, M., Hasegawa, H., Peter, L.R.: Bioaccumulation, biotransformation and trophic transfer of arsenic in the aquatic food chain. Environ. Res. **116**, 118–135 (2012)

35. Luvonga, C., Rimmer, C.A., Yu, L.L., Lee, S.B.: Analytical methodologies for the determination of organoarsenicals in edible marine species: a review. J. Agric. Food Chem. **68**(7), 1910–1934 (2020)
36. Taylor, V., Goodale, B., Raab, A., Schwerdtle, T., Reimer, K., Conklin, S., et al.: Human exposure to organic arsenic species from seafood. Sci. Total. Environ. **580**, 266–282 (2017)
37. Hu, Q., Liu, Y., Gu, X., Zhao, Y.: Adsorption behavior and mechanism of different arsenic species on mesoporous $MnFe_2O_4$ magnetic nanoparticles. Chemosphere **181**, 328–336 (2017)
38. Bednar, A.J., Garbarino, J.R., Ranville, J.F., Wildeman, T.R.: Presence of organoarsenicals used in cotton production in agricultural water and soil of the Southern United States. J. Agric. Food Chem. **50**(25), 7340–7344 (2002)
39. Mangalgiri, P.K., Adak, A., Blaney, L.: Organoarsenicals in poultry litter: detection, fate, and toxicity. Environ. Int. **75**, 68–80 (2015)
40. Fei, J., Wang, T., Zhou, Y., Wang, Z., Min, X., Ke, Y., et al.: Aromatic organoarsenic compounds (AOCs) occurrence and remediation methods. Chemosphere **207**, 665–675 (2018)
41. Saucedo-Velez, A.A., Hinojosa-Reyes, L., Villanueva-Rodríguez, M., Caballero-Quintero, A., Hernández-Ramírez, A., Guzmán-Mar, J.L.: Speciation analysis of organoarsenic compounds in livestock feed by microwave-assisted extraction and high performance liquid chromatography coupled to atomic fluorescence spectrometry. Food Chem. **232**, 493–500 (2017)
42. Lei, J., Peng, B., Min, X., Liang, Y., You, Y., Chai, L.: Modeling and optimization of lime-based stabilization in high alkaline arsenic-bearing sludges with a central composite design. J. Environ. Sci. Health Part A Tox. Hazard. Subst. Environ. Eng. **52**(5), 449–458 (2017)
43. Desheng, Q., Niya, Z.: Effect of arsanilic acid on performance and residual of arsenic in tissue of Japanese laying quail. Poult. Sci. **85**(12), 2097–2100 (2006)
44. Garbarino, J.R., Bednar, A.J., Rutherford, D.W., Beyer, R.S., Wershaw, R.L.: Environmental fate of roxarsone in poultry litter. I. degradation of roxarsone during composting. Environ. Sci. Technol. **37**(8), 1509–1514 (2003)
45. Nachman, K.E., Graham, J.P., Price, L.B., Silbergeld, E.K.: Arsenic: a roadblock to potential animal waste management solutions. Environ. Health Perspect. **113**(9), 1123–1124 (2005)
46. Liu, Y., Zhang, Z., Zhao, X., Wen, M., Cao, S., Li, Y.: Arsenic contamination caused by roxarsone transformation with spatiotemporal variation of microbial community structure in a column experiment. J. Groundw. Sci. Eng. **9**(4), 304–316 (2021)
47. Wang, F.M., Chen, Z.L., Zhang, L., Gao, Y.L., Sun, Y.X.: Arsenic uptake and accumulation in rice (Oryza sativa L.) at different growth stages following soil incorporation of Roxarsone and Arsanilic Acid. Plant Soil. **285**(1), 359–367 (2006)
48. Liu, Y., Wen, M., Wu, L., Cao, S., Li, Y.: Environmental behavior and remediation methods of roxarsone. Appl. Sci. **12**(15), 7591 (2022)
49. Mondal, R., Shanmughan, A., Murugeswari, A., Shanmugaraju, S.: Recent advances in fluorescence-based chemosensing of organoarsenic feed additives using luminescence MOFs, COFs, HOFs, and QDs. Chem. Commun. **59**(77), 11456–11468 (2023)
50. Iwuozor, K.O., Akpomie, K.G., Conradie, J., Adegoke, K.A., Oyedotun, K.O., Ighalo, J.O., et al.: Aqueous phase adsorption of aromatic organoarsenic compounds: a review. J. Water Process. Eng. **49**, 103059 (2022)
51. Emmanuel, S.S., Adesibikan, A.A.: A review on photocatalytic degradation of aromatic organoarsenic compounds in aqueous environment using nanomaterials. J. Chin. Chem. Soc. **71**(10), 1130–1153 (2024)
52. Zhou, S., Shang, H., Luo, J., Shen, M., Wang, Q., Zhang, S., et al.: Organoarsenic conversion to As(III) in subcritical hydrothermal reaction of livestock manure. J. Hazard. Mater. **402**, 123571 (2021)
53. Li, Y., Liu, Y., Zhang, Z., Fei, Y., Tian, X., Cao, S.: Identification of an anaerobic bacterial consortium that degrades roxarsone. MicrobiologyOpen. **9**(4), e1003 (2020)
54. Ji, Y., Shi, Y., Kong, D., Lu, J.: Degradation of roxarsone in a sulfate radical mediated oxidation process and formation of polynitrated by-products. RSC Adv. **6**(85), 82040–82048 (2016)

55. Chen, L., Li, H., Qian, J.: Degradation of roxarsone in UV-based advanced oxidation processes: a comparative study. J. Hazard. Mater. **410**, 124558 (2021)
56. Xu, J., Hu, M., Wang, F., Ding, Y., Wang, L.: Constructing PPy-encapsulated needle-like Fe_2O_3 nanoarrays on carbon cloth as electro-Fenton cathodes for high efficiency roxarsone degradation. New J. Chem. **44**(16), 6277–6284 (2020)
57. Ma, X., Liu, X., Zhao, Y., Shang, X., Zhang, Z., Lin, C., et al.: Enhanced removal of roxarsone and inorganic arsenic using ball-milled heating pad waste to activate peroxymonosulfate: critical role of non-radical pathways. Chem. Eng. J. **488**, 151100 (2024)
58. Ibrahim, R.K., Hayyan, M., AlSaadi, M.A., Hayyan, A., Ibrahim, S.: Environmental application of nanotechnology: air, soil, and water. Environ. Sci. Pollut. Res. **23**(14), 13754–13788 (2016)
59. Nasrollahzadeh, M., Sajadi, S.M., Sajjadi, M., Issaabadi, Z.: Chapter 1-An introduction to nanotechnology. In: Nasrollahzadeh, M., Sajadi, S.M., Sajjadi, M., Issaabadi, Z., Atarod, M. (eds.) Interface Science and Technology [Internet], p. 1–27. Elsevier. (An Introduction to Green Nanotechnology; vol. 28). Available from: https://www.sciencedirect.com/science/article/pii/B9780128135860000018. cited 23 Aug 2022
60. Jr, C.P.P., Owens, F.J.: Introduction to Nanotechnology, p. 408. John Wiley & Sons (2003)
61. O'Brien, T.: ArcGIS StoryMaps. Concentrated Animal Feeding Operations. Available from: https://storymaps.arcgis.com/stories/5def1eeaa079473a93778ba2397a5ee0 (2020). [cited 29 Aug 2022]
62. Liu, X., Zhang, W., Hu, Y., Hu, E., Xie, X., Wang, L., et al.: Arsenic pollution of agricultural soils by concentrated animal feeding operations (CAFOs). Chemosphere. **119**, 273–81 (2015)
63. Christen, K.: Chickens, manure, and arsenic. Environ. Sci. Technol. **35**(9), 184A-185A (2001)
64. Medicine C for V.: Arsenic-based Animal Drugs and Poultry. FDA [Internet]. Available from: https://www.fda.gov/animal-veterinary/product-safety-information/arsenic-based-animal-drugs-and-poultry. (2022) [cited 29 Aug 2022]
65. Jones, F.T.: A broad view of arsenic. Poult. Sci. **86**(1), 2–14 (2007)
66. Clark, S., Kimminau, E.: Critical review: future control of blackhead disease (Histomoniasis) in poultry. Avian Dis. **61**(3), 281–288 (2017)
67. Mondal, N.K.: Prevalence of Arsenic in chicken feed and its contamination pattern in different parts of chicken flesh: a market basket study. Environ. Monit. Assess. **192**(9), 590 (2020)
68. Nachman, K.E., Raber, G., Francesconi, K.A., Navas-Acien, A., Love, D.C.: Arsenic species in poultry feather meal. Sci. Total. Environ. **417–418**, 183–188 (2012)
69. Arai, Y., Lanzirotti, A., Sutton, S., Davis, J.A., Sparks, D.L.: Arsenic speciation and reactivity in poultry litter. Environ. Sci. Technol. **37**(18), 4083–4090 (2003)
70. Yao, L., Li, G., Dang, Z.: Major chemical components of poultry and livestock manures under intensive breeding. Ying Yong Sheng Tai Xue Bao J. Appl. Ecol. **17**(10), 1989–1992 (2006)
71. Fu, Q.L., Blaney, L., Zhou, D.M.: Natural degradation of roxarsone in contrasting soils: degradation kinetics and transformation products. Sci. Total. Environ. **607–608**, 132–140 (2017)
72. Hu, Y., Cheng, H., Tao, S., Schnoor, J.L.: China's ban on phenylarsonic feed additives, a major step toward reducing the human and ecosystem health risk from arsenic. Environ. Sci. Technol. **53**(21), 12177–12187 (2019)
73. Dr. Rajesh Singh.: Status of poultry production in India–Pashudhan praharee [Internet]. Available from: https://www.pashudhanpraharee.com/status-of-poultry-production-in-india/. (2019) [cited 24 Aug 2021]
74. Maneka, G.: Arsenic Kills Silently as you Relish Your Chicken and Mutton Dishes; High Time to Pick Safe Food. 2018th ed. (2018)
75. Huang, L., Yao, L., He, Z., Zhou, C., Li, G., Yang, B., et al.: Roxarsone and its metabolites in chicken manure significantly enhance the uptake of As species by vegetables. Chemosphere **100**, 57–62 (2014)
76. Yao, L., Huang, L., He, Z., Zhou, C., Lu, W., Bai, C.: Delivery of roxarsone via chicken diet→chicken→chicken manure→soil→rice plant. Sci. Total. Environ. **566–567**, 1152–1158 (2016)

77. Yao, L., Li, G., Dang, Z., He, Z., Zhou, C., Yang, B.: Arsenic speciation in turnip as affected by application of chicken manure bearing roxarsone and its metabolites. Plant Soil **316**(1), 117–124 (2009)
78. Yao, L., Huang, L., He, Z., Zhou, C., Li, G., Yang, B., et al.: External inorganic N source enhances the uptake of As species in garland chrysanthemum (C. coronarium) amended with chicken manure bearing roxarsone and its metabolites. J. Hazard Mater. **254–255**, 270–276 (2013)
79. Stolz, J.F., Perera, E., Kilonzo, B., Kail, B., Crable, B., Fisher, E., et al.: Biotransformation of 3-Nitro-4-hydroxybenzene Arsonic Acid (Roxarsone) and release of inorganic arsenic by clostridium species. Environ. Sci. Technol. **41**(3), 818–823 (2007)
80. Bednar, A.J., Garbarino, J.R., Ferrer, I., Rutherford, D.W., Wershaw, R.L., Ranville, J.F., et al.: Photodegradation of roxarsone in poultry litter leachates. Sci. Total. Environ. **302**(1–3), 237–245 (2003)
81. Zhu, X.D., Wang, Y.J., Liu, C., Qin, W.X., Zhou, D.M.: Kinetics, intermediates and acute toxicity of arsanilic acid photolysis. Chemosphere **107**, 274–281 (2014)
82. Goutam, S.P., Saxena, G., Roy, D., Yadav, A.K., Bharagava, R.N.: Green synthesis of nanoparticles and their applications in water and wastewater treatment. In: Saxena, G., Bharagava, R.N. (eds.) Bioremediation of Industrial Waste for Environmental Safety [Internet], pp. 349–379. Springer Singapore, Singapore. Available from: http://link.springer.com/https://doi.org/10.1007/978-981-13-1891-7_16. (2020) [cited 11 Sep 2021]
83. Lv, Y., Huang, S., Huang, G., Liu, Y., Yang, G., Lin, C., et al.: Remediation of organic arsenic contaminants with heterogeneous Fenton process mediated by SiO_2-coated nano zero-valent iron. Environ. Sci. Pollut. Res. **27**(11), 12017–12029 (2020)
84. Sun, T., Shi, Z., Zhang, X., Wang, X., Zhu, L., Lin, Q.: Efficient degradation of p-arsanilic acid with released arsenic removal by magnetic CeO_2–Fe_3O_4 nanoparticles through photo-oxidation and adsorption. J. Alloys Compd. **808**, 151689 (2019)
85. Sun, T., Zhao, Z., Liang, Z., Liu, J., Shi, W., Cui, F.: Efficient degradation of p-arsanilic acid with arsenic adsorption by magnetic CuO-Fe_3O_4 nanoparticles under visible light irradiation. Chem. Eng. J. **334**, 1527–1536 (2018)
86. Wu, Q., Ye, X., Lv, Y., Pei, R., Wu, M., Liu, M.: Lignin-based magnetic activated carbon for p-arsanilic acid removal: applications and absorption mechanisms. Chemosphere **258**, 127276 (2020)
87. Fu, W., Lu, D.L., Yao, H., Yuan, S., Wang, W., Gong, M., et al.: Simultaneous roxarsone photocatalytic degradation and arsenic adsorption removal by TiO2/FeOOH hybrid. Environ. Sci. Pollut. Res. Int. **27**(15), 18434–18442 (2020)
88. Chen, S., Deng, J., Ye, C., Xu, C., Huai, L., Li, J., et al.: Simultaneous removal of para-arsanilic acid and the released inorganic arsenic species by $CuFe_2O_4$ activated peroxymonosulfate process. Sci. Total. Environ. **742**, 140587 (2020)
89. You, N., Wang, X.F., Li, J.Y., Fan, H.T., Shen, H., Zhang, Q.: Synergistic removal of arsanilic acid using adsorption and magnetic separation technique based on Fe3O4@ graphene nanocomposite. J. Ind. Eng. Chem. **70**, 346–354 (2019)
90. Zhang, W., Liu, W., Li, G., Du, W., Lu, J., Song, J., et al.: Simultaneous adsorption and oxidation of para arsanilic acid by a highly efficient nanostructured Fe-Ti-Mn composite oxide. Chem. Eng. J. **407**, 127142 (2021)
91. Zhao, Z., Wu, P., Fang, Z., Zhang, X.: Selective sequestration of p-arsanilic acid from water by using nano-hydrated zirconium oxide encapsulated inside hyper-cross-linked anion exchanger. Chem. Eng. J. **391**, 123624 (2020)
92. Wen, Z., Xi, J., Lu, J., Zhang, Y., Cheng, G., Zhang, Y., et al.: Porous biochar-supported $MnFe_2O_4$ magnetic nanocomposite as an excellent adsorbent for simultaneous and effective removal of organic/inorganic arsenic from water. J. Hazard. Mater. **411**, 124909 (2021)
93. Tian, C., Zhao, J., Zhang, J., Chu, S., Dang, Z., Lin, Z., et al.: Enhanced removal of roxarsone by Fe_3O_4@3D graphene nanocomposites: synergistic adsorption and mechanism. Environ. Sci. Nano **4**(11), 2134–2143 (2017)

94. Su, S., Cao, C., Zhao, Y., Dionysiou, D.D.: Efficient transformation and elimination of roxarsone and its metabolites by a new α-FeOOH@GCA activating persulfate system under UV irradiation with subsequent As(V) recovery. Appl. Catal. B Environ. **245**, 207–219 (2019)
95. Wu, S., Yang, D., Zhou, Y., Zhou, H., Ai, S., Yang, Y., et al.: Simultaneous degradation of p-arsanilic acid and inorganic arsenic removal using M-rGO/PS Fenton-like system under neutral conditions. J. Hazard. Mater. **399**, 123032 (2020)
96. Kong, D., Wilson, L.D.: Synthesis and characterization of cellulose-goethite composites and their adsorption properties with roxarsone. Carbohydr. Polym. **169**, 282–294 (2017)
97. Zou, S.J., Ding, B.H., Chen, Y.F., Fan, H.T.: Nanocomposites of graphene and zirconia for adsorption of organic-arsenic drugs: performances comparison and analysis of adsorption behavior. Environ. Res. **195**, 110752 (2021)
98. Li, B., Wei, D., Li, Z., Zhou, Y., Li, Y., Huang, C., et al.: Mechanistic insights into the enhanced removal of roxsarsone and its metabolites by a sludge-based, biochar supported zerovalent iron nanocomposite: adsorption and redox transformation. J. Hazard. Mater. **389**, 122091 (2020)
99. Liu, B., Liu, Z., Wu, H., Pan, S., Cheng, X., Sun, Y., et al.: Effective and simultaneous removal of organic/inorganic arsenic using polymer-based hydrated iron oxide adsorbent: capacity evaluation and mechanism. Sci. Total. Environ. **742**, 140508 (2020)
100. Gu, B., Zhang, H., Ye, M., Zhou, T., Yi, J., Hu, Q.: Adsorption behavior of organoarsenicals over MnFe2O4-graphene hybrid nanocomposite: the role of organoarsenic chemical structures. Materials. **16**(24), 7636 (2023)
101. Li, M., He, Z., Zhong, H., Hu, L., Sun, W.: Multi-walled carbon nanotubes facilitated Roxarsone elimination in SR-AOPs by accelerating electron transfer in modified electrolytic manganese residue and forming surface activated-complexes. Water Res. **200**, 117266 (2021)
102. Hu, J., Tong, Z., Chen, G., Zhan, X., Hu, Z.: Adsorption of roxarsone by iron (hydr)oxide-modified multiwalled carbon nanotubes from aqueous solution and its mechanisms. Int. J. Environ. Sci. Technol. **11**(3), 785–794 (2014)
103. Liu, X., Liu, T., Zhou, H., Hu, W., Yao, B., Li, J., et al.: High-efficiency degradation of p-arsanilic acid and arsenic immobilization with iron encapsulated B/N-doped carbon nanotubes at natural solution pH. Sci. Total. Environ. **785**, 147152 (2021)
104. Fu, D., He, Z., Su, S., Xu, B., Liu, Y., Zhao, Y.: Fabrication of α-FeOOH decorated graphene oxide-carbon nanotubes aerogel and its application in adsorption of arsenic species. J. Colloid Interface Sci. **505**, 105–114 (2017)
105. Ke, M.K., Huang, G.X., Mei, S.C., Wang, Z.H., Zhang, Y.J., Hua, T.W., et al.: Interface-promoted direct oxidation of p-arsanilic acid and removal of total arsenic by the coupling of peroxymonosulfate and mn-fe-mixed oxide. Environ. Sci. Technol. **55**(10), 7063–7071 (2021)
106. Cao, S., Zhang, X., Huang, X., Wan, S., An, X., Jia, F., et al.: Insights into the facet-dependent adsorption of phenylarsonic acid on hematite nanocrystals. Environ. Sci. Nano **6**(11), 3280–3291 (2019)
107. Liu, K., Huang, Z., Dai, J., Jiang, Y., Yang, G., Liu, Y., et al.: Fabrication of amino-modified electrospun nanofibrous cellulose membrane and adsorption for typical organoarsenic contaminants: behavior and mechanism. Chem. Eng. J. **382**, 122775 (2020)
108. Acuña, K., Yañez, J., Suresh, R., Ramírez, E., Cuevas, J., Mansilla, H., et al.: Photocatalytic degradation of roxarsone by using synthesized ZnO nanoplates. Sol. Energy **157**, 335–341 (2017)
109. Liu, W., Ai, Z., Dahlgren, R.A., Zhang, L., Wang, X.: Adsorption and reduction of roxarsone on magnetic greigite (Fe3S4): indispensable role of structural sulfide. Chem. Eng. J. **330**, 1232–1239 (2017)
110. Zhu, S., Zhao, J., Zhao, N., Yang, X., Chen, C., Shang, J.: Goethite modified biochar as a multifunctional amendment for cationic Cd(II), anionic As(III), roxarsone, and phosphorus in soil and water. J. Clean. Prod. **247**, 119579 (2020)
111. Biswas, P., Wu, C.Y.: Nanoparticles and the environment. J. Air Waste Manag. Assoc. **55**(6), 708–746 (2005)

112. Synthesis of nanomaterials using various top-down and bottom-up approaches, influencing factors, advantages, and disadvantages: a review. Adv. Colloid Interface Sci. **300**, 102597 (2022)
113. Assaad, N., Sabeh, G., Hmadeh, M.: Defect control in Zr-based metal-organic framework nanoparticles for arsenic removal from water. ACS Appl. Nano Mater. **3**(9), 8997–9008 (2020)
114. Reddy, K.J., McDonald, K.J., King, H.: A novel arsenic removal process for water using cupric oxide nanoparticles. J. Colloid Interface Sci. **397**, 96–102 (2013)
115. Fu, W., Lu, D.L., Yao, H., Yuan, S., Wang, W., Gong, M., et al.: Simultaneous roxarsone photocatalytic degradation and arsenic adsorption removal by TiO_2/FeOOH hybrid. Environ. Sci. Pollut. Res. **27**(15), 18434–18442 (2020)
116. Jeon, I.Y., Baek, J.B.: Nanocomposites derived from polymers and inorganic nanoparticles. Materials. **3**(6), 3654–3674 (2010)
117. Sen, M.: Nanocomposite Materials [Internet]. Nanotechnology and the Environment. IntechOpen. Available from: https://www.intechopen.com/chapters/72636. (2020) [cited 9 May 2022]
118. Ashraf, S., Siddiqa, A., Shahida, S., Qaisar, S.: Titanium-based nanocomposite materials for arsenic removal from water: a review. Heliyon. **5**(5), e01577 (2019)
119. Siddiqui, S.I., Chaudhry, S.A.: Arsenic removal from water using nanocomposites: a review. Curr. Environ. Eng. **4**(2), 81–102 (2017)
120. Hornyak, G.L., Moore, J.J., Tibbals, H.F., Dutta, J.: Fundamentals of Nanotechnology, p. 786. CRC Press, Boca Raton (2018)
121. Ma, M.D., Wu, H., Deng, Z.Y., Zhao, X.: Arsenic removal from water by nanometer iron oxide coated single-wall carbon nanotubes. J. Mol. Liq. **259**, 369–375 (2018)
122. Chen, B., Zhu, Z., Ma, J., Yang, M., Hong, J., Hu, X., et al.: One-pot, solid-phase synthesis of magnetic multiwalled carbon nanotube/iron oxide composites and their application in arsenic removal. J. Colloid Interface Sci. **434**, 9–17 (2014)
123. Liu, H., Zuo, K., Vecitis, C.D.: Titanium dioxide-coated carbon nanotube network filter for rapid and effective arsenic sorption. Environ. Sci. Technol. **48**(23), 13871–13879 (2014)
124. Vadahanambi, S., Lee, S.H., Kim, W., Oh, I.K.: Arsenic removal from contaminated water using three-dimensional graphene-carbon nanotube-iron oxide nanostructures. Environ. Sci. Technol. **47** (2013)
125. Chen, R., Zhi, C., Yang, H., Bando, Y., Zhang, Z., Sugiur, N., et al.: Arsenic (V) adsorption on Fe_3O_4 nanoparticle-coated boron nitride nanotubes. J. Colloid Interface Sci. **359**(1), 261–268 (2011)
126. Arancibia-Miranda, N., Escudey, M., Pizarro, C., Denardin, J.C., García-González, M.T., Fabris, J.D., et al.: Preparation and characterization of a single-walled aluminosilicate nanotube-iron oxide composite: its applications to removal of aqueous arsenate. Mater. Res. Bull. **51**, 145–152 (2014)
127. Niu, H.Y., Wang, J.M., Shi, Y.L., Cai, Y.Q., Wei, F.S.: Adsorption behavior of arsenic onto protonated titanate nanotubes prepared via hydrothermal method. Microporous Mesoporous Mater. **122**(1), 28–35 (2009)
128. Jenewein, C., Cölfen, H.: Mesocrystals from platinum nanocubes. Nanomaterials **11**(8), 2122 (2021)
129. Yu, X., Wei, Y., Liu, C., Ma, J., Liu, H., Wei, S., et al.: Ultrafast and deep removal of arsenic in high-concentration wastewater: a superior bulk adsorbent of porous Fe_2O_3 nanocubes-impregnated graphene aerogel. Chemosphere **222**, 258–266 (2019)
130. Burt, J.L., Elechiguerra, J.L., Reyes-Gasga, J., Martin Montejano-Carrizales, J., Jose-Yacaman, M.: Beyond archimedean solids: star polyhedral gold nanocrystals. J. Cryst. Growth **285**(4), 681–691 (2005)
131. Pineda, T., Perales, O., Roman-Velazquez, F.: Effect of surface functionalization on the adsorption of arsenic using magnetite nanocrystals. Adsorpt. Sci. Technol. **31**, 807–822 (2013)
132. Xu, W.H., Wang, L., Wang, J., Sheng, G.P., Liu, J.H., Yu, H.Q., et al.: Superparamagnetic mesoporous ferrite nanocrystal clusters for efficient removal of arsenite from water. Cryst. Eng. Comm. **15**(39), 7895–7903 (2013)

133. Ji, Y., Qian, W., Yu, Y., An, Q., Liu, L., Zhou, Y., et al.: Recent developments in nanofiltration membranes based on nanomaterials. Chin. J. Chem. Eng. **25**(11), 1639–1652 (2017)
134. Pendergast, M.M., Hoek, E.M.V.: A review of water treatment membrane nanotechnologies. Energy Environ. Sci. **4**(6), 1946–1971 (2011)
135. Vandezande, P., Gevers, L.E.M., Vankelecom, I.F.J.: Solvent resistant nanofiltration: separating on a molecular level. Chem. Soc. Rev. **37**(2), 365–405 (2008)
136. Petersen, R.J.: Composite reverse osmosis and nanofiltration membranes. J. Membr. Sci. **83**(1), 81–150 (1993)
137. Min, L.L., Yuan, Z.H., Zhong, L.B., Liu, Q., Wu, R.X., Zheng, Y.M.: Preparation of chitosan based electrospun nanofiber membrane and its adsorptive removal of arsenate from aqueous solution. Chem. Eng. J. **267**, 132–141 (2015)
138. Chauhan, D., Dwivedi, J., Sankararamakrishnan, N.: Novel chitosan/PVA/zerovalent iron biopolymeric nanofibers with enhanced arsenic removal applications. Environ. Sci. Pollut. Res. **21**(15), 9430–9442 (2014)
139. Bahmani, P., Maleki, A., Daraei, H., Rezaee, R., Khamforoush, M., Dehestani Athar, S., et al.: Application of modified electrospun nanofiber membranes with α-Fe_2O_3 nanoparticles in arsenate removal from aqueous media. Environ. Sci. Pollut. Res. **26**(21), 21993–22009 (2019)
140. Talukder, M.E., Pervez, Md.N., Jianming, W., Gao, Z., Stylios, G.K., Hassan, M.M., et al.: Chitosan-functionalized sodium alginate-based electrospun nanofiber membrane for As (III) removal from aqueous solution. J. Environ. Chem. Eng. **9**(6), 106693 (2021)
141. Li, N., Xu, X., Sun, B., Xie, K., Huang, W., Yu, T.: Recent nanosheet-based materials for monovalent and multivalent ions storage. Energy Storage Mater. **25**, 382–403 (2020)
142. Zeng, J., Qi, P., Shi, J., Pichler, T., Wang, F., Wang, Y., et al.: Chitosan functionalized iron nanosheet for enhanced removal of As(III) and Sb(III): synergistic effect and mechanism. Chem. Eng. J. **382**, 122999 (2020)
143. Bangari, R.S., Yadav, V.K., Singh, J.K., Sinha, N.: Fe3O4-functionalized boron nitride nanosheets as novel adsorbents for removal of Arsenic(III) from contaminated water. ACS Omega **5**(18), 10301–10314 (2020)
144. Srivastava, R., Kommu, A., Sinha, N., Singh, J.K.: Removal of arsenic ions using hexagonal boron nitride and graphene nanosheets: a molecular dynamics study. Mol. Simul. **43**(13–16), 985–996 (2017)
145. Bangari, R.S., Singh, A.K., Namsani, S., Singh, J.K., Sinha, N.: Magnetite-coated boron nitride nanosheets for the removal of arsenic(v) from water. ACS Appl. Mater. Interfaces **11**(21), 19017–19028 (2019)
146. Koilraj, P., Takaki, Y., Sasaki, K.: Adsorption characteristics of arsenate on colloidal nanosheets of layered double hydroxide. Appl. Clay Sci. **134**, 110–119 (2016)
147. Liu, R., Liu, J.F., Zhang, L.Q., Sun, J.F., Jiang, G.B.: Low temperature synthesized ultrathin γ-Fe2O3 nanosheets show similar adsorption behaviour for As(III) and As(V). J. Mater. Chem. A. **4**(20), 7606–7614 (2016)
148. Njuguna, J., Ansari, F., Sachse, S., Zhu, H., Rodriguez, V.M.: 1-Nanomaterials, nanofillers, and nanocomposites: types and properties. In: Njuguna, J., Pielichowski, K., Zhu, H. (eds.) Health and Environmental Safety of Nanomaterials [Internet], p. 3–27. Woodhead Publishing. Available from: https://www.sciencedirect.com/science/article/pii/B9780857096555500015. (2014) [cited 12 May 2022]
149. Liang, Y., Jin, J., Xiang, Y., Wang, M., Xiong, J., Hou, J., et al.: Insights into the improving mechanism of defect-mediated As(V) adsorption on hematite nanoplates. Chemosphere **280**, 130597 (2021)
150. Peng, F., Luo, T., Yuan, Y.: Controllable synthesis of Mg–Fe layered double hydroxide nanoplates with specific Mg/Fe ratios and their effect on adsorption of As(V) from water. New J. Chem. **38**(9), 4427–4433 (2014)
151. Lin, Y.F., Chen, J.L., Xu, C.Y., Chung, T.W.: One-pot synthesis of paramagnetic iron(III) hydroxide nanoplates and ferrimagnetic magnetite nanoparticles for the removal of arsenic ions. Chem. Eng. J. **250**, 409–415 (2014)

152. Wang, J., Zhang, J., Chen, R., Yang, C., Xiang, L., Yi, M.: A vacuum calcination route to high-surface-area MgO nanoplates for superior arsenate adsorption and catalytic properties. Vacuum **158**, 231–235 (2018)
153. Moradlou, O., Dehghanpour Farashah, S., Masumian, F., Banazadeh, A.Z.: Magnetite nanoplates decorated on anodized aluminum oxide nanofibers as a novel adsorbent for efficient removal of As(III). Int. J. Environ. Sci. Technol. **13**(4), 1149–1158 (2016)
154. Siangproh, W., Chailapakul, O., Songsrirote, K.: Simple and fast colorimetric detection of inorganic arsenic selectively adsorbed onto ferrihydrite-coated silica gel using silver nanoplates. Talanta **153**, 197–202 (2016)
155. Acuña, K., Yáñez, J., Ranganathan, S., Ramírez, E., Pablo Cuevas, J., Mansilla, H.D., et al.: Photocatalytic degradation of roxarsone by using synthesized ZnO nanoplates. Sol. Energy **157**, 335–341 (2017)
156. Liu, G., Zheng, H., Jiang, Z., Zhao, J., Wang, Z., Pan, B., et al.: Formation and physicochemical characteristics of nano biochar: insight into chemical and colloidal stability. Environ. Sci. Technol. **52**(18), 10369–10379 (2018)
157. Ramanayaka, S., Vithanage, M., Alessi, D.S., Liu, W.J., Jayasundera, A.C.A., Ok, Y.S.: Nanobiochar: production, properties, and multifunctional applications. Environ. Sci. Nano **7**(11), 3279–3302 (2020)
158. Naghdi, M., Taheran, M., Brar, S., Rouissi, T., Verma, M., Surampalli, R., et al.: A green method for production of nanobiochar by ball milling- optimization and characterization. J. Clean. Prod. **164**, 1394–1405 (2017)
159. Jiang, M., He, L., Niazi, N.K., Wang, H., Gustave, W., Vithanage, M., et al.: Nanobiochar for the remediation of contaminated soil and water: challenges and opportunities. Biochar. **5**(1), 2 (2023)
160. Nath, B.K., Chaliha, C., Kalita, E.: Iron oxide permeated mesoporous rice-husk nanobiochar (IPMN) mediated removal of dissolved arsenic (As): chemometric modelling and adsorption dynamics. J. Environ. Manage. **246**, 397–409 (2019)
161. Yu, X., Han, X., Chang, C., Hu, Y., Xu, C. (Charles), Fang, S.: Corncob-derived activated carbon for roxarsone removal from aqueous solution: isotherms, kinetics, and mechanism. Environ. Sci. Pollut. Res. **27**(13), 15785–15797 (2020)
162. Wang, Y.J., Ji, F., Wang, W., Yuan, S.J., Hu, Z.H.: Removal of roxarsone from aqueous solution by Fe/La-modified montmorillonite. Desalination Water Treat. **57**(43), 20520–20533 (2016)
163. Zang, S., Zuo, Y., Wang, J., Liu, X., Gomez, M.A., Wei, L.: Adsorption removal of roxarsone, arsenite(III), and arsenate(V) using iron-modified sorghum straw biochar and its kinetics. Acta. Geochim. **40**(3), 409–418 (2021)
164. Joshi, T.P., Zhang, G., Koju, R., Qi, Z., Liu, R., Liu, H., et al.: The removal efficiency and insight into the mechanism of para arsanilic acid adsorption on Fe-Mn framework. Sci. Total. Environ. **601–602**, 713–722 (2017)
165. Wang, Y., Jiang, B., Wang, L., Feng, Z., Fan, H., Sun, T.: Hierarchically structured two-dimensional magnetic microporous biochar derived from hazelnut shell toward effective removal of p-arsanilic acid. Appl. Surf. Sci. **540**, 148372 (2021)
166. Yang, Y., Tao, S., Dong, Z., Xu, J., Zhang, X., Pan, G.: Adsorption of p-arsanilic acid on iron (hydr)oxides and its implications for contamination in soils. Minerals. **11**(2), 105 (2021)
167. Peng, Y., Wei, W., Zhou, H., Ge, S., Li, S., Wang, G., et al.: Iron humate as a novel adsorbent for p-arsanilic acid removal from aqueous solution. J. Dispers. Sci. Technol. **37**(11), 1590–1598 (2016)
168. Chen, W., Guo, G., Huang, L., Ouyang, L., Shuai, Q.: Facet-dependent adsorption of aromatic organoarsenicals on hematite: the mechanism and environmental impact. J. Hazard. Mater. **464**, 132976 (2024)
169. Joshi, T.P., Zhang, G., Jefferson, W.A., Perfilev, A.V., Liu, R., Liu, H., et al.: Adsorption of aromatic organoarsenic compounds by ferric and manganese binary oxide and description of the associated mechanism. Chem. Eng. J. **309**, 577–587 (2017)
170. Chen, W.R., Huang, C.H.: Surface adsorption of organoarsenic roxarsone and arsanilic acid on iron and aluminum oxides. J. Hazard. Mater. **227–228**, 378–385 (2012)

171. Xie, X., Cheng, H.: Adsorption and desorption of phenylarsonic acid compounds on metal oxide and hydroxide, and clay minerals. Sci. Total. Environ. **757**, 143765 (2021)
172. Cai, J., Wang, X., Zhou, Y., Jiang, L., Wang, C.: Selective adsorption of arsenate and the reversible structure transformation of the mesoporous metal–organic framework MIL-100(Fe). Phys. Chem. Chem. Phys. **18**(16), 10864–10867 (2016)
173. Li, B., Zhu, X., Hu, K., Li, Y., Feng, J., Shi, J., et al.: Defect creation in metal-organic frameworks for rapid and controllable decontamination of roxarsone from aqueous solution. J. Hazard. Mater. **302**, 57–64 (2016)
174. Tian, C., Zhao, J., Ou, X., Wan, J., Cai, Y., Lin, Z., et al.: Enhanced adsorption of p-arsanilic acid from water by amine-modified uio-67 as examined using extended x-ray absorption fine structure, x-ray photoelectron spectroscopy, and density functional theory calculations. Environ. Sci. Technol. **52**(6), 3466–3475 (2018)
175. Jung, B.K., Jun, J.W., Hasan, Z., Jhung, S.H.: Adsorptive removal of p-arsanilic acid from water using mesoporous zeolitic imidazolate framework-8. Chem. Eng. J. **267**, 9–15 (2015)
176. Zhu, K., Wu, J., Fan, R., Cao, Y., Lu, H., Wang, B., et al.: Selective adsorption and detection of p-arsanilic acid on MOF-on-MOF heterostructure induced by nitrogen-rich self-assembly template. Chem. Eng. J. **427**, 131483 (2022)
177. Lv, Y., Zhang, R., Zeng, S., Liu, K., Huang, S., Liu, Y., et al.: Removal of p-arsanilic acid by an amino-functionalized indium-based metal–organic framework: adsorption behavior and synergetic mechanism. Chem. Eng. J. **339**, 359–368 (2018)
178. Pang, D., Wang, C.C., Wang, P., Liu, W., Fu, H., Zhao, C.: Superior removal of inorganic and organic arsenic pollutants from water with MIL-88A(Fe) decorated on cotton fibers. Chemosphere **254**, 126829 (2020)
179. Xu, Y., Lv, J., Song, Y., Zhou, X., Tian, C., Hong, X., et al.: Efficient removal of low-concentration organoarsenic by Zr-based metal–organic frameworks: cooperation of defects and hydrogen bonds. Environ. Sci. Nano **6**(12), 3590–3600 (2019)
180. Li, Z., Liu, X., Jin, W., Hu, Q., Zhao, Y.: Adsorption behavior of arsenicals on MIL-101(Fe): the role of arsenic chemical structures. J. Colloid Interface Sci. **554**, 692–704 (2019)
181. Sarker, M., Song, J.Y., Jhung, S.H.: Adsorption of organic arsenic acids from water over functionalized metal-organic frameworks. J. Hazard. Mater. **335**, 162–169 (2017)
182. Jun, J.W., Tong, M., Jung, B.K., Hasan, Z., Zhong, C., Jhung, S.H.: Effect of central metal ions of analogous metal-organic frameworks on adsorption of organoarsenic compounds from water: plausible mechanism of adsorption and water purification. Chem. Weinh. Bergstr Ger. **21**(1), 347–354 (2015)
183. Lin, Z.J., Zheng, H.Q., Zeng, Y.N., Wang, Y.L., Chen, J., Cao, G.J., et al.: Effective and selective adsorption of organoarsenic acids from water over a Zr-based metal-organic framework. Chem. Eng. J. **378**, 122196 (2019)
184. Ding, W.Q., Labiadh, L., Xu, L., Li, X.Y., Chen, C., Fu, M.L., et al.: Current advances in the detection and removal of organic arsenic by metal-organic frameworks. Chemosphere **339**, 139687 (2023)
185. Tchinsa, A., Hossain, M.F., Wang, T., Zhou, Y.: Removal of organic pollutants from aqueous solution using metal organic frameworks (MOFs)-based adsorbents: a review. Chemosphere **284**, 131393 (2021)
186. Yang, T., Liu, Y., Wang, L., Jiang, J., Huang, Z., Pang, S.Y., et al.: Highly effective oxidation of roxarsone by ferrate and simultaneous arsenic removal with in situ formed ferric nanoparticles. Water Res. **147**, 321–330 (2018)
187. Fan, J., Ding, Z., Zhao, Z., Liu, J.: Degradation of p-arsanilic acid and simultaneous in-situ removal of arsenic species with ferrate(VI): kinetics, intermediate and degradation pathway. Chem. Eng. J. **350**, 453–462 (2018)
188. Yang, T., Wang, L., Liu, Y., Jiang, J., Huang, Z., Pang, S.Y., et al.: Removal of organoarsenic with ferrate and ferrate resultant nanoparticles: oxidation and adsorption. Environ. Sci. Technol. **52**(22), 13325–13335 (2018)
189. Xie, X., Cheng, H.: A simple treatment method for phenylarsenic compounds: oxidation by ferrate (VI) and simultaneous removal of the arsenate released with in situ formed Fe(III) oxide-hydroxide. Environ. Int. **127**, 730–741 (2019)

190. Xu, C., Luo, M., Zhao, J., Yin, J., Feng, C., Zhang, H., et al.: Ferrate(VI) combined with sulfur(IV) as an effective process for aromatic organoarsenic degradation: essential role of Fe(III) flocs. China Chem. Lett. 111538 (2025)
191. Kong, Y., Lu, F., Huang, Z., Nie, Y., Chen, Z., Ma, J.: Aluminum chloride induced ferrate activation for p-arsanilic acid removal: The significant synergism mechanism and enhanced role of reactive Fe(IV)/Fe(V). Sep. Purif. Technol. **343**, 127121 (2024)
192. Peng, Q., Xu, W., Qi, W., Hu, C., Liu, H., Qu, J.: Removal of p-arsanilic acid and phenylarsonic acid from water by Fenton coagulation process: influence of substituted amino group. Environ. Sci. Pollut. Res. Int. **28**(44), 63319–63329 (2021)
193. Qin, J., Li, H., Lin, C.: Fenton process-affected transformation of roxarsone in paddy rice soils: effects on plant growth and arsenic accumulation in rice grain. Ecotoxicol. Environ. Saf. **130**, 4–10 (2016)
194. Xie, X., Hu, Y., Cheng, H.: Rapid degradation of p-arsanilic acid with simultaneous arsenic removal from aqueous solution using Fenton process. Water Res. **89**, 59–67 (2016)
195. Zhang, A.Y., Huang, N.H., Zhang, C., Zhao, P.C., Lin, T., He, Y.Y., et al.: Heterogeneous Fenton decontamination of organoarsenicals and simultaneous adsorption of released arsenic with reduced secondary pollution. Chem. Eng. J. **344**, 1–11 (2018)
196. Hu, P., Liu, Y., Jiang, B., Zheng, X., Zheng, J., Wu, M.: High-efficiency simultaneous oxidation of organoarsenic and immobilization of arsenic in fenton enhanced plasma system. Ind. Eng. Chem. Res. **54**(33), 8277–8286 (2015)
197. Lunge, S., Singh, S., Sinha, A.: Magnetic iron oxide (Fe_3O_4) nanoparticles from tea waste for arsenic removal. J. Magn. Magn. Mater. **356**, 21–31 (2014)
198. Chen, Y., Lin, C., Zhou, Y., Long, L., Li, L., Tang, M., et al.: Transformation of roxarsone during UV disinfection in the presence of ferric ions. Chemosphere **233**, 431–439 (2019)
199. Chen, N., Wan, Y., Zhan, G., Wang, X., Li, M., Zhang, L.: Simulated solar light driven roxarsone degradation and arsenic immobilization with hematite and oxalate. Chem. Eng. J. **384**, 123254 (2020)
200. Zhou, J., Zhao, Z., Wang, Y., Ding, Z., Xu, X., Peng, W., et al.: BiOCl0.875Br0.125/ polydopamine functionalized PVDF membrane for highly efficient visible-light-driven photocatalytic degradation of roxarsone and simultaneous arsenic immobilization. Chem. Eng. J. **402**, 126048 (2020)
201. Tyutereva, Y.E., Sherin, P.S., Polyakova, E.V., Koscheeva, O.S., Grivin, V.P., Plyusnin, V.F., et al.: Photodegradation of para-arsanilic acid mediated by photolysis of iron(III) oxalate complexes. Chemosphere **261**, 127770 (2020)
202. Xu, J., Wu, Y., Ma, M., Luo, T., Xia, J., Zhang, X.: A novel transformation pathway of p-arsanilic acid in water by colloid ferric hydroxide under UVA light. Environ. Sci. Pollut. Res. Int. **29**(4), 5043–5051 (2022)
203. Czaplicka, M., Bratek, Ł, Jaworek, K., Bonarski, J., Pawlak, S.: Photo-oxidation of p-arsanilic acid in acidic solutions: kinetics and the identification of by-products and reaction pathways. Chem. Eng. J. **243**, 364–371 (2014)
204. Czaplicka, M., Jaworek, K., Bąk, M.: Study of photodegradation and photooxidation of p-arsanilic acid in water solutions at pH = 7: kinetics and by-products. Environ. Sci. Pollut. Res. **22**(21), 16927–16935 (2015)
205. Zheng, S., Jiang, W., Cai, Y., Dionysiou, D.D., O'Shea, K.E.: Adsorption and photocatalytic degradation of aromatic organoarsenic compounds in TiO_2 suspension. Catal. Today **224**, 83–88 (2014)
206. Nakajima, T., Xu, Y.H., Mori, Y., Kishita, M., Takanashi, H., Maeda, S., et al.: Combined use of photocatalyst and adsorbent for the removal of inorganic arsenic(III) and organoarsenic compounds from aqueous media. J. Hazard. Mater. **120**(1), 75–80 (2005)
207. Nasir, A., Khalid, S., Yasin, T., Mazare, A.: A review on the progress and future of TiO_2/ graphene photocatalysts. Energies **15**(17), 6248 (2022)
208. Kumbhakar, P., Pramanik, A., Biswas, S., Kole, A.K., Sarkar, R., Tiwary, C.S., et al.: 7-Green and sustainable methods of syntheses of photocatalytic materials for efficient application in dye degradation. In: Shah, M., Dave, S., Das, J. (eds.) Photocatalytic Degradation of Dyes

[Internet], p. 167–206. Elsevier. Available from: https://www.sciencedirect.com/science/article/pii/B9780128238769000044. (2021) [cited 1 Jan 2025]
209. Guzmán-Fierro, V.G., Moraga, R., León, C.G., Campos, V.L., Smith, C., Mondaca, M.A.: Isolation and characterization of an aerobic bacterial consortium able to degrade roxarsone. Int. J. Environ. Sci. Technol. **12**(4), 1353–1362 (2015)
210. Chen, J., Rosen, B.P.: Organoarsenical biotransformations by shewanella putrefaciens. Environ. Sci. Technol. **50**(15), 7956–7963 (2016)
211. Liu, D.F., Huang, X.N., Cheng, R., Min, D., Cheng, L., Zhao, F., et al.: Anaerobic respiration on nitarsone in aquatic environments by shewanella oneidensis MR-1 Lacking Known C·As lyases. ACS EST Water. (2020)
212. Zhang, W., Xu, F., Han, J., Sun, Q., Yang, K.: Comparative cytotoxicity and accumulation of Roxarsone and its photodegradates in freshwater Protozoan Tetrahymena thermophila. J. Hazard. Mater. **286**, 171–178 (2015)
213. Lateef, A., Darwesh, O.M., Matter, I.A.: Microbial nanobiotechnology: the melting pot of microbiology, microbial technology and nanotechnology. In: Lateef, A., Gueguim-Kana, E.B., Dasgupta, N., Ranjan, S. (eds.) Microbial Nanobiotechnology: Principles and Applications [Internet], p. 1–19. Springer, Singapore. Available from: https://doi.org/10.1007/978-981-33-4777-9_1. (2021) [cited Jan 1 2025]

Energy Storage Materials

High Grade Carbon Extraction from Tire Waste Upcycling for Energy Storage Capabilities

Himanshu Priyadarshi, Sivasankar Kakku, and Abhishek Sharma

Abstract This work presents a method for synthesizing tire char nanoparticles for use in electrodes meant for energy storage devices in pursuance of sustainable development goals. Empirical exploration has been carried out with a sharp emphasis on sustainability. A detailed empirical analysis explores the morphological, thermogravimetric, and optical characteristics of these nanoparticles, which are produced using a simple synthesis process. Characterization studies confirm the presence of graphitic peaks and a high carbon content. The nanoparticles (with size of ~ 20 nm) have been identified through crystallographic diffraction techniques and scanning electron microscopy image processing. Spectroscopic analysis reveals a band gap (in the tune of 3–4 eV), indicating potential conductivity improvements. Thermogravimetric analysis provides insights into the thermal stability of the samples. These findings support the suitability of the synthesized nanoparticles as active electrode materials for sustainable energy storage applications.

1 Introduction

Long-term energy security and the required planning for the same is essential for any society aiming for sustainable progress. A society's sovereignty and well-being are dependent on its ability to meet its energy demands locally. Since energy is the ability to do work, to ensure that the societal goals are accomplished, it is mandatory that every society is self-sufficient when it comes to achieving sufficient reserves of energy. Unfortunately, many countries remain heavily reliant on fossil fuels and

H. Priyadarshi
Department of Electrical Engineering, Manipal University Jaipur, Jaipur, India

S. Kakku · A. Sharma (✉)
Waste to Resources Laboratory, Department of Biotechnology and Chemical Engineering, Manipal University Jaipur, Jaipur, India
e-mail: abhishek.sharma@jaipur.manipal.edu

A. Sharma
Department of Chemical Engineering, BITS Pilani K. K. Birla Goa Campus, Goa, India

L. Ledwani and S. Srivastava (eds.), *Experimental and Computational Advances in Materials*, Engineering Materials, https://doi.org/10.1007/978-981-95-4180-5_6

do encourage drilling unabashedly, which has led to significant loss of life and a decline in quality of life. Ever-increasing awareness of the physiological, economic, ecological, and environmental consequences of abject dependency on fossil fuels can be understood from the standard texts comparing it to Frankenstein's monster. Sustainable energy storage solutions offer advantages at multiple levels which may be enumerated as follows: (i) demand-side energy management, (ii) optimal utilization of energy assets, (iii) energy self-reliance, (iv) full utilization of existing power grid infrastructure, and (v) cost savings by avoiding new infrastructure investments. The total power utilization by the energy consumer varies based on total connected load. Electricity distribution policies do encourage time-segregated consumption, with peak hours designated for essential applications, while non-essential consumption is discouraged. During off-peak hours, when vital applications are not being operational, discretionary domestic energy utilization is recommended. Prudent implementation of such demand-side management is only feasible with energy storage systems which are capable of buffering energy efficiently. Self-sufficiency, reliability, and security in energy assets are vital for economic growth. Energy pricing fluctuates due to sources and intermediaries involved in energy generation, transmission and distribution. While traditional power grids have limited ability to leverage energy price variations, modern smart grids, equipped with data-driven intelligence, can optimize energy pricing strategies, making intelligent decisions to maximize benefits. Historically, balancing supply–demand fluctuations relied on high-inertia alternators, but as these become obsolete, energy storage technologies are taking on this role. The ability to time-shift power consumption and supply ensures grid stability, even during blackouts. Therefore, the success of decentralized renewable energy systems in ensuring reliable power supply is intrinsically linked to advancements in energy storage technology.

2 Context of Sustainable Technology

Technological advancements are intended to enhance quality and longevity of all life forms within the compass and territory of Mother Earth. However, being driven by hedonistic psychology, it has often led to detrimental consequences rather than benefits. The hazardous effects of harmful but apparently innocuous technologies have resulted in numerous man-made disasters that continue to impact society. A prime example is the widespread use of petroleum-powered internal combustion (IC) engine vehicles. Despite its deep integration into modern supply chains and logistics, the negative impact of this technology is undeniable. The hazards of petroleum technology extend across its entire lifecycle, from extraction to emissions. Some of the key concerns are as follows: (i) tectonic plate and earthquakes' correlation with petroleum drilling; (ii) petroleum refineries cause significant environmental waste which is quite challenging to manage; (iii) countries with their economies dependent on petroleum face instability due to fluctuating crude oil prices; (iv) oil-rich regions, such as the Middle East, have been parched by conflicts; and (v) emissions from petroleum-powered vehicles contribute to air pollution, making urban air quality hazardous in densely populated areas.

It is reasonably questionable why previous generations of researchers failed to prognose the harmful effects of petroleum-powered transportation before widely advocating this technology. Petroleum has become so convolutedly interdigitated in daily life that fluctuations in its price directly impact the necessities of human life. Technology with such widescale negative consequences can hardly be considered conducive to human or environmental well-being. A truly sustainable technology is one which enables societal progress without unintended hazards.

Understanding these hazards requires examining the concept of corporate social responsibility (CSR) through the lens of the quadruple bottom line (QBL). Large-scale technologies are often elevated to the scale of global mass adoption due to the influence of corporate conglomerates which shape global policies. Scientific concerns regarding the environment and sustainability are often ignored due to myopic pursuit of profitability. Following the first industrial revolution (Industry 1.0), such corporations prioritized short-term profitability, overlooking sustainability, which led to severe environmental and social repercussions.

In recent times, forward-thinking corporate leaders have recognized that long-term profitability cannot come at the expense of core values like sustainability and respect for divinity of life. This realization has led to the establishment of the quadruple bottom line, which considers four key elements: (i) the planet, which provides vital resources; (ii) people, including both service providers and consumers; (iii) the purpose of the business, which should align with societal well-being; and (iv) profitability, which must be achieved sustainably. By adopting the holistic approach of QBL for CSR, business units can contribute to technological advancements that support both sustainable economic growth and environmental responsibility.

The 17 SDGs identified by United Nations in 2015 start with eradication of poverty (SDG1). Poverty in any corner of the world is a big threat to international peace and security, just as pain in any part of the body destroys the proper functioning of the mind. The population below poverty line must be fed properly and employed gainfully so that nobody suffers due to hunger or malnutrition. Zero hunger and adequate nutrition for all can be ensured only when (SDG2) practices of sustainable agriculture are scientifically adopted, promoted and canvased in a top-down approach from governance to governed. Scientific environment friendly agriculture will ensure good health and well-being for the inhabitants of Planet Earth (SDG3). Once the members at the wrong side of the poverty index threshold are taken care of to cross that threshold, it is equally important to ensure that they are progressive in the trajectories of their lives. This requires an attitude of lifelong learning which can be imparted by ensuring inclusive and equitable quality education (SDG4). Since the first source of education for a child is the mother, hence there has to be women empowerment (SDG5), if a society wants to reap the benefits of equitable and inclusive quality education amongst the inhabitants of Planet Earth. Planet Earth and most of the living beings as well nearly three-fourth water with respect to their composition, however availability of clean water with systematic water management (SDG6), along with affordable, reliable, sustainable and clean energy(SDG7) has emerged as one of the biggest challenges to be tackled with. Affordability of sustainable energy ensures that the populace in general has the opportunity of getting engaged in decent opportunities of work (SDG8). Decent avenues of employment ensure that the populace have sufficient purchase capacity, which will in turn, ensure that the demand–supply cycle is kept gyrational at a healthy speed of economic well-being (book by Sir John Maynard Kaynes). Promoting a culture of innovation with adaptive industrialization is crucial for generating employment opportunities. It is certainly true that innovation has to be cultured in the mindset of the intellectually inclined section of society, the responsibility of ensuring equitable industrialization and adaptive infrastructure wherein innovation can prosper (SDG9) lies with the government. Once there are opportunities, it also needs to be ensured that people from different sections of society have equal access to the employment opportunities commensurate to their qualifications in a meritocratic way (SDG10). The outcomes of this experiment design may be mapped with the following sustainable development goals:

S. No.	Sustainable development goal (SDG) in running head phrase	SDG	Correlation
1	No poverty	End poverty in all its forms everywhere	Direct
2	Zero hunger	End hunger, achieve food security and improved nutrition and promote sustainable agriculture	Direct
3	Good health and well-being	Ensure healthy lives and promote well-being for all at all ages	Direct
4	Quality education	Ensure inclusive and equitable quality education and promote lifelong learning opportunities for all	Direct
5	Gender equality	Achieve gender equality and empower all women and girls	Direct
6	Clean water and sanitation	Ensure availability and sustainable management of water and sanitation for all	Direct
7	Affordable and clean energy	Ensure access to affordable, reliable, sustainable and modern energy for all	Direct
8	Decent work and economic growth	Promote sustained, inclusive and sustainable economic growth, full and productive employment and decent work for all	Direct
9	Industry, innovation and infrastructure	Build resilient infrastructure, promote inclusive and sustainable industrialization and foster innovation	Direct
10	Reduced inequalities	Reduce inequality within and among countries	Direct
11	Sustainable cities and communities	Make cities and human settlements inclusive, safe, resilient and sustainable	Direct

(continued)

(continued)

S. No.	Sustainable development goal (SDG) in running head phrase	SDG	Correlation
12	Responsible consumption and production	Ensure sustainable consumption and production patterns	Direct
13	Climate action	Take urgent action to combat climate change and its impacts	Direct
14	Life below water	Conserve and sustainably use the oceans, seas and marine resources for sustainable development	Direct
15	Life on land	Protect, restore and promote sustainable use of terrestrial ecosystems, sustainably manage forests, combat desertification, halt and reverse land degradation, and halt biodiversity loss	Direct
16	Peace and justice strong institutions	Promote peaceful and inclusive societies for sustainable development, provide access to justice for all and build effective, accountable and inclusive institutions at all levels	Direct
17	Partnerships for the goals	Strengthen the means of implementation and revitalize the global partnership for sustainable development	Direct

3 Choice of Precursors for Sustainable Energy Storage Solutions

Researchers are working vigorously to explore the potential of synthesizing different components of energy storage applications by using industrial waste materials for sustainable production. Efficient waste management not only enhances profitability but also ensures compliance with environmental regulations, avoiding financial penalties and reputational damage. Various waste materials, including lignin (a by-product of cellulose refineries, paper and pulp industries), coal, agricultural residues, fungal waste, and used coffee grounds, have been identified as viable sources for contributing to different components of energy storage devices. A closer examination reveals that most of these waste materials are good starting point for deriving activated carbon which can be used for energy storage applications.

Mahmood et al. [1] recommend laser-scribing as a frugal technique for producing high-quality graphene from lignin (abundantly available by-product of cellulose biorefineries and the paper & pulp industry). Similarly, Chu et al. [2] emphasize the economic and environmental congeniality of producing highly porous graphene from crude polysaccharides, such as fungal waste, using a three-electrode system. This methodology tackles critical energy storage challenges, particularly in achieving rapid energy retrieval from sources to sinks. Advantageously, research [3] indicates that simple treatment of coconut shells can provide porous carbon suitable for ultracapacitor electrodes. By finetuning activation parameters (such as temperature and water flow rate) energy storage empiricists and enthusiasts have achieved enhanced conductivity, higher energy density, and durable cyclic stability, snubbing the need for additional conductivity-enhancing modifications.

Raw materials for energy storage device fabrication significantly impact end-user affordability [4, 5, 6,7, 8,9]. Securing high-quality precursors remains a major challenge. However, with the current emphasis on sustainable development goals and dependence on indigenous resources, waste-to-wealth conversion has become increasingly feasible through integrated problem-solving platforms. Research contributions on converting industrial waste into valuable energy materials are vital as these offer multiple benefits such as environmental protection, cost reduction, and better life-cycle management. By adopting this approach, industries can transform loss-making ventures into profitable operations with minimal investment in waste management simultaneously strengthening their corporate social responsibility (CSR) compliance through the quadruple bottom line framework.

Porous activated carbon is an important component for electrodes for the unit cell in energy storage devices [10, 11]. Carbon is one of the main components of agricultural waste, such as rice husk, black gram husk, and green gram husk [2, 12, 13, 14], which constitute nearly one-third of total crop yield. These materials, abundantly available in Asia, are often left unutilized beyond their occasional use as cattle feed, bio-fertilizer, or fuel. Many times, these are burnt very irresponsibly, causing smog and related health hazards for the whole population. On the brighter side, they present a promising opportunity for activated carbon production. Through adsorption techniques, such activated carbon can be engineered into electrodes for unit cell of energy storage device applications. Once these electrodes reach the end of their lifecycle, they can be repurposed for wastewater treatment through phenol adsorption on reactivated carbon, ensuring zero sludge formation. By effectively utilizing waste materials, industries can achieve profitability, aligning with sustainable energy solutions and advancing circular economy principles, without endangering the environment.

Another avenue for waste valorization is tire waste. Vehicle tires play an important role in modern civilization by facilitating transportation and the swift delivery of goods. With rampant industrialization, tire production has surged significantly over years. A study estimates that by 2027, the global tire industry will manufacture over 2.7 billion tires to cater to the market demand. Typically, tires have a lifespan of about four years, after which they are dumped, occupying vast amounts of land. With the increasing transportation, the volume of discarded tires will continue to rise. Annually, approximately 5 million tons of tires require proper disposal at a global level. Research indicates that currently, four billion tires are stockpiled in landfills, which is projected to reach 50 billion by 2030. China is accountable for 13 million tons of waste tires annually, with a growth rate of 6–8%, while Canada causes 42,118 tons of waste tires in 2019. Tires are manufactured from vulcanized rubber with sulfur bonds. This makes them highly averse to physical and chemical changes under normal environmental conditions. This non-biodegradable nature ensures durability but comes with a flip side of environmental pollution. Since three-fourth of each tire consists of void space, landfilling of waste tires is a major issue. Besides land consumption, discarded tires pose serious environmental and health risks. When left in open spaces, a tire becomes breeding space for insects, pests, and rodents. In places which lack awareness, stagnant water in dumped tires provides ideal conditions for mosquitoes, spreading diseases such as malaria. From a carbon-emission perspective, the production of a single tire releases 330 g of CO_2, contributing significantly to climate change. In many developing countries, millions of tires are burned daily to clear landfill space, causing severe air pollution [15, 16, 17].

Industries have been contemplating and exercising various techniques to mitigate the environmental impact of waste tires. While many methods transform tires into different physical forms, they do not eliminate them from the environment. Prevalent recycling and reuse techniques include tire retreading, rubber reclaiming, grinding, and mechanical lapping. Retreading may be useful for heavy vehicle tires, such as truck tires, extending their lifespan. Crumb rubber, a shredded tire product, may be used to manufacture reclaimed rubber. Since it releases toxic emissions, incineration must be avoided. On a hopeful note, pyrolysis has been reported to thermally decompose tires into oil, char, and gas without harmful emissions. Thus pyrolysis has been hailed as a promising sustainable solution.

Pyrolysis has been investigated for upcycling of waste tires into energy storage precursor materials, one of which is char. Pyrolyzed tire char has multiple applications which include the following: conversion into activated carbon (AC), water treatment, heavy metal removal, and energy storage [18, 19, 20, 21]. The properties of AC depend on the following two factors: (i) raw material and (ii) operating conditions, including temperature, residence time, heating rate, and reactor type [22]. Similarly, factors such as activation temperature, reagents, and impregnation ratio influence AC quality and yield.

4 Synthesis and Characterization of Tire Char Nanoparticles for Supercapacitor Applications

The synthesis of char nanoparticles has been done through following algorithm, which includes the following stages:

1. Grinding—Tire-char obtained from the waste-tire upcycling process has been finely ground.
2. Ultrasonication—The grinded tire char was subjected to ultrasonication using a high-energy homogenizer to enhance dispersion.
3. Solvent Dispersion—A precise amount of char was dispersed in a suitable solvent and agitated using a magnetic stirrer for 45 min at a controlled process temperature.

4. Surfactant Addition—A surfactant was introduced into the dispersion, followed by additional magnetic stirring for 30 min under the same temperature conditions.
5. Graphitic Nanoparticle Formation—The solution underwent further ultrasonication for 30 min to induce the formation of char-derived graphitic nanoparticles.
6. Centrifugation—The ultrasonicated product was placed in a centrifuge and spun for 15 min to separate denser nanoparticles from the lighter supernatant liquid.
7. Precipitate Collection—The supernatant was carefully pipetted out, leaving behind the settled nanoparticles at the bottom of the vial.
8. Filtration and Purging—The obtained sediment was filtered and treated with acetone to remove impurities and ensure sample purity.
9. Drying and Powdering—The purified precipitate was dried in a vacuum oven at 373 K for two hours and then finely powdered.
10. Cooling—After vacuum desiccation, the sample was allowed to cool under standard temperature and pressure (STP) conditions for one hour.
11. Final Grinding—The cooled sample was transferred to a clean mortar and ground further to obtain a uniform powder for subsequent investigations.

This systematic approach ensures the synthesis of high-quality char-derived nanoparticles with controlled properties for further analysis.

The powdered samples were characterized (Fig. 1) using powder X-ray diffraction (PXRD) with Rigaku's automated multipurpose X-ray Diffractometer (SMARTLAB model). This advanced system supports high-end material science research across various sample types, including powders, thin films, and bulk solids with amorphous, nano-crystalline, and polycrystalline structures. It is equipped with a Photon Max high-flux 9 kW rotating anode X-ray source and a HyPix-3000 high-energy, high-resolution multidimensional semiconductor detector. The HyPix-3000 supports 0D, 1D, and 2D measurement modes, eliminating the need for multiple detectors and facilitating seamless data collection. This detector provides two-dimensional powder diffraction patterns for comprehensive qualitative analysis. The system also includes a high-resolution feedback-based goniometer drive with a planar diffraction arm and fully automated transmission and reflection optics, enabled by its crossbeam optics technology. The instrument operates using para-focusing Bragg – Brentano geometry at a rated voltage of 40 kV and a current of 40 mA. Lattice parameters were determined using diffraction peaks (002) and (101), while the mean crystallite diameter was calculated using the Scherrer equation.

For further analysis, field emission scanning electron microscopy (FESEM) was utilized to examine surface morphology and elemental composition without requiring sample insulation. The JSM-7610FPlus FESEM (JEOL) is an ultra-high-resolution electron microscope equipped with semi-in-lens detectors and an in-the-lens Schottky field emission gun. This allows for high-resolution imaging across a range of probe currents, enabling the visualization of fine structures in diverse samples. The instrument's chamber is integrated with an Energy Dispersive X-ray

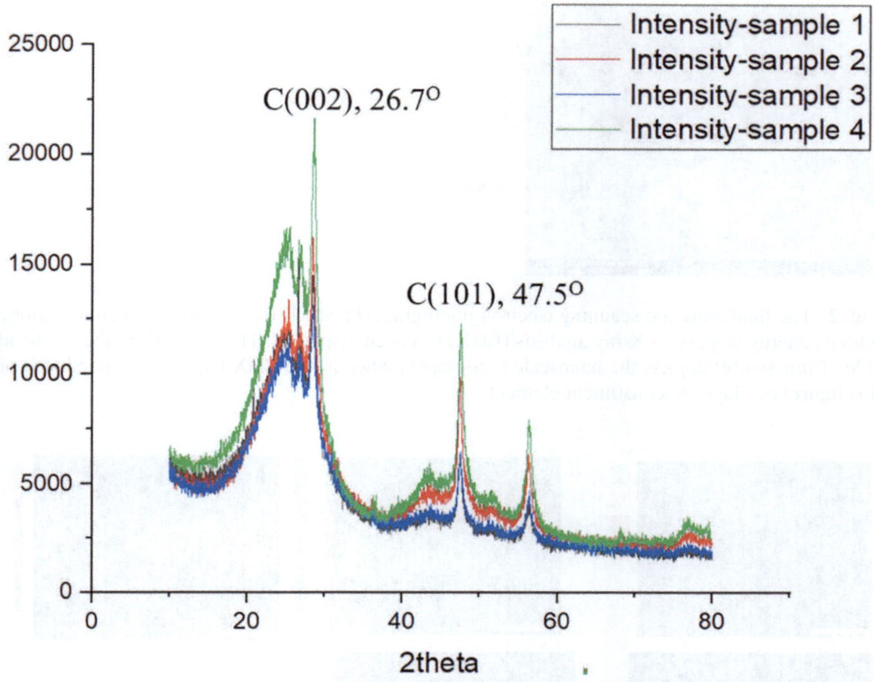

Fig. 1 The diffractograms of the sample specimens of tire-derived char nanoparticles derived from X-ray diffractometry techniques of the powder specimens reveal the presence of graphitic carbon

(EDAX) analyzer for elemental analysis, enhanced by its high probe current capability. High-power optics ensure a minimal probing diameter even at high probe currents.

During characterization, the gentle beam mode was applied to lower the landing voltage of electrons, reducing surface damage, charge accumulation, and beam-related artifacts. Energy-filtered imaging was employed for selective imaging of back-scattered and secondary electrons. The eucentric specimen stage minimized focus deviations while rotating, tilting, or shifting the sample. The JSM-7610FPlus FESEM provides exceptional spatial resolution, reaching 1 nm at 1 kV and 0.8 nm at 15 kV. The FESEM image has been displayed in Fig. 2, and the element-wise constitution has been depicted in Fig. 3.

Sample preparation for FESEM was supported by several auxiliary techniques:

1. Grinding and Polishing—meant for precise processing of nanocomposites.
2. Low-Speed Precision Cutting—applied to brittle or ductile materials such as electronic components and biomaterials, ensuring minimal deformation.
3. Automated Rotary-Pumped Sputtering Coater—used for gold, platinum, or carbon coating of non-conductive samples.

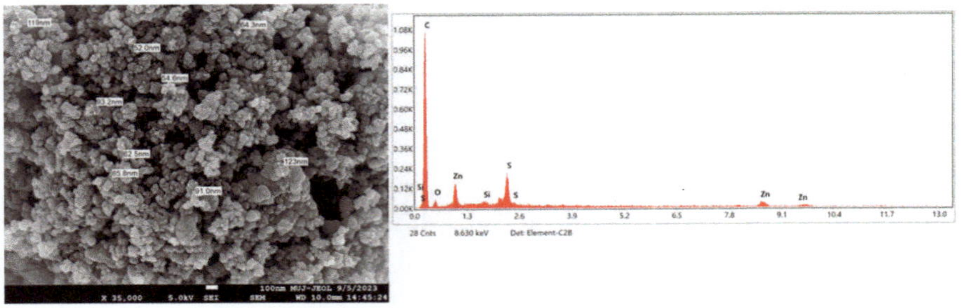

Fig. 2 The field emission scanning electron micrograph(FESEM) for one of the specimens along with its energy dispersive X-ray analysis (EDX) has been displayed. The FESEM (in the left hand side of this figure) depicts the nanoscale topography, whereas the EDX (in the right hand side of this figure) displays the constituent elements

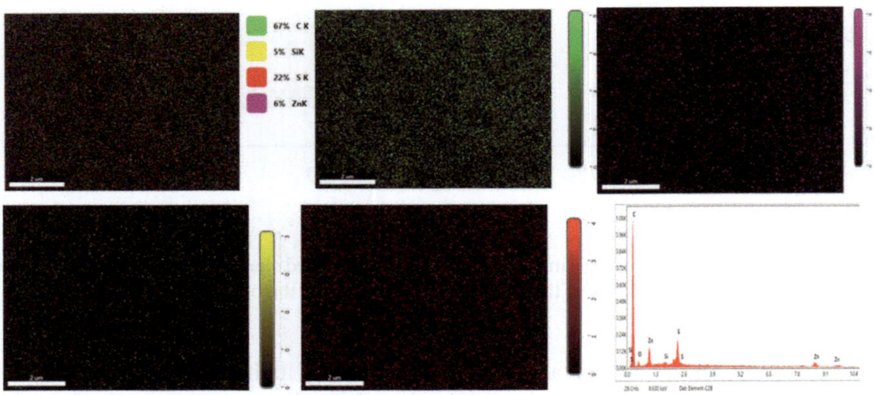

Fig. 3 The colored heatmap for elemental composition of the sample specimen displayed in figure 1 and 2. This specimen of the tire char nanoparticles has 67% carbon content, 22 percent sulfur, 6% zinc, and 5% silicon

These capabilities make the system highly suited for nanoscale imaging in material science research.

For spectral analysis, the sample's absorbance properties were examined using UV–Vis (ultraviolet–visible) spectroscopy. A 75 mg portion of the sample was dissolved in 100 mL of dimethylformamide (DMF), a suitable solvent, followed by sonication. The absorbance spectrum of the resultant solution was then compared with that of the solvent to analyze its optical characteristics. Based on the characterization results depicted in Fig. 4, it can be reasonably inferred that char nanoparticles and magnesium-doped silicon nanoparticles have been successfully synthesized. The nanoscale samples were analyzed using X-ray diffraction (XRD) to determine phase purity and estimate average particle size. Prior to FESEM characterization, the samples were gold-coated to prevent charging effects and potential sample damage.

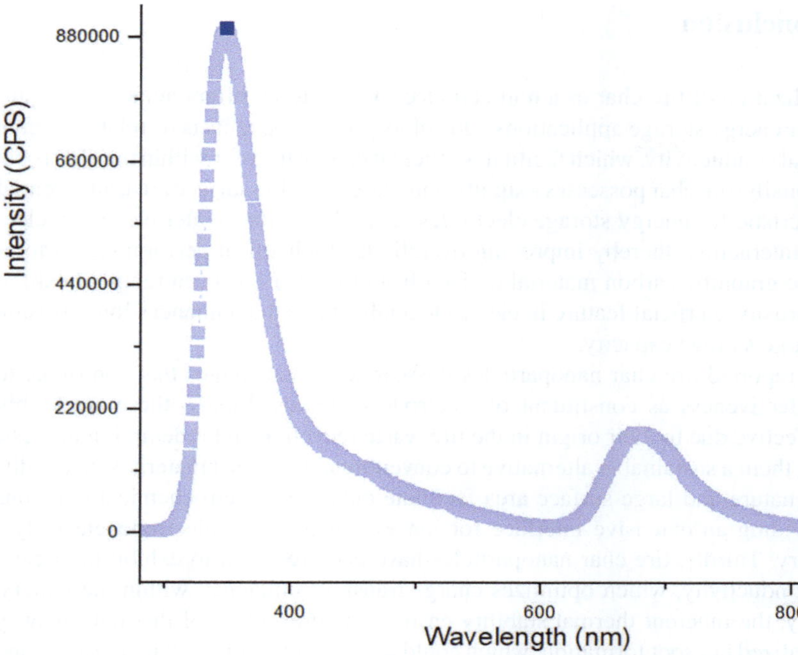

Fig. 4 The optical characterization also confirms high graphitization for conductivity modulation as is evident from the peak in this graph

This gold coating also ensures accurate topographical imaging, as FESEM relies on secondary electron emissions for image formation.

For spectral band-gap estimation, the sample was diluted in dimethylformamide (DMF) and analyzed accordingly. The crystalline phase integrity and average crystallite size were assessed through XRD. To reinforce the XRD-based size estimation, the results were compared with image-processing-derived size calculations from FESEM micrographs. Further validation can be obtained by cross-referencing these findings with transmission electron microscopy (TEM) images. The nanoscale dimensions confirm the potential applicability of these nanoparticles as electrode materials, with approximately 4% conductive carbon incorporated into the electrode.

Direct thermal analysis was conducted to determine the stoichiometric composition of electrode particles across varying temperatures. The samples are thermally quite stable and display weight loss of only 0.3% at 100 degree centigrade and nearly 9% up to 700 degree centigrade. Additionally, the surface area of the sample was measured using nitrogen adsorption isotherms, following appropriate purification through degassing under controlled vacuum conditions, temperature, and duration.

5 Conclusion

The utilization of tire char as a unit cell electrode material offers numerous advantages for energy storage applications. One of its primary benefits is its relatively high electrical conductivity, which facilitates efficient charge transfer within the electrode. Additionally, tire char possesses a significantly large specific surface area, an essential characteristic for energy storage electrodes, as it allows for greater electrode–electrolyte interaction, thereby improving overall electrochemical performance. Moreover, the graphitic carbon material derived from tire char has been reported exhibit high porosity, a crucial feature in electrode applications that enhances ion diffusion and charge storage capacity.

The reported tire char nanoparticles possess several attributes that contribute to their effectiveness as constituent of electrode materials. Firstly, these are highly cost-effective due to their origin in the tire waste recycling and redemption process, making them a sustainable alternative to conventional high-cost materials. Secondly, porous nature and large surface area facilitate enhanced electrochemical reactions by providing an extensive interface for ion exchange at the electrode–electrolyte boundary. Thirdly, tire char nanoparticles have been reported to exhibit high electrical conductivity, which optimizes charge transport efficiency within the battery. Fourthly, the inherent thermal stability ensures the prevention of thermal runaway and localized hot spot formation, which could lethally compromise battery safety and performance. Fifthly, these nanoparticles demonstrate high optoelectronic sensitivity, making them valuable in energy storage applications that involve opto-electronic elements. Moreover, these contribute to the scope and intent of energy storage device management through state diagnostics and prognostics, allowing for real-time monitoring and predictive maintenance of battery systems. All these factors significantly enhance overall battery performance, improving energy density, charge–discharge efficiency, and longevity.

Considering these advantages, tire char-derived nanoparticles hold immense potential in the scheme of sustainable energy storage solutions. This ability to convert waste material into a high-performance energy storage component aligns with cleaner energy production principles. Integrating tire waste-derived char in this manner can contribute to environmental conservation by reducing landfill waste simultaneously advancing next-generation energy storage technology.

Funding Sources The authors wish to express their gratitude to Department of Science and technology, Delhi, India (Grant: DST/TDT/WMT/End of life vehicles/2021/01 (G)/2).

References

1. Priyadarshi, H., Choudhary, H., Agarwal, S.N., Shrivastava, A., Singh, K., Akhtar, J.: Graphene based futuristic green batteries for energy harvesting. In: 2019 IEEE International Conference on engineering, Technology and Education (TALE), pp. 1–4. IEEE (2019)

2. Priyadarshi, H., Ahmed, G., Mishra, D., Srivastava, A.K., McGee, R., Shrivastava, A., Singh, K.: Sustainable graphene-based energy storage device technology: materials, methods, monitoring and digital twin. Crit. Rev. Solid State Mater. Sci. **49**(6), 1086–1119 (2024)
3. Xu, C., Xu, B., Gu, Y., Xiong, Z., Sun, J., Zhao, X.S.: Graphene-based electrodes for electrochemical energy storage. Energy Environ. Sci. **6**, 1388–1414 (2013)
4. Mahmood, F., Zhang, C., Xie, Y., Stalla, D., Lin, J., Wan, C.: Transforming lignin into porous graphene via direct laser writing for solid-state supercapacitors. RSC Adv. **9**(39), 22713–20 (2019). https://doi.org/10.1039/c9ra04073k
5. Chu, M., Li, M., Han, Z., Cao, J., Li, R., Cheng, Z.: Novel biomass-derived smoke-like carbon as a supercapacitor electrode material. R. Soc. Open Sci. **6**(7), 190132. https://doi.org/10.1098/rsos.190132
6. Conway, B.E.: Electrochemical Supercapacitors: Scientific Fundamentals and Technological Applications. Plenum Publishers, New York (1999)
7. Stoller, M.D., Park, S., Zhu, Y., An, J., Ruoff, R.S.: Graphene based ultracapacitors. Nanoletters **8**(10), 3498–3502 (2008)
8. Mi, J., Wang, X.R., Fan, R.J., Qu, W.H., Li, W.C.: Coconut-shell-based porous carbons with a tunable micro/mesopore ratio for high-performance supercapacitors. Energy Fuels **26**, 5321–5329 (2012)
9. Winter, M., Brodd, R.J.: What are batteries, fuel cells, and supercapacitors? Chem. Rev. **104**, 4245–4270 (2004)
10. Pandolfo, A.G., Hollenkamp, A.F.: Carbon properties and their role in supercapacitors. J. Power Sourc. **157**, 11–27 (2006)
11. Kosynkin, D.V., Higginbotham, A.L., Sinitskii, A., Lomeda, J.R., Dimiev, A., Price, B.K., Tour, J.M.: Longitudinal unzipping of carbon nanotubes to form graphene nanoribbons. Nature **458**(7240), 872–876 (2009)
12. Srihari, V., Das, A.: Comparative studies on adsorptive removal of phenol by three agro-based carbons: equilibrium and isotherm studies. Ecotoxicol. Environ. Safety **71**(1), 274–283 (2008)
13. Priyadarshi, H., Singh, K., Shrivastava, A.: Green technology solutions for energy storage devices. In: Energy Conversion and Green Energy Storage, pp. 117–132. CRC Press (2022)
14. Priyadarshi, H., Shrivastava, A., Singh, K.: Energy storage device fundamentals and technology. In: Electrical and Electronic Devices, Circuits and Materials, pp. 119–130. CRC Press (2021)
15. Boxiong, S., Chunfei, W., Cai, L., Binbin, G., Rui, W.: Pyrolysis of waste tyres: the influence of USY catalyst/tyre ratio on products. J. Anal. Appl. Pyrolysis **78**, 243–249 (2007)
16. Thomas, B.S., Gupta, R.C.: Properties of high strength concrete containing scrap tire rubber. J. Clean. Prod. **113**, 86–92 (2016)
17. Doja, S., Pillari, L.K., Bichler, L.: Processing and activation of tire-derived char: a review. Renew. Sustain. Energy Rev. **155**, 111860 (2022)
18. Sathiskumar, C., Karthikeyan, S.: Recycling of waste tires and its energy storage application of by-products–a review. Sustain. Mater. Technol. **22**, e00125 (2019)
19. Lin, Y.R., Teng, H.: Mesoporous carbons from waste tire char and their application in wastewater discoloration. Microporous Mesoporous Mater. **54**(1–2), 167–174 (2002)
20. Lahijani, P., Zainal, Z.A., Mohamed, A.R., Mohammadi, M.: Co-gasification of tire and biomass for enhancement of tire-char reactivity in CO_2 gasification process. Biores. Technol. **138**, 124–130 (2013)
21. González-González, R.B., González, L.T., Iglesias-González, S., González-González, E., Martinez-Chapa, S.O., Madou, M., Alvarez, M.M., Mendoza, A.: Characterization of chemically activated pyrolytic carbon black derived from waste tires as a candidate for nanomaterial precursor. Nanomaterials **10**(11), 2213 (2020)
22. Zerin, N.H., Rasul, M.G., Jahirul, M.I., Sayem, A.S.M.: End-of-life tyre conversion to energy: a review on pyrolysis and activated carbon production processes and their challenges. Sci. Total. Environ. **905**, 166981 (2023)
23. Deng, J., Bae, C., Denlinger, A., Miller, T.: Electric vehicles batteries: requirements and challenges. Joule (2020). https://doi.org/10.1016/j.joule.2020.01.013

24. Augustyn, V., McDowell, M.T., and Vojvodic, A. Toward an atomistic understanding of solid-state electrochemical interfaces for energy storage. Joule **2**(11), 2189–2193 (2018). https://doi.org/10.1016/j.joule.2018.10.014
25. Xu, L., Tang, S., Cheng, Y., Wang, K., Liang, J., Liu, C., Cao, Y.-C., Wei, F., Mai, L.: Interfaces in solid-state lithium batteries. Joule **2**, 1991–2015 (2018)
26. Priyadarshi, H., Ahmed, G., Singh, K., Shrivastava, A.: Two-dimensional modeling and analysis of lithium-ion cell for electric vehicle application. In: 2021 4th International Conference on Recent Developments in Control, Automation & Power Engineering (RDCAPE), pp. 548–552. IEEE (2021)
27. Priyadarshi, H., Singh, K., Shrivastava, A.: Insolation prediction for reliable energy storage using deep learning algorithms. In: 2021 International Conference on Recent Trends on Electronics, Information, Communication & Technology (RTEICT), pp. 416–419. IEEE (2021)

Photocatalytic Degradation of BTEX: Catalysts, Mechanisms, and Operational Parameters

Kanchan Guru, Anand G. Chakinala, and Praveen K. Surolia

Abstract Photocatalytic degradation of a group of volatile organic compounds BTEX (benzene, toluene, ethylbenzene and xylenes) exhibit significant environmental and health issues. We look at the most important parts of this potential remediation technique, starting with a comprehensive aspect of different photocatalysts. This chapter provides a comprehensive overview of the key aspects governing the photocatalytic removal of BTEX, beginning with a detailed examination of conventional photocatalysts—including titanium dioxide (TiO_2) and zinc oxide (ZnO)—as well as next-generation materials such as perovskite oxides, g-C_3N_4-based systems, bismuth-based compounds, metal–organic frameworks (MOFs), and porous materials like activated carbon, carbon nanotubes, zeolites, and clays. Here we discuss types of photocatalysts, nano structuring, doping, and development of heterojunctions, that are meant to improve light absorption, charge separation, and the availability of active sites. The chapter also goes into detail on how BTEX chemicals break down through photocatalytic processes. This covers how light absorption, electron–hole production, and the creation of very reactive oxygen species (ROS) like hydroxyl radicals ($OH^{\bullet}$) and superoxide radical ($O_2^{\bullet -}$) work together in a complex way and these ROS cause the oxidation pathways of BTEX, which turns them into harmless minerals. Lastly, we look at how important operational factors are for making BTEX photocatalytic systems work better like the initial BTEX content, catalyst loading, pH, temperature, light intensity. Overall, this chapter aims to serve as a valuable reference for researchers and practitioners developing environmentally sustainable photocatalytic technologies for BTEX abatement.

Keywords Volatile organic compounds (VOCs) · BTEX · Photocatalysis · Mechanism · Operational parameters · Environmental remediation

---

K. Guru · P. K. Surolia (✉)
Solar Energy Conversion and Nanomaterials Laboratory, Department of Chemistry, Manipal University Jaipur, Jaipur, Rajasthan, India
e-mail: praveenkumar.surolia@jaipur.manipal.edu

A. G. Chakinala (✉)
Chemical Reaction Engineering Laboratory, Department of Chemical Engineering, Manipal University Jaipur, Jaipur, Rajasthan, India
e-mail: anandgupta.chakinala@jaipur.manipal.edu

L. Ledwani and S. Srivastava (eds.), *Experimental and Computational Advances in Materials*, Engineering Materials, https://doi.org/10.1007/978-981-95-4180-5_7

1 Introduction

The rising trend in the emissions of volatile organic compounds (VOCs) in recent decades has resulted in a sequence of air pollution due to their mobility and instability [1]. VOCs, particularly those from industrial emissions of anthropogenic origin, are precursors of ozone and photochemical smog. The overwhelming contribution of VOCs to environmental pollution, along with their adverse effects on human health such as carcinogenic, mutagenic, and teratogenic agents, makes them particularly dangerous. Due to their active participation in photochemical processes and the destruction of ozone layers, aromatics are considered the principal pollutants [2]. The US Environmental Protection Agency (EPA) has set the maximum concentration for volatile organic compounds like benzene, toluene, ethylbenzene, and xylenes (BTEX) (Fig. 1) acceptable in drinking water at 5, 1000, 700, and 10,000 μg/L, respectively [3].

BTEX is one of the VOCs that pose a serious threat to the environment. BTEX is among the VOCs with a horrible eco-hazardous profile, so it is noteworthy. Illegal

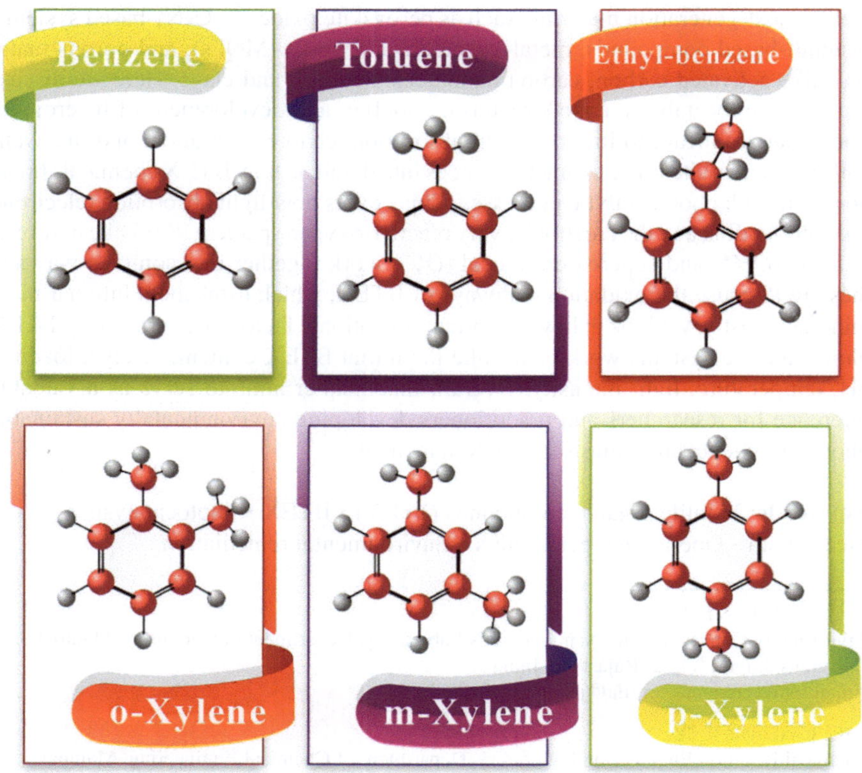

Fig. 1 Volatile organic compounds: benzene, toluene, ethylbenzene, o-, m-; and p-Xylene

dumping of industrial waste, poor sewage treatment and disposal, oil spills from tankers, accidental leaks from pipelines, municipal landfills, burning of coal and biomass, dumping in the ocean, and farming activities like using pesticides, painting equipment, and fertilizing, all release BTEX into groundwater and surface water [4]. Furthermore, naturally occurring in crude oil, diesel, gasoline, and other petroleum-based products are BTEX. They account for more than 60% of the VOCs present in urban environments. They are thus used as a normal baseline for the monitoring and assessment of the ambient VOCs levels and exposure. Most of the overall industrial emissions come from BETX using various structures with benzene rings, which also help research and comparison. The prospective technologies for eliminating aromatic molecules to safeguard the environment and human life are very much worth noticing. Absorption [5], adsorption [6], membrane separation [7], photocatalytic oxidation [8–10], catalytic oxidation [11], plasma destruction [12], and biological oxidation [13] are among the abatement technologies to break down BTEX over the past few decades. Among them, photocatalytic technology has attracted more interest because of its low energy, high efficiency, and harmlessness, which has wide application possibilities in many different industries [14–16]. Photocatalytic technology has become a major topic in the middle of the traditional energy crisis. Photocatalytic materials change light energy into chemical energy when exposed to UV or visible light. This makes it easier to make or break down organic molecules [17, 18]. In the past few years, BTEX detection and pretreatment technologies (extraction, separation, and enrichment) have come a long way. It's important to note that in the past several years, a lot of new materials, solvents, and other things have been created and used in the pretreatment and detection technology of BTEX, but there hasn't been a summary of all this yet [19].

Still, as far as we know, not much research has been done on the systematic study of the photocatalytic degradation of BTEX, focusing on how photocatalysis works and how BTEX photodegradation technology can be used in real life. The BTEX contaminants in the surroundings have attracted much attention. Reviewing previous remedial methods like natural attenuation and bioremediation used to BTEX compounds in water, Fayemiwo et al. detailed the benefits and drawbacks of each method. They also underlined newly developed nanotechnology-based technologies meant to eliminate harmful chemicals from water. New ideas that have emerged from recent BTEX remedial investigations including newly discovered bacteria, modern retrofit equipment, new materials and reagents, and joint removal technologies in recent years [20].

This chapter provides a comprehensive summary of the recent advances in photocatalytic technology for the degradation of BTEX. Like other VOCs with distinctive structures, BTEX can be studied in a systematic way to better understand the photocatalytic mechanism and make it more useful in real life. The chapter discusses widely studied photocatalysts such as TiO_2, g-C_3N_4, ZnO, and Bi-based oxides, along with their modifications to enhance visible-light response and degradation efficiency. Additionally, the influence of operational parameters on BTEX removal, kinetic modeling approaches, and the mechanistic pathways of photocatalytic degradation are summarized to support future research and practical implementation.

2 Mechanism of the Photocatalytic System

Under light irradiation, semiconductor photocatalysts having a suitable bandgap, help to completely mineralize dye and organic contaminants into non-toxic pieces. Organic compounds oxidatively break down into carbon dioxide and water when exposed to water, UV or visible light, and a semiconductor photocatalyst like TiO_2. Edmond Becquerel first discovered the photocatalyzed mechanism in 1839 during his research on natural photosynthesis [21]. Figure 2 offers a schematic of photocatalytic systems. Two necessary criteria must be satisfied to create successful and efficient photocatalysts. First, the existence of an electron with a high reduction potential in the conduction band (CB) and second, the valence band's (VB) high oxidation potential holes existence. Previous studies have suggested a photocatalytic mechanism whereby electrons acquire sufficient energy to transition from the VB to CB, thereby producing VB developing holes [22], when exposed to high-energy photons (E_{ph} > the bandgap of the photocatalysts). The commonly used term for this phase is photo-excitation phase [23].

$$photocatalyst + h\vartheta \rightarrow \text{photocatalyst}\left(e_{CB}^{-} + h_{VB}^{+}\right) \tag{1}$$

$$photocatalyst\left(e_{CB}^{-} + h_{VB}^{+}\right) \rightarrow photocatalyst + h\vartheta \tag{2}$$

$$h_{VB}^{+} + OH^{-} \rightarrow OH^{\bullet} \tag{3}$$

$$e_{CB}^{-} + O_2 \rightarrow O_2^{\bullet -} \tag{4}$$

$$O_2^{\bullet -} + H^{+} \rightarrow HO_2^{\bullet} \tag{5}$$

$$OH^{\bullet} + Organicpollutants \rightarrow Degradationproducts \tag{6}$$

Equation (1) depicts the photogeneration of the electron–hole pairs (EHP_S) in photocatalyst when they are exposed to light which results in the formation $\left(e_{CB}^{-}\right)$ and hole (h_{VB}^{+}) pairs. These EHPs can undergo radiative recombination and release thermal energy in just a few nanoseconds. Equation (2) illustrates the recombination of electron–hole pairs into the ground state. Trapping either electrons or holes helps to lower the presence of appropriate surface defect states or scavenger states, hence reducing the recombination. Furthermore, when pollutants are present in a fluid such as water, they bind to it instantly and move electrons and holes to the acceptor and

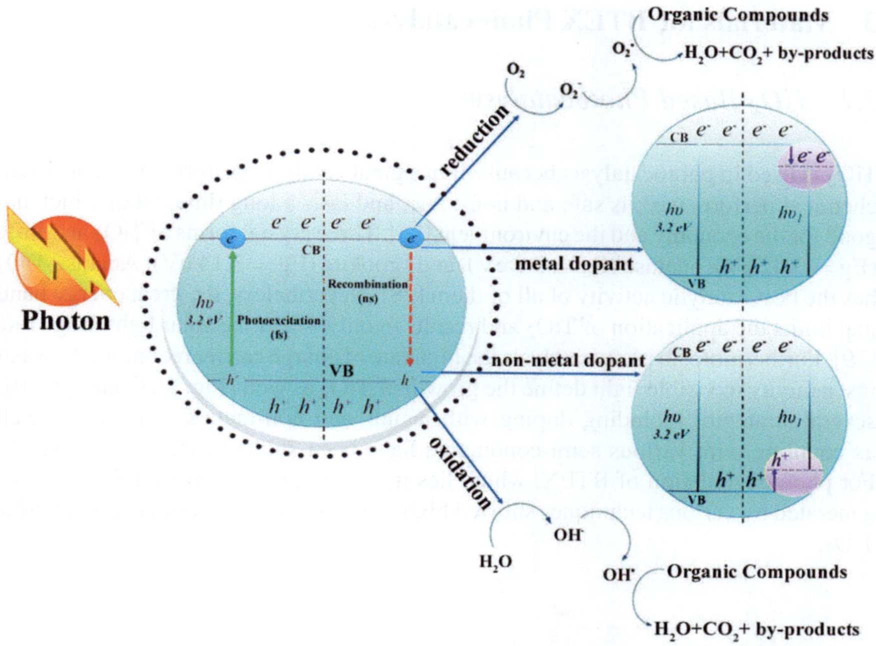

Fig. 2 Mechanism of semiconductor photocatalysis and doping of semiconductors with metal and non-metal dopants

donor molecules in line with the redox potential of each adsorbate [9, 24]. Reducing and oxidizing by the electron and holes, the different ROS are produced. By means of hole trapping the OH^- species transforms into free radicals ($OH^\bullet$) as in Eq. (3) [25]. Surface adsorbed molecular oxygen (O_2) can also absorb the photogenerated electrons as in Eq. (4). This generates superoxide radical anions ($O_2^{\bullet-}$) Sometimes the superoxide radicals ($O_2^{\bullet-}$) undergo protonation to produce hydroperoxyl radicals, hence possibly substituting hydroxyl radicals in the pathway of photocatalytic degradation (Eq. 5). These hydroxyl radicals help products to scavenge unwanted radicals, therefore limiting the recombination of photogenerated charge carriers [26]. These ROS further break pollutants, particularly via the ($H^\bullet$) radical and transform them into degradation products (CO_2 and H_2O) as in Eq. (6). Furthermore, when contaminants are present in a fluid such as water, they bond to it instantly and carry electrons and holes to the acceptor and donor molecules in line with each absorbates redox potential and shows the several phases of the typical photocatalytic degradation process.

3 Materials for BTEX Photocatalysis

3.1 TiO_2 Based Photocatalysts

TiO_2 is used in photocatalysis because it has great catalytic performance, consistent chemical performance, is safe and non-toxic, and lasts a long time, all of which are good for the economy and the environment [27]. Three crystal forms of TiO_2 are rutile ($Eg = 3.02$ eV), anatase ($Eg = 3.2$ eV), and brookite ($Eg = 3.14$ eV). Anatase TiO_2 has the best catalytic activity of all of them [28]. Nevertheless, the great energy band gap limits the application of TiO_2 and results in only 4% of the solar light usage rate [29]. Furthermore, in photocatalysis the high rate of charge carrier recombination and insensitivity to visible light define the growth of TiO_2 as well (Fig. 2). Consequently, several strategies including doping with metals and non-metals elements as well as coupling with various semi-conductors have been studied extensively (Fig. 3). For photodegradation of BTEX, which lies in Table 1, the modified TiO_2 catalysts generated by various techniques showed higher performance as compared to pristine TiO_2.

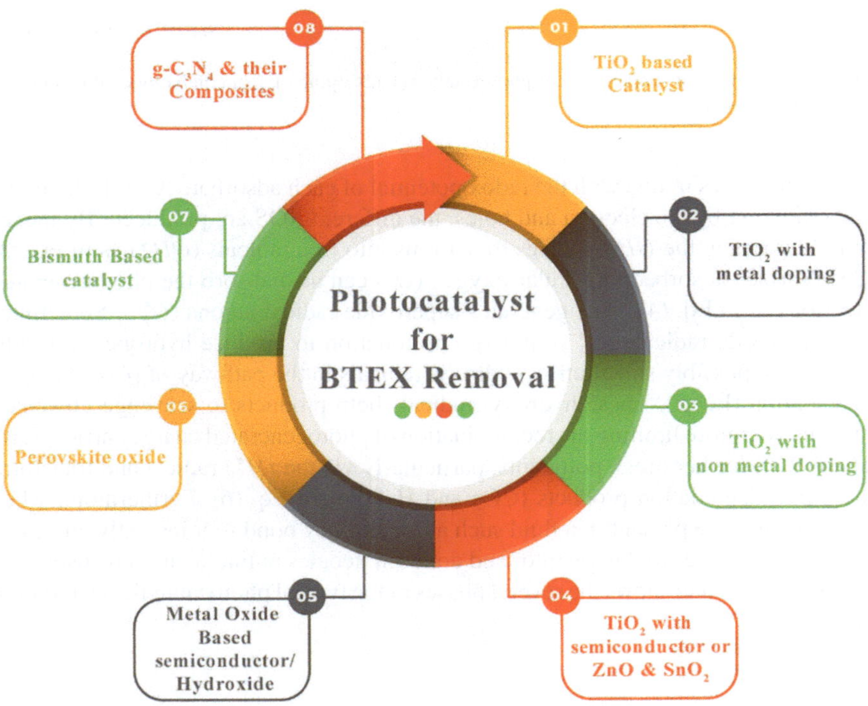

Fig. 3 Different photocatalysts for BTEX removal from wastewater

Table 1 Catalyst type and their modifications methods, targeted BTEX compounds, light source, performance and mechanism of photocatalysts and their effect on BTEX photocatalysis

Catalyst type	Catalyst and their modifications methods	Targeted BTEX compounds	Light source	Performance and mechanism	Key findings	References
I. Titanium Dioxide (TiO_2) & composites	Pure TiO_2	Toluene	UV light	Baseline performance	Pure TiO_2 (e.g., P25) shows high UV activity but limited visible light response and high e^-/h^+ recombination	[35]
	TiO_2 with Metal-doping ($Fe/Bi_2WO_6/TiO_2$-N heterostructure)	Benzene	Visible	Enhanced visible light utilization, improved charge separation, reduced recombination	Significantly enhanced benzene degradation efficiency compared to individual components	[36]
	TiO_2 with Metal-doping (Cu-doped TiO_2)	BTEX mixture	Visible	Enhanced visible light absorption, improved charge separation via Cu-TiO_2 interaction	Cu-TiO_2 showed higher BTEX degradation rates than undoped TiO_2, with optimal Cu loading	[37]
	TiO_2 with Non-metal doping (C-doped TiO_2)	BTEX	Visible	Reduced band gap, enhanced visible light absorption, increased surface defects	High degradation efficiency (e.g., 93.4% BTEX in 90 min) from paint thinners	[38]
	TiO_2 with Non-metal doping (N-Doped TiO_2)	Benzene/ toluene	Visible	Creation of oxygen vacancies, formation of Ti-N bonds, band gap narrowing	Enhanced photocatalytic activity for benzene and toluene degradation under visible light	[39]
II. Carbon-based materials & composites (with/ without TiO_2)	Graphene/rGO Composites (PIL@TiO_2/m-GO composites)	Benzene and toluene	Visible	High surface area for adsorption, facilitated charge transfer by m-GO, lowered band gap	97% degradation for both benzene and toluene, strong synergistic effect	[40]

(continued)

Table 1 (continued)

Catalyst type	Catalyst and their modifications methods	Targeted BTEX compounds	Light source	Performance and mechanism	Key findings	References
	Graphene Quantum Dots (GQDs)-TiO_2	Toluene	Visible	GQDs act as electron acceptors, sensitizers for visible light	Enhanced toluene degradation due to improved charge separation and visible light response	[41]
	Activated Carbon (AC) Composites (AC-TiO_2)	BTEX	UV–Vis light	High adsorption capacity of AC, synergistic adsorption-photocatalysis, improved mass transfer	Enhanced BTEX removal (adsorption + degradation) compared to pure TiO_2 or AC alone	[42]
	Carbon Nanotubes (CNTs) Composites (CNT-TiO_2)	BTEX	UV–Vis light	CNTs act as electron sinks, improve charge separation, increase surface area for adsorption	Improved BTEX degradation rates and mineralization compared to pure TiO_2	[43]
III. Zeolites & composites	Zeolite with TiO_2 (TiO_2/Zeolite (e.g., ZSM-5, clinoptilolite)	BTEX mixture	UV–Vis and solar	High adsorption capacity of zeolite pre-concentrates pollutants, improved dispersion and stability of TiO_2	Enhanced and more stable BTEX removal due to initial adsorption and subsequent degradation on active sites	[44]
IV. Biochar (BC) & composites	Biochar/TiO_2 Composites	Toluene and other VOCs	Visible light/ solar	High adsorption capacity of biochar, enhanced visible light absorption due to carbon, reduced e^-/h^+ recombination	Improved photocatalytic activity for organic pollutants (e.g., dyes, some VOCs), principle applicable to BTEX	[45]

(continued)

Table 1 (continued)

Catalyst type	Catalyst and their modifications methods	Targeted BTEX compounds	Light source	Performance and mechanism	Key findings	References
	TiO_2/g-C_3N_4/biochar ternary composite	BTEX mixture	Visible light	Synergistic effects of adsorption, visible light absorption, and charge transfer	High BTEX removal efficiency, demonstrating the benefit of multi-component systems involving biochar	[46]
V. Perovskite materials & composites	Perovskite (e.g., ABO_3), $BiVO_4$/TiO_2 heterojunction	Toluene	Visible light	Improved visible light absorption, efficient charge separation at the interface	Enhanced toluene degradation, demonstrating synergistic effects between the two semiconductors	[47]
	Perovskite (e.g., metal halides), Lead free halide perovskite heterojunction	BTEX	Visible light	Broad light absorption, high charge mobility	Emerging area for direct photocatalysis of VOCs; high potential for BTEX due to their optoelectronic properties	[48]
VI. Metal–organic frameworks (MOFs) & composites	MOF/TiO_2 Composites (MOF (MIL-125(Ti))/TiO_2 composite)	Benzene	UV–Vis	High adsorption capacity of MOF, photoactivity of Ti-MOF combined with TiO_2, precise active sites	Superior benzene degradation due to synergistic adsorption-photodegradation	[49]
	MOF alone or with composite	Aromatic VOCs including BTEX	UV–Vis/ Visible	High porosity for pre-concentration, tunable pore sizes, intrinsic photoactivity of some MOFs	MOFs demonstrate excellent adsorption and some intrinsic photocatalytic activity for VOCs, often enhanced in composites	[50, 51]

(continued)

Table 1 (continued)

Catalyst type	Catalyst and their modifications methods	Targeted BTEX compounds	Light source	Performance and mechanism	Key findings	References
VII. Graphitic carbon nitride ($g\text{-}C_3N_4$) & composites	Pure Graphitic Carbon Nitride ($g\text{-}C_3N_4$)	Benzene	Visible light	Visible light activity, polymeric structure, good stability	Showed promising activity for benzene degradation, often improved by modifications	[52]
	Graphitic Carbon Nitride ($g\text{-}C_3N_4$ /TiO_2) heterojunction	Toluene	Visible light	Efficient charge separation via Z-scheme or S-scheme, extended visible light absorption	Significantly enhanced toluene degradation compared to individual $g\text{-}C_3N_4$ or TiO_2	[46]

3.1.1 Doping with Metals

TiO_2 modified with noble metal doping can effectively enhance the photo generated separation of electron -hole pairs and improve the photocatalytic activity of catalyst TiO_2 [30]. In visible light, the absorption of TiO_2 can be promoted by noble metals [31]. However, noble metals are very expensive, so their use is less for photocatalysis. To overcome this challenge non-noble metals are used as a substitute for TiO_2 modification because non-noble metals are cheaper and easily available. Non-noble metals (such as Mn, Fe, Ce, Cu etc.) have the capability to narrow the band gap of TiO_2 and accelerate the response of TiO_2 to the visible light region [32]. Mono modified TiO_2 with Ag, Au and Pt modifiers has been prepared by chemical reduction method of photocatalysis [33]. The absorption spectra of mono-metallic modified TiO_2 photocatalysts showed that the absorption capacity of photocatalyst enhanced in visible light as compared to the pure TiO_2. Types of modifiers and number of modifiers also affect the photocatalytic efficiency of TiO_2. The catalytic oxidation of benzene on TiO_2 altered with base metals (Mn, Co, Cu, Ni, and Fe) was investigated by Huang et al. [50]. Base metals most certainly helped to drive the change from anatase phase to rutile one. The absorption of different metals changed differently, thus did the elimination efficiency of benzene. In another study, the degradation efficiency of toluene first raised and then dropped as the concentration of doped metal was increased. Like an electron acceptor, dopants trap electrons produced by the semiconductor and separate charge carriers if a specific amount of metal dopant is kept [34]. But when there are a lot of dopants, they can operate as a recombination center that stops holes from forming and even changes the shape of the catalyst crystal lenity, which makes the photocatalytic reactivity much lower.

3.1.2 Doping with Non-metals

Doping TiO_2 with non-metallic elements is a good idea because it contains a lot of raw materials, is cheap, and doesn't corrode easily. Doping TiO_2 photocatalyst with non-metallic components is a very good approach to lower the band gap and make it more responsive to visible light. Carbon (C), nitrogen (N), sulfur (S), and fluorine (F) are common nonmetallic doping elements on TiO_2 photocatalyst that work well to get rid of VOCs [53]. Carbon doping on TiO_2 can increase the crystallinity of TiO_2 and shift the absorption wavelength to the red shift [54]. Mahmood et al. improved the TiO_2 adsorption capability with (CQDs) [55]. Nitrogen exists with good stability and low ionization energy so it can reduce the band gap of photocatalyst TiO_2. Li et al. [56] investigated the NH-TiO_2 photocatalyst carbon quantum dots efficiently degrade the benzene. TiO_2 doped with NH exhibited deep photocatalytic activity because N–H co-doping lowered the charge carrier recombination. Different and interstitial crystal changes would occur on TiO_2 doped with carbon or nitrogen. According to Valentin et al. [57], the substitutional N was identified using X-ray photoelectron spectroscopy (XPS) study. The atomic radius of carbon and nitrogen is close to that of oxygen, and their electronegativity is also weaker than oxygen.

Hence, it is possible that in the TiO_2 lattice, nitrogen and carbon would take the position of oxygen. On the other hand, only a small amount of carbon and nitrogen could get through the lattice voids and subsequently connect with the atoms in the lattice. On the other hand, fluorine is more electronegative than oxygen. So, the F 2p states are below the O 2p VB. Adding fluorine to TiO_2 may cause the impurity states to show up in the band gap. The mixing of S's 3p orbit with the valence band might lower the prohibited bandwidth when S has different valence states. Doping S can also help separate pairs of electrons and holes. S-TiO_2 films glued to aluminum sheets worked very well to break down BTEX when exposed to visible light [58].

3.1.3 Composite of Semiconductors

Semiconductors are combined here with other catalysts that are called composite. Composites are made using the coupling effect between semiconductors like TiO_2 and catalyst which is nanomaterials. Composites work as a single semiconductor because there is difference in chemical potential between semiconductors and nanomaterials that result in bending of bands at the interface of junctions. There are 3 methods of composite synthesis based on the band gap of semiconductors- 1) narrow band gap semiconductors (CdS, C_3N_4). For example, the composites of CuS-CdS/TiO_2 produced a stepwise energy level structure and highly suppressed the recombining of electron–hole pairs [59]. 2) Second is wider band metal oxides (ZnO and SnO_2) to stop electron recombination. Shifted to SnO_2, the heterojunction developed between TiO_2, and the electrons produced by TiO_2 and SnO_2 (Eg = 3.8 eV). The charge-transfer mechanism improved to the performance [60]. 3) Third is combining the p-type semiconductor (CuO) on n-type TiO_2 to improve the efficiency of photocatalysis. Density functional theory (DFT) used to determine the energy band structure and density of TiO_2 and CuO-TiO_2 composites [61].

3.2 *Non-TiO_2 Photocatalysts*

Non-TiO_2 based photocatalysts are mainly mono-and bimetallic oxides, hydroxides, bismuth based and metal free materials. All materials are with different bandgaps and their operating parameters are based on different methods which were used in photocatalysis of BTEX (shown in Table 1).

3.2.1 Metal Oxide-Based Semiconductors

Metal oxides have a band gap, which allows them to absorb light (UV or visible) and produce pairs of holes and electrons. There are a lot of common metals like zinc, tin, and cerium in nature. The physicochemical properties of metal oxides (size, shape, morphology, composition, band gap) in different ways to make them work better as photocatalysts [51]. ZnO is a popular choice since it works well, is stable when exposed to light, and is safe for living things. It can absorb UV light because its band gap is about 3.2–3.37 eV. Like TiO_2, adding things like dopants or making composites can greatly improve its performance. SnO_2 also possesses photocatalytic properties and has been investigated to break down BTEX. WO_3 (Tungsten trioxide) is a promising photocatalyst that works in visible light. Ceramic foam filters with a WO_3/Pt coating have been proven to break down a lot of toluene when exposed to visible light. Iron Oxide (Fe_2O_3) can act as photocatalysts, especially when exposed to visible light [62]. CeO_2, or cerium dioxide, is another metal oxide that has been investigated as a photocatalyst. Different valence states of manganese oxide Mn_3O_4 can create heterojunctions that stop photogenerated electrons and holes from coming together again, which makes the photocatalytic activity better [63].

3.2.2 Hydroxide Photocatalysts

Degradation reactions are primarily driven by hydroxyl radicals ($OH^\bullet$), and the production of these radicals can be markedly increased by surfaces exhibiting high hydroxyl density. The generation of $OH^\bullet$ radicals is crucial in photocatalytic oxidation; consequently, hydroxides are commonly utilized as photocatalysts for the degradation of volatile organic compounds (VOCs). A series of $MSn(OH)_6$, where (M = Co, Cu, Fe, Mg, Mn, Zn) was synthesized and shown pr$_o$nounced visual absorptions [64]. Only $MgSn(OH)_6$ and $ZnSn(OH)_6$ exhibited a prominent and consistent efficacy in the degradation of benzene. The hydroxyl structure in these samples enhanced performance, facilitated the generation of active radicals ($OH^\bullet$ and $O_2^{\bullet -}$), and prevented the accumulation of stable by-products. Research indicates that ZnCr-LDHs (Layered Double Hydroxides) with a narrow band gap exhibits significant reactivity in photocatalytic reactions, possessing a substantial quantity of water molecules and hydroxyl ions. Water molecules and hydroxyl ions were essential to produce $OH^\bullet$ radicals during photocatalytic oxidation. The active species tests demonstrated that the primary impetus for the photocatalytic process in Au-supported ZnCr-LDHs was the hole and hydroxide radicals. This demonstrated that Au/ZnCr-LDHs displayed enhanced photocatalytic activity [65].

3.2.3 Perovskite Oxide as Photocatalyst

Perovskite (ABO_3), as a type of photocatalyst, has received attention due to its efficient absorption of solar energy, structural stability, and high energy conversion efficiency [66]. The ideal perovskite structure has A-site ions which are typically rare earth metal ions of larger radii like La^{3+}, Ce^{4+}, and alkaline earth metal ions such as Ca^{2+} and Sr^{2+}. These ions rest in the cuboctahedra voids formed by twelve oxygen atoms which form a 12-coordination sphere. The large ionic radius and high coordination number enable the A-site ions to stably occupy lattice positions region thus ensuring critical structural support for entire crystal framework [67]. The A-site ions also greatly impact on the electronic state and consequently the catalysis of B-site ions. The type and doping ratio of A-site ions can be altered, enabling control of the lattice parameters, oxygen vacancy concentration, and electronic structure of the catalyst which subsequently affects catalytic performance [47]. Typically, the B-site ions are formed by relatively small radius transition metal ions such as Mn^{3+}, Fe_3^{++}, Co_3^{+}, and Ni^{2+} which are positioned at the center of an octahedron of six oxygen atoms, forming an octahedral structure with sixfold coordination to the oxygen atoms [68].

The valence state and type of B-site ions have a critical impact on the activity of the catalyst. Due to having many electron orbits and changeable oxidation states, their transition metal ions can provide numerous active sites in catalytic reactions and undergo electron transfer and chemical reaction processes. Manganese-containing perovskite catalysts showed high performance regarding the catalytic oxidation of volatile organic compounds, especially toluene. $CaTiO_3$ perovskites were studied for toluene photodegradation. $CaTiO_3$ perovskite has a highly absorption surface with is gone upto 400 mm approximately [69].

3.2.4 Bismuth Based Photocatalyst

In the development of non-TiO_2 photocatalysts, bismuth-based compounds became the focus of investigation in recent years owing to their peculiar, layered structure and appropriate band gap. The bonding of the 6 s orbital of Bi to the 2p orbital of O causes the tilt of valence band to increase, permitting greater separation and migration of electron and hole pairs during photocatalytic processes. Bismuth-based photocatalysts have strong photocatalytic performance. A typical binary metal oxide, Bi_2WO_6, exhibits an orthorhombic Aurivillius-type structure characterized by alternating layers of bismuth oxide. These layered structures are found in some complex metal oxides as a specific stacking arrangement where layers are stacked on the top of each other. This arrangement provides these materials with unique properties, mainly in ferroelectrics and photocatalysis. The general formula for these compounds is (Bi_2O_2), ($A_{m-1}B_mO_{3m+1}$), where 'm' is the number of perovskite layers between each bismuth oxide layer. In the case of Bi_2WO_6, m = 1. This is illustrated as $[Bi_2O_2]^{2+}$ and $(WO_6)_2$ octahedral. Bi_2WO_6 has a narrow band gap of energy 2.5–2.8 eV, therefor its absorption of the visible light spectrum is significant. It was shown that the modifications to Bi_2WO_6 with platinum metal resulted in an increased amount of oxygen vacancies [70]. The reduction of the oxygen vacancies above the conduction band may be able to lower the band gap of in turn biasing depletion regions of Bi_2WO_6 and increase the production of ($O_2^{\bullet -}$) which should help mineralization of gaseous toluene. The addition of platinum helped to enhance the separation of photogenerated charge carriers. Liu et al. added carbon quantum dots (CQDs) onto Bi_2WO_6 and were successfully aiding the removal of toluene under visible light, causing red shifts in light absorption and an increase in biocompatibility for the composites [71].

Bismuth oxyhalides BiOX (X = F, Cl, Br, I) show unique optical and electrical properties with a tetragonal matlockite structure. BiOCl is a layered semiconductor photocatalyst that consists of alternate $[Bi_2O_2]^{2+}$ layers and Cl. BiOCl exhibits a band gap of 3.4 eV which explains its low sensitivity to visible light. In the Bi-W-Cl-O system, three stages were prepared by altering the reaction conditions: BiOCl, BiOCl/Bi_2WO_6 composites, and Bi_2WO_6 [72]. The UV–vis DRS spectra showed that Bi_2WO_6 had higher light efficiency relative to BiOCl. The photocatalytic activity of the three samples for benzene under visible light was ranked as follows: Bi_2WO_6 > BiOCl/Bi_2WO_6 composites > BiOCl. Dong et al. reported the successful formation of surface oxygen vacancies (OVs) on BiOCl crystal via solvothermal treatment with hot ethylene glycol reduction. The inclusion of OVs enhanced the adsorption and activation of O_2, H_2O, and toluene molecules which led to more free radicals being produced with greater oxidative potential. The internal electric field developed during the h^+/e^- transfer cycle facilitated the h^+/e^- separation which increases photocatalytic activity. BiOI is a potential semiconductor with a layered crystalline structure and a suitable bandgap. Nevertheless, BiOI's practical uses are limited due to low photocatalytic efficiency with high recombination rates of charge carriers and poor electrical conductivity [73]. Wang et al. used a solvothermal approach combined with thermal decomposition to develop a new composite photocatalyst I-deficient $BiO_{1.2}\ I_{0.6}/WO_3$. The incorporation of iodine vacancies and WO_3 greatly improved the photocatalytic performance of BiOI [74].

3.2.5 Graphite-Nitride Based Photocatalyst

Graphitic carbon nitride (g-C_3N_4) has a band gap often in the range 2.6 to 2.8 eV. This number renders it an exceptional photocatalyst due to its band gap being appropriate for the absorption of visible light, which represents a substantial segment of the solar spectrum. g-C_3N_4 is non-toxic, affordable, metal-free, and stable. It is made from common precursor materials that include the most abundant elements, carbon and nitrogen. Zhao et al. found that pure g-C_3N_4 has a lower catalytic activity for benzene compared to TiO_2. g-C_3N_4 has limited use because of its low specific surface area and a high rate of photocarrier recombination so the g-C_3N_4 is often combined with other semiconductors to boost its effectiveness. The $H_3PW_{12}O_{40}$/g-C_3N_4 film-coated optical fiber photoreactor was custom-designed and tested for removing aromatics like benzene, toluene, and m-xylene under simulated sunlight. The $H_3PW_{12}O_{40}$/g-C_3N_4 film showed much better photocatalytic activity, as well as improved stability and recyclability. A new α-Fe_2O_3/Cu/g-C_3N_4 nanocomposite was created using a simple method and showed excellent photocatalytic performance and stability for removing air ethylbenzene [75].

3.3 Porous Materials for BTEX Adsorption

BTEX can be removed through the adsorption process as well, which is now widely studied. High surface area of the materials is vital parameter for better adsorption capacity which can be enhanced by introducing the porosity in them. Some of the materials tested for the BTEX adsorption are activated carbon, biochar, carbon nanotube, zeolites, metal–organic frameworks etc. (Fig. 4).

3.3.1 Activated Carbon

Activated carbon (AC) has wide specific surface area (600–1400m^2 g^{-1}), well-developed pore structure (0.5–1.4cm^{3}g^{-1}), and strong capability to adsorb volatile organic compounds (VOCs) (10–600 mg g^{-1}), so activated carbon (AC) is regarded to be a versatile adsorbent. Lignocellulosic-biomass, Coal, wood, coconut shell, peat, agricultural waste, food waste and lignite are examples of carbonaceous materials that are often used as precursors in industrial air conditioning (AC). The cost of producing these materials ranges from $1000–1500 t^{-1} [76]. According to Romero-Anaya et al., it is even possible to make it in the shape of a pellet, granule, powder, or spherical once the carbonization and activation processes have been completed. AC has been widely used for environmental applications, including the treatment of wastewater, the purification of air, and the remediation of soil, particularly for the removal of volatile organic compounds (VOCs) like BTEX [77].

The adsorption circumstances and the physicochemical parameters of the adsorbent both have an impact on the adsorption efficacy of AC on VOCs. AC seems to be acceptable for application in the adsorption of VOCs at room temperature, low to

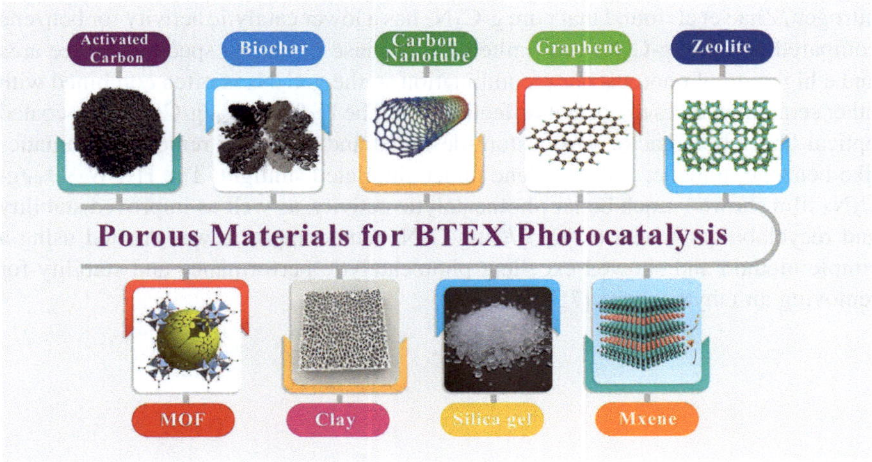

Fig. 4 Different porous materials for BTEX photocatalysis

medium concentration, and in an environment of nitrogen dioxide. When it comes to the adsorption capacity of VOCs, the high surface area and pore volume of AC have a beneficial impact. The chemical functional groups that are present on the surface of the AC are particularly important for the adsorption of certain VOCs [78].

3.3.2 Biochar

Biochar is a representative member of the carbon family, which is viewed as a viable alternative to commercial AC due to its abundant feedstocks, and efficient-low cost. Biochar is made under softer pyrolysis conditions (slow pyrolysis and relatively low temperature of < 700 °C) in an inert environment as opposed to AC. Biochar may be produced from a variety of abundant carbon-rich sources, including wood, forestry and agricultural waste, fruit byproducts, etc. [76]. Compared to AC ($1000 to $1500t^{-1}$), its production cost of $\$20t^{-1}$ is significantly lower [79]. Raw materials and manufacturing circumstances have a significant impact on the characteristics of biochar. significant lignin and mineral content raw materials often create significant amounts of biochar, and as pyrolysis temperature and duration increase, the mineral content may decrease. According to Hsi et al., the biochar that results from carbonization is a disordered elementary graphitic crystallite with a simple pore structure [80].

3.3.3 Carbon Nanotubes

A carbon nanotube (CNT) is made of a graphene sheet that has been wrapped up into a cylinder containing sp^2 hybridized carbon atoms. There are two types of carbon nanotubes: single-walled carbon nanotubes (SWCNT) and multi-walled carbon nanotubes (MWCNT). These depend on how the graphene cylinders are arranged [81]. The most popular ways to make it are arc discharge, laser ablation, and chemical vapor deposition (CVD). Compared to the CVD process, the arc discharge and laser treatment procedures prefer a greater yield [82]. CNT is a new type of nanomaterial that has special properties, such as being able to conduct electricity, changing light, and being strong. CNT is also better at getting rid of trace impurities from liquids and gases because it has a higher surface area, is naturally hydrophobic, and is very stable at high temperatures [83]. However, it is not common to use CNTs to get rid of VOCs.

3.3.4 Graphene

Graphene with a two-dimensional (2D) structure is made up of a sheet of carbon atoms arranged in hexagons that share sp^2 hybridized orbitals with three other carbon atoms. Common ways to prepare them are exfoliation, hydrothermal self-assembly, chemical vapor deposition, and nanotube slicing [84]. Graphene has a very high theoretical specific surface area and a very high electrical conductivity (3000 to 5000W m K^{-1}), which is much higher than copper. It also has a lot of mechanical strength [85]. Graphene has amazing physicochemical properties, so graphene has been used in electronics sectors, sensors, photonics, energy storage devices, biomedicine field, and cleaning up the environment through photocatalysis.

There are two most common derivatives of graphene, those are graphene oxide (GO) and reduced graphene oxide (rGO). The first one is what happens when graphene is oxidized with various groups that include oxygen, such carboxylic, hydroxyl, and epoxide groups. Chemical treatment or thermal annealing can remove the functional groups from GO to make the latter. Yu et al. looked at how well benzene and toluene adsorbed onto GO and rGO at room temperature, 50 ppm, and in a N_2 environment. GO and rGO have surface areas of 236.4 and 292.6 m^2g^{-1}, respectively. The rGO has greater adsorption capabilities for benzene and toluene (276.4 and 304.4 mg g^{-1}) than the GO (216.2 and 240.6 mg g^{-1}). It might be because rGO is more hydrophobic, has less oxygen, and has more defect sites [86].

3.3.5 Zeolite

Zeolite is a crystalline aluminosilicate framework and is made up of an endless number of three-dimensional (3D) arrangements of TO_4 tetrahedra, where T is Al or Si. The oxygen atoms are shared by the two tetrahedrons, which make crosslinking in space that makes channels and cavities with regular sizes that can hold tiny organic molecules [87, 88]. The structural formula for zeolite is $A_{(x/q)}$ $[(AlO_2)x$ $(SiO_2)_y]n(H_2O)$, where A represents Ca, Na, K, Ba, Sr, and other cations. It has 16–21% water by weight. Zeolite is a popular chemical sieve, adsorbent, and catalyst because it has great features including being water-repellent, having a huge surface area (250–800 m^2g^{-1}), being able to change its porosity, and not catching fire. Changing the Si/Al ratio can change the texture of zeolite [89]. Carbon-based materials generally have problems with being flammable and hard to regenerate. Zeolite's better hydrothermal and chemical stability lets it solve these challenges. Also, zeolite can completely desorb at a temperature as low as 150 °C, while carbon-based materials need a temperature of more than 300 °C [90].

3.3.6 Metal–organic Frameworks (MOFs)

The first people to find metal organic frameworks (MOFs) were Hoskins and Robson in 1989 [91]. They are a new type of crystalline hybrid porous material. Several types of porous MOFs are made up of metal ions or clusters that are connected to organic ligands in organized one, two, or three-dimensional frameworks such as MIL-101(Fe), MIL-125(Ti), UiO-66(Zr). There are several ways to make MOF, including the evaporation solvent approach, the diffusion method, the hydrothermal or solvent-thermal method, the ultrasonic and microwave method, and the microwave method. It is important to note that the structure of MOF may be easily changed by choosing the right organic ligands. MOF has been of attention all over the world for the last 20 years because of its unique qualities, such as its very large surface area (up to $3000 m^2 g^{-1}$), exceptional thermal stability (>400°C), customizable pore structure, and easy functionalization [92]. Researchers have looked at the many possible uses of MOFs in gas storage, separations, heterogeneous catalysis, and sensing. The open metal sites on the pore surfaces of MOFs can let different VOCs stick to them better. MOF, on the other hand, keeps its permanent structure and crystalline order even after being regenerated [93]. Different kinds of MOFs, such as the MIL series, the UiO series, and the ZIF series, have been made to get rid of VOCs. Vellingiri et al. looked at how well different kinds of MOFs adsorbed toluene under normal settings. We assessed the balanced adsorption capabilities of all the MOFs in the following order: ZIF-67 (224 mg g^{-1}) > UiO-66 (166 mg g^{-1}) > MOF-199 (159 mg g^{-1}) > MIL-101 (98.3 mg g^{-1}). ZIF-67 has the highest maximum adsorption capacity, which might be because it has the largest surface area ($1401 m^2 g^{-1}$) [94].

MOF is the best adsorbent for VOCs since its pore structure can be changed, and it has amazing physical and chemical capabilities. In general, MOF may hold more VOCs than regular adsorbents like AC and zeolite. The modification methods may be used in several ways to improve the hydrophobic property and adsorptive selectivity of MOF. There are further problems that make it hard to use in industry, such being weak dispersive forces because they have a lot of empty space and not enough open metal sites that are good for coordination and catalysis [95]. Also, using MOF to adsorb VOCs is still too expensive because it costs a lot to synthesize.

In conclusion, photocatalytic materials are important for converting light energy into chemical energy during the photocatalytic breakdown of BTEX. They can help to create electron–hole pairs that lead to active oxygen species. Semiconductors with a narrow band gap allow electrons to be excited from the valence band to the conduction band with relatively low energy from photons. However, the lifespan of photo-generated electrons is very short. Research shows that both TiO_2-based and non-TiO_2 photocatalytic materials perform differently in BTEX removal. Compared to semiconductor monomers or doping elements, heterojunction compositions have shown remarkable performance. With this, it warrants further exploration into developing heterojunctions using semiconductors with different bandgaps. Because of the difference between the conduction and valence bands, band bending is present at the heterojunction interface. This can provide a built-in field that can help take the electrons and holes in opposite directions which would decrease the photocarrier recombination rate. As a result, more reactive oxygen species could be produced so the appropriate semiconductors and heterojunction design is important for the photocatalytic degradation of BTEX.

4 Effect of Parameters on Photocatalytic BTEX Removal

4.1 Operating Reactor System

Both batch and continuous flow reactors can be utilized for photocatalytic removal of BTEX (Benzene, Toluene, Ethylbenzene, Xylene) with each having its pros and cons. Batch reactors are commonly found in laboratories or for small scale processes. In these reactors, a photocatalyst along with the contaminated water is placed in a vessel, irradiated, and allowed to react until a satisfactory level of BTX removal is accomplished. While offering great control over reaction parameters and flexibility for research, batch systems suffer from high operational costs, major idle time between cycles, and challenges in scaling up to industrial applications due to the control needed for catalyst separation. However, for batch systems tailored to high throughput or constant flow reactors like fixed bed reactors PMRs with immobilized catalysts or photocatalytic membrane reactors are best suited for large scale real-world scenarios of BTX removal from wastewater. These systems offer increased throughput and consistent product quality because they allow for polluted water to continuously enter, with treated water being continuously expelled. Although restrictions are due to mass transfer, the need for strong catalyst immobilization, or domination by continuous flow systems exist, the added efficiency of flexibility and scalability make them optimal for long term environmental restoration projects [96].

4.2 Flow Rate (Residence Time) and Initial Concentration of BTEX

The photocatalytic reaction of BTEX includes two processes: Photocatalytic oxidation and photocatalytic reduction. Photocatalytic oxidation/heterogeneous reaction of BTEX has also 3 different processes: adsorption, surface reaction and desorption process. In the surface reaction of BTEX, active species of oxidization are available which oxidize BTEX molecules into smaller molecules and break BTEX into CO_2 and H_2O. Initially increasing BTEX concentration increases photocatalytic degradation kinetics by supplying more target molecules for active species. Beyond a certain extent, nevertheless, the small number of active species is inadequate to completely oxidize all BTEX. This results in the accumulation of intermediate and byproducts that subsequently fight for active sites on the catalyst surface and stop more mineralization, therefore lowering the general efficiency of BTEX removal. This emphasizes a critical optimization point for effective photocatalytic treatment [97].

Particularly in continuous flow reactors, the flow rate directly affects the mass transfer properties and the residence time, that is, the average time a molecule stays in the reactor. Less time allows the BTEX molecules to come into touch with the UV light and the photocatalyst. This directly reduces the reaction time accessible for degradation, therefore lowering the BTEX conversion every pass over the reactor.

Reducing the total removal efficiency increases the flow rate lowers the fixed reactor volume. Ironically, in fixed-bed or baffled reactors particularly, raising the flow rate can improve turbulence and mixing. This enhanced fluid dynamics helps to lower external mass transfer restrictions, therefore assuring that more BTEX molecules reach the catalytic surface where reactions take place. This has advantages to a certain extent. The presence of a reduced flow rate enables a longer residence period, which in turn provides a greater number of opportunities for BTEX molecules to interact with the active species and undergo degradation. The elimination efficiency of BTEX is typically increased because of this. In situations when the flow rate is insufficient, particularly in immobilized catalyst systems, mass transfer may become a limiting issue. This is especially true in situations where the supply of new BTX to the catalyst surface is limited or when by-products build close to the surface [98].

4.3 *Type of Photocatalyst Used for BTEX Photocatalysis*

The selection of photocatalytic materials significantly influences the efficiency of photocatalytic activity of BTEX. Titanium dioxide (TiO_2) is highly recommended photocatalysts because of its stability, potential oxidizing capacity and non-toxicity. Modification of TiO_2 also used with different type of doping such as transition metals (e.g. Mn, Fe, V, Cr, Ni, Co), noble metals (Ag, Au, Pt) or integration of these modifications with other semiconductors like AgCl, ZIF-8, ZIF-67, Zr-MOF, Bi_2WO_6. These are capable for visible light absorption, reduce electron–hole recombination, and enhance overall photocatalytic efficiency. Ag/AgCl and Ag/AgCl/Bi_2WO_6 are the examples which have improved the visible light photocatalytic efficiency for the photocatalytic degradation [99, 100].

Different crystalline forms of TiO_2 (e.g., anatase, rutile, brookite) have different photocatalytic efficiencies for BTEX photo-degradation. Anatase is typically a more reactive phase of the TiO_2. Different nanoparticles with higher surface area (e.g., TiO_2 nanoparticles, ZIF-67, MIL-10, ZIF-8, Zr-MOF) offer more active sites for adsorption and reactivity which have resulted in enhanced BTEX degradation efficiency. Porous structures (high porous materials) can additionally enhance photocatalytic activity because of high absorption capacity [101].

4.4 *Light Source and Intensity*

For a photocatalyst to be effective, it must absorb photons that are equal to or greater than its band gap. For semiconductors with a wide bandgap such as titanium dioxide (TiO_2), the use of ultraviolet light is common; however, researchers are focusing on developing photocatalysts that can be activated using visible light so that solar energy can be utilized more efficiently. At lower levels of illumination, the rate of photocatalytic degradation reacts proportionally (first-order kinetics) to the illumination level and may increase in degree. The reasons for these phenomena are due to an increase in the flow of photons which results in an increased rate of creation of electron–hole pairs. With higher levels of illumination, either the rate may become proportional to the square root of the intensity (half-order kinetics) or not dependent at all on light intensity (zero-order kinetics). Saturation happens when there is too much charge carrier being produced compared to their reaction with adsorbed species ions. This will lead to accelerated recombination where electrons and holes combined [102].

4.5 Temperature

While photocatalytic reactions are a little less sensitive to temperature as compared to traditional thermal reactions, temperature still does play a role. It can facilitate kinetic limitations of charge carrier systems and adsorption processes. However, distinction processes within a system may cause an accelerator effect with excessive heating. Processes functioning below optimal ranges suffer from reduced oxygen solubility pivotal for ROS formation, oxygen species deficit plunge oxidizing electron–hole recombination; both of which degrade performance efficiency in excess. Optimal temperature range is often considered between 20 and 80 °C [103].

4.6 pH of the Solution (for Aqueous Systems)

pH has the potential to employ a significant impact on the surface charge of the photocatalyst. The adsorption behavior of BTEX molecules, and the production of reactive oxygen species (ROS) such as hydroxyl radicals ($OH^{\bullet}$) and superoxide radicals ($O_2^{\bullet-}$) has great impact on photocatalysis process of BTEX. The ideal pH for the breakdown of BTEX varies based on the photocatalyst that is being used and the properties of the BTEX compounds that are being used. According to some studies, the presence of a basic medium has the potential to accelerate the generation of ($OH^{\bullet}$) radicals, while the presence of acidic circumstances may enhance degradation for specific photocatalysts [104].

5 Demonstration of BTEX Photocatalysis at Industry Scale

5.1 Pilot-Scale Air Purification Systems

The photocatalytic process can purify the gaseous emission from industrial exhaust through the elimination of BTEX. An industrial air purification system was set up using a pilot-scale fixed bed photoreactor. In this reactor, TiO_2 nanoparticles were immobilized on a glass fiber mesh. The process showed an impressive removal of 500 ppm of the toluene degraded in a continuous flow system under UV-A irradiation [105]. In another study, toluene elimination was done using a photocatalytic reactor using a honeycomb monolith covered with TiO_2. In this system, effective toluene removal was observed [106, 107].

5.2 *Industrial Wastewater Treatment*

Some case studies have proven that photocatalytic technology can be applied to treat industrial wastewater. A photocatalytic reactor designed using TiO_2 nanoparticles was employed at pilot scale to treat the benzene polluted wastewater. The study demonstrated effective treatment, catalyst deactivation and requirements for the catalyst recycling [108]. In another study, photocatalyst was immobilized on a substrate and xylene was degraded in a fixed-bed reactor. The immobilization eliminates the expensive separation process of the catalysts and its recycling, adeal for the industrial application [109].

6 Challenges in Photocatalysis of BTEX

There are several problems with photocatalytic degradation of BTEX from wastewater and air. Most of these problems come from the fact that catalysts don't work well, such as when they don't absorb light well, when electron–hole recombination happens too quickly, or when they become inactive because they are dirty or poisoned. The procedure is made further harder by the creation of harmful intermediates, the difficulty of transferring BTEX to the catalyst surface, and the difficulty of finding the best operating settings. Wastewater has a lot of problems, such a complicated matrix with toxins that fight one another, ionic interference, turbidity, and the fact that it is hard to separate catalysts. When cleaning the air, low BTEX levels, high humidity, and high flow rates can all be problems. To get past these problems, we need to make new, very effective, and stable photocatalysts, as well as better reactor designs and a full understanding of how reactions work in different environmental situations [110].

7 Conclusion and Future Aspects of BTEX Photocatalysis

Photocatalysis is one of the most promising techniques due to its low energy consumption and less hydroxide based photocatalysts, semiconductors, TiO_2 modified with metals, non-metals and semiconductors which are improving photocatalytic activity of BTEX through photocatalytic oxidation of BTEX VOCs and improve its catalytic life. Metal oxides have a wide band gap and high light absorption capacity, so they are the efficient photocatalyst, other than metal oxides, hydroxides demonstrate excellent photocatalytic activity due to this hydroxyl radicals which lead towards high photocatalytic performance. g-C_3N_4 has shown the worthy photocatalytic activity due to its high visible light absorption capacity. Bismuth (Bi) based photocatalyst and their derivatives have also shown noteworthy performance due to its high visible light absorption capacity.

MOF-based composites, particularly when applied to cost-effective microporous materials such as biochar, clays, or zeolites, represent a promising alternative to traditional single adsorbents like activated carbons or zeolites. The absorption of VOCs is controlled mainly by the textural properties and the surface chemistry of porous adsorbents. The high specific surface area and pore volume, more so in micropores, of the serial carbons are beneficial to physical adsorption. Chemical adsorption refers to surface functional groups of adsorbents and polarity, molecular mass, melting point and mass of absorbents.

Moreover, the parameters also affect the photocatalysis of BTEX such as operation mode, initial concentration of pollutants, flow rate, temperature of reactors, reactor systems, pH and light intensity. Photocatalytic activity also depends on the mechanism of photocatalyst, the breakdown processes vary in specifics due to the positioning and quantity of methyl and ethyl groups. The integration with other technologies and the regeneration of deactivated catalysts are highly significant for practical applications. Despite advancements in photocatalytic degradation of BTEX, numerous challenges remain to be addressed to comply with severe emission requirements cost-effectively. In the future, we believe that effort should concentrate on the following elements.

- Development of new photocatalysts like quantum dots (CQDs), perovskite materials and graphitic carbon nitride g-C_3N_4 which have wide band gap and efficient absorption capacity of visible light.
- Integration of noble metals to semiconductors like TiO_2 and ZnO to improve their Surface plasmon resonance (SPR) and high absorption of visible light to improve photocatalytic activity.
- Development of 2D materials such as MXenes, graphene and their composites to enhance the surface area for charge separation and reduce recombination to improve photocatalytic activity.
- Design of novel photoreactors such as flow-through photoreactors which can be utilized for continuous flow for large scale degradation of BTEX in air and water streams such as drains of refinery.
- Most of the tests were conducted in a laboratory with high concentrations and simplified compositions, which most likely led to performance and mechanisms quite different from actual operating conditions. Therefore, further studies should focus on the photodecomposition of mixed volatile organic compounds (VOCs) having high emissions from industrial activities and greatly diluted pollutants from indoor spaces.

- For low concentration and non-continuous pollution, the design of materials that can effectively adsorb BTXs pollutants and thereafter photo-catalytically decompose them is very favorable.
- Using computational modeling such as density functional theory (DFT) and other modeling techniques to understand the properties of materials, reaction mechanism, rate of reaction, adsorption energy, calculating active sites, energy gap, designing of photocatalytic materials and their effect on photocatalysis, catalyst deactivation and regeneration capacity.
- In Green synthesis and synthesis of cost effective photocatalyst, the goal is to design energy-efficient photocatalysts with minimal environmental impact and hazardous waste production such as activated carbon from agricultural waste, bio-waste and food-waste which are the prominent materials for BTEX photocatalysis.
- Valorization of BTEX or selective oxidation of BTEX may also be explored because the primary goal of BTEX degradation is to mineralize it into CO_2 and H_2O, the controlled and specific conditions required to transform BTEX into valuable chemicals while paradoxically aiming for complete mineralization in air purification, aligning this selective approach with circular economy principles.

The future of BTX photocatalysis is promising, since it is progressing towards systems that are more efficient, durable, and intelligently constructed. The use of the entire spectrum of solar light, the achievement of improved charge separation, the guarantee of long-term stability, and the incorporation of these technological developments into solutions that are both practicable and scalable for cleaner air and water will be the primary focuses.

References

1. He, C., Cheng, J., Zhang, X., Douthwaite, M., Pattisson, S., Hao, Z.: Recent advances in the catalytic oxidation of volatile organic compounds: a review based on pollutant sorts and sources. Chem. Rev. **119**, 4471–4568 (2019)
2. Liu, Y., Song, M., Liu, X., Zhang, Y., Hui, L., Kong, L., Zhang, Y., Zhang, C., Qu, Y., An, J.: Characterization and sources of volatile organic compounds (VOCs) and their related changes during ozone pollution days in 2016 in Beijing. China. Environ. Pollut. **257**, 113599 (2020)
3. Yu, B., Yuan, Z., Yu, Z., Xue-song, F.: BTEX in the environment: an update on sources, fate, distribution, pretreatment, analysis, and removal techniques. Chem. Eng. J. **435**, 134825 (2022)
4. Speight, J.G.: Heavy oil recovery and upgrading. Gulf Professional Publishing (2019)
5. Fent, K.W., Toennis, C., Sammons, D., Robertson, S., Bertke, S., Calafat, A.M., Pleil, J.D., Wallace, M.A.G., Kerber, S., Smith, D.: Firefighters' absorption of PAHs and VOCs during controlled residential fires by job assignment and fire attack tactic. J. Expo. Sci. Environ. Epidemiol. **30**, 338–349 (2020)
6. Choi, W.-S., Hwang, M.J., Kwon, Y.-J., Jung, S.-H., Youn, Y., Balathanigaimani, M.S., Lee, J.-W., Hwang, K.-J., Shim, W.-G.: Radio-frequency thermal plasma treated activated carbon impregnated with Mn and Ag for volatile organic compounds adsorption. J. Nanosci. Nanotechnol. **20**, 4423–4427 (2020)
7. Liu, Y., Feng, X., Lawless, D.: Separation of gasoline vapor from nitrogen by hollow fiber composite membranes for VOC emission control. J. Memb. Sci. **271**, 114–124 (2006)

8. Surolia, P.K., Tayade, R.J., Jasra, R.: V: Effect of anions on the photocatalytic activity of Fe (III) salts impregnated TiO_2. Ind. Eng. Chem. Res. **46**, 6196–6203 (2007)
9. Surolia, P.K., Lazar, M.A., Tayade, R.J., Jasra, R.: V: Photocatalytic degradation of 3, 3′-Dimethylbiphenyl-4, 4′-diamine (o-Tolidine) over nanocrystalline TiO2 synthesized by sol−gel, solution combustion, and hydrothermal methods. Ind. Eng. Chem. Res. **47**, 5847–5855 (2008)
10. Surolia, P.K., Tayade, R.J., Jasra, R.: V: Photocatalytic degradation of nitrobenzene in an aqueous system by transition-metal-exchanged ETS-10 zeolites. Ind. Eng. Chem. Res. **49**, 3961–3966 (2010)
11. Napruszewska, B.D., Michalik-Zym, A., Dula, R., Bielańska, E., Rojek, W., Machej, T., Socha, R.P., Lityńska-Dobrzyńska, L., Bahranowski, K., Serwicka, E.M.: Composites derived from exfoliated Laponite and Mn-Al hydrotalcite prepared in inverse microemulsion: a new strategy for design of robust VOCs combustion catalysts. Appl. Catal. B **211**, 46–56 (2017)
12. Kim, E.H., Chun, Y.N.: VOC decomposition by a plasma-cavity combustor. Chem. Eng. Process. **104**, 51–57 (2016)
13. Padhi, S.K., Gokhale, S.: Biological oxidation of gaseous VOCs–rotating biological contactor a promising and eco-friendly technique. J. Environ. Chem. Eng. **2**, 2085–2102 (2014)
14. Surolia, P.K., Tayade, R.J., Jasra, R.: V: TiO2-coated cenospheres as catalysts for photocatalytic degradation of methylene blue, p-nitroaniline, n-decane, and n-tridecane under solar irradiation. Ind. Eng. Chem. Res. **49**, 8908–8919 (2010)
15. Sharma, K., Vaya, D., Prasad, G., Surolia, P.K.: Photocatalytic process for oily wastewater treatment: a review. Int. J. Environ. Sci. Technol. **20**, 4615–4634 (2023)
16. Vaya, D., Surolia, P.K.: Semiconductor based photocatalytic degradation of pesticides: an overview. Environ. Technol. Innov. **20**, 101128 (2020)
17. Mishra, S., Chakinala, N., Chakinala, A.G., Surolia, P.K.: Photocatalytic degradation of methylene blue using monometallic and bimetallic Bi-Fe doped TiO_2. Catal. Commun. **171**, 106518 (2022)
18. Mishra, S., Chakinala, N., Chakinala, A.G., Surolia, P.K.: Photocatalytic performance of Bi/Zn co-doped sol–gel synthesized TiO2 nanoparticles. J. Chem. Technol. Biotechnol. **99**, 531–539 (2024)
19. Njoku, V.O.N., Arinze, C., Chizoruo, I.F., Blessing, E.N.: A review: effects of air, water and land dumpsite on human health and analytical methods for determination of pollutants. Anal. Meth. Environ. Chemistr. J. **4**, 80–106 (2021)
20. Fayemiwo, O., Moothi, K., Daramola, M.: BTEX compounds in water–future trends and directions for water treatment. Water Sa. **43**, 602–613 (2017)
21. Mandade, P.: Introduction, basic principles, mechanism, and challenges of photocatalysis. In: Handbook of Nanomaterials for Wastewater Treatment, pp. 137–154. Elsevier (2021)
22. Dong, S., Cui, L., Tian, Y., Xia, L., Wu, Y., Yu, J., Bagley, D.M., Sun, J., Fan, M.: A novel and high-performance double Z-scheme photocatalyst ZnO-SnO_2-Zn2SnO4 for effective removal of the biological toxicity of antibiotics. J. Hazard. Mater. **399**, 123017 (2020)
23. Connelly, K.A., Idriss, H.: The photoreaction of TiO2 and Au/TiO2 single crystal and powder surfaces with organic adsorbates: emphasis on hydrogen production from renewables. Green Chemistr. **14**, 260–280 (2012)
24. Kumar, M., Surolia, P.K., Mishra, S., Guru, K., Sethia, G., Bhatt, H.B., Chakinala, A.G., Prasad, G.: Sustainable sunlight and UV-Driven photocatalytic degradation of methylene blue by employing cellulose/GO/TiO_2–Bi composite material. Arab J. Sci. Eng. 1–14 (2024)
25. Saini, P., Ahmadizamani, K., Chakinala, N., Mukherjee, S., Sethia, G., Chakinala, A.G., Surolia, P.K.: Bismuth doped g-C3N4 composites for enhanced photocatalytic degradation of ciprofloxacin. J. Mol. Struct. **1321**, 140013 (2025)
26. Crane, B.R., Arvai, A.S., Ghosh, S., Getzoff, E.D., Stuehr, D.J., Tainer, J.A.: Structures of the N ω-hydroxy-l-arginine complex of inducible nitric oxide synthase oxygenase dimer with active and inactive pterins. Biochemistry **39**, 4608–4621 (2000)
27. Mishra, S., Chakinala, N., Chakinala, A.G., Surolia, P.K.: Bimetallic Bi/Zn decorated hydrothermally synthesized TiO_2 for efficient photocatalytic degradation of nitrobenzene. Catal. Commun. **172**, 106538 (2022)

28. Fischer, K., Gawel, A., Rosen, D., Krause, M., Abdul Latif, A., Griebel, J., Prager, A., Schulze, A.: Low-temperature synthesis of anatase/rutile/brookite TiO_2 nanoparticles on a polymer membrane for photocatalysis. Catalysts **7**, 209 (2017)
29. Ma, X., Dang, R., Liu, Z., Yang, F., Li, H., Guo, T., Luo, J.: Facile synthesis of heterogeneous recyclable Pd/CeO2/TiO_2 nanostructured catalyst for the one pot hydroxylation of benzene to phenol. Chem. Eng. Sci. **211**, 115274 (2020)
30. Li, X.-S., Ma, X.-Y., Liu, J.-L., Sun, Z.-G., Zhu, B., Zhu, A.-M.: Plasma-promoted Au/TiO_2 nanocatalysts for photocatalytic formaldehyde oxidation under visible-light irradiation. Catal. Today **337**, 132–138 (2019)
31. Xu, M., Wang, Y., Geng, J., Jing, D.: Photodecomposition of NOx on Ag/TiO_2 composite catalysts in a gas phase reactor. Chem. Eng. J. **307**, 181–188 (2017)
32. Chen, M., Wang, H., Chen, X., Wang, F., Qin, X., Zhang, C., He, H.: High-performance of Cu-TiO_2 for photocatalytic oxidation of formaldehyde under visible light and the mechanism study. Chem. Eng. J. **390**, 124481 (2020)
33. Wysocka, I., Markowska-Szczupak, A., Szweda, P., Ryl, J., Endo-Kimura, M., Kowalska, E., Nowaczyk, G., Zielińska-Jurek, A.: Gas-phase removal of indoor volatile organic compounds and airborne microorganisms over mono-and bimetal-modified (Pt, Cu, Ag) titanium (IV) oxide nanocomposites. Indoor Air **29**, 979–992 (2019)
34. Rao, Z., Shi, G., Wang, Z., Mahmood, A., Xie, X., Sun, J.: Photocatalytic degradation of gaseous VOCs over Tm3+-TiO_2: revealing the activity enhancement mechanism and different reaction paths. Chem. Eng. J. **395**, 125078 (2020)
35. You, J.-W., Vikrant, K., Lim, D.-H., Verma, S., Kukkar, D., Anwer, H., Kim, K.-H.: Photocatalytic potential of a titanium dioxide–supported platinum catalyst against VOCs with complicated composition under varying humidity conditions. J. Clean. Prod. **371**, 133487 (2022)
36. Lyulyukin, M., Kovalevskiy, N., Bukhtiyarov, A., Kozlov, D., Selishchev, D.: Kinetic aspects of benzene degradation over TiO_2-N and composite Fe/Bi_2WO_6/TiO_2-N photocatalysts under irradiation with visible light. Int. J. Mol. Sci. **24**, 5693 (2023)
37. Bathla, A., Younis, S.A., Kim, K.-H., Li, X.: TiO 2-based catalytic systems for the treatment of airborne aromatic hydrocarbons. Mater. Horiz. **10**, 1559–1579 (2023)
38. Wu, X.D., Zhong, N.F., Yuan, R.H., Che, X.H., Wang, Y., Ye, X.Y.: Fabrication of C-doped TiO2 nanosphere for BTEX degradation of paint thinners with visible light irradiation. In: Materials Science Forum, pp. 177–183. Trans Tech Publ (2022)
39. Hasanova, S.A.: Compared the efficiency of TiO_2 and N-Doped TiO_2 to degrade BTEX. Adv. Physic. Res. **3**, 123–128 (2021)
40. Shajari, S., Kowsari, E., Seifvand, N., Boorboor Ajdari, F., Chinnappan, A., Ramakrishna, S., Saianand, G., Dashti Najafi, M., Haddadi-Asl, V., Abdpour, S.: Efficient photocatalytic degradation of gaseous benzene and toluene over novel hybrid PIL@ TiO_2/m-GO composites. Catalysts **11**, 126 (2021)
41. Tang, J., Zhu, J., Liu, L., Xia, L., He, Z., Wang, D., Xu, X., Song, S.: Coupling urchin-like TiO_2 nanospheres with nitrogen and sulfur co-doped graphene quantum dots for visible-light-induced degradation of toluene. Chem. Eng. J. **482**, 148813 (2024)
42. Li, Y., Zhang, H., Tao, B., Zhang, J., Li, J.: Adsorption and adsorption-photocatalytic degradation of VOCs based on carbon materials: a review. Available at SSRN 4245544 (2022)
43. Mukherjee, A.: Carbon nanotubes-based sensors for environmental applications. Carbon Nanotube-Based Sens. 197–235 (2024)
44. Xu, L., Guo, P., Xu, J., Shen, B., Zhao, Z.: Regulation of TiO_2/ZSM-5 catalyst for enhanced photocatalytic toluene oxidation: intensified light absorption, charge separation and toluene adsorption. Colloids Surf. A Physicochem. Eng. Asp. **690**, 133832 (2024)
45. Li, D., Fang, Y., Lu, J., Sun, J., Zhao, X., Hou, N., Xing, J.: Enhanced biodegradation of PAHs by biochar and a TiO_2@ biochar composite under light irradiation: photocatalytic mechanism, toxicity evaluation and ecological response. Chem. Eng. J. **458**, 141495 (2023)
46. Al Mamari, S., Al Marzouqi, F., Al Nabhani, A., Kim, Y., Selvaraj, R.: Hollow TiO_2/g-C_3N_4 nanocomposite for photodegradation of volatile organic carbons under visible-light. Toxicol. Environ. Chem. **104**, 211–231 (2022)

47. Yuan, J., Li, G., Liu, X., Yang, Y., Yu, F., Cao, J., Fei, Z., Ma, J., Nazeeruddin, M.K., Dyson, P.J.: Catalytic oxidation of BTX (benzene, toluene, and xylene) using metal oxide perovskites. Adv. Funct. Mater. **34**, 2401281 (2024)
48. Xiao, Z., Li, J., Mai, X., Yang, J., Zhu, M.: Halide perovskite-based nanomaterials for the detection and photocatalytic removal of gaseous pollutants. Catal. Sci. Technol. (2024)
49. Zhou, Z., Dontsova, T.A.: Modification Methods to Enhance the Performance of TiO_2 in Photocatalysis (2023)
50. Nikhar, S., Chakraborty, M.: Assessing the photodegradation efficiency of benzene, toluene, and xylene (BTX): a comparative investigation using activated charcoal (AC), zeolitic imidazolate framework-8 (ZIF-8), and zirconium metal–organic framework (Zr-MOF). Water Sci. Technol. **90**, 3193–3209 (2024)
51. Theka, T.J., Swart, H.C., Motaung, D.E.: Recent trends, advances, and challenges in MOFs-derived metal oxide semiconductor-based sensors for BTEX detection: a review. Inorg Chem Commun. 112884 (2024)
52. Kumar, N., Kumari, M., Ismael, M., Tahir, M., Sharma, R.K., Kumari, K., Koduru, J.R., Singh, P.: Graphitic carbon nitride (g–C3N4)–assisted materials for the detection and remediation of hazardous gases and VOCs. Environ. Res. **231**, 116149 (2023)
53. Tryba, B., Wozniak, M., Zolnierkiewicz, G., Guskos, N., Morawski, A., Colbeau-Justin, C., Wrobel, R., Nitta, A., Ohtani, B.: Influence of an electronic structure of N-TiO2 on its photocatalytic activity towards decomposition of acetaldehyde under UV and fluorescent lamps irradiation. Catalysts **8**, 85 (2018)
54. Kavitha, R., Devi, L.G.: Synergistic effect between carbon dopant in titania lattice and surface carbonaceous species for enhancing the visible light photocatalysis. J. Environ. Chem. Eng. **2**, 857–867 (2014)
55. Mahmood, A., Wang, X., Shi, G., Wang, Z., Xie, X., Sun, J.: Revealing adsorption and the photodegradation mechanism of gas phase o-xylene on carbon quantum dots modified TiO_2 nanoparticles. J. Hazard. Mater. **386**, 121962 (2020)
56. Li, M., Song, W., Zeng, L., Zeng, D., Xie, C., Yang, Q.: Mechanistic study of N-H-and H–N-codoping of a TiO_2 photocatalyst for efficient degradation of benzene under visible light. RSC Adv. **10**, 2757–2766 (2020)
57. Di Valentin, C., Pacchioni, G.: Trends in non-metal doping of anatase TiO2: B, C. N and F. Catal Today. **206**, 12–18 (2013)
58. Jo, W.-K., Kang, H.-J.: Aluminum sheet-based S-doped TiO_2 for photocatalytic decomposition of toxic organic vapors. Chin. J. Catal. **35**, 1189–1195 (2014)
59. Xin, Y., Chen, Q., Zhang, G.: Construction of ternary heterojunction CuS-CdS/TiO2 nanobelts for photocatalytic degradation of gaseous toluene. J. Alloys Compd. **751**, 231–240 (2018)
60. Paul, N., Purkayastha, D.D., Krishna, M.G.: TiO_2/SnO_2 and SnO_2/TiO_2 heterostructures as photocatalysts for degradation of stearic acid and methylene blue under UV irradiation. Superlattices Microstruct. **129**, 105–114 (2019)
61. Zhou, W., Shen, B., Wang, F., Zhang, X., Zhao, Z., Si, M., Guo, S.: Enhanced photocatalytic degradation of xylene by blackening TiO_2 nanoparticles with high dispersion of CuO. J. Hazard. Mater. **391**, 121642 (2020)
62. Muruganandam, L., ALSharyani, A.K.: Use of zinc oxide nanorods in the photocatalytic degradation activity of tricosane hydrocarbon from produced water under visible light irradiations. Desalin. Water Treat. **317**, 100288 (2024)
63. Wang, Q.: Environmental Catalysts for Remediation of Volatile Organic Compounds. Hong Kong University of Science and Technology (Hong Kong) (2018)
64. Huang, D., Fu, X., Long, J., Jiang, X., Chang, L., Meng, S., Chen, S.: Hydrothermal synthesis of MSn (OH) 6 (M= Co, Cu, Fe, Mg, Mn, Zn) and their photocatalytic activity for the destruction of gaseous benzene. Chem. Eng. J. **269**, 168–179 (2015)
65. Fu, S., Zheng, Y., Zhou, X., Ni, Z., Xia, S.: Visible light promoted degradation of gaseous volatile organic compounds catalyzed by Au supported layered double hydroxides: influencing factors, kinetics and mechanism. J. Hazard. Mater. **363**, 41–54 (2019)

66. Soni, A., Mishra, S., Vaya, D., Surolia, P.K.: Role of Ag and g-C3N4 over CaTiO3 for effective photocatalytic degradation of nitrobenzene. Inorg. Chem. Commun. **159**, 111862 (2024)
67. Huang, Z., Guan, B., Li, P., Liu, T., Ma, X., Xiong, R., Nian, J., Li, J.: Degradation performance and mechanism of VOCs in asphalt by alkali metal-doped La2Ce2O7-based fluorite through pyroelectric and photocatalytic synergy. Constr. Build. Mater. **443**, 137842 (2024)
68. Liang, C., Li, C., Zhu, Y., Du, X., Yao, C., Ma, Y., Zhao, J.: Recent advances of photocatalytic degradation for BTEX: materials, operation, and mechanism. Chem. Eng. J. **455**, 140461 (2023)
69. Cerón-Urbano, L., Aguilar, C.J., Diosa, J.E., Mosquera-Vargas, E.: Nanoparticles of the perovskite-structure $CaTiO_3$ system: the synthesis, characterization, and evaluation of its photocatalytic capacity to degrade emerging pollutants. Nanomaterials **13**, 2967 (2023)
70. Zhang, S., Pu, W., Chen, A., Xu, Y., Wang, Y., Yang, C., Gong, J.: Oxygen vacancies enhanced photocatalytic activity towards VOCs oxidation over Pt deposited Bi2WO6 under visible light. J. Hazard. Mater. **384**, 121478 (2020)
71. Liu, Y., Zhu, C., Sun, J., Ge, Y., Song, F., Xu, Q.: In situ assembly of CQDs/Bi_2 WO 6 for highly efficient photocatalytic degradation of VOCs under visible light. New J. Chem. **44**, 3455–3462 (2020)
72. Cheng, J., Shi, S., Tang, T., Tian, S., Yang, W., Zeng, D.: Controllable topological transformation from BiOCl hierarchical microspheres to Bi_2WO6 superstructures in the Bi–W–Cl–O system. J. Alloys Compd. **643**, 159–166 (2015)
73. Dong, X., Cui, W., Wang, H., Li, J., Sun, Y., Wang, H., Zhang, Y., Huang, H., Dong, F.: Promoting ring-opening efficiency for suppressing toxic intermediates during photocatalytic toluene degradation via surface oxygen vacancies. Sci. Bull. (Beijing) **64**, 669–678 (2019)
74. Wang, S., Guan, Y., Zeng, R., Zhang, Z., Liu, L., Li, Z., An, W., Fu, Y.: Plate-like WO3 inserting into I-deficient BiO1. 2I0. 6 microsphere for highly efficient photocatalytic degradation of VOCs. J. Taiwan Inst. Chem. Eng. **105**, 96–103 (2019)
75. Parvari, R., Ghorbani-Shahna, F., Bahrami, A., Azizian, S., Assari, M.J., Farhadian, M.: A novel core-shell structured α-Fe_2O_3/Cu/g-C_3N_4 nanocomposite for continuous photocatalytic removal of air ethylbenzene under visible light irradiation. J. Photochem. Photobiol. A Chem. **399**, 112643 (2020)
76. Zhao, X., Li, X., Zhu, T., Tang, X.: Adsorption behavior of chloroform, carbon disulfide, and acetone on coconut shell-derived carbon: experimental investigation, simulation, and model study. Environ. Sci. Pollut. Res. **25**, 31219–31229 (2018)
77. Romero-Anaya, A.J., Lillo-Rodenas, M.A., Linares-Solano, A.: Factors governing the adsorption of ethanol on spherical activated carbons. Carbon N Y. **83**, 240–249 (2015)
78. Amitay-Rosen, T., Leibman, A., Nir, I., Zaltsman, A., Kaplan, D.: The effects of aging on the dynamic adsorption of hazardous organic vapors on impregnated activated carbon. J. Occup. Environ. Hyg. **12**, 130–137 (2015)
79. Suzuki, R.M., Andrade, A.D., Sousa, J.C., Rollemberg, M.C.: Preparation and characterization of activated carbon from rice bran. Bioresour. Technol. **98**, 1985–1991 (2007)
80. Hsi, H.-C., Horng, R.S., Pan, T.-A., Lee, S.-K.: Preparation of activated carbons from raw and biotreated agricultural residues for removal of volatile organic compounds. J. Air Waste Manag. Assoc. **61**, 543–551 (2011)
81. Raphey, V.R., Henna, T.K., Nivitha, K.P., Mufeedha, P., Sabu, C., Pramod, K.: Advanced biomedical applications of carbon nanotube. Mater. Sci. Eng. C **100**, 616–630 (2019)
82. Smalley, R.E.: Crystalline ropes of metallic carbon nanotubes. In: Supercarbon: Synthesis, Properties and Applications, pp. 31–40. Springer (1998)
83. Na, C.-J., Yoo, M.-J., Tsang, D.C.W., Kim, H.W., Kim, K.-H.: High-performance materials for effective sorptive removal of formaldehyde in air. J. Hazard. Mater. **366**, 452–465 (2019)
84. Lu, J., Do, I., Drzal, L.T., Worden, R.M., Lee, I.: Nanometal-decorated exfoliated graphite nanoplatelet based glucose biosensors with high sensitivity and fast response. ACS Nano **2**, 1825–1832 (2008)
85. Allahbakhsh, A., Arjmand, M.: Graphene-based phase change composites for energy harvesting and storage: state of the art and future prospects. Carbon N Y. **148**, 441–480 (2019)

86. Yu, L., Wang, L., Xu, W., Chen, L., Fu, M., Wu, J., Ye, D.: Adsorption of VOCs on reduced graphene oxide. J. Environ. Sci. **67**, 171–178 (2018)
87. Guru, K., Chakinala, N., Chakinala, A.G., Surolia, P.K.: Recent advances in catalyst development for enhanced p-Xylene production via toluene methylation. Chem. Bio. Eng. Rev. e70021
88. Guru, K., Chakinala, N., Chakinala, A.G., Surolia, P.K.: Catalyst modification strategies and reactor engineering for enhanced p-Xylene production via toluene methylation: a review. Chem. Select **10**, e00585 (2025)
89. Nien, K.C., Chang, F.T., Chang, M.B.: Adsorption of mesitylene via mesoporous adsorbents. J. Air Waste Manag. Assoc. **67**, 1319–1327 (2017)
90. Su, Y.-C., Kao, H.-M., Wang, J.-L.: Mesoporous silicate MCM-48 as an enrichment medium for ambient volatile organic compound analysis. J. Chromatogr. A **1217**, 5643–5651 (2010)
91. Hoskins, B.F., Robson, R.: Infinite polymeric frameworks consisting of three dimensionally linked rod-like segments. J. Am. Chem. Soc. **111**, 5962–5964 (1989)
92. Yang, K., Sun, Q., Xue, F., Lin, D.: Adsorption of volatile organic compounds by metal–organic frameworks MIL-101: influence of molecular size and shape. J. Hazard. Mater. **195**, 124–131 (2011)
93. Zhao, Y.-T., Yu, L.-Q., Xia, X., Yang, X.-Y., Hu, W., Lv, Y.-K.: Evaluation of the adsorption and desorption properties of zeolitic imidazolate framework-7 for volatile organic compounds through thermal desorption-gas chromatography. Anal. Meth. **10**, 4894–4901 (2018)
94. Vellingiri, K., Kumar, P., Deep, A., Kim, K.-H.: Metal-organic frameworks for the adsorption of gaseous toluene under ambient temperature and pressure. Chem. Eng. J. **307**, 1116–1126 (2017)
95. Zhu, L., Meng, L., Shi, J., Li, J., Zhang, X., Feng, M.: Metal-organic frameworks/carbon-based materials for environmental remediation: a state-of-the-art mini-review. J. Environ. Manage. **232**, 964–977 (2019)
96. Khodadadian, F., De Boer, M.W., Poursaeidesfahani, A., Van Ommen, J.R., Stankiewicz, A.I., Lakerveld, R.: Design, characterization and model validation of a LED-based photocatalytic reactor for gas phase applications. Chem. Eng. J. **333**, 456–466 (2018)
97. Cheng, Z., Li, C., Chen, D., Chen, J., Zhang, S., Ye, J., Yu, J., Dionysiou, D.D.: A novel array of double dielectric barrier discharge combined with TiCo catalyst to remove high-flow-rate toluene: performance evaluation and mechanism analysis. Sci. Total Environ. **692**, 940–951 (2019)
98. Simantiraki, F., Gidarakos, E.: Comparative assessment of compost and zeolite utilisation for the simultaneous removal of BTEX, Cd and Zn from the aqueous phase: batch and continuous flow study. J. Environ. Manage. **159**, 218–226 (2015)
99. Silva, A.C.A., Ribeiro, W.S.M., Rezende, T.K. de L., Barbosa, A.I. dos S.: TiO_2 Doped with noble metals and transition metals: a review. Int. J. Metallurg. Alloys **9**, 1–12 (2023)
100. Ghamarpoor, R., Fallah, A., Jamshidi, M.: A review of synthesis methods, modifications, and mechanisms of ZnO/TiO_2-based photocatalysts for photodegradation of contaminants. ACS Omega **9**, 25457–25492 (2024)
101. Eddy, D.R., Permana, M.D., Sakti, L.K., Sheha, G.A.N., Solihudin, Hidayat, S., Takei, T., Kumada, N., Rahayu, I.: Heterophase polymorph of TiO_2 (Anatase, Rutile, Brookite, TiO_2 (B)) for efficient photocatalyst: fabrication and activity. Nanomaterials. **13**, 704 (2023)
102. Emmanuel, S.S., Olawoyin, C.O., Ayodele, I.D., Oluwole, O.J.: Emerging nanosemiconductors for photocatalytic degradation of mono-aromatic volatile organic compounds (BTEX): a pragmatic review. J. Organomet. Chem. **996**, 122767 (2023)
103. Chen, Y.-W., Hsu, Y.-H.: Effects of reaction temperature on the photocatalytic activity of TiO_2 with Pd and Cu cocatalysts. Catalysts **11**, 966 (2021)
104. Bahmani, M., Bitarafhaghighi, V., Badr, K., Keshavarz, P., Mowla, D.: The photocatalytic degradation and kinetic analysis of BTEX components in polluted wastewater by UV/H2O2-based advanced oxidation. Desalination Water Treat. **52**, 3054–3062 (2014)
105. Hosseini, S., Moghaddas, H., Masoudi Soltani, S., Kheawhom, S.: Technological Applications of Honeycomb Monoliths Prepared via Extrusion and Coating of Structured Supports in Environmental Processes: A Review (2019)

106. Ku, Y., Chen, J.-S., Chen, H.-W.: Decomposition of benzene and toluene in air streams in fixed-film photoreactors coated with TiO_2 catalyst. J. Air Waste Manag. Assoc. **57**, 279–285 (2007)
107. Tomašić, V., Jović, F., Gomzi, Z.: Photocatalytic oxidation of toluene in the gas phase: modelling an annular photocatalytic reactor. Catal. Today **137**, 350–356 (2008)
108. Rabahi, A., Assadi, A.A., Nasrallah, N., Bouzaza, A., Maachi, R., Wolbert, D.: Photocatalytic treatment of petroleum industry wastewater using recirculating annular reactor: comparison of experimental and modeling. Environ. Sci. Pollut. Res. **26**, 19035–19046 (2019)
109. Chen, Z., Kuschk, P., Reiche, N., Borsdorf, H., Kästner, M., Köser, H.: Comparative evaluation of pilot scale horizontal subsurface-flow constructed wetlands and plant root mats for treating groundwater contaminated with benzene and MTBE. J. Hazard. Mater. **209**, 510–515 (2012)
110. Garcia, P.M.P., Ortiz, N.: Solar photodecomposition for removing BTEX compounds from groundwater contaminated by gasoline station activities. J. Geosci. Environ. Protect. **13**, 325–346 (2025)

Applications of Perovskite Materials in Energy Storage and Efficiency Systems

Rashi Nathawat, Sanjay Kumar, and Satyapal S. Rathore

Abstract Perovskite materials, generally in the form of ABX_3, are promising candidates for next-generation energy storage and efficiency applications. This chapter presents an in-depth investigation of perovskite structures, synthetic routes, and their novel properties. It examines single-and multi-component perovskites from structural, optical, dielectric, magnetic, and multiferroic perspectives. Methods for the preparation of various types of perovskites using solid-state reactions, sol–gel, and hydrothermal methods are presented, and their effects on material properties are described. This chapter explains the use of perovskites in energy technologies, particularly photovoltaics, optoelectronics, and sensors, and highlights their potential for energy storage and conservation. Although there has been impressive progress in perovskite-based technologies, such as the rapid increase in solar cell efficiency, challenges such as material instability and environmental concerns due to lead-based perovskites have also been studied. This chapter provides a comprehensive overview of the role of perovskites in tackling global energy problems and summarizes the research required to address existing problems, thereby driving the development of future energy technologies.

1 Introduction

Energy makes every day of modern life possible. Energy is the backbone of every facet of our modern existence, whether it is maintaining the lights in our buildings, keeping our transportation machines in rotation, or providing the stomach of the industrial factories that keep us progressing as a cohesive unit. Internationally, the need for energy is likely to continue rising, motivated by population expansion, urban growth, and the boosting of dependence on digital structures. As we progress through

---

R. Nathawat (✉) · S. Kumar
Functional Ceramics and Smart Materials Lab, Department of Physics, Manipal University Jaipur, Jaipur, India
e-mail: rashi.nathawat@jaipur.manipal.edu

S. S. Rathore
Department of Physics, Cluster University of Jammu, Jammu, India

L. Ledwani and S. Srivastava (eds.), *Experimental and Computational Advances in Materials*, Engineering Materials, https://doi.org/10.1007/978-981-95-4180-5_8

the twenty-first century, one of the most significant challenges is discovering energy sources that are sustainable, efficient, and environmentally friendly.

Energy is a fundamental requirement for modern society, and the worldwide demand is projected to increase by 50% between 2018 and 2050 [1]. Fossil fuels covered 84% of the worldwide energy supply in 2020, but renewables only accounted for 11% [2]. This highlights the need for sources that can provide viable alternatives to power the rapidly approaching future. Energy can manifest as mechanical, thermal, chemical, electrical, or nuclear forms, and it may originate from fossil fuels, nuclear reactions, or renewable source. The role of each state or region in the international energy landscape is heterogeneous, depending on natural resources, technological development, and policy architecture. For example, nations that hold fossil fuel resources, such as the Middle East is a large example, have always cornered a lot of the fossil fuel market, and countries that have enough water have demanded hydroelectricity, such as Norway and Canada. In contrast, rapidly growing economies rely on solar and wind energy to meet their energy needs while minimizing environmental damage.

Energy is the vein of modern economies, driving daily production and consumption activities [3, 4]. The onset of the twenty-first century, the so-called era of green economic development, is characterized by an important transition from the use of fossil fuels to the adoption of renewable energy sources to combat the negative consequences of depletion and climate change caused by fossil fuels. The United Nations Sustainable Development Goals (SDGs) also support this transformation, especially SDG 7, which aims to "ensure access to affordable, reliable, and sustainable energy for everyone" [5, 6]. Countries have come to recognize that sustainable energy development and energy security are the two essential ingredients that can pave the way for each nation to pursue sustainable targets through strategic routes in the upcoming century [4]. As agreed and integrated into Vietnam's policies, both elements have been allies along a strategic road for Vietnam.

The BRICS nations have recently reaffirmed their collective commitment to achieving the targets outlined in the 2030 Sustainable Development Goals agenda, which further highlights the importance of sustainable development for their economic growth paths [7]. The global shift towards renewables is due to the demands of climate change mitigation, net-zero energy goals, and energy security in the future [8, 9]. The renewable energy transition presents both opportunities and challenges for states. While energy transition offers a path towards economic prosperity, energy independence, and a lower environmental footprint, it requires significant capital expenditure on infrastructure, technology, and policies [10, 11].

Brazil, Germany, and India have adopted numerous approaches to augment their renewable energy capacities, which can serve as varied templates to expedite this transition [5]. The problem is that the transition is not happening quickly enough for us to be on track to meet global climate targets which means that we still need to see a lot more ambitious policy changes and more integrated approaches [5].

One of the most talked-about areas in renewable energy technologies is 'perovskite materials.' Perovskites, named after a mineral with a similar crystal structure, are a type of material known for their synchronized optical and electronic features. Their

Fig. 1 The perovskite configuration of $CaTiO_3$ and Lev Perovski, a Russian mineralogist [13]

high light absorption, low production costs, and low fabrication complexity make them particularly attractive for photovoltaic applications (e.g., solar cells). In just over ten years, perovskite solar cells have accomplished efficiency rates that, in some cases, surpass those of conventional silicon-based solar cells. These promising materials have many applications in energy storage. storage. Their exceptional ionic conductivity, stability under heat, and adjustable electronic properties make them highly appealing for application in energy storage systems, including batteries, supercapacitors, and fuel cells..

Perovskites have already demonstrated their ability to significantly enhance the energy density, charge-discharge rates, and longevity of storage systems. Perovskite-based photovoltaics now compete with conventional silicon-based technologies, with record power conversion efficiencies greater than 25% (NREL 2021). Owing to their versatility and effectiveness, perovskite materials have emerged as crucial materials in the fight against global energy problems that we are currently facing and have the potential to transform energy generation and energy storage technology.

The efficiency of these perovskite materials is not the only important factor. This low-temperature method for creating flexible solar cells opens new avenues for lightweight, portable, and wearable solar technology. Moreover, their tunable band gaps allow tandem solar cells to operate, in which various layers absorb a larger part of the solar spectrum and maximize energy conversion. The ability to adjust these properties positions perovskites as a revolutionary development in solar energy and a formidable competitor in various optoelectronic applications, such as light-emitting diodes (LEDs), photodetectors, and lasers.

As global energy demands rise and the environmental impacts of fossil fuel use become more pronounced, advanced materials like perovskites are set to be pivotal in the future development and commercialization of these technologies. They are sanctuaries of hope as we move toward a clean, efficient, and universally accessible energy future—one that coordinates the inventive character of scientific research alongside the lived conditions of people and the planet.

2 Perovskite Materials

Perovskite materials have quickly established themselves as important in the domains of materials science and energy technology. The main reason for their rapid rise in prominence is the remarkable structural versatility and excellent physical properties of perovskite materials. In 1839, Rose identified the first perovskite mineral, known as calcium titanate ($CaTiO_3$). It was named after the Russian mineralogist Lev Perovsky, giving rise to a large family of compounds that share the same crystal structures, see in Fig. 1. They are described by the general formula ABX_3 [12, 13].

Although initially referring to the mineral perovskite, perovskite has come to encompass a large family of synthetic and naturally occurring inorganic and organic-inorganic hybrid materials that share an unusual crystal structure that allows for extensive chemical substitution and property tuning through synthesis.

Perovskite materials have been a topic of interest for over 50 years because of their ferroelectric, piezoelectric, and superconducting properties and their important roles in capacitor technology and high-temperature superconductivity. The introduction of organic-inorganic halide perovskites, like $CH_3NH_3PbI_3$, as nearly perfect materials for photovoltaic uses in the early 2000s transformed the dynamics of this field of materials. However, this could be about to change, because for the last 30 years, hybrid peroskites have shown remarkable optoelectronic properties that have brought their performance (solar cell efficiency) to extremes by breaking the 4% barrier in 2009, followed by more than a 25% efficiency in only slightly more than a decade, making them serious challengers to traditional silicon solar cells. Beyond photovoltaics, the beneficial properties of perovskites, specifically their high ionic mobility, tunable electronic structures, redox activity, and structural flexibility, have made them appealing options for various energy storage uses [14, 15].

Perovskite oxides such as LLTO ($Li_{1/3}La_{1/3}TiO_{1/3}O_2$) and $Na_{1/2}Bi_{1/2}TiO_3$ have been successfully integrated as electrolytes or electrode materials in conventional Li-ion and Na-ion battery systems [16]. These exhibit enhanced conductivity and thermal stability, which are crucial for establishing safer and more efficient solid-state battery systems. Consequently, their catalytic and electrochemical properties make them the most desirable candidates for redox flow batteries that can connect electricity networks with energy stored at the grid level by enhancing energy conversion efficiency and long-term operational stability. Notably, perovskite oxides such as $LaNiO_3$ and $BaTiO_3$ can exhibit pseudo-capacitive properties that can also be implemented in supercapacitors (enabling rapid charge-discharge processes and frequent high power densities) [17].

Moreover, perovskites are excellent for oxygen ion conductivity and perform catalytic reactions in fuel cells and water-splitting apparatuses under hydrogen-powered energy systems, which will remain vital for hydrogen production and storage. Other perovskites exhibit phase transitions and shape memory effects, highlighting the potential for new applications of this family of materials in thermal and mechanical energy storage. However, these applications are not practical because of a myriad of challenges, including the long-term environmental stability of halide perovskites, toxicity of lead content, scalability of synthesis, and material degradation under operating stress. Nonetheless, ongoing advances in materials engineering, dopant design, and interface engineering have incrementally mitigated these bottlenecks.

Owing to continued investigation and interdisciplinary team efforts, perovskite materials are poised to enhance established energy-storage techniques and enable emerging technologies for energy sustainability and resilience on a global scale.

3 Structure of the Perovskite Materials

The wide range of structures and compositions found in perovskite materials (with the general formula ABX_3) allows for numerous applications. The ideal perovskite structure is cubic this means cubic corners will have an A cation, centre with B Cation, and face centres will have X anions [18, 19].

In the perovskite structure, the divalent A cation is positioned at the corner, where it is encircled by 12 oxygen anions located at the (0, 0, 0) corner site. Meanwhile, the tetravalent B cations are found at the (1/2, 1/2, 1/2) center of the structure, nestled within the core of the oxygen octahedral. Oxygen ions at (1/2, 1/2, 0) face center location of cubic scalor lattice A third, well-known description of BO_6 octahedra is a 3D framework with regular corner connectivities. The coordination number of atom A was 12. Ideal perovskites are constructed according to the structure shown in a three-dimensional (3D) octahedron corner-sharing BO_6 framework (with all B-O-B angles of 180°). In perovskites, structural distortion can be described by two elements: one that influences the dimensions of the a and b ionic factors, and the other being the electronic configuration of the metal ions.

Perovskite has two favored structural distortions: tilting of BO_6 octahedra and off-centering of B ions within the BO_6 octahedra. The first is a dispersive phase transition, and the second is an order-disorder phase transition. However, in many perovskites, this ideal structure is distorted from its decimal configuration owing to the ionic radii, temperature, and pressure. The Goldschmidt tolerance factor is utilized to forecast the compatibility of specific ions in forming a stable perovskite structure, taking into account the geometric limitations of the ions being combined [20–22].

$$t = \frac{r_A + r_O}{\sqrt{2}(r_B + r_O)} \tag{1}$$

The ionic radii of cations A and B are denoted by r_A and r_B, respectively, while r_O represents the ionic radius of the oxygen anions, measured in units.

In the perovskite, t $\sim$ 1.0, the material exhibited an ideal cubic perovskite structure with the highest symmetry and a stable form. For values between 0.9 and 1.0, the structure remains relatively stable but becomes distorted, with orthorhombic or rhombohedral polymorphs appearing more frequently. For other values between 0.75 and 0.9, the distortions become more pronounced, but the material can still have a perovskite-type phase. However, once the tolerance value is lower than 0.75 or larger than approximately 1.0, the structure is typically very different from a normal perovskite framework, resulting in the formation of other crystal structures, such as hexagonal or other non-perovskite structures.

Perovskites exhibit a variety of structural distortions, as they can take on tetragonal, orthorhombic, and rhombohedral forms [23]. For example, $BaTiO_3$ can be cubic, tetragonal, or rhombohedral (depending on the temperature [23]), and some structures even adopt more than one perovskite under different external conditions. It is worth noting that, although perovskites are commonly oxides, they may also have halide-type anions or even more complex ones, such as BH_4 [24–26]. This adjustability permits the incorporation of about 90% of the metallic elements from the Periodic Table into the A and/or B positions [27]. Owing to their structural variations and compositions, these materials exhibit a wide range of properties and applications, from solar cells to superconductors and catalysts [27–29]. Recent developments in machine learning and integrative approaches have offered new strategies to search for and predict new perovskite materials, facilitating material design and optimization to some extent [29].

4 Types of Perovskite Materials

4.1 Single Perovskite

In this section, we introduce the different types of single perovskites. The basic types of single perovskites are oxide perovskite, halide perovskite, hybrid perovskite, and lead-free perovskite (Fig. 2).

4.1.1 Oxide Perovskite

Perovskite oxides, represented by the formula ABO_3, consist of a wide range of materials where A and B are cations of differing sizes. The ideal structures of the oxide perovskites are shown in Fig. 3 [30]. Such materials display a broad diversity of properties and thus have attracted much attention in diverse areas such as energy conversion and storage, catalysis, and electronics [31, 32]. Perovskites have flexible structures that can be altered by changing their composition with various elements to modulate their physicochemical properties [33].

Examples of oxide perovskites include $BaTiO_3$, $PbTiO_3$, $SrTiO_3$ [34]. These materials are known to exhibit multiple structural phases, including tetragonal, orthorhombic, and rhombohedral, which can lead to diverse physical effects and

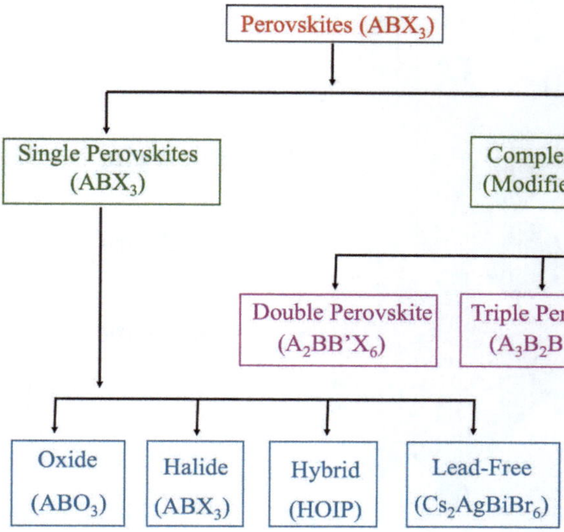

Fig. 2 A diagrammatic flowchart illustrating the various categories of perovskite materials

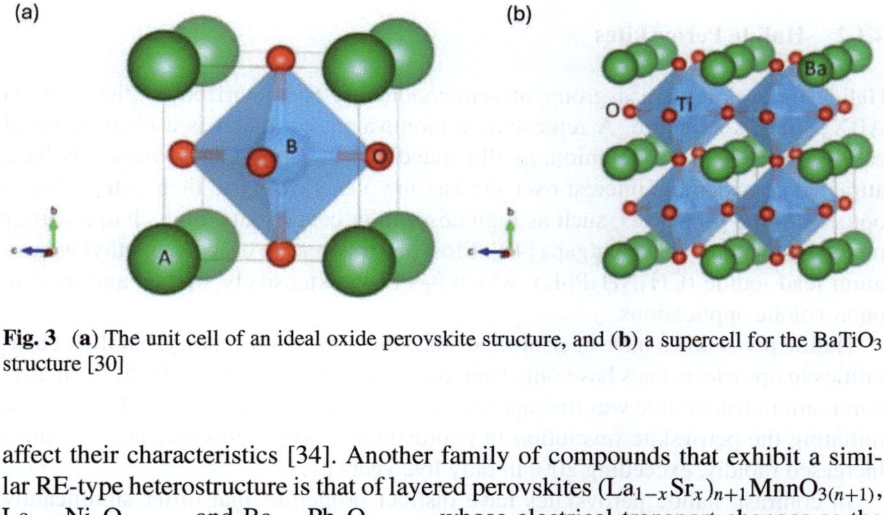

Fig. 3 (**a**) The unit cell of an ideal oxide perovskite structure, and (**b**) a supercell for the $BaTiO_3$ structure [30]

affect their characteristics [34]. Another family of compounds that exhibit a similar RE-type heterostructure is that of layered perovskites $(La_{1-x}Sr_x)_{n+1}MnnO_{3(n+1)}$, $La_{n+1}Ni_nO_{3(n+1)}$, and $Ba_{n+1}Pb_nO_{3(n+1)}$, whose electrical transport changes as the number of integrated perovskite slabs increases, the behavior shifts from semiconducting to metallic [33].

Notably, oxide perovskites are the most diverse in terms of applications (e.g., solid oxide fuel cells, cathode materials, and solid electrolytes) [35, 36]. They are attractive for these applications because of their ability to conduct oxide ions and electrons and their thermal stability. Moreover, perovskite oxides have potential applications in photocatalysis, thermoelectrics, and dielectrics [37, 38].

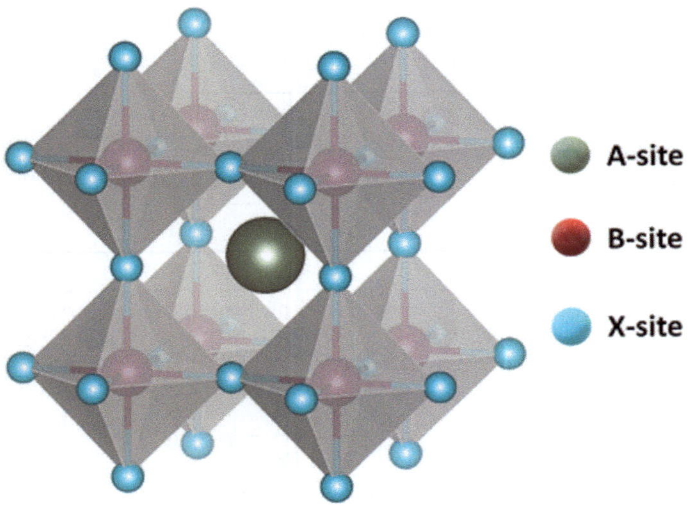

Fig. 4 A schematic structure of the halide perovskite ABX_3 [39]

4.1.2 Halide Perovskites

Halide perovskites are a group of semiconductors characterized by the formula ABX_3. In this structure, A represents a monovalent cation, B is a divalent metal cation, and X is a halide anion, as illustrated in Fig. 4 [39]. These materials have attracted considerable interest over the last few years owing to their extraordinary optoelectronic properties, such as high absorption coefficients, high charge-carrier mobilities, and tunable bandgaps [40]. Most perovskites are based on methyl ammonium lead iodide ($CH_3NH_3PbI_3$), which has been extensively studied and used in photovoltaic applications.

Halide perovskites have been around since the nineteenth century, but their capabilities in optoelectronics have only been realized in the last decade. In 2009, methylammonium lead iodide was first applied as a sensitizer in dye-sensitized solar cells, initiating the perovskite revolution in photovoltaics [41]. Cell efficiency has since increased rapidly, exceeding 20% in only five years [41].

In contrast, halide perovskites have distinct properties that differ significantly from those of common semiconductors. For instance, although they are crystalline, they exhibit low thermal conductivity and high thermal expansion coefficients [42]. Moreover, the ease with which these materials can be processed into high-quality thin films and single crystals is attractive for the cost-effective creation of high-performance optoelectronic devices [42]. As research in this field expands, halide perovskites are being increasingly utilized in various applications, including light-emitting diodes, lasers, X-ray detectors, and memory devices [43].

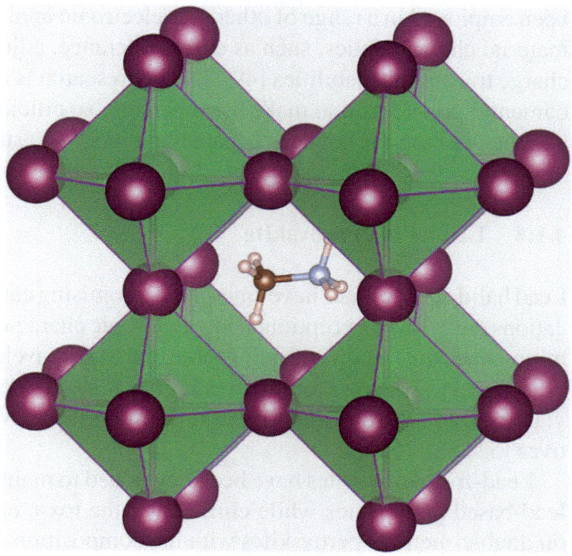

Fig. 5 The hybrid perovskite structure of the $CH_3NH_3PbI_3$ with atoms representation [45]

4.1.3 Hybrid Perovskite

Hybrid perovskites are distinct from other oxide perovskites because they consist of organic and inorganic components that form a perovskite crystal structure. These materials have been extensively studied in optoelectronics, primarily because of their excellent photovoltaic performance. Among the organics, methylammonium lead iodide ($CH_3NH_3PbI_3$ or $MAPbI_3$) stands out, with photovoltaic devices with power conversion efficiencies greater than 20% [44]. Figure 5 illustrates the crystal structure of the hybrid perovskite $CH_3NH_3PbI_3$ [45].

Hybrid perovskites have had a fascinating history in the field of optoelectronics, beginning in the early 2010s when their promise for solar cell applications was first acknowledged. Since then, research in this area has proliferated, studying a wide variety of compositions, structures, and uses beyond photovoltaics. For instance, formamidinium lead iodide ($FAPbI_3$) has emerged as a novel choice for high-performance solar cells [46].

Hybrid perovskites have shown remarkable performance, yet, they suffer from stability and toxicity issues. To alleviate these concerns, scientists have investigated low-dimensional perovskites, such as 2D arrangements, which have higher stability in the presence of moisture than their 3D structures [47]. Furthermore, research has been conducted to find lead-free replacements by investigating other ns^2 electronic configuration materials, where "n" denotes the outermost shell and "s^2" indicates two paired electrons in the s-orbital [48]. Beyond their use in solar cells, hybrid perovskites have been employed in a range of other optoelectronic applications due to their distinctive material characteristics, such as defect tolerance, adjustable bandgap, and excellent charge transport capabilities [49]. Further research is ongoing to understand the fundamental principles that make them perform so efficiently, such as polaronic effects and the interactions between organic and inorganic sublattices [50, 51].

4.1.4 Lead-Free Perovskite

Lead halide perovskites have emerged as promising candidates for photovoltaic applications due to their exceptional optoelectronic characteristics, such as a high photoluminescence quantum yield, adjustable emission wavelength, and easy solution-based processing [52]. However, its toxicity has raised concerns regarding its commercial viability and environmental impact, leading to the development of lead-free alternatives [52, 53].

Lead-free perovskites have been developed to maintain the excellent properties of lead-based perovskites while eliminating the toxic element. One approach focuses on double-metallic perovskites with the composition $A_2BB'X_6$, in which the B sites of the crystal lattice is sequentially occupied by metal cations with single and triple positive charges [53]. For example, this class of materials has been synthesized using a mechanochemical method for $Cs_2AgInBr_6$. Other strategies include tin-containing perovskites [54] and the incorporation of chirality into hybrid, lead-free perovskites [55].

Despite advancements in creating lead-free perovskites, challenges related to stability, phase transitions, and photoelectric conversion efficiency persist [56]. However, the use of these materials is not limited to photovoltaics; they can also be used in photodetectors, light-emitting diodes, and other optoelectronic devices [52, 53].

4.2 Complex Perovskite

Another class of materials is complex perovskites, which follow a general formula of ($A_3B_2B'O_9$) or ($A_2BB'O_6$), where sites B and B' are occupied by different cations within a three-dimensional framework. Such materials can exhibit a diverse set of physical properties, including ferroelectricity, magnetism, and high dielectric constants, which have valuable applications in multiple electronic and energy devices. The flexibility of the perovskite lattice enables ionic substitution to tailor its properties, provoking considerable interest in its application in sensors, capacitors, fuel cells, and photovoltaic devices.

These perovskites are divided into three parts: double perovskites, triple perovskites, and layered perovskites. All detailed information about these perovskites is discussed below.

4.2.1 Double Perovskite

A newly emerging category of sophisticated materials, known as double perovskites, has recently attracted interest in various optoelectronic applications as potential alternatives to lead-based perovskites. These materials are part of the broader perovskite family. Goldschmidt's early work on structure–composition relationships paved the way for the ubiquitous application of the perovskite crystal structure, a versatile ceramic host [57]. This adaptability has facilitated the creation of materials exhibiting a variety of characteristics, ranging from dielectrics to superconductors.

The general formula for double perovskites is $A_2BB'X_6$, which provides further versatility in customizing material properties compared to their single perovskite counterparts Evans et al. [58]. This structural adaptability permits the insertion of diverse components at multiple sites within the same material system (i.e., A, B, B′, and X sites), thus facilitating multiple functionalities. An example is $Cs_2AgBiBr_6$, which has been investigated as a lead-free option for photovoltaic applications and has interesting optoelectronic properties [59]. Instances include Ba_2AgIO_6, which shows promise as a novel semiconductor with a visible band gap, as well as the iodide double perovskite family Cs_2NaLnI_6 (Ln = Ce, Nd, Gd, Tb, and Dy) [60].

In summary, double perovskites are an exciting development in the field of materials science, providing enhanced stability and tunable characteristics compared to single perovskite counterparts. This intense research and development activity in recent years has been motivated by the potential applications of these new systems in a wide range of technologies, from photovoltaics to optoelectronics and energy-related technologies.

4.2.2 Triple Perovskite

Triple perovskites are an extension of the widely investigated perovskite family, which originated from the fundamental science of structure-composition relationships established by Goldschmidt [57]. The simple ABX_3 formula for perovskites has been generalized to include more complex compositions, such as triple perovskites.

The typical structure of triple perovskites is represented as $A_3B'B''X_9$. These compounds are usually composed of A, a monovalent cation such as $CH_3NH_3^+$, with B′ and B″ being distinct metal cations, and X serving as a halide anion. A well-known example of a triple perovskite is $Ba_3CoSb_2O_9$, which features a hexagonal crystal structure [61, 62]. This framework provides greater compositional versatility than classical perovskites and can be decorated with various inorganic and organic elements [63]. They also offer significantly more complexity than well-known double perovskites, which can confer interest with respect to selecting material properties and functionalities.

In contrast to standard perovskites, which are generally represented by the chemical formula ABX_3 and exhibit a disordered arrangement of the B cation, triple perovskites possess a three-dimensional lattice of corner-sharing BX_6 octahedra with an orderly distribution of B' and B" cations. This specific arrangement can give rise to unique electronic and optical characteristics in the material. A-site organic-inorganic hybrid triple perovskites are depicted as organic cations occupying one of the A-sites in an all-inorganic composition using rare earths [25]. The perovskite structure is distinguished by the diversity of molecular anions that can be hosted in its lattice, such as BH_4 in complex hydride perovskites, enabling a library of compositions and properties [25].

4.2.3 Lyered Perovskite

Two-dimensional (2D) perovskites, also known as layered perovskites, hold great potential for use in optoelectronic devices. Based on their three-dimensional (3D) structures, these materials have a stack of layered structures that allow for greater stability and tunable properties. These materials follow the general formula A_2BX_4, where A denotes a larger cation, B is a metal cation, and X is a halide anion [64].

The 2D perovskite crystal structure consists of inorganic layers made up of corner-sharing BX_6 octahedra, which are positioned between layers of organic spacer cations. This staggered configuration yields a naturally constructed quantum-well system with substantial exciton binding energies [65]. By controlling the number of layers in a non-zero dimension, for example, they exhibit different physical properties that can be used to tune their optoelectronic properties [65].

Among the key advantages of 2D perovskites over their 3D analogs is their superior environmental stability, most notably towards moisture [65, 66]. This enhanced stability has attracted more research interest in their potential applications in solar cells, light-emitting diodes, photodetectors, and lasers [66].

In their recent study, Liao et al. [67] examined a variety of 2D perovskite structures, including the Ruddlesden-Popper (RP) phase, the Dion-Jacobson (DJ) phase, and the phase characterized by alternating cations within the interlayer space (ACI). Each of these structures has similar functions and prospective uses. For instance, DJ phase perovskites have been reported to achieve a desirable balance between photovoltaic and stability in perovskite solar cells [68].

Recent studies have identified various forms of all-organic 2D perovskites, including the Choi-Loh van der Waals phase, which expands the range beyond traditional metal-based perovskites [64]. Furthermore, embedding dopants, such as Mn(II) ions, into 2D perovskites has led to the creation of tunable, emissive materials [69].

In conclusion, layered perovskites offer significant advantages for a variety of optoelectronic uses. Their distinct structure, enhanced stability, and adaptable properties make them strong contenders for numerous technological applications, leading to substantial research and development efforts.

5 Synthesis of the Perovskite Materials

This section will explore the potential methodologies for synthesizing perovskite materials in their polycrystalline form. While perovskite materials can be synthesized through various techniques, our focus will be on the primary methods, including the solid-state reaction method, sol-gel method, and hydrothermal method. The fundamental details are discussed in the subsequent section.

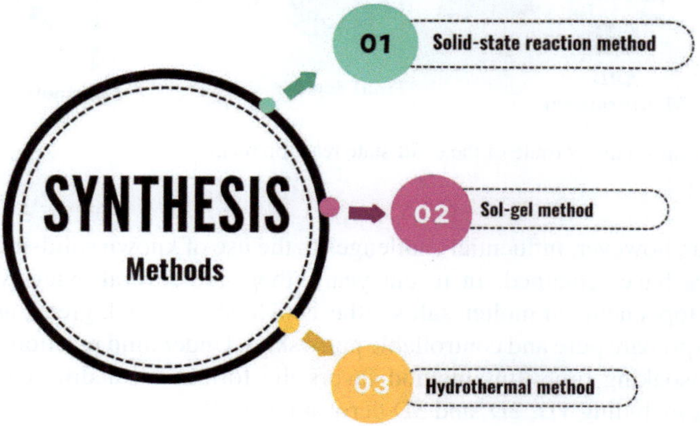

5.1 Solid-State Reaction Method

The solid-state reaction is a commonly used method for synthesizing perovskite precursors. This is usually carried out by first mixing and grinding stoichiometric amounts of the precursor materials, followed by calcination at high temperatures.

The solid-state approach is a general synthesis method for perovskites; however, it consists of multiple processes. Initially, precursor materials, which are typically metal oxides or carbonates, are independently weighed and mixed in stoichiometric proportions. These are then ground together, typically in a mortar and pestle or ball milling, to ensure homogeneous mixing and enhance the surface area of the reactants. This mixture is pressed into pellets that are then heated at temperatures on the order of 800–1200 °C for several hours or days, with intermediate grinding and re-pressing steps to allow for improved reactivity and homogeneity throughout the fuel [70, 71]. The solid-state synthesis route is represented by a flow chart in Fig. 6.

The three-step grinding method allows for the formation of $LaBO_3$ perovskites, where B is either Mn or Fe. First, combine lanthanide oxides with alkaline earth carbonates. Finally, thermal treatment, heating them at 400°C for 3 h. It involves solid-state synthesis (SSR), high-energy ball milling (HEBM), and low-energy ball milling (LEBM) using water. With each step, the perovskite crystals become better. The final step, laser desorption of a Bi, increases the surface area of the perovskite particles dramatically [72].

Fig. 6 General synthesis route of the solid-state reaction method

Thus far, however, influential challenges to the use of known solid-state synthesis approaches have remained. In recent years, there are several extensive methods, including topochemical molten salt synthesis (TMSS), a rapid, green, and effective method to prepare pure and controllable perovskite. Under mild reaction temperature and short soaking time, this method favors the formation of different perovskite structures, including 1D, 2D, and 3D perovskites [73].

Overall, solid-state reactions serve as the primary route for the synthesis of perovskites, and research remains active to develop better techniques to optimize the reaction conditions in the future. The synthesis route used affects the final properties of the perovskite; thus, many synthesis methodologies are being developed and refined.

5.2 *Sol-Gel Method*

The sol-gel technique is highly regarded for producing perovskite materials, offering numerous benefits such as low-temperature processing, high purity, and exceptional control over both composition and structure [74]. Perovskite materials are synthesized in multiple sequential steps. Initially, metal salts or alkoxides are dissolved in a suitable solvent, and chelating agents are added to regulate the hydrolysis rate. Next, hydrolysis and condensation reactions occur, where the precursors form a colloidal suspension called sol. In the second gelation stage, the sol condensed and polymerized to form a gel. This gel was aged to cross-link the internal truss network, which was then dried to eliminate residual solvents. Subsequently, the dried gel was calcined at elevated temperatures to facilitate crystallization into the desired perovskite phase. The sol-gel synthesis route is shown in Fig. 7.

In this approach, $Sr(NO_3)_2$ and $TiCl_3$ precursors can be used, for example, for $SrTiO_3$ perovskite synthesized using the Pechini sol-gel technique [75]. This technique, further demonstrated by spark plasma sintering, yields apparent ceramics with

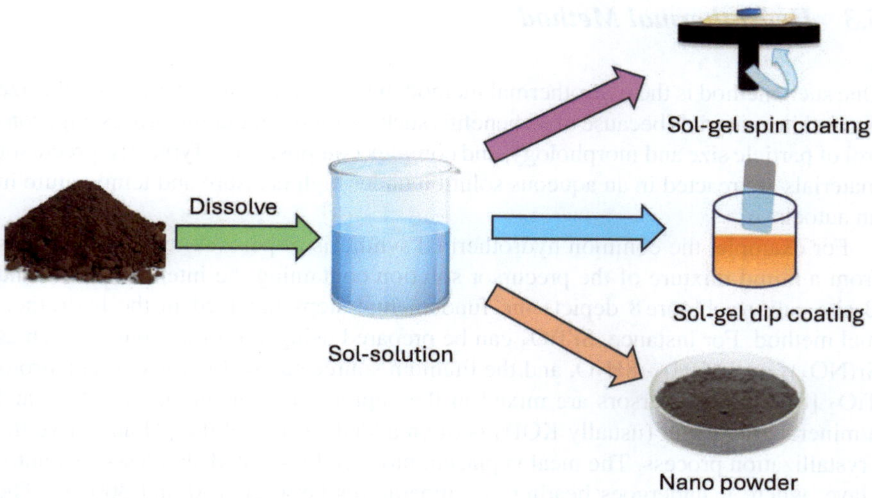

Fig. 7 Synthesis route of the sol-gel method

excellent physicomechanical performance and hydrolytic stability, making it highly favorable for radioactivity waste isolation applications. Recent investigations have extended the sol-gel approach to develop more sophisticated architectures. As an example, porous spheres composed of perovskite titanates like $SrTiO_3$, $BaTiO_3$, and $CaTiO_3$ can be synthesized by employing a combination of sol-gel and hydrothermal methods [76]. This strategy offers tunable particle sizes and inner structures, allowing the fabrication of materials with enhanced photocatalytic activity.

In recent years, this sol-gel method has been applied beyond oxides to non-oxide compounds, such as MAX phase materials. V_2GeC and Cr_2GeC were prepared using sol-gel chemistry with conventional or microwave heating methods [77]. This progress paves the way for broader access to advanced materials. Gel cracking and precursor reactivity are common issues in sol–gel synthesis. Cracking often occurs due to rapid solvent evaporation or excessive shrinkage during drying, whereas precursor reactivity can lead to uncontrolled hydrolysis or inhomogeneity in the gel network. These challenges can be mitigated by controlled drying, pH adjustment, chelating agent use, or precursor ratio optimization [78].

The sol-gel process is constantly evolving, and many new strategies have been developed to obtain more advanced catalytic formulations and multifunctional inorganic materials [79]. Agile and efficient methods for evaluating and aerating perovskite devices at the microscale would also extend progress in the same direction, complementing advances in artificial intelligence for perovskite discovery and synthesis [80] and leading to further perovskite materials and device innovation and economic and sustainable upscaling of energy devices and catalysis.

5.3 Hydrothermal Method

One such method is the hydrothermal method, which is a common way to synthesize perovskite materials because of its benefits, such as low-temperature processing, control of particle size and morphology, and complex compositions. Typically, precursor materials are reacted in an aqueous solution under high pressure and temperature in an autoclave.

For example, the common hydrothermal synthesis of perovskite materials starts from a liquid mixture of the precursor solution containing the intended A-site and B-site cations. Figure 8 depicts the fundamental steps involved in the hydrothermal method. For instance, $SrTiO_3$ can be prepared using strontium sources such as $Sr(NO_3)_2$ or $Sr(OH)_2 \cdot 8H_2O$, and the titanium source can be hydrous or anhydrous TiO_2 [81]. The precursors are mixed in the appropriate stoichiometric ratios, and a mineralizing agent (usually KOH) is often added to control the pH and drive the crystallization process. The meal is placed into a Teflon-coated stainless-steel autoclave, where it undergoes heating at temperatures between 150 and 300 °C. The

Fig. 8 A step by step synthesis process of the hydrothermal method

duration of this heating process, which can last from several hours to multiple days, is influenced by the specific characteristics of the perovskite composition and the desired properties to be achieved.

Advancements in membrane technology have led to the achievement of its expected performance through various methods, including hydrothermal synthesis, which emerged from recent progress in the microwave-assisted hydrothermal (MHT) technique. This strategy leverages the advantages of microwave heating in the hydrothermal synthesis process to achieve a shorter reaction time and better control of the particle morphology. Moreover, the MHT method synthesis of $NaTaO_3$ nanocrystals exhibited higher photocatalytic activity for overall water splitting than those synthesized by conventional hydrothermal (CHT) methods [82].

The hydrothermal method has been proven to prepare a variety of compositions and performances of perovskite materials. To take another example, $SrTiO_3$, $BaTiO_3$, and $CaTiO_3$ porous spheres were produced through a combination of the sol-gel and hydrothermal techniques in which the basic building blocks were single-crystalline nanocubes [76]. These materials exhibited excellent photocatalytic oxygen evolution and photovoltaic behaviour in CdSe-sensitized solar cells. Similarly, in another study, a one-step hydrothermal method was used to fabricate $LaFeO_3$ oxide, where the produced material was reported to present favorable surface properties for steam methane reforming [83].

More complex perovskite structures, including double perovskites, have recently been synthesized. The symmetrical double-perovskite bismuth oxide $(K_{1.00})(Ba_{1.00})_3(Bi_{0.89}Na_{0.4}O_{12}$ has been hydrothermally synthesized and shows superconductivity with a transition temperature of $T_C = 31.5\,K$ [84]. Likewise, a simple perovskite-type bismuth oxide with the composition $(Ba_{0.62}K_{0.38})(Bi_{0.92}Mg_{0.}Ni_3O_8$ was synthesized via a simple hydrothermal method, showing a large shielding volume fraction with a high muon transition temperature of $T_N \sim 30K$ [85].

6 Properties of the Perovskite Materials

This section will explore the intriguing and distinctive characteristics of perovskite materials. Our focus will be on their structural, optical, dielectric, magnetic, and multiferroic properties, as these aspects are crucial for their use in energy storage, photovoltaic applications, light-emitting devices, and sensors.

6.1 Structural Properties

Perovskite materials are a major focus of research in materials science due to their distinctive structural characteristics and diverse applications. This class of materials follows a general ABX_3 formula where A and B are considered cations while X is the anion, whose ratios can be adjusted to modify various properties [86]. Functional properties such as ferroelectricity, piezoelectricity, and multiferroicity of perovskites are widely utilized in microelectronic devices [78].

Perovskites possess structural properties that can be finely tuned and adapted to form a variety of low-dimensional variants at the atomic level [87]. This structural adaptability drives their use in optoelectronic and photonic applications. Halide perovskites, known for their distinctive opto-electrical characteristics, have been shown to be promising materials for use in solar cells, LEDs, and electronic screens [86, 88].

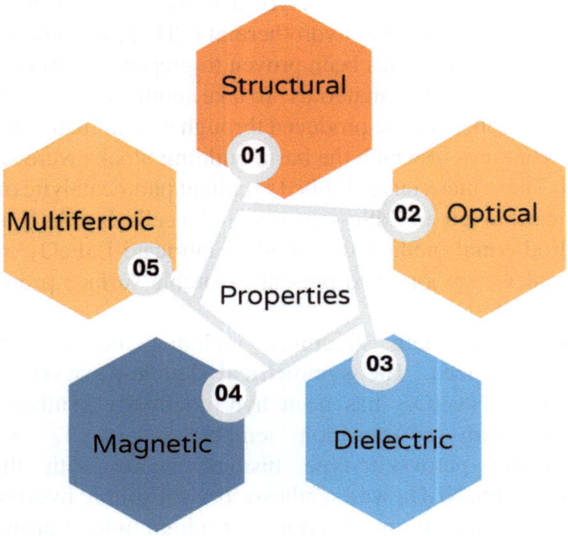

Structural phase transitions in perovskites are essential for their optical and electronic characteristics. One common example is leadhalide perovskites, which undergo multiple phase transitions as a function of temperature, affecting their performance in photovoltaic devices [89]. Likewise, bismuth ferrite ceramics doped with rare earths have complex structural phase diagrams that are important for their structure-property relationships [90]. The temperature-dependent photoluminescence of methylammonium lead iodide ($CH_3NH_3PbI_3$) perovskite, also known as MAPI, showed a redshift in the PL wavelength with cooling, indicating a decreasing bandgap opposite to that of typical semiconductors. Discontinuities in the shift identified orthorhombic to tetragonal and tetragonal to cubic phase transitions at 160 $\pm$ 5 K and 310 $\pm$ 5 K, respectively [91].

Table 1 summarizes the types of perovskite materials, their structural formulas, and important structural features. Halide double perovskites ($A_2MI_{III}X_6$) are characterized by their iodide structures and unique phase transitions, which hold promise for optoelectronic applications. Lead halide perovskites ($APbX_3$, where A = MA or FA, and X = I or Br) have raised much interest in the literature because of such temperature-dependent phase transitions. These transitions give rise to material properties such as negative thermal expansion, which render these materials ideal candi-

Table 1 Summary of different perovskite types and their key structural properties

Perovskite type	Key structural properties	Structural formula	References
de Perovskites	Temperature-dependent phase transitions, negative thermal expansion	$APbX_3$ (A = MA, FA; X = I, Br)	[89]
Perovskite Oxide Nanotubes	Low-dimensional nanostructures	Various	[78]
Bismuth Ferrite	Complex phase diagrams, morphotropic phase boundaries	$BiFeO_3$ (rare-earth doped)	[90]
Halide Double Perovskites	Iodide-based structures, phase transitions	$A_2MI_{III}X_6$	[60]
Cubic Perovskites	Lattice constant as a key parameter	Various	[92]

dates for use in solar cells and thermoelectrics. Bismuth ferrite ($BiFeO_3$),especially when doped with rare-earth elements, exhibits intricate phase diagrams and morphotropic phase boundaries, which are significant for improving its ferroelectric and multiferroic characteristics. Cubic perovskites have diverse compositions but are typically defined by their lattice constants, a key structural parameter that strongly affects their physical behavior. Finally, perovskite oxide nanotubes are low-dimensional nanostructures characterized by surface- and dimension-dependent properties, providing new opportunities for nanoelectronics and catalysis. Together, these studies highlight the extraordinary structural versatility and functional tunability of the perovskite family.

6.2 *Optical Properties*

Perovskite materials have garnered significant interest for their potential use in solar energy technology due to their exceptional optical characteristics. These materials, with their sharp optical absorption edges and high absorption rates, are considered strong contenders for solar cell applications [93].

The optical characteristics of perovskite materials, especially $CH_3NH_3Sn_xPb_{1-x}I_3$ groups, were systematically examined [94]. Perovskite materials have a tunable bandgap, enabling the optimization of light absorbance over a wide range [95]. Such dopant exposure is especially beneficial for tandem solar cell structures, as perovskite-on-perovskite devices can yield efficiencies exceeding 35% [94]. From chemical management, the capability to engineer the bandgap has led to the realization of perovskite solar cells with different colors (for example, yellow and red), which can potentially be developed for building integrated photovoltaics [96].

This was particularly because the absorption coefficient (α) of $CH_3NH_3PbI_3$ was found to be linked to the significant surface roughness of the perovskite layers created through solution processing. Experimental data correlating the direct and indirect bandgaps with experimental measurements verified that the hybrid perovskites possessed a direct bandgap, and the indirect nature was negligible. Moreover, for hybrid perovskites of the $APbX_3$ type, the A-site cation is important for determining the absorption strength through hydrogen bonding interactions [97].

To conclude, the optical characteristics of perovskite materials make them exceptionally suitable for use in solar energy applications. Owing to their high absorption coefficients, tunable bandgaps, and long charge carrier diffusion lengths, perovskites have rapidly improved solar cell efficiencies, with certified values of 25.2% [98]. However, many challenges must be solved for the commercialization of perovskite-based photovoltaic technologies [99], especially stability and scalability.

6.3 Dielectric Properties

The interest in perovskite criteria materials, both single and complex systems, has emerged owing to their varied dielectric permittivity behavior and potential applications in many fields. This review summarizes various perovskite structures, their dielectric properties, functional features, and potential applications.

Single perovskites $(Bi_{1/2}Na_{1/2})TiO_3$ have shown promise for increasing piezoelectric The addition of small amounts of $Bi(Mg_{2/3}Nb_{1/3})O_3$ to $(Bi_{1/2}Na_{1/2})TiO_3$ produces adds to the piezoelectric performance of it, with the best property for d_{33} and kt at x = 0.7% and x = 0.9% respectively. Temperature-dependent dielectric constant measurements revealed the typical relaxor ferroelectric behavior of these compounds.

Even more interesting dielectric properties can be achieved for complex perovskites such as double and quadruple perovskites. For example, the double perovskite $PbBaBiNbO_6$ exhibits ferroelectric-type behavior and non-exponential conductivity relaxation [100]. The activation energy was further reduced by approximately 1 eV at the B-site Bi^{5+}, indicating that it is an excellent candidate for bulk photovoltaic effects and solar applications. In addition, the same phenomenon was observed in $Bi(Ga_xFe_{1-x})O_3$–$PbTiO_3$ (BGF–PT), where two ferroelectric phases, rhombohedral and tetragonal, coexist, and the substitution of Ga in the lattice leads to an enhanced electrical resistivity and dielectric constant and a decreased dielectric loss [101].

Interestingly, some perovskite materials have dielectric properties that are responsive to temperature. For example, the metal-free perovskite $(H_2dabco)(NH_4)[BF_4]_3$ shows a temperature-dielectric response at approximately 333 K, ascribed to the ordered—disordered transformation of its organic "spherical" cations [102]. This behavior indicates that these devices may be useful as thermal sensors. Likewise, the organic-inorganic layered perovskite (cyclohexyl methylammonium)$_2$CdCl$_4$ exhibits a reversible first-order phase transition at approximately 342 K with a significant variation in the dielectric constant and significant dielectric anisotropy [103]. Some non-perovskite materials, such as langesite and spinel ferrite family compounds, also exhibit interesting dielectric properties [104–108].

Therefore, perovskite materials possess various dielectric properties, including ferroelectricity, piezoelectricity, and temperature-dependent properties. These features render them promising candidates for photovoltaics, memory devices, and thermal sensors. Further investigations need to be carried out to improve these properties and to search for new perovskite compositions to perform their role in diverse applications.

6.4 *Magnetic Properties*

Perovskite materials have diverse structural and functional properties and have drawn increasing attention in the field of magnetism. Materials characterized by the general formula ABO_3 display a diverse range of magnetic behaviors, including ferromagnetism, antiferromagnetism, and intricate spin arrangements.

The composition and structure of perovskite materials significantly affect their magnetic properties. One obvious example is rare-earth-doped bismuth ferrite ceramics, which exhibit enhanced functional responses owing to the stabilization of distinct ordering patterns of bismuth ions after introducing rare earths into the Bi^{3+} site [90]. Adjusting these parameters significantly improves the electronic and magnetic characteristics, especially at the structural morphotropic phase boundaries (Fig. 9).

Sc_2NiMnO_6 is a fully ordered magnetodielectric double perovskite that displays successive magnetic transitions at 35 and 17 K, which are assigned to single-k and 2-k magnetic orders, respectively [109]. The ST deviation observed, which indicates

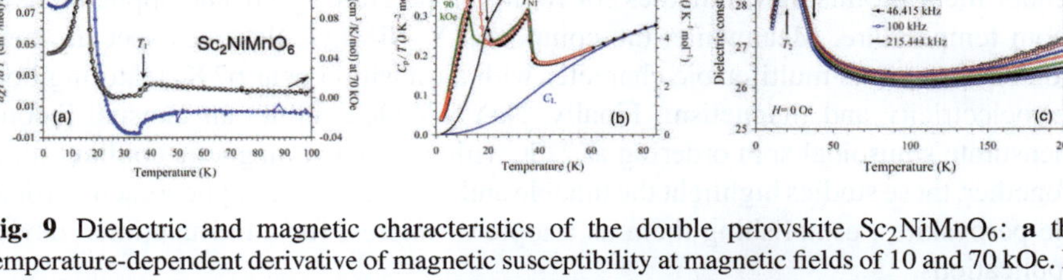

Fig. 9 Dielectric and magnetic characteristics of the double perovskite Sc_2NiMnO_6: **a** the temperature-dependent derivative of magnetic susceptibility at magnetic fields of 10 and 70 kOe. In this illustration, T_1 and T_2 denote the transition temperatures. **b** Heat capacity measurements under various magnetic fields (0, 70, and 90 kOe). **c** The dielectric constant ϵ' as a function of temperature across different frequencies [109]

Table 2 Overview of different perovskites and their magnetic properties

Compound	Transition temperature(s)	Magnetic behavior	References
La_2NiMnO_6	$T_C = 275\,K$	Para–ferro transitions	[111]
Sc_2NiMnO_6 s	$T_1 = 35\,K, T_2 = 17\,K$	Sequential magnetic transitions	[109]
La_2CoMnO_6	$T_C = 220\,K$	Para–ferro transitions	[111]
$NaYNiWO_6$	$T_N = 21\,K$	Incommensurate sinusoidal spin ordering	[112]
Y_2NiMnO_6	$T_C = 67\,K$	Multiferroic	[113]

the charge-coupling effect, points to the potential of 3d double perovskites with small A-site cations for use in magnetoelectric applications.

The particle size can also affect the magnetic properties of perovskite materials. Utilization of new nanoscale approaches has revealed new magnetic and electronic effects in perovskite manganites that are dramatically different from those of the corresponding bulk and films [110]. The behavior that varies with size, as discussed here, is crucial not only for basic research but also for the advancement of future electronic and magnetic nanodevices.

Therefore, the magnetic properties of earth-abundant perovskite materials are dynamic and versatile, providing a wide variety of opportunities for study by magnetism researchers. Spin ordering of complexity and size-dependent behavior are only a few lessons learned from these materials that continue to excite and inspire new paths in fundamental research and possible applications of these materials.

Table 2 shows the magnetic properties of different perovskite-based materials, which demonstrate a great diversity of behaviors and transition temperatures. For instance, Sc_2NiMnO_6 exhibits two magnetic transitions at $T_1 = 35\,K$ and $T_2 = 17\,K$, highlighting the complexity of the spin physics in these honeycomb systems. In contrast, both La_2NiMnO_6 and La_2CoMnO_6 exhibit a more conventional paramagnetic-to-ferromagnetic transition at approximately 275 K and 220 K, respectively, according to the study by Mao et al. (2013). Such relatively high Curie temperatures render them promising candidates for room-temperature spintronic applications at room temperature. Meanwhile, the compound Y_2NiMnO_6 deserves special attention because of its multiferroic character with a transition near 67 K, featuring both ferroelectricity and magnetism. Finally, $NaYNiWO_6$ exhibits an unusual incommensurate sinusoidal spin ordering at 21 K with a complex magnetic configuration. Together, these studies highlight the tunable and diverse magnetic phenomena in double perovskites, establishing them as a topic of interest for multifunctional device applications.

6.5 Multiferroic Properties

Over the past several decades, multiferroic perovskite materials have attracted tremendous interest in material design strategies and applications owing to the distinctive phenomenon of presenting more than one ferroic order, such as ferroelectricity, ferromagnetism, and ferroelasticity, simultaneously. These materials exhibit rich physics and new device concepts; therefore, they are of great interest, particularly for researchers working in the fields of material science and condensed matter physics [114].

Perovskite oxides, characterized by the general formula ABO_3, possess a range of functional properties, including ferroelectricity, piezoelectricity, and multiferroicity. These attributes are essential for diverse applications, particularly in microelectronic devices, with feature sizes shrinking to nanoscale dimensions [78].

In recent decades, numerous multiferroic materials have been identified, yet perovskite manganites, known for their magnetically induced ferroelectricity, have garnered significant interest due to their strong multiferroic properties and the diverse physical mechanisms they encompass [43]. Recently, new multiferroic materials comprising double-perovskite structures have been studied in detail. For example, Bi_2NiMnO_6 and Bi_2FeCrO_6 have been synthesized in bulk and thin-film forms, exhibiting multiferroic properties at low and room temperatures, respectively [115]. The Bi^{3+} ion at the A site produces ferroelectric structural distortion in these compounds, whereas the B-site configuration of transition metal ions gives rise to ferromagnetism following the Kanamori-Goodenough criterion.

Another exciting finding is the multiferroic behavior of double perovskite Y_2NiMnO_6. Showcasing a high polarization of 145 μC/m2, a reasonably high multiferroic transition temperature of 67 K, and a strong magnetoelectric coupling of ~21%, this material is unique. The ferroelectricity of Y_2NiMnO_6 is thought to be caused by the interaction between Ni^{2+}/Mn^{4+} charge ordering and spin ordering [113]. The examples of practical applications for Y_2NiMnO_6 , such as spintronic devices, magnetic sensors, and neuromorphic computing. The hybrid multiferroicity at nanoscale has also been investigated by synthesizing composite nanofibers of $NiFe_2O_4$–$Pb(Zr_{0.52}Ti_{0.48})O_3$ and $CoFe_2O_4$–Pb $(Zr_{0.52}Ti_{0.48})O_3$ via solar-gel methods, electrospinning. These nanofibers exhibit inherent ferroelectric and ferromagnetic properties, enabling the development of novel multiferroic nanostructures and devices.

In this section, we provide some background on this method, applied to the 3D multiferroic family: lead-free perovskite. Researchers have successfully synthesized (R-3AP)$RbBr_3$ by introducing homochirality into 3D ferroelectric (Rac-3AP)$RbBr_3$, which exhibits both ferroelasticity, good thermal stability, reversible second-harmonic generation switching characteristics, and a structural phase transition at 401 K. Here, we discuss the recent developments in ABO_3 multiferroic perovskite materials and explore their capabilities in multifunctional nanoelectronic devices. Owing to their distinctive electron spin, transfer, polarization, and magnetoelectric coupling properties, these materials can be effective candidates for data storage and information processing, as well as unlocking the mysteries of artificial intelligence via neuromorphic computers by providing diverse functionalities.

7 Applications

In this part, we also explore the possible uses or applications of perovskite materials. Perovskites were first discovered as new compounds over a century ago, but it is only recently that they have gained popularity because of their remarkable properties and tunability [57]. These materials have been used in various fields such as, energy storage and conversion systems, solar cell, light emitting device, sensor.

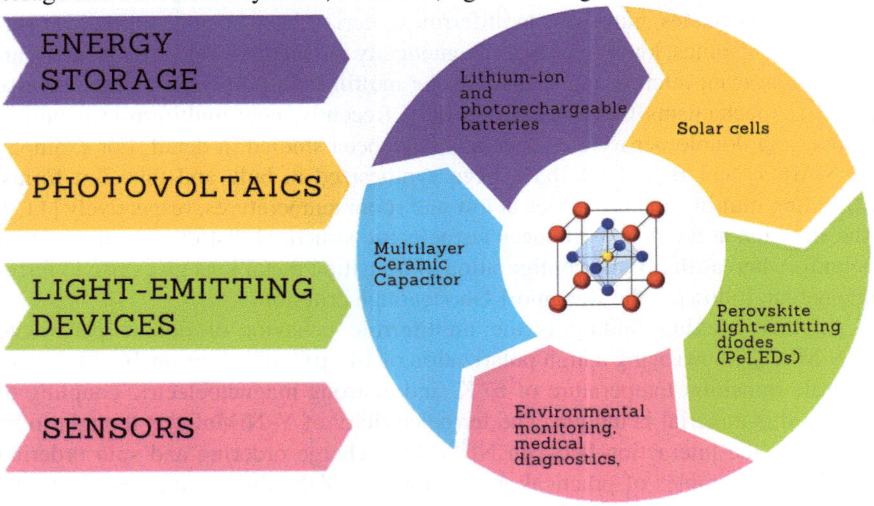

7.1 Energy Storage and Conservation Systems

Due to their adaptable characteristics, perovskite materials have become essential in a range of energy storage and conversion systems. The use of perovskites for energy applications was started more than 60 years ago, in the work of Professor John B. Goodenough focused on perovskite-type oxides to develop random access memory [116]. Since then, these materials have become important components of energy conversion and storage technologies.

Perovskites, which have the general formula ABO_3, have been used in many important energy storage and conversion applications. The latter include, for instance, oxygen-transporting membranes for oxy-fuel combustion, solid oxide fuel cells for efficient power generation, electrocatalysts for oxygen reduction and evolution reactions, and solar cells for solar energy harvesting [117]. Recently, halide perovskites originally designed for solar cells have been applied to energy storage systems, such as lithium-ion batteries and photorechargeable batteries [118]. Integration of energy conversion and storage technologies makes use of the potential of perovskite materials. Perovskite solar cells have even been integrated with batteries and supercapacitors, and employed for photovoltaic driven catalysis [119].

In addition to the potential for cost savings, lightweight, high-performance materials, such as the use of novel systems, benefit electric vehicles and portable electronic devices, which use machine learning [92]. Perovskite materials are central to next-generation energy technologies. They offer broad versatility due to unique characteristics such as mixed electronic and ionic conductivities, tunable composition and morphology, and nanostructurability [116, 120].

Recently, high-entropy perovskites (HEPs), a new high-entropy class of perovskite frameworks with an inherent high atomic disorder in their crystalline structure, have exhibited impressive operational stability and performance [121]. As energy conversion and storage has become an increasingly high-demand topic, perovskites can indeed make strides towards overcoming global energy and climate change challenges [120].

7.2 *Photovoltaics Applications*

Perovskite materials have been recognized as one of the most promising materials for next generation photovoltaic applications owing to their remarkable optoelectronic properties and their rapid development in power conversion efficiency (PCE). Perovskite solar cells (PSCs) first discovered in 2009 and showing a PCE of 3.81% have made incredible progress and reached a certified PCE of 24.2% in nearly a decade [122].

The reported perovskite solar cell efficiency of 21.2% is useful for comparison with other established photovoltaic technologies. Monocrystalline silicon cells have achieved certified efficiencies of approximately 26.8%, whereas polycrystalline silicon devices have reached approximately 24.4%. Cadmium telluride (CdTe) solar cells have efficiencies close to 22.1%, and copper indium gallium selenide (CIGS) cells have efficiencies of approximately 23.6%. These figures are based on data compiled by the National Renewable Energy Laboratory (NREL 2025) and provide a broader context for assessing the performance of perovskite solar cells. Table 3 summarize the PCE values across real materials [123].

PSC typically consists of a perovskite absorber layer sandwiched between electron and hole transport layers. Functionalized fullerenes act like charge transport layers, interfacial layers, and addi- tives in these devices, ultimately playing crucial roles in PCE and device stability [124]. Due to the large quantities of organic semiconductors available, PSCs can be deposited either in a regular (n-i-p) configuration or as an inverted (p-i-n) structure, the latter of which are becoming more common due to their low cost of manufacture and ease of fabrication, as well as their low hysteresis seen in measurements [125].

Table 3 Certified efficiencies of various photovoltaic technologies

S. No.	Photovoltaic technology	Material example	Certified PCE (%)	References
1	Perovskite Solar Cells	$CH_3NH_3PbI_3$-based	21.2	[123]
2	Monocrystalline Silicon	Si	26.8	[123]
3	Polycrystalline Silicon	Si	24.4	[123]
4	Cadmium Telluride (CdTe)	CdTe	22.1	[123]
5	Copper Indium Gallium Selenide (CIGS)	$Cu(In,Ga)Se_2$	23.6	[123]

The advantages of perovskite materials as the next generation, low-cost, flexible, mass manufacturable photovoltaic technologies include appropriate bandgaps, solution processability, and high absorption coefficients. Such unique properties have enabled their utilization in various applications beyond traditional solar panels, including tandem solar cells, building-integrated photovoltaics, space applications, and photovoltaic-driven catalysis [119].

Although they show significant potential, PSCs face challenges such as lead toxicity and long-term stability [122]. Meanwhile, material and device engineering (e.g., the use of low-dimensional perovskites for enhanced stability and alternative, less toxic compositions) are being conducted to address these issues [126, 127].

In conclusion, perovskite materials represent a significant advancement in photovoltaic technology, offering the potential for solar cells that are both highly efficient and cost-effective. These materials have been developed for commercial use due to their adaptability and ability to enhance performance across a wide range of applications, from conventional solar panels to innovative integrated energy systems [88]. As research continues to address challenges related to stability and toxicity, perovskite solar cells are anticipated to play a crucial role in the future of photovoltaic technologies.

7.3 Light-Emitting Devices

Perovskite light-emitting diodes (PeLEDs) have arisen into promising candidates for future display and illumination applications, receiving intensive interest recently due to their remarkable optoelectronic features [128]. PeLEDs have gone from their first bright electroluminescence election demo in 2014, reaching EQEs over 20% by 2018, to efficiencies near 36% [129] in four years. Because of their high color purity, solution processability, and flexibility potential, these devices have been used in numerous applications [130].

Light-emitting devices utilize perovskite materials such as 3D bulk perovskites, 2D and quasi-2D perovskites, as well as colloidal perovskite nanocrystals [131]. Representatives include $CsPbBr_3$ & Cs_4PbBr_6, which presented $EQE_{max} = 22.3\%$ (Luminance=10,050 cd m^{-2}; [32]) and $FAPbI_3$ allowing operational lifetimes reaching a record of 1.9×10^6 hours under low current densities [50]. These materials can emit light across the entire visible spectrum, enabling high color reproduction to achieve a performance comparable to or even superior to that of the best commercial display devices [132].

PeLEDs hold promise for future applications as they can potentially overcome the drawbacks of existing technologies. They can be highly efficient, low-cost, and versatile devices for displays and lighting [130]. However, these issues must be resolved for commercial viability in terms of operational and spectral stability and lead toxicity. Although PeLEDs have faced challenges in terms of performance and stability, recent progress in materials engineering, device architecture, and passivation strategies has achieved promising results, bringing PeLEDs closer to practical use [7, 50, 98, 131].

7.4 Sensors

Perovskite materials, with their exceptional properties and flexibility, are seen as highly promising for use in sensor applications. Perovskites are being used in chemical sensors and have shown promise for the desired response at room temperature. For example, $CsPbBr_2I$ (CPBI) perovskites respond rapidly to various compounds at concentrations as low as 1 ppm [98]. Self-powered devices show enhanced photocurrents when exposed to both reducing and oxidizing gas molecules, enabling the detection of diverse environmentally and medically important compounds.

Remarkably, perovskite nanomaterials have exhibited a remarkable detection ability by responding to extensive types of chemical and biological entities in solid and solution states. They have demonstrated exceptional effectiveness in identifying small molecules, including O_2, NO_2, and CO_2. The excellent sensing capability of perovskites is ascribed to their unique structural characteristics, such as a good direct bandgap, low defect density, and high sensitivity. Additionally, to mitigate the potential toxicity of Pb, studies have shifted towards using lead-free halide perovskite materials for humidity sensing.

Perovskite materials are important for next-generation sensors because of their ability to form highly sensitive, compact, and robust devices. A flexible sensor utilizing mechanoluminescent-perovskite technology has been created, capable of detecting pressures as low as 11 kPa. It features a wide sensing range and a sensitivity of 0.095kPa^{-1} [133]. The self-powered sensor, which has signal stability over 1000 cycles, indicates its dependable performance for practical application scenarios such as structural health monitoring. Moreover, PQDs have been applied to develop dual-function light-sensing/display devices, paving the way for the realization of non-contact interactive screens and communication ports [134]. Continuing advancements in research in this area will place perovskite-based sensors at the forefront of many applications, such as environmental monitoring, medical diagnostics, and wearable technology.

8 Conclusion

In summary, this study provides a comprehensive overview of perovskite materials and their applications in energy storage and efficiency systems. Perovskites, with their versatile ABX_3 structure, exhibit a wide range of compositions and properties, including single and complex structures. This study explores various synthesis methods and examines the structural, optical, dielectric, magnetic, and multiferroic properties of these materials. Perovskites show great promise for advancing energy technologies, particularly in improving the efficiency and performance of solar cells, batteries, light-emitting devices, and sensors. However, challenges such as stability, toxicity (in lead-based perovskites), and scalability must be addressed for widespread commercial adoption. Overall, this study underscores the importance of perovskite materials in addressing global energy challenges and highlights the need for continued research to overcome the existing limitations and fully realize their potential in energy applications.

References

1. Osaki, K.: US energy information administration (EIA). In: 2019 Edition US Annual Energy Outlook Report (aeo2019). Haikan Gijutsu 61(8):32–43 (2019)
2. Newell, R., Raimi, D., Villanueva, S., Prest, B., et al.: Global energy outlook 2021: pathways from Paris. Resour. Future **8**, 39 (2021)
3. Saleh, H.M., Hassan, A.I.: The challenges of sustainable energy transition: a focus on renewable energy. Appl. Chem. Eng. **7**(2), 2084 (2024)
4. Batra, G.: Renewable energy economics: achieving harmony between environmental protection and economic goals. Social Sci. Chronicle **2**(2), 1–32 (2023)
5. Ishola, A.: Iot applications in sustainability and sustainable community development. World J. Adv. Res. Rev. (2024)
6. Parra, C., Kirschke, J., Ali, S.H.: Ensure access to affordable, reliable, sustainable and modern energy for all. In: Mining, Materials, and the Sustainable Development Goals (SDGs), pp. 61–68. CRC Press, Boca Raton
7. Zhang, X., Khaskheli, M.B., Shen, C., Haider Jafri, M.A., Shamsi, I.H.: A comparative study of economic activity and sustainable development in China and three south Asian countries. Discov. Sustain. **6**(1), 170 (2025)
8. Gyani, R., Chandel, S.S.: Towards developing an effective energy management system for decentralized solar mini-grids in a remote location in india—a case study. In: International

Conference on Signal, Machines, Automation, and Algorithm, pages 157–171. Springer, Berlin (2024)

9. Zakari, A., Musibau, H.O.: Sustainable economic development in OECD countries: does energy security matter? Sustain. Dev. **32**(1), 1337–1353 (2024)
10. Tang, X., Li, X.: Exploring carbon neutrality potential in the petroleum industry: the role of green monetary policies as motivators. Resour. Policy **86**, 104230 (2023)
11. Wu, L.,Broadstock, D.C.: Does economic, financial and institutional development matter for renewable energy consumption? Evidence from emerging economies. Int. J. Econ. Policy Emerg. Econ. **8**(1), 20–39 (2015)
12. Kay, H.F., Bailey, P.C.: Structure and properties of catio3. Acta Crystallogr. A **10**(3), 219–226 (1957)
13. Khirade, P.P., Raut, A.V.: Synthesis, structure, physical properties and applications. in: Recent Advances in Multifunctional Perovskite Materials, p. 3 (2022)
14. Franz, A., Többens, D., Schorr, S.: Structural investigations of ch3nh3pbi3. Chemistry **9**, 139–175 (1980)
15. Gesi, K.: Effect of hydrostatic pressure on the structural phase transitions in ch3nh3pbx3 (x= cl, br, i). Ferroelectrics **203**(1), 249–268 (1997)
16. Thakur, A., Singh, D., Kaur Gill, S.: Numerical simulations of 26.11% efficient planar ch3nh3pbi3 perovskite nip solar cell. Mater. Today Proc. **71**, 195–201 (2002)
17. Jain, A., Wang, Y.G. Shi, L.N.: Recent developments in batio3 based lead-free materials for energy storage applications. J. Alloys Compounds 928:167066 (2022)
18. Jiang, L.Q., Guo, J.K., Liu, H.B., Zhu, M., Zhou, X., Wu, P., Li, C.H.: Prediction of lattice constant in cubic perovskites. J. Phys. Chem. Solids **67**(7), 1531–1536 (2006)
19. Galasso, F.S.: Perovskites and high-t c superconductors (1990)
20. Goldschmidt, V.M.: Crystal structure and chemical correlation. Ber. Dtsch. Chem. Ges. **60**, 1263–1296 (1927)
21. Peña, M.A., Fierro, J.L.G.: Chemical structures and performance of perovskite oxides. Chem. Rev. **101**(7), 1981–2018 (2001)
22. Wang, W., Tade, M.O., Shao, Z.: Research progress of perovskite materials in photocatalysis- and photovoltaics-related energy conversion and environmental treatment. Chem. Soc. Rev. **44**(15), pp. 5371–5408 (2015)
23. Megaw., H.D., Solids. crystal structure of barium titanium oxide and other double oxides of the perovskite type. Trans. Faraday Soc. **42**, A224–A231 (1946)
24. Ley, M.B., Paskevicius, M., Schouwink, P., Richter, B., Sheppard, D.A., Buckley, C.E, Jensen, T.R.: Novel solvates m (bh 4) 3 s (ch 3) 2 and properties of halide-free m (bh 4) 3 (m= y or gd). Dalton Trans. **43**(35), 13333–13342 (2014)
25. Schouwink, P., Ley, M.B., Tissot, A., Hagemann, H., Jensen, T.R., Smrčok, L., Černý, R.: Structure and properties of complex hydride perovskite materials. Nat. Commun. **5**(1), 5706 (2014)
26. Schouwink, P., Sadikin, Y., Van Beek, W., Černý, R.: Experimental observation of polymerization from bh4- to b12h122- in mixed-anion a3bh4b12h12 (a= rb+, cs+). Int. J. Hydrogen Energy **40**(34), 10902–10907 (2015)
27. Zhu, J., Li, H., Zhong, L., Xiao, P., Xuelian, X., Yang, X., Zhao, Z., Li, J.: Perovskite oxides: preparation, characterizations, and applications in heterogeneous catalysis. ACS Catal. **4**(9), 2917–2940 (2014)
28. Ali, N., Shehzad, N., Uddin, S., Ahmed, R., Jabeen, M., Kalam, A., Al-Sehemi, A.G., Alrobei, H., Kanoun, M.B., Khesro, A., et al.: A review on perovskite materials with solar cell prospective. Int. J. Energy Res. **45**(14), 19729–19745 (2021)
29. Li, S., Luo, J., Liu, J., Tang, J.: Self-trapped excitons in all-inorganic halide perovskites: fundamentals, status, and potential applications. J. Phys. Chem. Lett. **10**(8), 1999–2007 (2019)
30. Aïssa, B., Ali, A., El-Mellouhi, F.: Oxide and organic-inorganic halide perovskites with plasmonics for optoelectronic and energy applications: a contributive review. Catalysts **11**(9), 1057 (2021)

31. Deng, Q., Lin, B.: Automated machine learning structure-composition-property relationships of perovskite materials for energy conversion and storage. Energy Mater **1**, 100006 (2021)
32. Lin, R., Xu, J., Wei, M., Wang, Y., Qin, Z., Liu, Z., Wu, J., Xiao, K., Chen, B., Park, S.M., et al.: All-perovskite tandem solar cells with improved grain surface passivation. Nature **603**(7899), 73–78 (2022)
33. Mitzi, D.B., Feild, C.A., Harrison, W.T.A., Guloy, A.M.: Conducting tin halides with a layered organic-based perovskite structure. Nature **369**(6480), 467–469 (1994)
34. Peng, B., Hu, Y., Murakami, S., Zhang, T., Monserrat, B.: Topological phonons in oxide perovskites controlled by light. Sci. Adv. **6**(46), eabd1618 (2020)
35. Istomin, S.Ya., Antipov, E.V.: Cathode materials based on perovskite-like transition metal oxides for intermediate temperature solid oxide fuel cells. Russ. Chem. Rev. **82**(7), 686 (2013)
36. Malavasi, L., Fisher, C.A.J., Saiful Islam, M.: Oxide-ion and proton conducting electrolyte materials for clean energy applications: structural and mechanistic features. Chem. Soc. Rev. **39**(11), 4370–4387 (2010)
37. Atta, N.F., Galal, A., El-Ads, E.H.: Perovskite nanomaterials–synthesis, characterization, and applications. Perovskite Mater. Synthesis Character. Prop. Appl. **1**, 3–5 (2016)
38. Fuertes, A.: Nitride tuning of transition metal perovskites. APL Mater. **8**(2) (2020)
39. Wu, H.-S., Murti, B.-T., Singh, J., Yang, P.-K., Tsai, M.-L.: Prospects of metal-free perovskites for piezoelectric applications. Adv. Sci. **9**(12), 2104703 (2022)
40. Sheng, M., Zhu, H., Wang, S., Liu, Z., Zhou, G.: Accelerated discovery of halide perovskite materials via computational methods: a review. Nanomaterials **14**(13), 1167 (2024)
41. Stoumpos, C.C., Kanatzidis, M.G.: Halide perovskites: poor man's high-performance semiconductors. Adv. Mater. **28**(28), 5778–5793 (2016)
42. Handa, T., Yamada, T., Nagai, M., Kanemitsu, Y.: Phonon, thermal, and thermo-optical properties of halide perovskites. Phys. Chem. Chem. Phys. **22**(45), 26069–26087 (2020)
43. Dong, S., Liu, J.-M.: Recent progress of multiferroic perovskite manganites. Mod. Phys. Lett. B **26**(09), 1230004 (2012)
44. Obraztsov, P.A., Lyashenko, D., Chizhov, P.A., Konishi, K., Nemoto, N., Kuwata-Gonokami, M., Welch, E., Obraztsov, A.-N., Zakhidov, A.: Ultrafast zero-bias photocurrent and terahertz emission in hybrid perovskites. Commun. Phys. **1**(1), 14 (2018)
45. Eames, C., Frost, J.M., Barnes, P.R.F., O'regan, B.C., Walsh, A., Saiful Islam, M.: Ionic transport in hybrid lead iodide perovskite solar cells. Nat. Commun. **6**(1), 7497 (2015)
46. Wang, S., Xiao, W., Wang, F.: Structural, electronic, and optical properties of cubic formamidinium lead iodide perovskite: a first-principles investigation. RSC Adv. **10**(54), 32364–32369 (2020)
47. Grancini, G., Nazeeruddin, M.K.: Dimensional tailoring of hybrid perovskites for photovoltaics. Nat. Rev. Mater. **4**(1), 4–22 (2019)
48. Ganose, A.M., Savory, C.N., Scanlon, D.O.: Beyond methylammonium lead iodide: prospects for the emergent field of ns 2 containing solar absorbers. Chem. Commun. **53**(1), 20–44 (2017)
49. Jeon, T., Kim, S.J., Yoon, J., Byun, J., Hong, H.R., Lee, T.-W., Kim, J.-S., Shin, B., Kim, S.O.: Hybrid perovskites: effective crystal growth for optoelectronic applications. Adv. Energy Mater. **7**(19), 1602596 (2017)
50. Guo, B., Lai, R., Jiang, S., Lian, Y., Ren, Z., Li, P., Cao, X., Xing, S., Wang, Y., Li, W., et al.: Efficient and ultra-stable perovskite light-emitting diodes. *arXiv preprint* arXiv:2204.07829 (2022)
51. Park, J.-G., Le, M.D., Jeong, J., Lee, S.: Structure and spin dynamics of multiferroic bifeo3. J. Phys. Condensed Matter **26**(43), 433202 (2014)
52. Li, X., Gao, X., Zhang, X., Shen, X., Lu, M., Wu, J., Shi, Z., Colvin, V.L., Hu, J., Bai, X., et al.: Lead-free halide perovskites for light emission: recent advances and perspectives. Adv. Sci. **8**(4), 2003334 (2021)
53. Yin, H., Xian, Y., Zhang, Y., Li, W., Fan, J.: Structurally stabilizing and environment friendly triggers: double-metallic lead-free perovskites. Solar RRL **3**(9), 1900148 (2019)
54. Hu, S., Smith, J.A., Snaith, H.J., Wakamiya, A.: Prospects for tin-containing halide perovskite photovoltaics. Precis. Chem. **1**(2):69–82 (2023)

55. Zhong, W.-H., Chen, H.-R., Li, Z.-M., Zhu, J.-Y., Shi, C.-H., Cao, Q.-L., Zhao, J.-J., Chen, L.-Z.: 1d chiral enantiomer lead-free perovskites induced chiralopical activity and photoelectric response. Inorg. Chem. **62**(43), 17985–17992 (2023)
56. Wang, H., Zhang, C., Huang, W., Zou, X., Chen, Z., Sun, S., Zhang, L., Li, J., Cheng, J., Huang, S., et al.: Research progress of abx 3-type lead-free perovskites for optoelectronic applications: materials and devices. Phys. Chem. Chem. Phys. **24**(45), 27585–27605 (2022)
57. Bhalla, A.S., Guo, R., Roy, R.: The perovskite structure–a review of its role in ceramic science and technology. Mater. Res. Innov. **4**(1), 3–26 (2000)
58. Evans, H.A., Mao, L., Seshadri, R., Cheetham, A.K.: Layered double perovskites. Ann. Rev. Mater. Res. **51**(1), 351–380 (2021)
59. Li, Q., Wang, Y., Pan, W., Yang, W., Zou, B., Tang, J., Quan, Z.: High-pressure band-gap engineering in lead-free cs2agbibr6 double perovskite. Angew. Chem. Int. Ed. **56**(50), 15969–15973 (2017)
60. Kent, G.T., Morgan, E., Albanese, K.R., Kallistova, A., Brumberg, A., Kautzsch, L., Wu, G., Vishnoi, P., Seshadri, R., Cheetham, A.K.: Elusive double perovskite iodides: structural, optical, and magnetic properties. Angewandte Chemie Int. Ed. **62**(32), e202306000 (2023)
61. Kumar, S., Nathawat, R., Yogi, A.K., Rathore, S.S.: Relaxor ferroelectric-like glassy dielectric dispersion in the triple perovskite Ba3cosb2o9. Phys. Chem. Chem. Phys. **26**(16), 12580–12586 (2024)
62. Kumar, S., Nathawat, R., Yogi, A.K., Rathore, S.S.: Field dependent magnetic and dielectric properties of the triangular lattice antiferromagnet: Ba3cosb2o9. J. Alloys Compounds 182702 (2025)
63. Gebhardt, J., Rappe, A.M.: Mix and match: organic and inorganic ions in the perovskite lattice. Adv. Mater. **31**(47), 1802697 (2019)
64. Choi, H.S., Lin, J., Wang, G., Wong, W.P.D., Park, I.-H., Lin, F., Yin, J., Leng, K., Lin, J., Loh, K.P.: Molecularly thin, two-dimensional all-organic perovskites. Science **384**(6691), 60–66 (2024)
65. Wang, H., Fang, C., Luo, H., Li, D.: Recent progress of the optoelectronic properties of 2d ruddlesden-popper perovskites. J. Semicond. **40**(4), 041901 (2019)
66. Chu, Z., Chu, X., Zhao, Y., Ye, Q., Jiang, J., Zhang, X., You, J.: Emerging low-dimensional crystal structure of metal halide perovskite optoelectronic materials and devices. Small Struct. **2**(6), 2000133 (2021)
67. Liao, C.-H., Arafat Mahmud, Md., Ho-Baillie, A.W.Y.: Recent progress in layered metal halide perovskites for solar cells, photodetectors, and field-effect transistors. Nanoscale **15**(9), 4219–4235 (2023)
68. Niu, T., Xue, Q., Yip, H.-L.: Advances in Dion-Jacobson phase two-dimensional metal halide perovskite solar cells. Nanophotonics **10**(8), 2069–2102 (2021)
69. Adhikari, S.D., Pradhan, N.: Recent developments of Mn(ii)-doped 2d-layered and 2d platelet perovskite nanostructures. Front. Mater. **7**, 206 (2020)
70. Kohlmann, H.: Looking into the black box of solid-state synthesis. Eur. J. Inorg. Chem. **2019**(39–40), 4174–4180 (2019)
71. Tsujimoto, Y., Yamaura, K., Takayama-Muromachi, E.: Oxyfluoride chemistry of layered perovskite compounds. Appl. Sci. **2**(1), 206–219 (2012)
72. Heidinger, B., Royer, S., Alamdari, H., Giraudon, J.-M., Lamonier, J.-F.: Reactive grinding synthesis of labo3 (b: Mn, Fe) perovskite; properties for toluene total oxidation. Catalysts **9**(8), 633 (2019)
73. Li, L., Deng, J., Chen, J., Xing, X.: Topochemical molten salt synthesis for functional perovskite compounds. Chem. Sci. **7**(2), 855–865 (2016)
74. Guo, X., Zhang, Q., Ding, X., Shen, Q., Chunchun, W., Zhang, L., Yang, H.: Synthesis and application of several sol-gel-derived materials via sol-gel process combining with other technologies: a review. J. Sol-Gel. Sci. Technol. **79**, 328–358 (2016)
75. Belov, A.A., Shichalin, O.O., Papynov, E.K., Buravlev, I.Y., Kolodeznikov, E.S., Kapustina, O.V., Azon, S.A., Kondrikov, N.B., Fedorets, A.N., Tananaev, I.G.: Hybrid sol–gel and spark plasma sintering to produce perovskite-like srtio3 ceramics for radioactive waste isolation. J. Compos. Sci **7**(10), 421 (2023)

76. Jia Hong Pan, Chao Shen, Irina Ivanova, Na Zhou, Xingzhu Wang, Wee Chong Tan, Qing-Hua Xu, Detlef W Bahnemann, and Qing Wang. Self-template synthesis of porous perovskite titanate solid and hollow submicrospheres for photocatalytic oxygen evolution and mesoscopic solar cells. *ACS Applied Materials & Interfaces*, 7(27):14859–14869, 2015
77. Siebert, J.P., Patarakun, K., Birkel, C.S.: Mechanistic insights into the nonconventional sol–gel synthesis of max phase m2gec (m= v, cr). Inorganic Chem. **61**(3), 1603–1610 (2022)
78. Zhu, X., Liu, Z., Ming, N.: Perovskite oxide nanotubes: synthesis, structural characterization, properties and applications. J. Mater. Chem. **20**(20), 4015–4030 (2010)
79. Esposito, S.: "traditionalâŁž sol-gel chemistry as a powerful tool for the preparation of supported metal and metal oxide catalysts. Materials **12**(4), 668 (2019)
80. Liang, J., Tingting, W., Wang, Z., Yunduo, Yu., Linfeng, H., Li, H., Zhang, X., Zhu, X., Zhao, Yu.: Accelerating perovskite materials discovery and correlated energy applications through artificial intelligence. Energy Mater. **2**(3), 200016 (2022)
81. Lencka, M.M., Riman, R.E.: Hydrothermal synthesis of perovskite materials: thermodynamic modeling and experimental verification. Ferroelectrics **151**(1), 159–164 (1994)
82. Shi, J., Liu, G., Wang, N., Li, C.: Microwave-assisted hydrothermal synthesis of perovskite NaTaO3 nanocrystals and their photocatalytic properties. J. Mater. Chem. **22**(36), 18808–18813 (2012)
83. Gómez-Cuaspud, J.A., Vera-López, E., Carda-Castelló, J.B., Barrachina-Albert, E.: One-step hydrothermal synthesis of LaFeo3 perovskite for methane steam reforming. Reaction Kinetics Mech Catal **120**, 167–179 (2017)
84. Rubel, M.H.K., Takei, T., Kumada, N., Mozahar Ali, M., Miura, A., Tadanaga, K., Oka, K., Azuma, M., Yashima, M., Fujii, K., et al.: Hydrothermal synthesis, crystal structure, and superconductivity of a double-perovskite bi oxide. Chem. Mater. **28**(2), 459–465 (2016)
85. Rubel, M.H.K., Takei, T., Kumada, N., Mozahar Ali, M., Miura, A., Tadanaga, K., Oka, K., Azuma, M., Magome, E., Moriyoshi, C., et al.: Hydrothermal synthesis, structure, and superconductivity of simple cubic perovskite (Ba0.62K0.38)(Bi0.92Mg0.08)O3 with Tc 30k. Inorg. Chem. **56**(6), 3174–3181 (2017)
86. Abiram, G., Thanihaichelvan, M., Ravirajan, P., Velauthapillai, D.: Review on perovskite semiconductor field-effect transistors and their applications. Nanomaterials **12**(14), 2396 (2022)
87. Liu, J., Chen, K., Khan, S.A., Shabbir, B., Zhang, Y., Khan, Q., Bao, Q.: Synthesis and optical applications of low dimensional metal-halide perovskites. Nanotechnology **31**(15), 152002 (2020)
88. Zhang, P., Li, M., Chen, W.-C.: A perspective on perovskite solar cells: Emergence, progress, and commercialization. Front. Chem. **10**, 802890 (2022)
89. Masoumeh Keshavarz, Martin Ottesen, Steffen Wiedmann, Michael Wharmby, Robert Küchler, Haifeng Yuan, Elke Debroye, Julian A Steele, Johan Martens, Nigel E Hussey, et al. Tracking structural phase transitions in lead-halide perovskites by means of thermal expansion. *Advanced materials*, 31(24):1900521, 2019
90. Arnold, D.C.: Composition-driven structural phase transitions in rare-earth-doped bifeo3 ceramics: a review. IEEE Trans. Ultrasonics Ferroelectrics Frequency Control **62**(1), 62–82 (2015)
91. Thomson, S.: Observing phase transitions in a halide perovskite using temperature dependent photoluminescence spectroscopy. Edinburgh Instruments, Livingston (2018)
92. Zhang, Y., Xiaojie, X.: Machine learning lattice constants for cubic perovskite compounds. ChemistrySelect **5**(32), 9999–10009 (2020)
93. Zuo, C., Bolink, H.J., Han, H., Huang, J., Cahen, D., Ding, L.: Advances in perovskite solar cells. Adv. Sci. **3**(7), 1500324 (2016)
94. Anaya, M., Correa-Baena, J.P., Lozano, G., Saliba, M., Anguita, P., Roose, B., Abate, A., Steiner, U., Grätzel, M., Calvo, M.E., et al.: Optical analysis of CH3NH3SnxPb1-xi3 absorbers: a roadmap for perovskite-on-perovskite tandem solar cells. J. Mater. Chem. A **4**(29), 11214–11221 (2016)

95. Shao, J.-Y., Li, D., Shi, J., Ma, C., Wang, Y., Liu, X., Jiang, X., Hao, M., Zhang, L., Liu, C., et al.: Recent progress in perovskite solar cells: material science. SCIENCE CHINA Chem. **66**(1), 10–64 (2023)
96. Lee, K.-T., Jay Guo, L., Park, H.J.: Neutral-and multi-colored semitransparent perovskite solar cells. Molecules **21**(4), 475 (2016)
97. Fujiwara, H., Kato,M., Tamakoshi, M., Miyadera, T., Chikamatsu, M.: Optical characteristics and operational principles of hybrid perovskite solar cells. Physica status solidi (A) **215**(12), 1700730 (2018)
98. Cheng, X., Yang, S., Cao, B., Tao, X., Chen, Z.: Single crystal perovskite solar cells: development and perspectives. Adv. Func. Mater. **30**(4), 1905021 (2020)
99. Elumalai, N.K., Arafat Mahmud, Md., Wang, D., Uddin, A.: Perovskite solar cells: progress and advancements. Energies **9**(11), 861 (2016)
100. Panda, N., Parida, B.N., Padhee, R., Choudhary, R.N.P.: Dielectric and electrical properties of the double perovskite pbbabinbo 6. J. Electron. Mater. **44**, 4275–4282 (2015)
101. Cheng, J.-R., Li, N., Eric Cross, L.: Structural and dielectric properties of ga-modified bifeo 3–pbtio 3 crystalline solutions. J. Appl. Phys. **94**(8), 5153–5157 (2003)
102. Chu, L.-L., Zhang, T., Zhang, W.-Y., Shi, P.-P., Gao, J.-X., Ye, Q., Da-Wei, F.: Three-dimensional metal-free molecular perovskite with a thermally induced switchable dielectric response. J. Phys. Chem. Lett. **11**(5), 1668–1674 (2020)
103. Lv, X.-H., Liao, W.-Q., Wang, Z.-X., Li, P.-F., Mao, C.-Y., Ye, H.-Y.: Design and prominent dielectric properties of a layered phase-transition crystal:(cyclohexylmethylammonium) 2cdcl4. Crystal Growth Design **16**(7), 3912–3916 (2016)
104. Rathore, S.S., Mishra, S.K., Vitta, S.: Low temperature dielectric dispersion and relaxor like behavior in multiferroic ba3nbfe3si2o14. J. Appl. Phys. **111**(7) (2012)
105. Satyapal Singh Rathore and Satish Vitta: Effect of divalent Ba cation substitution with Sr on coupled 'multiglass' state in the magnetoelectric multiferroic compound Ba3NbFe3Si2O14. Sci. Rep. **5**(1), 9751 (2015)
106. Rathore, S.S., Nathawat, R., Vitta, S.: A 'mixed' dielectric response in langasite Ba3 NbFe3Si2O14. Phys. Chem. Chem. Phys. **23**(1), 554–562 (2021)
107. Sharma, V.K., Nathawat, R., Rathore, S.S.: Structural, thermal and dielectric studies of mixed spinel cofe2o4. Mater. Today: Proc. **46**, 2193–2196 (2021)
108. Sharma, V.K., Nathawat, R., Rathore, S.S.: Dielectric properties correlation with microstructure in ABi4Ti4O15 (A= Sr, Ba) bismuth layered ferroelectrics. Mater. Adv. **3**(12), 4890–4898 (2022)
109. Yi, W., Princep, A.J., Guo, Y., Johnson, R.D., Khalyavin, D., Manuel, P., Senyshyn, A., Presniakov, I.A., Sobolev, A.V., Matsushita, Y., et al.: Sc2nimno6: a double-perovskite with a magnetodielectric response driven by multiple magnetic orders. Inorganic Chem. **54**(16), 8012–8021 (2015)
110. Zhang, T., Wang, X.P., Fang, Q.F., Li, X.G.: Magnetic and charge ordering in nanosized manganites. Appl. Phys. Rev. **1**(3) (2014)
111. Mao, Y., Parsons, J., McCloy, J.S.: Magnetic properties of double perovskite La2BMnO6(B= Ni or Co) nanoparticles. Nanoscale **5**(11), 4720–4728 (2013)
112. Ravi Shankar, P.N., Orlandi, F., Manuel, P., Zhang, W., Shiv Halasyamani, P., Sundaresan, A.: A-site and b-site cation ordering induces polar and multiferroic behavior in the perovskite nalnniwo6 (ln= y, dy, ho, and yb). Chem. Mater. **32**(13), 5641–5649 (2020)
113. Su, J., Yang, Z.Z., Lu, X.M., Zhang, J.T., Gu, L., Lu, C.J., Li, Q.C., Liu, J.-M., Zhu, J.S.: Magnetism-driven ferroelectricity in double perovskite y2nimno6. ACS Appl. Mater. Interfaces **7**(24), 13260–13265 (2015)
114. Prellier, W., Singh, M.P., Murugavel, P.: The single-phase multiferroic oxides: from bulk to thin film. J. Phys. Condensed Matter **17**(30), R803 (2005)
115. Shimakawa, Y., Azuma, M., Ichikawa, N.: Multiferroic compounds with double-perovskite structures. Materials **4**(1), 153–168 (2011)
116. Sun, C., Alonso, J.A., Bian, J.: Recent advances in perovskite-type oxides for energy conversion and storage applications. Adv. Energy Mater. **11**(2), 2000459 (2021)

117. Zhu, L., Ran, R., Tade, M., Wang, W., Shao, Z.: Perovskite materials in energy storage and conversion. Asia-Pac. J. Chem. Eng. **11**(3), 338–369 (2016)
118. Zhang, L., Miao, J., Li, J., Li, Q.: Halide perovskite materials for energy storage applications. Adv. Func. Mater. **30**(40), 2003653 (2020)
119. Bati, A.S.R., Zhong, Y.L., Burn, P.L., Nazeeruddin, M.K., Shaw, P.E., Batmunkh, M.: Next-generation applications for integrated perovskite solar cells. Commun. Mater. **4**(1), 2 (2023)
120. Kostopoulou, A., Brintakis, K., Nasikas, N.K., Stratakis, E.: Perovskite nanocrystals for energy conversion and storage. Nanophotonics **8**(10), 1607–1640 (2019)
121. Wang, Y., Liu, J., Song, Y., Yu, J., Tian, Y., Robson, M.J., Wang, J., Zhang, Z., Lin, X., Zhou, G., et al.: High-entropy perovskites for energy conversion and storage: design, synthesis, and potential applications. Small Methods **7**(4), 2201138 (2023)
122. Ferguson, V., Silva, S.RP., Zhang, W.: Carbon materials in perovskite solar cells: prospects and future challenges. Energy Environ. Mater. **2**(2), 107–118 (2019)
123. National Renewable Energy Laboratory (NREL): Best research-cell efficiencies chart. https://www.nrel.gov/pv/cell-efficiency.html (2025). Accessed Aug 2025
124. Xing, Z., Li, S.-H., Yang, S.: Targeted molecular design of functionalized fullerenes for high-performance and stable perovskite solar cells. Small Struct. **3**(6), 2200012 (2022)
125. Liu, T., Chen, K., Qin, H., Zhu, R., Gong, Q.: Inverted perovskite solar cells: progresses and perspectives. Adv. Energy Mater. **6**(17), 1600457 (2016)
126. Hussain, I., Tran, H.P., Jaksik, J., Moore, J., Islam, N., Jasim Uddin, M.: Functional materials, device architecture, and flexibility of perovskite solar cell. Emergent Mater. **1**, 133–154 (2018)
127. Dai, X., Ke, X., Wei, F.: Recent progress in perovskite solar cells: the perovskite layer. Beilstein J. Nanotechnol. **11**(1), 51–60 (2020)
128. Qin, F., Min, L., Po, L., Sun, S., Bai, X., Zhang, Yu.: Luminescence and degeneration mechanism of perovskite light-emitting diodes and strategies for improving device performance. Small Methods **7**(10), 2300434 (2023)
129. Yang, D., Zhao, B., Yang, T., Lai, R., Lan, D., Friend, R.H., Di, D.: Toward stable and efficient perovskite light-emitting diodes. Adv. Funct. Mater. **32**(9), 2109495 (2022)
130. Hou, S., Liu, J., Shi, F., Zhao, G.-X., Tan, J.-W., Wang, G.: Recent advances in silver nanowires electrodes for flexible organic/perovskite light-emitting diodes. Front. Chem. **10**, 864186 (2022)
131. Yi, Yu., Tang, Y., Wang, B., Zhang, K., Tang, J.-X., Li, Y.-Q.: Red perovskite light-emitting diodes: Recent advances and perspectives. Laser Photonics Rev. **17**(2), 2200608 (2023)
132. Zhao, X., Andrew Ng, J.D., Friend, R.H., Tan, Z.-K.: Opportunities and challenges in perovskite light-emitting devices. ACS Photonics **5**(10), 3866–3875 (2018)
133. Shohag, Md.A.S., Eze, V.O., Carani, L.B., Okoli, O.I.: Fully integrated mechanoluminescent devices with nanometer-thick perovskite film as self-powered flexible sensor for dynamic pressure sensing. ACS Appl. Nano Mater. **3**(7), 6749–6756 (2020)
134. Tong-Yu, J., Fan, C.-C., Liang, B.-D., Liu, C.-D., Jin, M.-L., Chai, C.-Y., Zhang, W.: Chirality triggered biferroicity in a 3d rubidium based perovskite. Adv. Func. Mater. **34**(28), 2316747 (2024)

Multifunctional Nanomaterials

Synthesis and Antimicrobial Activity of Transition Metal-based Biogenic Nanoparticles

Sriparna Ray and Rishabh Basrur

Abstract The chemical synthesis of nano particles may involve harmful chemicals which may be toxic towards humans and the environment. Hence, different techniques for the manufacture of nanoparticles through environmentally friendly methods have grown in leaps and bounds in the last few years. *Plectranthus amboinicus* has been used for many years as ayurvedic medicines for treating bacterial infections. Herein, the green synthesis of different transition metal-based nanoparticles will be discussed, utilizing phytochemicals from *P. amboinicus.* In addition, their spectroscopic analysis, with the help of UV-Vis and FTIR spectroscopy, as well as XRD techniques will be discussed. The utility of these biogenic nanoparticles has been based on their effectiveness on the restriction of bacterial growth. Each of the nanoparticles were tested against multiple bacteria, for example, Gram-negative *E. coli* and *Klebsiella pneumonia* as well as Gram-positive *Streptococcus aureus* and *Bacillus subtilis.* The zone of inhibition found for each bacterium was used as the criteria for measuring the effectiveness of the antibacterial nanoparticles.

Keywords Biogenic nanoparticles · Antimicrobial activity · Transition metal complexes · Earth abundant transition metal · *Plectranthus amboinicus*

1 Introduction

Nanoparticles are a collection of materials with characteristics that are different from their bulk and their molecular levels. It is defined by the size of the particles which range from 1 to 100 nanometres. The technology involving nanoparticles has great potential which is why there has been a great increase in the recent investment and

S. Ray (✉) · R. Basrur
Catalytic Applications Laboratory, Department of Chemistry, School of Physical and Biological Sciences, Faculty of Science, Technology and Architecture, Manipal University Jaipur, Jaipur 303007, Rajasthan, India
e-mail: sriparna.ray@jaipur.manipal.edu; sriparna.ray@gmail.com

L. Ledwani and S. Srivastava (eds.), *Experimental and Computational Advances in Materials*, Engineering Materials, https://doi.org/10.1007/978-981-95-4180-5_9

development in this field. It can be used in various field like electronics, biomedical, pharmaceuticals, cosmetics, catalysts, and materials [1].

Biomedical nanoparticles are of two types: inorganic and organic. Inorganic nanoparticles have been found to be successful in clinical studies and are being developed for various medical applications including ablations of tumours, and anaemia treatments. Organic nanoparticles have shown positive results in the field and are being designed for medical applications like vaccinations [2].

The formation of iron, copper, silver, nickel, zinc, etc. nanoparticles using the extract of *Plectranthus amboinicus* is reported by various research groups. Different analytical techniques were utilized on the obtained nanoparticles to get an understanding of the shape, size and bonding in the nanoparticles. Techniques involving spectroscopic analysis like, UV-visible (UV-vis) and Fourier-transform infrared (FTIR) spectroscopies, Scanning electron microscopic (SEM) analysis, X-ray diffraction (XRD) analysis, etc. have been utilized for complete characterization of the nanoparticles [3–5].

Among the different applications of nanoparticles, one emerging area is their utilization in inhibiting disease-causing microbes. Hence, the antibacterial properties of all the aforementioned nanoparticles were explored and used to draw a conclusion towards their utilization in pharmaceutical sector.

2 *Plectranthus amboinicus*

Indian Borage and Ajwain is the common name of *P. amboinicus* in India. It is found all over India and some parts of Africa. Figure 1 shows the general appearance of the plant. It is used in home remedies for cough, cold and fever [6]. Several reports have exemplified the utility of the phytochemicals present in *P. amboinicus*. Apigenin, luteolin, and salvigenin are some of the flavonoids present in it [7]. Different parts of this plant have exhibited different utility. Apart from antibacterial activity, the medicinal use of this plant includes anti-inflammatory activity, antidiabetic activity, etc.

2.1 Preparation of The Leaf Extract

The leaves of the *P. amboinicus* were taken and thoroughly washed with distilled water. The leaves were then cut into small pieces and boiled in distilled water for several hours till the water turned into a green coloured solution. The change in the colour indicated the formation of the extract. The solution is then filtered to remove the pieces of the leaves. The filtrate then obtained is the leaf extract. Through FTIR spectrum, major peaks ranging from 3433.29 to 1064.71 cm^{-1} were observed, which provided the evidence of the different organic functional groups present in the phytochemicals. Notable among them are 3433.29 and 3417.86 cm^{-1} which

Fig. 1 *P. amboinicus* plant

corresponded to O–$H_{stretch}$. This indicated the existence of alcoholic –OH or phenolic groups. The 2924.09 and 2862.36 cm^{-1} peaks corresponded to C–$H_{stretch}$ of acyclic alkanes. The C=O stretching vibration at 1616.35 cm^{-1} indicated the presence of aromatic ketones or carboxyl groups. The peaks at 1411.89 and 1064.71 cm^{-1} were indicating the presence of aliphatic aldehydes and primary aliphatic amines [3, 5].

3 Preparation of the Nanoparticles

The preparation of the nanoparticles with the individual transition metals were quite similar in procedure. A detailed schematic diagram is shown in Fig. 2. To the leaf extract, a water-soluble solution of the required metal was added and stirred for 8 to 12 h until a colour change is observed. This solution was then centrifuged, and the supernatant liquid was discarded, to obtain the respective nanoparticles.

3.1 Manganese Nanoparticles and their Applications

Revathi et al. reported that the manganese-based nanoparticles were synthesized by mixing the freshly prepared solutions of $KMnO_4$ (5 mL) and the leaf extract (25 mL) [8]. On storing this mixture under ambient conditions, the manganese ions are gradually reduced, which could be monitored by the colour change from pale pink to brown colour. Consequently, the mixture was centrifuged and the nanoparticles washed with distilled water a number of times to purify the manganese-based NPs. Interestingly, the bioactive phytochemicals were screened using various tests including Meyer's test and Wagner's test for alkaloids, Froth test for saponins, Bradford's test for proteins,

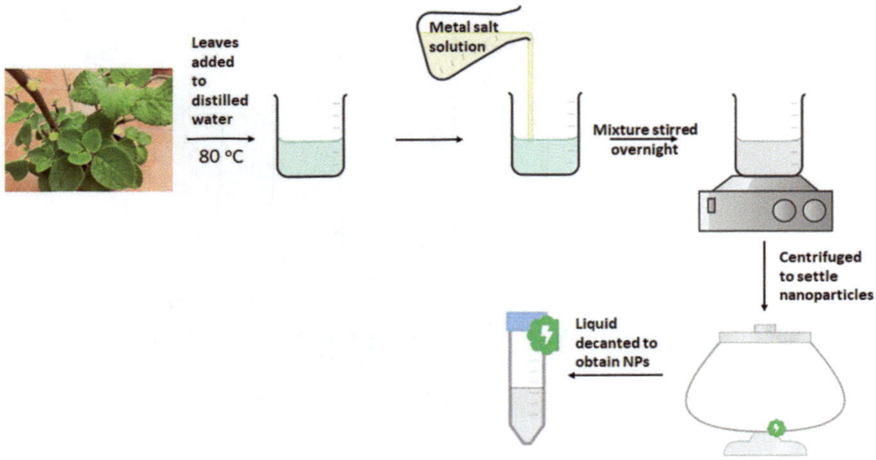

Fig. 2 Synthetic strategy of transition metal-based nanoparticles

Fehlings test, Molisch's test and Barfoed's test for carbohydrates, Shibita's reaction test and Pew's test for flavonoids, test for ketones and Fehlings test for free reducing sugars. The presence of these phytochemicals is extremely essential for reducing the manganese ions to their zerovalent forms in the nanoparticles, through the biosynthesis process.

After the synthesis, the manganese-based nanoparticles were thoroughly characterized utilizing UV-visible and FTIR spectroscopy, SEM, EDX and XRD analysis. The bioactivity of the Mn NPs was successfully checked using agar well diffusion assay and the zone of inhibition was measured to find their efficacy against specific bacterial cells. To prepare the petri dishes for the bioactivity analysis experiment, the first step was to pour nutrient agar medium into petri dishes, in which the test bodies were injected into with the help of a micropipette. This was allowed to dry and grow. The result is tabulated in Table 1.

Table 1 Antibacterial activity of the manganese nanoparticles in terms of zone of inhibition

Bacteria	Zone of inhibition (mm)
S. aureus	25
S. pneumonia	23
B. subtilis	21
E. coli	21
K. pneumonia	22
P. aeruginosa	21
P. mirabilis	19

Table 2 Antibacterial activity of the iron nanoparticles in terms of zone of inhibition

Bacteria	Zone of inhibition (mm)
E. coli	15
Klebsiella	18
Proteus	16
S. typhi	12
S. aureus	14

3.2 *Iron Nanoparticles and their Applications*

To prepare the Fe_2O_3 nanoparticles, 1M $FeCl_3$ solution was taken. Plant extract (20 mL) was added to the ferric chloride solution [3]. The solution was then stored until a colour change occurred to a dark brown colour. To obtain the nano particles the solution is centrifuged. The prepared nanoparticles were brownish in colour.

The iron nanoparticles were analysed by X-ray diffraction, FTIR spectroscopy, UV-visible spectroscopy and scanning electron microscope. The XRD analysis shows peak at (0 1 2), (1 0 4), (1 1 0), (1 1 3), (0 2 4), (1 1 6), (2 1 4), (4 1 4) and (1 1 9). This concluded that the structure of the nanoparticles was found to be rhombohedral. By the help of Debye-Scherer's equation the average particle size was around 80 to 100 nm.

In the FTIR spectrum, the band observed at 3460 cm^{-1} corresponded to a stretch in the O–H–O vibration, the band at 2935 cm^{-1} corresponds to CH_2 while 2436 cm^{-1} is due to the presence of a cyano group. 1625 cm^{-1} is present due to the water that has been absorbed by the nanoparticle. 1085 cm^{-1} is due to the presence of a C–O–C vibration. In the UV-Vis spectrum the absorption at 200 and 250 nm is due to the occurrence of Fe-O formation. In addition, through SEM, the morphology of the particles were observed, where the particle distribution was non-homogeneous with small spherical shapes distributed in the entire domain.

The iron nanoparticles were treated against Gram-negative *E. coli*, *Klebsiella*, *Proteus* and *S. typhi* and Gram-positive *S. aureus*. Their antimicrobial activity is tabulated in Table 2.

3.3 *Copper Nanoparticles*

To prepare the copper nanoparticles, 0.1 M $CuNO_3$ solution was taken similarly and reacted with 20 mL of leaf extract [4]. The copper nanoparticles were analysed by X-ray diffraction, FTIR spectroscopy, UV-visible spectroscopy and scanning electron microscope. The XRD analysis showed peaks at (110), (002), (111), (202), (020), (021), (113), (311) (220), (312) and (044). This indicates the copper nano particles had a crystalline structure. The average particle size was between 20 to 35 nm using Debye-Scherer's equation. From the FTIR spectrum, prominent peaks can be seen

Table 3 Antibacterial activity of the copper nanoparticles in terms of zone of inhibition

Bacteria	Zone of inhibition (mm for 50 mg/mL)
E. coli	4.45 ± 0.24
K. pneumonia	4.20 ± 0.22
P. aeruginosa	3.95 ± 0.20
S. aureus	4.10 ± 0.21
S. pyogenes	3.55 ± 0.17
B. subtilis	3.70 ± 0.18

at 3467, 1694, 1100, 1471, 850, 582, 529, and 691 cm^{-1}. Similar to that of the iron NPs, the band at 3467 cm^{-1} corresponded to the OH group, the band at 1694 cm^{-1} is caused by an amide group. The peak at 1471 cm^{-1} is owed to the presence of a $C \equiv N$ group. The peak at 1100 cm^{-1} corresponded to the presence of a carboxylic group. The peaks below 1000 cm^{-1} correlate to the metal-oxygen bonding. The UV-Vis spectrum of the formed Cu nanoparticles shows a peak at 398 nm. This is due to the small sized particles of CuO formed. In the SEM, the morphology of the copper nanoparticles was studied. It was observed that spherical, circular, and oval shaped nanoparticles were presented. This showed that the Cu nanoparticles had high surface energy.

The copper nanoparticles were tested against *E. coli*, *S. aureus*, *Bacillus subtilis*, *S. pyogenes*, *Pseudomonas aeruginosa* and *Klebsiella pheumoniae* (Table 3).

3.4 Silver Nanoparticles

The silver nanoparticles were synthesized in a similar manner, using 0.1 mM silver nitrate solution (90 mL) with 10 mL of *P. amboinicus* leaf extract [5]. The silver nanoparticles were analysed by X-ray diffraction, FTIR spectroscopy, UV-visible spectroscopy and scanning electron microscope. The XRD spectrum exhibited peaks at (111), (200), (220), and (311) which correspond to a cubic structure of metallic silver ion which further reveal that the NPs were composed of pure crystalline silver. The size of the particles was found to be an average of 17.3 nm. The FTIR spectrum of the silver nanoparticles showed peaks similar to that of the peaks on the FTIR of the plant extract but with a small shift towards the higher wavenumbers. Peaks are seen at 3444.87, 2962.66, 1627.92, 1431.18, and 1095.57 cm^{-1}. These peaks correspond to O–H, C–H, C=O, and aliphatic amines respectively. UV-visible spectrum of the silver nanoparticles showed a sharp peak between 400 and 420 nm, signifying the presence of Ag nanoparticles. The SEM image of the Ag nanoparticles were evenly shaped and spherical.

The silver nanoparticles were tested against Gram-positive *Pseudomonas aeruginosa*, and Gram-negative *Enterobacter sp.*, *Klebsiella*, and *E. coli*, and their antibacterial characteristics are presented in Table 4.

Table 4 Antibacterial activity of the silver nanoparticles in terms of zone of inhibition

Bacteria	Zone of inhibition (mm for 50 mg/mL)
P. aeruginosa	22.00 ± 1.00
Enterobacter sp.	11.66 ± 1.52
E. coli	13.00 ± 1.00
K. pneumonia	14.00 ± 1.00

Prabakaran *et al.* also had synthesized similar silver nanoparticles from silver nitrate solution and leaf extract of *P. amboinicus* [9]. They, too, reported good potency of these NPs against *Pseudomonas aeruginosa, Escherichia coli* and *Staphylococcus aureus.* In addition, these NPs were found to inhibit the proliferation of B16F10 murine melanoma cells, thereby exhibiting anticancer properties. Another notable application of industrial importance was its utility as a biosensor for detecting triethylamine, which is considered as a toxic impurity.

3.5 Nickel Nanoparticles

R. Ramesh and co-workers had successfully synthesized nickel nanoparticles by stirring 1 M nickel nitrate solution with 20 mL freshly prepared extract of *P. amboinicus* for 30 min [10]. The product was centrifuged and washed thoroughly with distilled water and dried in hot air oven. The resultant compound turned dark green in colour after the calcination process. Detailed characterization of the nanoparticles was carried out using XRD, SEM and EDAX analysis. While XRD exhibited the crystalline nature of NPs with average size of 100 nm, SEM and EDAX confirmed their elemental composition. The confirmation for Ni-O bonds came from FTIR spectroscopy (452 cm^{-1}) and UV-visible spectroscopy (250–370 nm).

These nickel-based nanoparticles were used for anti-bacterial and antifungal studies, utilizing the agar disc diffusion method. Positive tests were obtained with respect to bacteria *Proteus* and *Candida* fungal strain.

3.6 Zinc Nanoparticles

A solution of zinc acetate dihydrate in distilled water was added to the *P. amboinicus* leaf extract and stirred to obtain the zinc oxide nanoparticles [11]. The XRD analysis exhibited the crystalline wurtzite structure for the zinc oxide nanoparticles, with an average size of 42 nm, which matched with prior literature reports. The EDX studies clearly proved the pure presence of Zn and O elements in the biosynthesized NPs. The FTIR spectroscopy confirmed the presence of Zn–O bond at 600–400 cm^{-1}. The presence of flavonoids or terpenoids from the *P. amboinicus* leaf extract was also visible through the distinct peaks at 3406 and 2925 cm^{-1}.

Table 5 Antibacterial activity of the zinc nanoparticles in terms of zone of inhibition

Bacteria	Zone of inhibition (mm)
B. subtilis	14
E. coli	15
S. epidermidis	11

In order to explore the antimicrobial activity against Gram-positive and Gram-negative bacteria, minimum inhibitory concentration (MIC) values and zone of inhibition were obtained using either the broth microdilution or the agar well diffusion methods. The bioactivity was investigated against *E. coli*, *Bacillus subtilis* and *Staphylococcus epidermidis* (Table 5).

The large surface area present in the biosynthesized zinc oxide nanoparticles made it ideal to inhibit the proliferation of bacterial growth, Due to the strong interaction with the microbial cell, the ZnO nanoparticles, thus synthesized, can treat the common Gram-negative bacteria.

4 Conclusion and Future Scope

All the biogenic nanoparticles had been effective against a number of bacterial species. While all the nanoparticles were effective against bacteria, according to the results discussed in the above reports, the manganese nanoparticles were most effective against the bacteria. This can be rationalized by the fact that manganese-based nanoparticles are known for their small size, thereby leading to distinctive characteristics and interactions with the bacterial cells [12, 13]. These NPs can easily penetrate the cell membrane, which enhances their ability to inhibit the growth and proliferation of the microbes. Additionally, the positively charged manganese ions can interact destructively with the negatively charged cell membrane surface.

Various research groups are exploring transition metal-based nanoparticles, which are being synthesized using the green biosynthetic pathway. In fact, utilizing the earth abundant transition metals are being more impactful due to their sustainable nature. It is noteworthy that chemical catalysis facilitated by various transition metal-based nanoparticles is also a very active arena of research [14–16]. Both rare earth metals as well as earth abundant transition metals are being active explored in their nanoparticle moieties and various chemical utility and bioactivity are being investigated for further utilization in the industrial and pharmaceutical sectors.

References

1. Biswas, P., Wu, C.-Y.: J. Air & Waste Manag. Assoc. **55**, 708 (2005)
2. Anselmo, A.C., Mitragotri, S.: Bioeng. Transl. Med. **1**, 10 (2016)

3. Ramesh, R., Yamini, V., Rajkumar, D., John Sundaram, S., Lakshmi, D., Liakath Ali Khan, F.: Mater. Today: Proc. **36**, 453 (2021)
4. Velsankar, K., Vinothini, V., Sudhahar, S., Krishna Kumar, M., Mohandas, S.: Appl. Nanosci. **10**, 3953 (2020)
5. Sreelakshmy, S., Thangapandiyan, S.: Asian J. Pharm. Clin. Res. **12**, 153 (2019)
6. Arumugam, G., Swamy, M.K., Sinniah, U.R.: Molecule **21**(4), 369 (2016)
7. Kumar, P., Nitin Kumar, S.: A. J. Physiol. Biochem. Pharmacol. **10**(2), 55 (2020)
8. Revathi, N., Sankarganesh, M., Raja, J.D., Rajakanna, J., Senthilkumar, O.: Inorg. Chem. Commun. **154**, 110935 (2023)
9. Prabakaran, L., Sathyaraj, W.V., Yesudhason, B.V., Subbaraj, G.K., Atchudan, R.: Chemosensors **11**, 373 (2023)
10. Ramesh, R., Yamini, V., John Sundaram, S., Liakath Ali Khan, F., Kaviyarasu, K.: Mater Today Proc. **36**(2), 268 (2021)
11. Porkalai, V., Lavanya, N., Nithya, R., Sagayaraj, R.: Asian J. Chem. **36**(2), 301 (2024)
12. Velho-Pereira, S., Parmekar, M.V.: Chem. Select **9**, e202402247 (2024)
13. Islam, B., Samir, D.: J. Microbiol. Infect. Dis. **14**(4), 146 (2024)
14. Garge, S.S., Joglekar, P., Ray, S.: Bulg. Chem. Commun. **54**, 50 (2022)
15. Choudary, B.M., Madhi, S., Chowdari, N.S., Kantam, M.L., Sreedhar, B.: J. Am. Chem. Soc. **124**, 14127 (2002)
16. Chahdoura, F., Pradel, C., Gómez, M.: Chem. Cat. Chem. **6**, 2929 (2014)

Surface Functionalized Materials

Surface Engineering in Antibacterial Paints: Innovations for Revolutionizing Paints and Coating Technologies

Himanshi Saini and Lalita Ledwani

Abstract The ever-increasing requirement for pathogenic-resistant atmospheres and sanitation has propelled the progress of antimicrobial coatings across all areas and has not been limited to healthcare facilities. These antibacterial paints and coatings have now become a crucial revolution in surface protection technologies. The chapter explores all the possible roles of surface engineering in enhancing the functionality and efficiency of antimicrobial paint, along with taking care of the sustainability criteria. It delves foremost into the fundamental mechanism of antimicrobial action, along with types of antimicrobial agents used to incorporate the antimicrobial properties. A comprehensive discussion follows up those deals with different surface engineering techniques for antimicrobial coatings, such as chemical method, physical method, bio-inspired and nanomaterial-assisted strategies that contribute towards improving pathogenic resistance, durability and environmental compatibility. Key applications of these coatings and paints in healthcare environments, food storage areas, marine industries, commercial and residential areas are highlighted, alongside a discussion on testing standards and protocols for performance assessment. The chapter is focused on the most recent developments on coatings, where case studies and empirical examples and the psychological aspect of functions are illustrated to put the functional efficacy of antimicrobial paints into perspective. A comparison of performance metrics of surface-engineered coatings with conventional paints is also addressed. Finally, it also provides insights into the future emerging materials and technologies in antimicrobial coatings.

Keywords Smart coatings · Antibacterial paints · Surface engineering · Coating technology · Surface modification

H. Saini · L. Ledwani (✉)
Department of Chemistry, Manipal University Jaipur, Jaipur, India
e-mail: lalita.ledwani@jaipur.manipal.edu

L. Ledwani and S. Srivastava (eds.), *Experimental and Computational Advances in Materials*, Engineering Materials, https://doi.org/10.1007/978-981-95-4180-5_10

1 Introduction

The swift growth of antimicrobial-resistant microorganisms alongside the growing demand for sterile and hygienic surroundings have fuelled substantial advancements in the arena of surface engineering. Amongst all the promising innovations in the surface engineering field, the development of surface coatings with bacterial resistance has emerged in recent years. These antibacterial paints and coatings play an acute crucial part in preventing microbial contamination in the immediate environment, which in turn enhances public health primarily, and principally reduces the spread of communicable diseases [1]. Predominantly microbial contamination of surfaces, especially in communal areas, has been an enduring challenge in healthcare, food processing, as well as public spaces. Traditional paints and coating are essentially aesthetically pleasing but can serve as reservoirs for bacteria, fungi, and other microbes, allowing their transmission and proliferation [2]. This predicament has directed to the advancement of coating technologies leading to the development of antibacterial paints. These coatings are designed to help in inhibition of microbial growth which successively creates a hygienic environment. In hospitals and healthcare facilities, antimicrobial coatings have played crucially in reducing hospital-acquired infections (HAIs). These HAIs are highly accountable for increasing mortality rates worldwide, however this can be reduced by a noteworthy amount using antibacterial coatings. Common pathogens namely *Staphylococcus aureus*, *Escherichia coli*, and *Pseudomonas aeruginosa*, contribute to the transmission of diseases, thus accenting the need for antimicrobial interventions [3]. Apart from healthcare, antibacterial paints and coatings are fetching importance in residential and commercial buildings as well. As discussed earlier, the rise in antibiotic-resistant bacteria and global pandemics, there is a budding awareness about the need for antimicrobial protection and coatings in areas of high frequency touch, such as walls, doorknobs, and countertops. Therefore, from last decade the construction industry has responded by integrating antibacterial coatings into contemporary building designs, suggestively decreasing the risk of microbial contamination whilst improving indoor air quality. These coatings are mostly advantageous in areas where recurrent cleaning and decontamination are required, such as healthcare facilities, schools, restaurants, and public transportation [4].

The efficiency of antibacterial coatings resides in their ability to inactivate, inhibit or destroy the growth of microbes through various mechanisms. Some of the antibacterial coatings integrate biocidal agents within those aids in killing microbes upon interaction. These agents, such as silver nanoparticles, copper, and zinc oxide, chiefly interact with cell membranes of microbes, instigating structural damage of cell and its lysis [5]. In other kind of formulations, antibacterial agents are progressively and steadily released from the matrix of the paint, thus providing continual antimicrobial activity over a stretched period of time. These formulations have common agents like quaternary ammonium compounds (QACs), triclosan, and iodine-based reagents [6]. Apart from that certain class of paints operate over photocatalysts such as titanium dioxide (TiO_2) to further generate reactive oxygen species (ROS) when and where exposed to light. These generated ROS excellently degrade organic matter and damages the cellular components of microbes. Other techniques are also incorporated in the class such as hydrophobic coatings which prevent microbial adhesion by forming water-repellent surfaces, which further inhibits the formation of biofilms [7]. Also, there are some advanced coatings that are particularly designed to alter the surface pH upon microbial contact, generating an inhospitable environment for proliferation of the bacteria. The above discussed mechanisms allow researchers to develop coatings with improved performance, guaranteeing long-lasting antimicrobial protection.

Since the chapter focuses on surface engineering and its integration with antibacterial coatings, it is important to discuss that the formulation of such highly effective paints has resurfaced the interdisciplinary approaches, which includes, material science, nanotechnology, and microbiology altogether, to generate coatings with enhanced antimicrobial properties. The integration of the above-mentioned foci in the surface engineering field has brought momentous advancements to the antibacterial paint's formulation and effectiveness in the environment. Articulating about nanomaterials, such as silver and copper nanoparticles, these have transformed antimicrobial coatings by providing a significantly higher surface area-to-volume ratios along with increased reactivity [8]. These nanoparticles due to their composition and characteristics produces a broad-spectrum antimicrobial activity and enduring stability in formulation of paint. Despite the efficiency of nanoparticle-based paint, the hunt for eco-friendly alternatives have sustained predominantly due to the search of a sustainable and environment friendly paint. This has led to the incorporation of natural antimicrobial agents, such as chitosan, natural oils, and polyphenols. These bio-based coatings are now becoming a sustainable alternative which does not compromises with the antimicrobial efficacy of the paint. Apart from that some modern coatings parade self-cleaning properties, which actively repelling contaminants and microbes. One such example is a class of hydrophilic paints along with photocatalytic functionality. The photocatalytic activity assists in the breakdown of organic matter and also helps with reducing microbial adhesion to the surface, further extending the lifespan of the paint [9]. Additionally, researchers are currently focusing on incorporating the mechanical properties as well along with sustained antimicrobial property. This will safeguard that antimicrobial functionality remains intact despite any abrasion, excessive cleaning, and even harsh environmental exposure [10].

Despite all the advancements discussed as of now in antibacterial paint technology, several challenges are yet to be addressed in order to optimize their overall effectiveness and worldwide adoption. A major requirement of biocidal/antibacterial agents in paint formulations is that it must obey with regulatory guidelines to ensure safety for both individuals as well as the environment. This is mandatory because some antimicrobial compounds have raised alarming concerns due to their toxicity and likely effect on the ecosystems. Also, prolonged exposure to these antibacterial agents may as well contribute towards the development of resistant strains [11]. Nevertheless, to battle this, scientists are now focusing on evolving multi-modal strategies that would eventually reduce the probability of development of these resistant strains. The research also targets on maintaining the functionality over long periods, requiring careful formulation. Additionally, even though antibacterial properties are the primary focus and need of the hour but, it is also important for the developed paints to retain their colour, texture, and other aesthetic properties. Correspondingly, during large scale adoption, making these coatings affordable is also a crucial consideration, as it has much higher production cost compared to conventional paints [12]. The chapter discusses about how the application of surface engineering in the field of antibacterial paints is reforming the coatings industry, providing operative solutions to combat microbial contamination in various environments and settings. The chapter also briefly discusses about nanotechnology, biomaterials, and other smart coatings that are paving the way for next-generation antimicrobial coatings that are vitally enhancing health and sanitation whilst maintaining the sustainability factor. Despite the discussed challenges, continued innovation have driven the development of more effective, durable, and environmentally friendly antibacterial paints, in due course shaping and changing the future of surface protection in private and public spaces.

2 Fundamentals of Antibacterial Paints

As discussed vividly in the introduction section antibacterial paints are advanced type of coatings that are articulated to inhibit or minimize bacterial growth on painted surfaces whilst maintain sustainability, mechanical aspects and other aesthetically pleasing criterions. They are commonly used in areas that have hygiene-sensitive environments most likely hospitals, institutes, kitchens/food-processing areas, and other public or residential spaces where maintaining cleanliness is essential due to frequent contact [13]. The fundamental aspects of antibacterial paints are outlined below.

2.1 Mechanisms of Antibacterial Action

The functionality of Antibacterial paints works through various mechanisms, which including contact-killing, release of biocidal, and anti-adhesion properties. These beforehand mentioned mechanisms are usually influenced by the type of composition of the matrix and also its surface modifications of the paint.

- **Contact-Killing Mechanism**: Surface-engineered antibacterial paints incorporate metallic nanoparticles such as silver, copper, and zinc oxide or other polymeric biocides. These agents ultimately disrupt the bacterial membranes after they come in contact while minimizing the leaching of antibacterial agents during the killing of bacteria. This mechanism ensuring that the paint last longer and maintains its effectiveness. These nanoparticles and agents have the ability to generate ROS which further augments their antibacterial properties, as they tend to disrupt bacterial membrane causing cell death ultimately. A phenomenal example of this is showed in a study where it has been demonstrated that nano silver coatings can hinder over 99% of pathogenic bacterial growth, hence making them highly effective for healthcare and public spaces [14].

- **Biocide Release**: while some paints rely on contact killing, others antibacterial paints rely on the controlled release of antimicrobial agents over the time. Surface engineering techniques especially help regulate the release. This regulation is done by encapsulating biocides within microcapsules or polymeric matrices. This slow-release mechanism thus confirms extended antibacterial action. It also reduces the frequency of reapplication and enhances durability. One such example is copper-based biocidal coatings which have been found to maintain antimicrobial activity for several months, providing an extended protection period [15].
- **Anti-Adhesion Mechanism**: Another mechanism involves modifying the surface topography, which eventually can prevent microbial attachment and formation of biofilm. Some commonly used coatings are hydrophobic coatings, self-cleaning surfaces, and zwitterionic polymers which effectively reduces bacterial adhesion. For instance, superhydrophobic coatings, creates a water-repellent effect, making it difficult for bacteria adhesion on the surface. Such coatings are bio-inspired by leaves of lotus, and are particularly useful in clinics/healthcare facilities and food processing industries where it is crucially required [16].

2.2 Surface Engineering Strategies for Antibacterial Paints

Antibacterial paints along with surface engineering plays a prominent role in enhancing antibacterial performance. It can be done through various mechanisms, this includes nano structuring, chemical functionalization, and hybrid coatings and are mentioned below:

- **Nano structuring**: The incorporation of nanomaterials, such as titanium dioxide (TiO_2) and graphene oxide, enhances the surface reactivity of paints, thereby leading to promotion of antimicrobial properties. These nanostructures generate reactive oxygen species (ROS) under light exposure, effectively killing bacteria. The produced nanoparticles further increase the overall surface area which is accessible for bacterial interaction. This eventually tames antimicrobial efficiency. Advanced techniques such as atomic layer deposition (ALD) allow for precise control over the nanostructure and thickness of these coatings, optimizing their performance [17].
- **Chemical Functionalization**: Surface coatings can be functionalized with antibacterial moieties such as quaternary ammonium compounds (QACs), chitosan, and peptides as mentioned earlier in the manuscript. The functional groups that are incorporated in the paint matrix via additional agents interact with bacterial membranes. Upon contact with the membrane it leads to cell lysis, and it is also interesting to note that the covalent bonds incorporated by surface engineering allows steady attachment of these antibacterial agents for increasing the

efficacy. This further reducing their depletion over time. A recent development includes the use of antimicrobial peptides (AMPs) derived from natural sources, which provide effective antibacterial action with minimal toxicity to human cells [18].

- **Physical surface coating**: Physical surface coatings can provide effective measures preventing bacterial attachment and biofilm formation in a variety of ways. Micro- and nano-patterned surface disruption of bacterial attachment occurs due to the change of topography, providing an unfavourable environment to microbial life, and may be species specific or broadly protective [19]. The hydrophobicity and hydrophilicity adjustment alter the surface wettability that hinders bacterial colonizing, the hydrophobic surfaces defy the presence of water-based free-living bacteria suspension, and the hydrophilic surfaces create a water-revolving shield to block the direct contact. The microorganism retention and attachment to surfaces could be affected by control of surface roughness whereby, smoother surfaces have the tendency to reduce bacterial attachment and particular roughness patterns have the capacity to either trap or repel microorganism. These physical processes synergize with chemical and photocatalytic processes. Chemical coatings include antimicrobial agents or present inhospitable surfaces, and photocatalytic surfaces produce the reactive oxygen species during light exposure. Physical, chemical and photocatalytic techniques in combination have a good potential to generate a multifactor approach to preventing the colonization of bacteria on surfaces and biofilm formation in them [20].
- **Hybrid coating**: Hybrid coating that are formed by inorganic and organic agents tends to improve its durability and instigate multi-functional performance of the resulting single formulation coating. For instant, silver nanoparticles embedded in paint matrices due to its alternating effects provide a sustained antibacterial effect. It also enhances mechanical properties of the formulated paint. The synergy between different organic and inorganic agents creates a robust antibacterial surface resistant towards environmental degradation. Hybrid coatings incorporating both photoreactive TiO_2 and silver have shown excellent results in self-cleaning applications, further extending the longevity of these coatings [21].

2.3 Characterization and Performance Evaluation

In order to study the antibacterial paints efficiency, it can be accessed through surface characterization and microbiological testing.

- **Surface Morphology Analysis**: Techniques such as scanning electron microscopy (SEM) and atomic force microscopy (AFM) evaluate surface roughness and bacterial adhesion. High-resolution imaging obtained from the analysis provides insights into the bacterial interaction with the modified surfaces. These images aids in the optimization of paint formulations for better efficacy. Studies have precisely demonstrated that nanostructured surfaces with optimal roughness parameters significantly reduce bacterial adhesion, which in turn offers improved protection against bacterial attack [22].

- **Chemical Composition Analysis**: Fourier-transform infrared spectroscopy (FTIR) and X-ray photoelectron spectroscopy (XPS) is helpful in determining the chemical bonding and possible functional groups responsible for antibacterial mechanism of the paint. These techniques can also help in assessing the stability of the formulated paint and distribution of biocidal agents present in the coating matrix. Chemical mapping through the help of XPS has been predominantly useful in determining the dispersion of silver, copper and other similar kind of particles in polymer matrices [23].
- **Microbiological Testing**: Standardized tests for instance ISO 22196 and ASTM E2180 help in measuring bacterial reduction rates and its long-term efficacy under different kind of environmental conditions. Additional testing, namely zone inhibition assays and live/dead fluorescence staining, provides supplementary validation of antibacterial performance of the paint. Recent studies have also indicated that engineered coatings in which TiO_2 and CuO nanoparticles are present tend to exhibit a 5-log reduction in bacterial populations just after few hours of application [24].

2.4 *Factors Influencing Antibacterial Performance*

There are several factors that impact the antibacterial efficacy of engineered surfaces. A few factors are stated below briefly:

- **Surface Roughness**: Surfaces that have a smooth interface tend to have reduced bacterial adhesion compared to that of rough interfaces. Engineered micro- and nanostructures can be designed to prevent bacterial colonization as they tend to form smooth surface. Research has also shown that micro-patterned surfaces that mimic shark like skin exhibit a significant reduction in bacterial attachment which in turn reduce bacterial biofilm formation. This phenomenon has been leveraged in especially medical and marine applications [25].
- **Hydrophobicity and Hydrophilicity**: The wettability similar to that of surface roughness of the surface influences bacterial attachment to that particular surface. Hydrophobic surfaces have a tendency to repel water as well as bacteria, while hydrophilic surfaces can be engineered in a way that it ultimately prompts its antimicrobial interactions. Superhydrophobic coatings with nano-texturing have also shown remarkable bacterial resistance by preventing retention of moisture, which is a key factor in proliferation of microbial colonies [26].
- **Environmental Conditions**: The effectiveness of antibacterial coatings along with the influence of surface may also vary depending on its immediate environment. Humidity, temperature, and exposure to contaminants effect antibacterial performance profoundly. In environments with high-humidity, antibacterial paints

incorporating moisture-responsive polymers have established enhanced antimicrobial activity. It is prominently due to activation of biocides upon exposure to moisture [27].

- **Durability and Wear Resistance**: Recent research have shown multi-layer coatings incorporating biocides namely graphene and fluoropolymers has provided a breakthrough in developing wear-resistant antibacterial surfaces. These kinds of antibacterial paints have major industrial applications. This is an important aspect that are to be incorporated in antibacterial paints as the paint must maintain their efficacy over extended periods, even under mechanical stress [28].

3 Materials Used in Antibacterial Surface Engineering

Antibacterial surface engineering is emerging as a new field of the materials science discipline because there is a need to avoid the spread of harmful pathogens on high-touch frequency surfaces. Incorporation of antibacterial material into engineered surfaces is necessary in most applications from healthcare devices and health buildings to consumer products and public facilities. The present chapter provides a comprehensive review of the variety of materials used for antibacterial surface modification, how they work, and the mechanism of their operation. Antimicrobial property of a surface largely relies on the material employed [29].

Metal compounds, polymers, ceramics, and composite materials find applications to confer antimicrobial property. Among them, metals like silver, copper, and zinc have always been recognized to exhibit strong antibacterial effects. Silver, for example, has been highly studied to observe that it releases ions that break bacterial cell membranes and cause disruption in essential enzyme functions. Just like copper has a wide range of antimicrobial efficacy by creating reactive oxygen species that destroy bacterial proteins and DNA, zinc, when added in the form of an oxide, not only imparts antimicrobial effect but also increases toughness in coatings. Such metals are generally embedded in a matrix or used in the form of a thin film to develop surfaces that kill bacteria upon contact reliably [30].

Besides metallic agents, polymers of different kinds have been designed to be antibacterial. Cationic polymers, like quaternary ammonium salts, possess positive functional groups that electrostatically interact with the negatively charged bacterial cell walls. Cell membrane disruption and cell killing ensue from this interaction. The benefit of polymers is that they can be designed to produce coatings having controlled release functions, thereby providing long-duration antimicrobial action. Some polymers can also be designed to include other antibacterial agents to produce synergistic functions that have greater overall effectiveness. Ceramic materials are also being discovered, and they have very significant applications in antibacterial surface engineering [31].

Ceramics like titanium dioxide (TiO_2) are attracting attention with their photocatalytic nature. Under ultraviolet light, TiO_2 generates reactive oxygen species, which have been reported to be cytotoxic to bacteria in quantities. Apart from being the mechanism for the antibacterial action, photocatalysis is also self-cleaning surface technology. The long-term stability and durability of ceramics render them most appropriate for applications in environments where they need to endure severe chemicals and heat. Other than this, development in the area of nanotechnology has made it possible for nanoceramics to be synthesized with increased surface area, which also increases their interface with bacterial cells and optimizes the production of antimicrobial species [25].

One of the thrilling new families of materials in the field of antibacterial surface engineering is the family of composite materials. Composites are made up of two or more dissimilar materials combined to obtain a set of properties not possible from an individual component. Such as polymer-metal composite materials, for instance, that can leverage the antibacterial activity of metals without compromising the plasticity and formability of polymers. In such composites, metal nanoparticles are dispersed in a polymer matrix in a manner that the metal nanoparticles are evenly distributed, and metal ions are released in a controlled fashion. This controlled release is required to offer long-term activity and prevent cytotoxicity to human cells. Analogously, ceramic-polymer composites may offer improved mechanical strength and chemical resistance with antibacterial activity. Processing of composite materials likely requires careful regard to parameters such as particle size, distribution, and surface functionalization, which are likely to directly influence the antimicrobial character of the final product [32].

Bio-inspiration and integration of bio-based materials into antibacterial surface engineering has been among the emerging strategies. They mimic nature's approach to antibacterial action. Insect wing-bio-inspired micro- and nano-architectured surfaces, for example, have been shown to physically disrupt bacterial cells mechanically. Biomimetic surfaces rely not on chemical entities and are therefore a promise to the challenges of bacterial resistance. The bio-degradable polymers made from renewable feedstocks such as chitosan also exhibit in-built antibacterial activity. Chitosan, a derivative of the deacetylation of the protein constituent chitin found in shellfish, is inherently biocompatible, adhesive to bacterial cell walls, and biodegradable. Usage of chitosan as a coating would find maximum applicability in food-packaging and in biomedical-devices where the need would be focused on environmental safety as well as biocompatibility [33]. Along with chitosan, other examples of bio-based materials used to engineer antibacterial surfaces include cellulose, alginate, silk fibroin, and polyhydroxyalkanoates (PHA) because of their general biocompatibility, biodegradability, and growth inhibitory effects on bacteria.

Antibacterial properties of materials are regulated by a series of significant parameters. Roughness and morphology of a surface are the regulators of the interaction of bacteria with a surface. Micro- and nano-structured surfaces were found to inhibit bacterial cell adhesion through reduced contact area and hence enhanced antibacterial performance. Chemical structure and surface functionalization also regulate the interaction between a surface and microbial cells. For instance, hydrophilic surfaces are more resistant to bacterial adhesion than hydrophobic surfaces. The selection of the material must also take into consideration the use conditions in the target application environment where the coating is going to be applied. Temperature, humidity, and UV exposure may affect the stability and release profile of antibacterial agents. Optimization of material properties to the target application environment is therefore the secret to achieving long-term antimicrobial performance [34].

One of the greatest challenges to surface antibacterial engineering is how to achieve the desired effective antimicrobial activity while addressing problems of concern generated by toxicity and environmental impact. Bacteria are highly inhibited by metals such as copper and silver, but environmental and cellular toxicities must be well managed. This is addressed by controlling the concentration and release rate of metal ions through appropriate formulation and surface modification. In most instances, the materials are designed to exhibit a controlled time-release of antibacterial agents over extended periods, with antibacterial protection without reaching toxic concentrations. Besides this, as composite materials have been introduced, incorporating antibacterial agents is now simple in a manner that the exposure is minimized but with a very high antimicrobial activity [35].

4 Techniques and Materials in Antibacterial Surface Engineering

The techniques employed to incorporate antibacterial materials into coatings are just as important as the materials themselves. Sputtering, sol–gel coating, and layer-by-layer deposition are just a few of the methods that are widely used to deposit antibacterial agents onto surfaces. Sputtering, for instance, can sputter thin, uniform layers of metal onto substrates with precise control over film thickness and uniformity of the antibacterial coat. Sol–gel processing is equally effective in incorporating ceramic nanoparticles into coatings because it can create a homogeneous film with good adhesion properties. Layer-by-layer assembly, on the other hand, has the advantage of building multilayered coatings with compositional and thickness regulation, which is beneficial in tailoring the release kinetics of antibacterial agents [36].

Other than their direct antibacterial action, the majority of them possess additional useful properties which complement the entire functionality of the coatings. As an example, TiO_2, besides expressing antibacterial effect via photocatalysis, facilitates the self-cleaning property which can be valuable in preserving the visual appearance of a surface over extended periods. In the same way, composite material made of metals mixed with polymers may offer enhanced mechanical strength, enhanced durability, and abrasion, making them appropriate for high-traffic areas. Such material's multiple functionalities are among the factors leading to its utilization in many different industries as it can allow the development of coatings with many different performance attributes at the same time [37].

Most of the antibacterial surface engineering materials are multivariated and diversified and involve metals, ceramics, polymers, composites, and bio-inspired materials. All such forms of materials entail some in-built weaknesses and advantages. Metal complexes of silver, copper, and zinc exhibit great antibacterial efficacy with ion release and production of reactive oxygen species, while cationic polymers lyse bacterial cell walls by electrostatic attraction. TiO_2-type ceramics use photocatalytic property to decompose organic contaminants and prevent microbial proliferation. Composite materials and bio-products also increase the possibilities of design, where engineers can use coatings to undertake special functions and serve as environmental protectants and safety agents [38].

The choice and inclusion of these materials need a harmonious combination of antibacterial activity, stability, and biocompatibility. Sputtering, sol–gel processing, and layer-by-layer assembly are some of the advanced fabrication methods that are critical in allowing efficient inclusion of such materials in coatings with controlled release kinetics and long-term stability. With continued growth in the area of research, achieving sustainable and multifunctional antibacterial materials can fuel future progress in surface engineering to realize ultimately cleaner and safer environments for a wide range of applications.

5 Innovative Techniques in Antibacterial Paint Development

Antibacterial paints are now a part of daily life, a pre-emptive action for the reduction of the spread of harmful bacteria in settings ranging from hospitals and public transport to schools and food. With increasingly escalating antibiotic resistance and bacteria still proving uncontrollable on high-contact surfaces, innovation in paint technology is necessary to enable public hygiene and protection. Most prevalent among these are to include nanotechnology in paint. Copper, silver, and zinc oxide nanoparticles are most frequently mentioned as having a strong antimicrobial action. Nanoparticles have been found to release ions that interfere with bacterial cell membrane and intracellular processes and thereby cause cell death. Silver nanoparticles, for example, prove to be very effective even at low concentrations and have thus become the blue-eyed boy of antibacterial coatings. Their very small size allows them a large surface area-to-volume ratio, which helps them to readily interact with bacterial cells, and with controlled release systems for extended antibacterial activity. Recent research has focused on the creation of homogeneous dispersion of nanoparticles in the paint matrix so that they are not in agglomerates and hence the functional properties are intact and even uniform antimicrobial activity is assured [39].

Another novel method exploits photocatalytic material, specifically titanium dioxide (TiO_2). TiO_2 photocatalyzes the generation of reactive oxygen species (ROS) upon exposure to ultraviolet (UV) light, which is lethal to bacteria. The photocatalytic effect not only microbicides the microbes but also makes the paint self-cleaning because organic dirt is minimized by the generated ROS. Researchers have attempted to modify TiO_2 to promote its activation into the visible light, desirable for indoor use because exposure to UV will be minimal. Methods like doping of TiO_2 with nitrogen or carbon atoms have been found to alter its spectrum of light absorption and, in turn, its antibacterial efficiency under usual circumstances [40].

Polymer carriers are another important innovation in antibacterial paint technology. Through the incorporation of antibacterial agents into polymer matrices, developers obtain controlled release profiles for long-term antimicrobial activity. Quaternary ammonium compounds (QACs) are cationic polymers that have been utilized widely because they can destabilize bacterial cell walls through electrostatic forces. The incorporation of QACs in a polymer network allows for gradual release of the active ingredient such that antibacterial activity is prolonged long after the initial coat of paint has been dried. Also, the polymer matrices can be designed to respond to specific environmental stimuli, i.e., changes in pH levels, which would indicate bacterial development. This intelligent release system maximizes the function of the paint by only allowing its antibacterial action when the need arises [41].

Layer-by-layer (LbL) assembly is a very fine method of creating antibacterial coatings. The sequential deposition of oppositely charged materials results in multilayer films, which are structurally and functionally tunable. At every layer, the structure and function can be tuned with or without adding some antibacterial agents, i.e., metal ions or antimicrobial peptides, apart from contributing other functionalities such as adhesion and toughness. LbL modularity offers the opportunity for scientists to engineer coatings with tailored release characteristics and synergistic antibacterial activity. For instance, a multi-layer assembly may consist of an outer nanoparticle-enriched layer of silver ions together with inner layers containing QAC, resulting in total protection against bacterial contamination. Electrospinning is another promising technology that has been applied in the production of antibacterial paints. Electrospinning employs high-voltage fields to produce ultrafine polymer fibers that can be incorporated with antibacterial agents. The resulting nonwoven fiber mats possess a high surface area that favors effective interaction with bacterial cells. When incorporated into paint formulations, electrospun fibers help provide a porous network that not only optimizes antibacterial performance but also enhances mechanical properties of the coating. Progress in electrospinning has lately centered on making fibers with defined diameters and morphologies that can be engineered to optimize exposure of antibacterial agents and general paint performance. In addition to these methods, bio-inspired methods have attracted a lot of attention as a method to prevent bacterial colonization without the use of chemicals. Nature offers many examples of bacteria-resistant surfaces which act naturally, for example, the micro- and nano-structured wings of some insects. These biological structures have been replicated by researchers to obtain texturized surfaces that physically disrupt bacterial cell membranes when in contact. These biomimetic surfaces' function based on mechanical forces rather than chemical interactions, and the likelihood of resistance by bacteria is minimized. In addition, such surfaces can be fabricated using low-cost fabrication methods such as soft lithography, making them suitable for bulk applications in commercial and public use [42].

These hybrid coatings that integrate multiple antibacterial strategies also lead the research today. By combining metal nanoparticles, photocatalytic films, polymer resin matrices, and bio-inspired coatings into one formula, researchers are able to develop multifunctional coatings with enhanced and synergistic antibacterial activities. A hybrid antibacterial coating, for instance, can employ silver nanoparticles to inactivate bacteria upon contact, TiO_2 to self-cleaning over long durations after irradiation, and a polymer matrix for sustained release of secondary antimicrobial agents. Such a symbiotic strategy not only lowers the mentioned above overall efficiency of the coating but also stabilizes its antibacterial activity in changing environmental conditions.

Application of these new methods is also accompanied by many challenges that need to be overcome to enable their mass-scale use. One of the major challenges is the release of antibacterial agents at effective but non-harmful concentrations because very high concentrations would undermine human health and the environment, too. Developments in controlled release technology have led the way in reducing these risks, allowing tuning of antibacterial agent levels within the paint matrix. In addition, long-term stability and durability of such coatings are still areas of major research interest. Formulations must be developed by researchers that will remain effective as antibacterials over extended periods, even under wear, weathering, and repeated cleaning regimes. Scalability is yet another critical element of business promise of new antibacterial paints. Techniques like electrospinning and layer-by-layer assembly, while as effective as they are in the laboratory, need to be scalable in order to apply them on an industrial scale. Research still remains centered on scalable production processes that will make antibacterial coatings affordable without diluting effectiveness. Collaboration between industry and academia is driving progress in this area, with pilot schemes holding out the prospect of mass production of high-tech antibacterial paints [43].

In general, new technology to produce antibacterial paint is changing the landscape of surface cleaning and public sanitation. With the blend of nanotechnology, photocatalysis, polymer science, layer-by-layer assembly, electrospinning, and bio-inspiration, scientists are designing functional coatings with prolonged, effective antibacterial activity. All these technologies aside from making health care, mass transit, and consumer product manufacturing safer, represent the promise of other technologies engineered to address anticipated challenges in bacterial resistance and nature conservation. Ultimately, successful application of these novel methods into useable commercial products will be central to improving world public health and safety as this technology continues its development.

6 Testing and Characterisation of Antibacterial Paints

As it is clear that antibacterial paints are specifically designed to prevent growth of microorganism on surfaces. These coatings thus create an essential in environments where hygiene is paramount. In order to guarantee their performance and durability in long term and harsh environments as well, these paints must be tested and characterized thoroughly. This includes a complete set of laboratory and field testing that measures antimicrobial activity, physical and chemical properties, adhesion, durability, and environmental stability. There we explain the procedures employed to test antibacterial paints, provide comments on approaches to characterization utilized, and place the issues encountered in efforts at standardization of tests.

One of the aims of testing antibacterial paints is to establish the efficiency with which they kill or inhibit bacteria. Microbiological testing is the foundation upon which this decision is made. Standard tests like the JIS Z 2801 and ASTM E2180 protocols are also routinely used in measuring antibacterial activity. A test involves the addition of a fixed bacterial inoculum concentration on the paint surface. After a specified contact time, incubation leads to an estimation of bacteria surviving by the culture technique. A lower number of colony-forming units (CFUs) of bacteria on the test surface than on a control surface validates the antimicrobial activity of the paint. These tests do not only supply quantitative estimates of activity but assist with comparisons across formulations [44].

In conjunction with microbiological assays, characterization of these antibacterial paints is also done through contemporary spectroscopic and microscopic methods. Transmission electron microscopy (TEM) and scanning electron microscopy (SEM) are frequently used to examine the distribution and morphology of the antibacterial agents, i.e., nanoparticles, within the paint system. SEM provides high-resolution microscopic images of the surface morphology and demonstrates paint and bacteria interaction behavior at the microscale. Compared to SEM, TEM, however, gives information regarding the nanostructures embedded in the matrix. Energy-dispersive X-ray spectroscopy (EDX), which is typically employed with these imaging methods, provides elemental analysis of the coating and verifies the occurrence and distribution of metals such as silver, copper, or zinc [45].

Fourier-transform infrared (FTIR) and Raman spectroscopy are crucial to recognize chemical bonds and functional groups of antibacterial paint. With the help of the spectroscopic investigation, scientists ensure efficient integration of antibacterial substances into the matrix of the paint and chemical alteration during curing or air exposure. By studying vibrational modes of the molecules, scientists can trace stability of antibacterial compounds with time. This is highly relevant for those paints that will continue to offer antimicrobial functionality even after continued use [46].

Physical characteristics of the paint should be examined as an integral part of characterization as well. The physical characteristics that are crucial in appearance and operation performance include gloss, colour, viscosity, and drying time. Physical characteristics are determined under strict control using standardized tests, such that the antibacterial function is not detrimental to the usability of the paint. For example, hydrophobicity of the surface is measured by contact angle measurements, which ultimately determines bacterial adhesion. A resulting large contact angle is generally used to justify that a hydrophobic surface that can prevent bacterial colonization. In addition, abrasion resistance, adhesion strength, and flexibility tests are performed to determine the long-term stability of the antibacterial coating under actual application conditions [47].

Environmental stability is also an extremely important parameter that is extensively screened. Antibacterial paint is subjected to various conditions of stress such as temperature, humidity, ultraviolet (UV) light, and chemical exposure. Accelerated weathering test such as QUV exposure mimics the long-term exposure of environmental conditions on the performance of the paint. The tests predict the future life of the antibacterial activity and that the paint will remain effective in the future. On certain occasions, paints are exposed to cyclic ageing tests, which replicate artificially the natural process of weathering, to evaluate their outdoor performance. Standardization of test procedures will be required for the comparison of paints with antibacterial properties that have been supplied by different producers. International standards, such as ISO 22196, provide guidelines for evaluating the antibacterial properties of plastics and other porous materials. These guidelines are frequently adapted for application to paint, following necessary modifications. Reproducibility at laboratories is an issue, yet. Differences in test conditions, including inoculum concentration, contact time, and recovery procedures, can cause differences in reported antibacterial effectiveness. To prevent these issues, collaborative research and inter-laboratory comparisons are carried out. These activities are designed to harmonize testing procedures and develop consensus protocols that provide consistent and comparable results [3].

The incorporation of real-time monitoring methods has also improved the antibacterial paint characterization science. Biosensors and real-time PCR analysis can be utilized to detect rapidly bacterial DNA on the surfaces of paints with real-time indication of antimicrobial effectiveness. Such methodologies could be of most utility in facilities like hospitals or food preparation zones where quick determination of microbial contamination is crucial. Besides, time-lapse microscopy allows scientists to directly view the dynamic interaction between the coated surface and the bacteria and offer information on the mechanisms of inactivation. Nonetheless, there are a number of challenges in antibacterial paint characterization and testing. One of them is the possibility of bacteria becoming resistant to the antimicrobial agents employed in the paints. Long-term trials are needed to determine if repeated exposure to antibacterial surfaces could induce adaptive microbial community change. Further, the environmental and safety implications of discharging antimicrobial agents, specifically metal ions, into the environment must be thoroughly evaluated. A lot of toxicology testing must be required by regulators to ensure antibacterial paints will not harm human health or ecosystems. A second challenge is projecting laboratory test results to in-use performance. While controlled testing is helpful, the nuances that occur in practice—i.e., varying levels of soil, biofilm, and wear—may affect the antibacterial performance of a paint. Field trials and extended monitoring in actual conditions of use are needed to confirm laboratory results and ensure that the antibacterial performance is sustained under varying conditions. Computational simulation and modelling advances provide promising technologies to supplement laboratory testing. The models can forecast the release kinetics of the antimicrobial agent, the bacterial cell-surface interaction of the coated film, and the aging degradation of the coating. By combining experiment data with simulations, scientists could optimize paint formulations more effectively and minimize trial-and-error testing needs [48].

In short, testing and characterization of antibacterial paints constitute a complex procedure encompassing microbiological testing, sophisticated imaging, spectroscopic methods, and environmental and mechanical testing. With each method comes specific information on the performance and lifespan of the coatings to ensure high standards of modern uses. Standardization challenges, long-term effectiveness, and safety in the environment initiate continuous research and cooperation among scientists, manufacturers, and regulators. As further test methods become available, they will be instrumental in future development of antibacterial paint that is effective and sustainable to protect public health.

7 Challenges and Future Perspectives

As of now, it has been vividly quantified that these antibacterial paints have emerged as a groundbreaking innovation in surface engineering. Thus, offering protection against harmful microbes and contributing to improved hygiene across multiple industries. These antibacterial coatings have played a very valuable role not only in healthcare but also at food processing units, residential places, and public organizations, where preventing bacterial growth is difficult but yet essential. However, despite the above stated numerous benefits, antibacterial paints still face significant challenges worldwide that are hindering in their prevalent adoption.

These challenges assort from regulatory and environmental concerns to cost effectiveness and potential bacterial resistance due to continuous exposure. Additional to these challenges, research and developers' efforts endure to focus on improving the efficacy along with increasing longevity, and sustainability of antibacterial coatings in inbuilt environment. The following sections explores the key challenges that are associated with antibacterial paints and discusses its future perspectives alongside. It effectively highlights the potential innovations and opportunities in the field.

7.1 Challenges in Antibacterial Paint Development and Use

7.1.1 Bacterial Resistance and Long-Term Efficacy

Development of bacterial resistance is one of the risks and is the most persistent concerns regarding antibacterial paints and the coating. This phenomenon is similar to when a bacteria develop resistance towards antibiotics after being exposed to the reagent for a longer period. Similarly, prolonged and frequent exposure to antibacterial coatings may as well lead to adaptation and added resistance mechanisms in immediate residing microbial populations. This resistance development phenomenon may render the effectiveness of certain antimicrobial agents over time, for instance silver nanoparticles or quaternary ammonium compounds. A few studies have proposed that some of the bacteria can develop resistance by modifying their cell wall structure, articulating efflux pumps, or by producing biofilms, which eventually shield them from antibacterial agents. Moreover, over abusing of antibacterial coatings at non-critical places may hasten the resistance development. This however can be tackled by necessitating strict usage guidelines and periodical assessment of coating efficacy. Scientists are currently pursuing work on multifunctional coatings that combine antibacterial mechanism with anti-fouling and self-cleaning properties to alleviate the risk of resistance development. Continuous innovation in antibacterial mechanisms is required to mitigate this issue, including the development of coatings that disrupt bacterial communication i.e. quorum sensing rather than just directly killing the bacteria upon contact [49].

7.1.2 Environmental and Toxicological Concerns

While it is prominently stated that antibacterial paints contribute to better hygiene but at the same time, they also pose potential environmental and health risks. This is due to the use of metal-based nanoparticles and synthetic biocides in the paint matrix. Nanoparticle leaching is one such problem in which antibacterial paints that rely on silver, copper, or zinc oxide like nanoparticles, can leach into the environment over time period. This can lead to bioaccumulation especially in aquatic ecosystems. This will ultimately affect marine life, and the food chain involved with it. Another such health risk is when some biocidal agents, such as triclosan and quaternary ammonium compounds (QACs) use has linked to hormonal disruption and skin irritation in human beings. Likewise continuous exposure to these chemicals in interior environments or closed spaces raises concerns about long-term toxicity due to altered air quality index. However, governments/agencies globally are enforcing strict environmental regulations on usage of antimicrobial agents used in consumer products such as antibacterial coatings. For example, the European Chemicals Agency (ECHA) and the U.S. Environmental Protection Agency (EPA) have imposed limits on certain biocides used in consumer products, necessitating reformulation of antibacterial paints. In order to subside the health concerns, researchers and academicians are continuously investigating bio-based agents alternatives, such as essential oils, chitosan, and plant-derived polyphenols, which are reasonably eco-friendly when compared to the traditional metal-based coatings [50].

7.1.3 Cost and Market Barriers

Despite their numerous advantages and benefits, antibacterial paints remain significantly more expensive than traditional/conventional paints, limiting their implementation in price-sensitive markets. The high price of these paints is due to incorporation of nanoparticles, specialized polymers, and self-cleaning mechanisms increases its overall manufacturing expenditures. For instance, while multiscale large hospitals and corporate institutions can have the funds for antibacterial coatings, small businesses and households may find them economically impractical unfeasible. Also, it is on interesting to note that many potential users are unaware of the benefits of antibacterial paints. It therefore reduces market demand and slows down its large-scale production. In order to surge affordability and market penetration of these specialized coatings, mass production procedures and substitute low-cost antimicrobial agents have to be explored. For instance, in near future, biodegradable antimicrobial coatings derived from either agricultural waste or natural polymers may offer a cost-effective solution than currently available smart coatings with metals and nanoparticles [51].

7.1.4 Compatibility with Existing Coating Systems

Apart from the before mentioned challenges, another significant challenge is ensuring that antibacterial agents do not compromise with the actual mechanical properties or adhesion performance of paints. Numerous antibacterial coatings must be integrated into existing decorative, industrial, and protective paint systems. This amalgamation requires extensive compatibility testing. Durability concerns raised conspicuously as some antimicrobial additives may as well alter viscosity, gloss, and color stability of the paint, thus leading to issues with product performance. Certain biocidal agents may also interfere with the paint's ability to adhere to substrates, altering the adhesion properties. This issue is especially in areas with high-moisture or chemically aggressive environments. It is also seen that unlike traditional paints, antibacterial coatings may require recurrent reapplication. This is done in order to maintain their effectiveness, raising major concerns about cost-effectiveness and maintenance of the paint. An alternative to tackle these challenges, hybrid coating formulations are developed that integrate antibacterial properties while preserving conventional paint performance characteristics [52].

7.2 Future Perspectives and Innovations in Antibacterial Paints

Despite discussing about the challenges, there are ongoing research and technological advancements that are paving its way for next-generation antibacterial coating classes. The future of antibacterial paints resides in the progress of smarter, more sustainable, and much higher efficient formulations that address and tackles both microbial threats and environmental concerns.

7.2.1 Smart and Responsive Antibacterial Coatings

The next emerging antibacterial paint classes are incorporating stimuli-responsive materials that can either activate or deactivate based on the immediate environmental conditions. One such example is pH-sensitive coatings. These paints release antibacterial agents only when a microbial biofilm begins to form in its vicinity. It reduces unnecessary release of chemical rand impact the environmental. Another instance of these smart paints are light-activated antimicrobial surfaces. Photocatalytic coatings, that too particularly TiO_2-based paints, use UV or visible light that triggers the antimicrobial activity. This mechanism makes them suitable for health facilities and public spaces. One more class is the self-healing antibacterial paints which is inspired by biological systems. These bio-inspired coatings have the potential to repair minor damages and cracks, which generically ensuring long-term antimicrobial efficacy and also provide mechanical strength. Such smart coatings will significantly enhance the sustainability and efficiency of antibacterial paints whilst minimizing inadvertent side effects and concerns rising from conventional antibacterial paints [53].

7.2.2 Sustainable and Biodegradable Antibacterial Paints

Since the beginning of modernization, an expected shift towards biodegradable and naturally derived antimicrobial agents is observed in order to address environmental concerns. For instance, chitosan-based antibacterial paints that are derived from cutaneous shells. The derived chitosan has natural antimicrobial and biodegradable properties which makes it an excellent green alternative of synthetic biocides. Similarly essential oil-infused coatings with prominent ingredients like tea tree oil, eucalyptus, and thyme extracts provide antibacterial properties without conducing to environmental toxicity. Certain protein-based coatings such as soy and casein-derived polymers offer inherent antimicrobial benefits. These are additionally biodegradable and VOC-free making them eco-friendly alternative while allowing antibacterial paints to comply with global sustainability goals and regulatory frameworks [42].

7.2.3 Integration with Artificial Intelligence (AI) and Nanotechnology

An unexplored phase of paint formulation is the use of AI and nanotechnology which is expected to revolutionize antibacterial paint development in the coming decades. AI-assisted formulation of paints using machine learning algorithms can optimize and enhance antibacterial efficacy whilst minimizing material costs and environmental impact. Scientists are also exploring nanotextured engineered surfaces that can physically disrupt bacterial membranes due to its structural characteristics, thus eliminating the need for chemical biocides. Graphene based coatings are also a promising alternative for superior class paints as it exhibits exceptional antimicrobial properties, along with superior mechanical strength and heat resistance. The combination of all the AI technology, nanotechnology, along with sustainable chemistry is all set to transform the landscape of antibacterial and its incorporation in surface engineering, leading to the development of highly efficient and environmentally friendly coatings [54].

8 Conclusion

The investigation of different aspects of antibacterial paints and surface engineering presented in this chapter accentuates the key role that these coatings play in modern materials science and public health. The perspective on the topic is rooted in the belief that environmentally sustainable, effective, and technologically advanced antibacterial coatings have now become an integral part of construction and industrial applications and will sustain in the future. Commencement with the basic details of antibacterial paints, these coatings are not just reflexive surface treatment, but they assist as an active defence mechanism against any contamination. From a perspective of surface engineering, the progressions in coating modification techniques have been noteworthy. The capability to tailor surface properties at both the nano and micro levels using surface engineering allows for the accurate control over the coating's adhesion properties, wettability, and antimicrobial activity. After the comprehensive study of the current status of antibacterial coatings, it is quite clear that impending research must endure to focus on optimizing the coating's durability and antibacterial efficiency, ensuring that all the properties including its mechanical efficiency remain active over extended periods without degradation. Traditional antibacterial agents have demonstrated strong efficacy but come with environmental and toxicity concerns. However, the shift towards bio-based materials, is a promising step towards eco-friendly solutions to the problem. However, it is important that it ensures that these alternatives offer comparable performance while remaining cost-effective. Along with the other criterions, it is also important that the testing and characterization must not be overstated. Another aspect is by guaranteeing scientific validation and regulatory compliance, the commercial and large-scale success of antibacterial coatings worldwide will be assured. It is important that the development of standardized and globally recognized testing protocols will boost the credibility and marketability of these paints. The forthcoming phase of innovation must prioritize sustainability, recyclability, and non-toxic of the paint formulations. Furthermore, the incorporation of AI-driven formulation techniques and advanced computational modelling will accelerate the design of next-generation antibacterial coatings. In the overall conclusion of the chapter, it can be stated that it has provided a comprehensive outline of the surface engineering field and its relation with antibacterial paints and coatings, illustrating both its current state and future potential. It can also be firmly stated through the literature that antibacterial paints represent a ground-breaking step in surface engineering field. The future of antibacterial surface engineering lies in collaboration, along with innovation and a steadfast commitment to sustainability in reference to the environment—a vision that will eventually transform how surface protection is approached in various industries.

References

1. Jariya, S.A.I., Manivannan, N., Ali, B.M., Narayanan, T.S.N.S., Ravichandran, K.: Engineering the surface of titanium to improve its bioactivity and antibacterial activity through a multifunctional coating approach. New J. Chem. **47**(12), 5843–5862 (2023). https://doi.org/10.1039/d2nj04485d
2. Atul, S.C., Ilaiyavel, S., Jaishree, K.T.: Anti-bacterial properties of wear resistant thermal diffusion coatings. Trans. Indian Instit. Metals **73**(11), 2911–2917 (2020). https://doi.org/10.1007/s12666-020-02094-5
3. Hoque, J., Akkapeddi, P., Ghosh, C., Uppu, D.S.S.M., Haldar, J.: A biodegradable polycationic paint that kills bacteria in vitro and in vivo. ACS Appl. Mater. Interf. **8**(43), 29298–29309 (2016). https://doi.org/10.1021/acsami.6b09804
4. Uppu, D.S.S.M., Haldar, J.: Lipopolysaccharide neutralization by cationic-amphiphilic polymers through pseudoaggregate formation. Biomacromol **17**(3), 862–873 (2016). https://doi.org/10.1021/acs.biomac.5b01567
5. Guo, J., Xu, Q., Zheng, Z., Zhou, S., Mao, H., Wang, B., Yan, F.: Intrinsically antibacterial poly (ionic liquid) membranes: the synergistic effect of anions. ACS Macro Lett. **4**(10), 1094–1098 (2015). https://doi.org/10.1021/acsmacrolett.5b00609
6. Hoque, J., Konai, M.M., Gonuguntla, S., Manjunath, G.B., Samaddar, S., Yarlagadda, V., Haldar, J.: Membrane active small molecules show selective broad spectrum antibacterial activity with no detectable resistance and eradicate biofilms. J. Med. Chem. **58**(14), 5486–5500 (2015). https://doi.org/10.1021/acs.jmedchem.5b00443
7. Wei, Z., Teng, S., Fu, Y., Zhou, Q., Yang, W.: Micro/nano-structured electrospun membranes with superhydrophobic and photodynamic antibacterial performances. Progr. Org. Coat. **164**, 106703 (2022). https://doi.org/10.1016/j.porgcoat.2021.106703
8. Ghodake, V.S., Deshmukh, S.P., Delekar, S.D.: Metal oxide nanocomposites: design and use in antimicrobial coatings. In: Elsevier eBooks, pp. 549–599 (2022). https://doi.org/10.1016/b978-0-323-85705-5.00011-7
9. Kumar, S.S.A., Bashir, S., Ramesh, K., Ramesh, S.: A comprehensive review: super hydrophobic graphene nanocomposite coatings for underwater and wet applications to enhance corrosion resistance. FlatChem **31**, 100326 (2021). https://doi.org/10.1016/j.flatc.2021.100326
10. Gupta, S., Puttaiahgowda, Y.M., Nagaraja, A., Jalageri, M.D.: Antimicrobial polymeric paints: an up-to-date review. Polym. Adv. Technol. **32**(12), 4642–4662 (2021). https://doi.org/10.1002/pat.5485
11. Dey, R., Mukherjee, S., Barman, S., Haldar, J.: Macromolecular nanotherapeutics and antibiotic adjuvants to tackle bacterial and fungal infections. Macromol. Biosci. **21**(11), 182 (2021). https://doi.org/10.1002/mabi.202100182
12. Mahanta, U., Khandelwal, M., Deshpande, A.S.: Antimicrobial surfaces: a review of synthetic approaches, applicability and outlook. J. Mater. Sci. **56**(32), 17915–17941 (2021). https://doi.org/10.1007/s10853-021-06404-0
13. Park, D., Wang, J., Klibanov, A.: One-step, painting-like coating procedures to make surfaces highly and permanently bactericidal. Biotechnol. Progr. **22**(2), 584–589 (2006). https://doi.org/10.1021/bp0503383
14. Haldar, J., An, D., De Cienfuegos, L.Á., Chen, J., Klibanov, A.M.: Polymeric coatings that inactivate both influenza virus and pathogenic bacteria. Proceed. Natl. Acad. Sci. **103**(47), 17667–17671 (2006). https://doi.org/10.1073/pnas.0608803103
15. Haldar, J., Weight, A.K., Klibanov, A.M.: Preparation, application and testing of permanent antibacterial and antiviral coatings. Nat. Protocols **2**(10), 2412–2417 (2007). https://doi.org/10.1038/nprot.2007.353
16. Bastarrachea, L.J., Goddard, J.M.: Self-healing antimicrobial polymer coating with efficacy in the presence of organic matter. Appl. Surf. Sci. **378**, 479–488 (2016). https://doi.org/10.1016/j.apsusc.2016.03.198

17. Hung, Y., McLandsborough, L.A., Goddard, J.M., Bastarrachea, L.J.: Antimicrobial polymer coatings with efficacy against pathogenic and spoilage microorganisms. LWT **97**, 546–554 (2018). https://doi.org/10.1016/j.lwt.2018.07.046
18. Ungureanu, C., Pirvu, C., Mindroiu, M., Demetrescu, I.: Antibacterial polymeric coating based on polypyrrole and polyethylene glycol on a new alloy TiAlZr. Progr. Org. Coat. **75**(4), 349–355 (2012). https://doi.org/10.1016/j.porgcoat.2012.07.015
19. Cheng, Y., Feng, G., Moraru, C.I.: Micro-and nanotopography sensitive bacterial attachment mechanisms: a review. Front. Microbiol. **10**, 191 (2019)
20. Kreve, S., Dos Reis, A.C.: Bacterial adhesion to biomaterials: what regulates this attachment? A review. Jpn. Dent. Sci. Rev. **57**, 85–96 (2021)
21. Nautiyal, A., Qiao, M., Ren, T., Huang, T., Zhang, X., Cook, J., Bozack, M.J., Farag, R.: High-performance engineered conducting polymer film towards antimicrobial/anticorrosion applications. Eng. Sci. **4**, 70–78 (2018)
22. Vahabi, H., Jouyandeh, M., Cochez, M., Khalili, R., Vagner, C., Ferriol, M., Movahedifar, E., Ramezanzadeh, B., Rostami, M., Ranjbar, Z., Hadavand, B.S., Saeb, M.R.: Short-lasting fire in partially and completely cured epoxy coatings containing expandable graphite and halloysite nanotube additives. Progr. Org. Coat. **123**, 160–167 (2018). https://doi.org/10.1016/j.porgcoat.2018.07.014
23. Chen, D., Yin, Z., Wu, F., Fu, H., Kundu, S.C., Lu, S.: Orientational behaviors of silk fibroin hydrogels. J. Appl. Polym. Sci. **134**(32), 50 (2017). https://doi.org/10.1002/app.45050
24. Dizaj, S.M., Lotfipour, F., Barzegar-Jalali, M., Zarrintan, M.H., Adibkia, K.: Antimicrobial activity of the metals and metal oxide nanoparticles. Mater. Sci. Eng. C **44**, 278–284 (2014). https://doi.org/10.1016/j.msec.2014.08.031
25. Perelshtein, I., Applerot, G., Perkas, N., Guibert, G., Mikhailov, S., Gedanken, A.: Sonochemical coating of silver nanoparticles on textile fabrics (nylon, polyester and cotton) and their antibacterial activity. Nanotechnology **19**(24), 245705 (2008). https://doi.org/10.1088/0957-4484/19/24/245705
26. Saji, V.S.: Progress in rust converters. Progr. Org. Coat. **127**, 88–99 (2018). https://doi.org/10.1016/j.porgcoat.2018.11.013
27. He, B., Zhao, J., Ou, Y., Jiang, D.: Biofunctionalized peptide nanofiber-based composite scaffolds for bone regeneration. Mater. Sci. Eng. C **90**, 728–738 (2018). https://doi.org/10.1016/j.msec.2018.04.063
28. Xie, M., Gao, M., Yun, Y., Malmsten, M., Rotello, V.M., Zboril, R., Akhavan, O., Kraskouski, A., Amalraj, J., Cai, X., Lu, J., Zheng, H., Li, R.: Antibacterial nanomaterials: mechanisms, impacts on antimicrobial resistance and design principles. Angew. Chem. Int. Edn. **62**(17), 7345 (2023). https://doi.org/10.1002/anie.202217345
29. Querido, M.M., Paulo, I., Hariharakrishnan, S., Rocha, D., Pereira, C.C., Barbosa, N., Bordado, J.M., Teixeira, J.P., Santos, R.G.D.: Auto-disinfectant acrylic paints functionalised with triclosan and isoborneol—antibacterial assessment. Polymers **13**(13), 2197 (2021). https://doi.org/10.3390/polym13132197
30. Yang, L., Wu, Y., Yang, F., Wang, W.: The effect of antibacterial and waterproof coating prepared from hexadecyltrimethoxysilane and nano-titanium dioxide on wood properties. Front. Mater. **8**, 579 (2021). https://doi.org/10.3389/fmats.2021.699579
31. Akhavan, O., Ghaderi, E.: Toxicity of graphene and graphene oxide nanowalls against bacteria. ACS Nano **4**(10), 5731–5736 (2010). https://doi.org/10.1021/nn101390x
32. Li, P., Poon, Y.F., Li, W., Zhu, H., Yeap, S.H., Cao, Y., Qi, X., Zhou, C., Lamrani, M., Beuerman, R.W., Kang, E., Mu, Y., Li, C.M., Chang, M.W., Leong, S.S.J., Chan-Park, M.B.: A polycationic antimicrobial and biocompatible hydrogel with microbe membrane suctioning ability. Nat. Mater. **10**(2), 149–156 (2010). https://doi.org/10.1038/nmat2915
33. Krishnamoorthy, M., Hakobyan, S., Ramstedt, M., Gautrot, J.E.: Surface-initiated polymer brushes in the biomedical field: applications in membrane science, biosensing, cell culture, regenerative medicine and antibacterial coatings. Chem. Rev. **114**(21), 10976–11026 (2014). https://doi.org/10.1021/cr500252u

34. Morones, J.R., Elechiguerra, J.L., Camacho, A., Holt, K., Kouri, J.B., Ramírez, J.T., Yacaman, M.J.: The bactericidal effect of silver nanoparticles. Nanotechnology **16**(10), 2346–2353 (2005). https://doi.org/10.1088/0957-4484/16/10/059
35. Duncan, R., Gaspar, R.: Nanomedicine(s) under the microscope. Mol. Pharm. **8**(6), 2101–2141 (2011). https://doi.org/10.1021/mp200394t
36. Liu, S., Zeng, T.H., Hofmann, M., Burcombe, E., Wei, J., Jiang, R., Kong, J., Chen, Y.: Antibacterial activity of graphite, graphite oxide, graphene oxide, and reduced graphene oxide: membrane and oxidative stress. ACS Nano **5**(9), 6971–6980 (2011). https://doi.org/10.1021/nn202451x
37. Akhavan, O., Ghaderi, E., Esfandiar, A.: Wrapping bacteria by graphene nanosheets for isolation from environment, reactivation by sonication, and inactivation by near-infrared irradiation. J. Phys. Chem. B **115**(19), 6279–6288 (2011). https://doi.org/10.1021/jp200686k
38. Mao, H.Y., Laurent, S., Chen, W., Akhavan, O., Imani, M., Ashkarran, A.A., Mahmoudi, M.: Graphene: promises, facts, opportunities, and challenges in nanomedicine. Chem. Rev. **113**(5), 3407–3424 (2013). https://doi.org/10.1021/cr300335p
39. Bryaskova, R., Philipova, N., Bakov, V., Georgiev, N.: Innovative antibacterial polymer coatings. Appl. Sci. **15**(4), 1780 (2025). https://doi.org/10.3390/app15041780
40. Verma, J., Khanna, A.S., Sahney, R., Bhattacharya, A.: Super protective anti-bacterial coating development with silica–titania nano core–shells. Nanoscale Adv. **2**(9), 4093–4105 (2020). https://doi.org/10.1039/d0na00387e
41. Manna, J., Begum, G., Kumar, K.P., Misra, S., Rana, R.K.: Enabling antibacterial coating via bioinspired mineralization of nanostructured ZnO on fabrics under mild conditions. ACS Appl. Mater. Interf. **5**(10), 4457–4463 (2013). https://doi.org/10.1021/am400933n
42. Shu, H., Chen, P., Yang, R.: Advances in antibacterial polymer coatings synthesized via chemical vapor deposition. Chem. Bio-Eng. **1**(6), 516–534 (2024). https://doi.org/10.1021/cbe.4c00043
43. Morin-Crini, N., Lichtfouse, E., Torri, G., Crini, G.: Applications of chitosan in food, pharmaceuticals, medicine, cosmetics, agriculture, textiles, pulp and paper, biotechnology, and environmental chemistry. Environ. Chem. Lett. **17**(4), 1667–1692 (2019). https://doi.org/10.1007/s10311-019-00904-x
44. Cloutier, M., Mantovani, D., Rosei, F.: Antibacterial coatings: challenges, perspectives, and opportunities. Trends Biotechnol. **33**(11), 637–652 (2015). https://doi.org/10.1016/j.tibtech.2015.09.002
45. Hoque, J., Akkapeddi, P., Yadav, V., Manjunath, G.B., Uppu, D.S.S.M., Konai, M.M., Yarlagadda, V., Sanyal, K., Haldar, J.: Broad spectrum antibacterial and antifungal polymeric paint materials: synthesis, structure–activity relationship, and membrane-active mode of action. ACS Appl. Mater. Interf. **7**(3), 1804–1815 (2014). https://doi.org/10.1021/am507482y
46. Mutschlechner, M., Gstir, R., Schöbel, H., Rössler, A., Lass-Flörl, C., Bach, K.: From process to product: exploring microbial diversity in paints. J. Coat. Technol. Res. **22**, 481–490 (2024). https://doi.org/10.1007/s11998-024-00990-9
47. Song, B., Zhang, E., Han, X., Zhu, H., Shi, Y., Cao, Z.: Engineering and application perspectives on designing an antimicrobial surface. ACS Appl. Mater. Interf. **12**(19), 21330–21341 (2020). https://doi.org/10.1021/acsami.9b19992
48. Kook, M., Kaur, H., Danilian, D., Rosenberg, M., Kisand, V., Ivask, A.: Durability of photocatalytic ZnO-based surface coatings and preservation of their antibacterial effect after simulated wear. J. Coat. Technol. Res. **21**(3), 1005–1016 (2024). https://doi.org/10.1007/s11998-023-00868-2
49. Lee, J., Lee, C., Lee, J., Kim, S., Cho, H., Kim, M., Cho, W., Joo, J., Kim, C.: Cerium oxide–polysulfone composite separator for an advanced alkaline electrolyzer. Polymers **12**(12), 2821 (2020). https://doi.org/10.3390/polym12122821
50. Querido, M.M., Paulo, I., Hariharakrishnan, S., Rocha, D., Barbosa, N., Gonçalves, D., Santos, R.G.D., Bordado, J.M., Teixeira, J.P., Pereira, C.C.: Development and in vitro validation of antibacterial paints containing chloroxylenol and terpineol. Toxics **10**(7), 343 (2022). https://doi.org/10.3390/toxics10070343

51. Lin, H., Jin, T., Dmytruk, A., Saito, M., Yazawa, T.: Preparation of a porous ITO electrode. J. Photochem. Photobiol. Chem. **164**(1–3), 173–177 (2004). https://doi.org/10.1016/j.jphotochem.2003.12.023
52. Thatikonda, S.K., Goswami, D., Dobbidi, P.: Effects of CeO_2 nanoparticles and annealing temperature on the microwave dielectric properties of $MgTiO_3$ ceramics. Ceram. Int. **40**(1), 1125–1131 (2013). https://doi.org/10.1016/j.ceramint.2013.06.113
53. Kitade, S., Asuka, K., Akiba, I., Sanada, Y., Sakurai, K., Masunaga, H.: Shear-induced pre-crystallization structures of long chain branched polypropylene under steady shear flow near the melting temperature. Polymer **54**(1), 246–257 (2012). https://doi.org/10.1016/j.polymer.2012.10.048
54. Yılmaz, G.E., Göktürk, I., Ovezova, M., Yılmaz, F., Kılıç, S., Denizli, A.: Antimicrobial nanomaterials: a review. Hygiene **3**(3), 269–290 (2023). https://doi.org/10.3390/hygiene3030020

Computational Approaches to Epoxy Polymers as Corrosion Inhibitors for Metal Surfaces: A Review

Saraswati Kumari Ola, Ishita Chopra, Triveni, and Veena Dhayal

Abstract In this review focus on computational methods used to study corrosion inhibition along with epoxy polymer behaviour when interacting with metals. The analysis of epoxy polymers regarding their reactivity and adsorption behaviour as well as electronic properties utilized DFT, MD simulations, MC simulations and quantum chemical descriptors. The research demonstrates that corrosion inhibition efficiency predictions depend on E_{HOMO}, E_{LUMO}, chemical hardness/softness quantum descriptors with Fukui indices, molecular electrostatic potential (MEP) and dual local descriptors. Computational simulations demonstrated both stable interface positions and strong surface binding of the polymers with Fe(110) surfaces through the use of specific functional groups like oxygen or nitrogen to enhance protective film capabilities. These results demonstrate the ability of computational approaches to enhance corrosion inhibitor optimization before implementing such systems in industry.

Keywords Computational · Corrosion · Metal · Surface · Simulation

1 Introduction

Corrosion research to link laboratory investigations with practical solutions which protect materials that need to survive longer in various industrial systems [1]. The field explores how materials and their surroundings along with their basic weaknesses

S. K. Ola
Department of Chemistry, B K Birla Institute of Higher Education, Pilani, Jhunjhunu, Rajasthan, India

I. Chopra · V. Dhayal (✉)
Department of Chemistry, Manipal University Jaipur, Jaipur, India
e-mail: veena.dhayal@jaipur.manipal.edu

Triveni
Department of Chemistry, Smt. Indramani Mandelia Shiksha Niket, Pilani, Jhunjhunu, Rajasthan, India

L. Ledwani and S. Srivastava (eds.), *Experimental and Computational Advances in Materials*, Engineering Materials, https://doi.org/10.1007/978-981-95-4180-5_11

work together [2]. The scientific community conducts analyses on corrosion mechanisms which include uniform distortion and localized damage and galvanic failure and stress corrosion cracking together with erosion-corrosion events [3]. Examinations of materials analyze substances across different environments which include atmospheric exposure as well as corrosive chemical and marine solutions [4]. Corrosion functions as a widespread natural occurrence because it refers to material deterioration which advances at a gradual pace through chemical, electrochemical or environmental elements [5]. Extensive research exists to study corrosion phenomena since it creates major economic losses and safety concerns in a variety of industries [6]. Studies of corrosion work to identify natural material-degrading processes and formulate methods for both prevention and control purposes [7]. The evaluation of corrosion rates alongside corrosion product identification depends on electrochemical measurement techniques together with microscopic and spectroscopic and surface analytical procedures [8]. Research findings enable development of protective coatings together with inhibitors and alloy modifications which improve material resistance [9].

The industrial significance of epoxy polymers and their composites remains high because these materials combine multiple properties which include mechanical, thermal, electrical, dielectric, rheological and anticorrosive characteristics [10]. The coating of materials with epoxy polymers occurs across many industrial sectors including aeronautics alongside automotive and construction fields plus packaging while applying landfill techniques for radioactive waste disposal and serving as corrosion protective coatings [11]. Studies of corrosion involve practical uses beyond theoretical experiments [12]. Studies of corrosion play a critical role for industries that operate across transportation and infrastructure and energy production together with biomedical sectors to ensure materials' endurance and safety [13]. Corrosion research scientists have developed two primary innovations which include self-healing coatings and advanced corrosion-resistant alloys [8]. Environmental elements that consist of temperature and humidity together with pollutants actively participate in corrosion while requiring multiple specialists from material science and engineering and chemistry and environmental science to work together [14]. A complete evaluation of corrosion complexities becomes possible through this entire systems method [15]. Computational methods represent essential tools researchers use to study corrosion and corrosion inhibition through their economical evaluation of both inhibitor molecule characteristics and the performance of these molecules regarding metallic surface interactions [16].

Computational techniques including molecular dynamics (MD) simulations, Monte Carlo (MC) simulations and density functional theory (DFT) for analyzing and predicting material behavior [17]. The atomistic resolution of metal surface-inhibitor molecule interactions provided by MD simulations makes these simulations highly powerful for corrosion study [18]. The investigation of adsorption thermodynamics in corrosion inhibition requires MC simulations as a highly effective computational method [18]. The quantum mechanical program DFT enables researchers to determine precisely how corrosion inhibitor systems behave on an electronic scale

[19]. The advancement of corrosion processes understanding and materials degradation requires computational methods as fundamental instruments [20]. Computer simulations along with models and data analysis techniques provided through these methods enable researchers to investigate complex material-environment interactions thus revealing information about corrosion mechanisms [21] and both the speed of corrosion and methods to minimize it [22]. Scientists have established various molecular modelling approaches and quantum chemical processes which determine how inhibitor molecular characteristics affect inhibition efficiency [23]. According to Obot and his team in their review density functional theory computations play a vital role in corrosion research [24]. Researchers at Moradi et al. [25] showed that improving Monte-Carlo simulation methods effectively determines system reliability through multiple component analysis at any correlation level. Hsissou et al. [26] examined the influence of significant corrosion model parameter modifications upon times to onset of corrosion through Monte Carlo method simulations. The reduction of concrete chloride diffusion coefficient values would create substantial increases in the time before corrosion initiation happens [27]. MD simulation research by Harris and coworkers proves itself as an effective tool to understand molecular level corrosion inhibition mechanisms [28].

This study aims to conduct a critical assessment along with a summary of all computational methods which investigate corrosion inhibition processes. This study evaluates both benefits and weaknesses of computational methods to develop comprehensive understanding about how these approaches enhance corrosion inhibitor development processes including evaluation and optimization steps. The study utilised Monte Carlo simulation and dual local philicity alongside dual Fukui function and dual local softness and local electrophilicity and Fukui indices to understand the mechanism behind two epoxy polymer and metallic surface interactions.

2 Computational Methods in Corrosion Studies and Applications

2.1 *Global Quantum Chemical Descriptors Calculations*

Quantum chemical descriptor calculations at the global scale help researchers understand how reactions proceed as well as serve as an essential investigative tool for molecule properties. The physical characteristics exhibit a clear relationship with molecular structure [29]. The Global Quantum Chemical Descriptors (GQCDs) of two ethylenic epoxy polymers received theoretical calculations through density functional theory (DFT) at the B3LYP/6-311G(d,p) level of theory [30]. The GQCD computational workload occurred within the Gaussian 03 software suite. A B3LYP/6-311G(d,p) basis set coupled with the DFT method delivered global quantum chemical descriptors through E_{HOMO} and E_{LUMO} [31]. Molecular orbitals E_{HOMO} and

E_{LUMO} from polymers exhibit direct relationships with two fundamental phenomena: ionization energy (IE = $-E_{HOMO}$) and electron affinity (EA = $-E_{LUMO}$) [32].

The shape of the energy gap arises from the difference between orbital energy levels of E_{HOMO} and E_{LUMO} (Eq. 1). The formula for chemical hardness (η) appears in Eq. 2 while chemical softness (σ) obtains its value through Eq. 3. The coating behavior electronegativity ($\varkappa$) gets evaluated through this mathematical expression (4) [33]. The electronic character ω functions as a descriptor to assess electrophilic properties by providing relative ranking for compound electrophilic behaviour [34]. According to Eq. (5) the parameter ω represents the energy loss measured during donor–acceptor electron transfers. Researchers calculated the Fe-metal (Fe-CS) work function through a quantum mechanical method that applied Eq. (6) [35].

$$\Delta E = E_{LUMO} - E_{HOMO} \tag{1}$$

$$\eta = \frac{\Delta E}{2} = \frac{E_{LUMO} - E_{HOMO}}{2} \tag{2}$$

$$\sigma = \frac{1}{\eta} = -\frac{2}{E_{HOMO} - E_{LUMO}} \tag{3}$$

$$\chi = \frac{1 + A}{2} \tag{4}$$

$$\omega = \frac{\chi^2}{2\eta} \tag{5}$$

$$\text{Fe - CS} = \frac{\chi_{Fe} - \chi_{inh}}{2(\eta_{Fe} + \eta_{inh})} \tag{6}$$

The calculation utilized the absolute electronegativity values χ_{Fe} and χ_{int} for iron and polymer while absolute hardness values η_{Fe} and η_{int} were used for both substances. The evaluation of Fe-metal (Fe-CS) work function occurred by implementing theoretical $\chi_{Fe} = 7.0$ eV and $\eta_{Fe} = 0$ eV values [36]. The Fukui indices determined the local reactivity during the optimization process [37]. The studied molecules show their areas of reactivity because this analysis reveals both nucleophilic and electrophilic attack responsibility for each specific atomic area [38]. Experimental findings regarding carbon steel coating behavior exposed to marine conditions were collected using electrode-based assessment methods [39]. Results showed outstanding protection of the carbon steel material and precursor [40]. The metal protection capabilities of ethylenic polyepoxide polymers TGETBAE (triglycidyl ether of tribisphenol A and ethylene/methylene dianiline) and HGTMDAE (hexaglycidyl trimethylene dianiline and ethylene/methylene dianiline) show dramatic improvements. Metal protection efficiency of these formulations emerges in TGETBAE/MDA, HGTMDAE/MDA, TGETBAE/MDA/NP, and

HGTMDAE/MDA/NP under all conditions without (NP: absence) and with (NP: presence) natural phosphate [17].

To verify the locations where epoxy polymers adhere we evaluated global quantum chemical descriptor calculations using Gaussian software version 03 W for our tri and hexafunctional epoxy polymers because they show exceptional adhesive properties between their six aromatic nuclei and one ethylenic radical [41]. The electronic basis has enabled researchers to study several GQCDs for epoxy polymers when calculating these descriptors [42]. The DFT method with the B3LYP/6-311G(d,p) basis set performed the calculations of different GQCDs [43]. A complete analysis of global quantum chemical descriptors occurred to uncover the reactive nature of electronegative sites [44]. The research analyzes TGETBAE and HGTMDAE molecular sites with electropositive and electronegative properties which it presents in Table 1. Both epoxy polymers show their optimized geometric structure and E_{LUMO}/E_{HOMO} electron distribution patterns in Figs. 1 and 2 [30].

As indicated in Figs. 1 and 2 the HOMO electronic density of TGETBAE extends across its aromatic rings along with epoxy groups while HGTMDAE's HOMO distribution emphasizes the ethylenic ring [45]. The LUMO distributions settle on epoxy ring surfaces for TGETBAE and span aromatic and epoxy ring surfaces for HGTMDAE. The energetic values belonging to highest occupied molecular orbitals (E_{HOMO}) establish how effectively compounds donate electrons [46]. Chemical species that possess high HOMO energy values exhibit increased electron donation characteristics toward acceptor species that contain low-energy unoccupied molecular orbitals [47]. A molecule takes electrons more easily if its lowest unoccupied molecular orbital (LUMO) energy level remains low. A molecule's electron-accepting power depends directly on its LUMO energy level value [17].

Table 1 The research analyzed GQCDs calculations of ethylenic polymers; permission from Elsevier

Property	TGETBAE (0)	TGETBAE (−1)	TGETBAE (+1)	HGTMDAE (0)	HGTMDAE (−1)	HRTMDAE (+1)
EHOMO	−5.876	−7.243	−7.21	−6.499	−6.499	−6.344
ELUMO	−2.536	−4.168	−5.003	−4.625	−4.631	−4.789
ΔE_{gap} (eV)	3.340	3.075	2.207	1.874	1.868	1.555
IP (eV)	5.876	7.243	7.210	6.499	6.499	6.344
EA (eV)	2.536	4.168	5.003	4.625	4.631	4.789
μ (Debye)	7.777	37.450	21.259	3.143	8.438	9.122
η (eV)	1.670	1.537	1.103	0.937	0.934	0.777
σ (eV)	0.598	0.598	0.650	0.906	1.070	1.287
$\varkappa$ (eV)	4.206	5.705	6.106	5.562	5.565	5.566
ω (eV)	5.296	10.588	16.90	16.507	16.578	19.935
Fe-metal (eV)	0.836	0.421	0.405	0.767	0.768	0.922

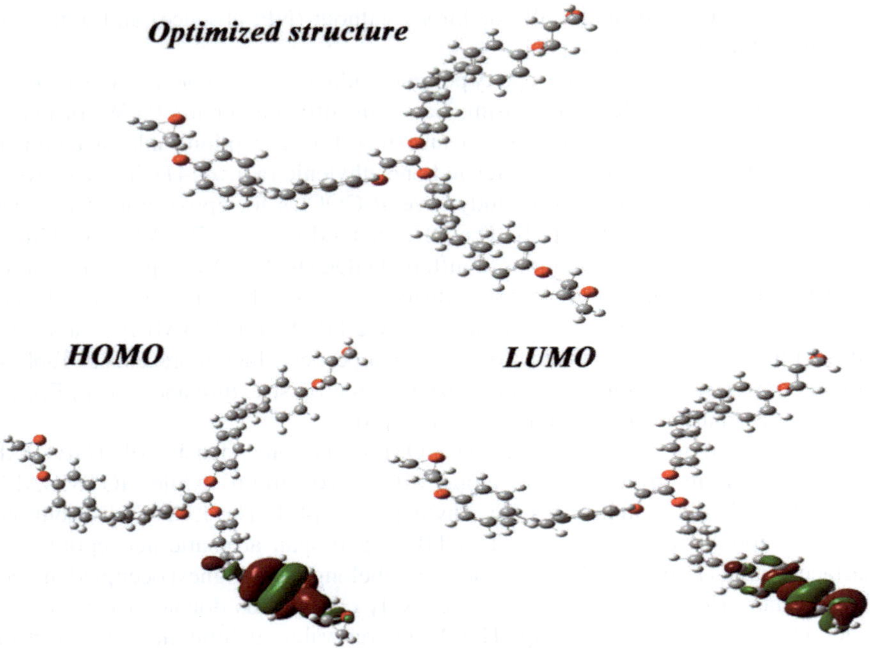

Fig. 1 A computational analysis evaluated the structural characteristics alongside HOMO–LUMO orbital distributions of TGETBAE; permission from Elsevier

When the gap energy (ΔE_{gap}) becomes lower the coating molecule shows better adsorption toward the metal surface [48]. Analysis of Table 1 reveals that HGTMDAE molecules with respective charges of neutral, positive or negative possess lower gap energy values than TGETBAE (HGTMDAE ΔE_{gap} < TGETBAE ΔE_{gap}). The inhibitory effectiveness of HGTMDAE against TGETBAE matrix appears to be high. We observed that HGTMDAE as a positively charged matrix possesses a gap energy of 1.55 eV which stands lower than both the neutral and negative charges. The identified energy values for these molecules amount to 1.868 and 1.874 eV. Table 1 presents the hardness (η) and softness (σ) values obtained through calculation. The analysis concludes that both TGETBAE and HGTMDAE matrices display good metal surface interaction [49].

The HGTMDAE matrix demonstrates better chemical reactions towards steel surface compared to TGETBAE matrix because its softness values for neutral, negative and positive forms have increased to 1.067, 1.070 and 1.287 eV^{-1} [50]. The determined hardness of HGTMDAE matrix at neutral and negatively charged and positively charged states reached 0.937, 0.934 and 0.777 eV respectively. Table 1 reveals HGTMDAE achieves higher electrophilicity index (χ) measurements than TGETBAE does. The electrophilicity of HGTMDAE reaches values of 16.507 eV for neutral conditions and 16.578 eV under negative conditions with a maximum value

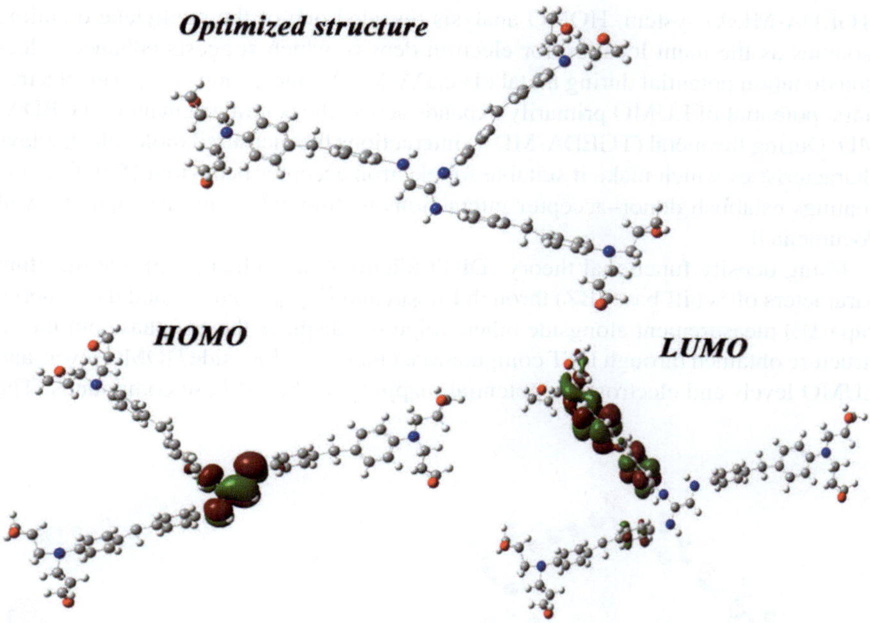

Fig. 2 A computational analysis evaluated the structural characteristics alongside HOMO–LUMO orbital distributions of HGTMDAE; permission from Elsevier [17]

of 19.935 eV during positive charging. These measurements show precise electron-accepting behavior for HGTMDAE. The work functions of Fe metal between both TGETBAE and HGTMDAE matrices fall under 3.6 eV. Molecules display a natural inclination to offer electrons to metallic surfaces when measured through this parameter [51].

The degree of inhibition effectiveness increases as the electron donation capability of the matrix material increases toward the metal surface. The electron transfer performance from neutral HGTMDAE matrix with positive and negative charges surpasses that of TGETMDAE matrix according to the sequence ΔN(HGTMDAE) > ΔN(TGETBAE). The HGTMDAE matrix exhibits well-formed film quality based on this study result. Research on epoxy polymers containing ethylenic nuclei interacting with TGETBAE and HGTMDAE matrices (in neutral, negatively charged, and positively charged forms) confirms that electronic effects govern epoxy-diamine system behaviors and properties [50]. Literature demonstrates that coating efficiency of macromolecules decreases as functional atom electronegativities rise among molecules with subtle differences. The sequence of increasing efficiency among the atoms starts with O followed by N and then S while Se comes next followed by P [52].

Figure 3 displays the optimized molecular structure and visualizes HOMO and LUMO orbitals and ESP distributions of the epoxy resin-methylene dianiline

(TGEDA-MDA) system. HOMO analysis revealed one of the methylene dianiline moieties as the main location for electron density which suggests enhanced electron donation potential during metal (TGEDA-MDA) interactions [53]. The electrostatic potential of LUMO primarily expands across the central segment of TGEDA-MD. During the metal (TGEDA-MDA) interactions the measured molecule displays characteristics which make it suitable for electron acceptor behaviour [54]. Organic coatings establish donor–acceptor interactions with metallic surfaces that are well documented.

Using density functional theory (DFT) Sliem et al. studied electronic quantum parameters of Schiff base (BZ) through E_{HOMO} and E_{LUMO} analysis and their energy gap (ΔE) measurement alongside others. Figure 4 displays the gas phase optimized structure obtained through DFT computational methods alongside HOMO levels and LUMO levels and electrostatic potential mapping for Schiff base compounds. The

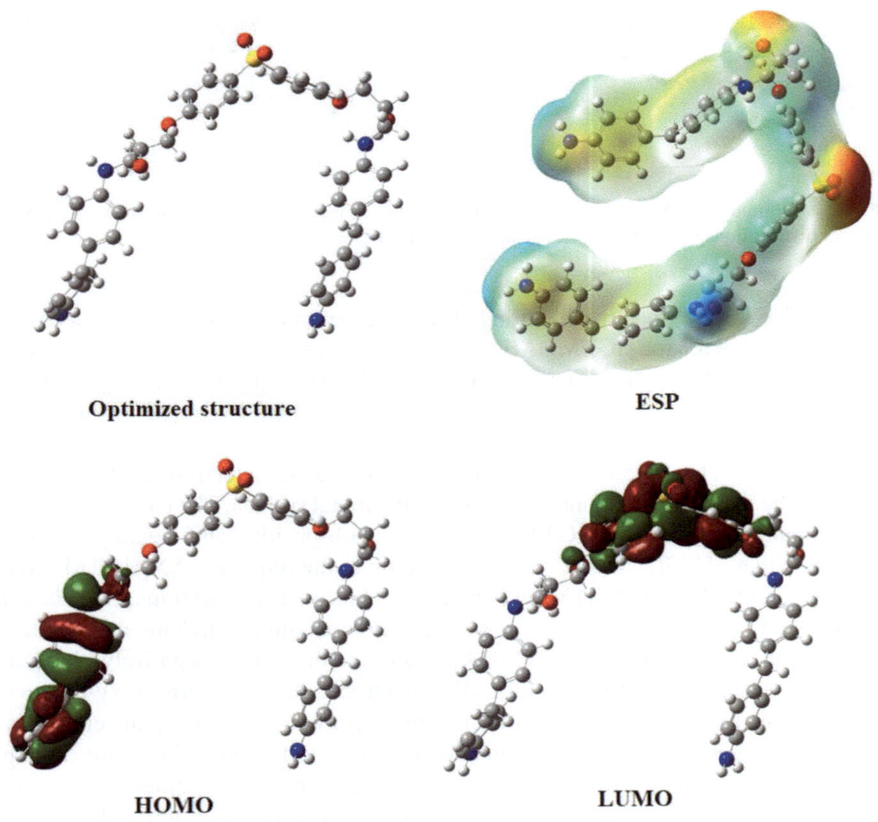

Fig. 3 The optimized structure for TGEDA-MDA exists alongside the HOMO, LUMO surfaces as well as ESP data; permission from Elsevier

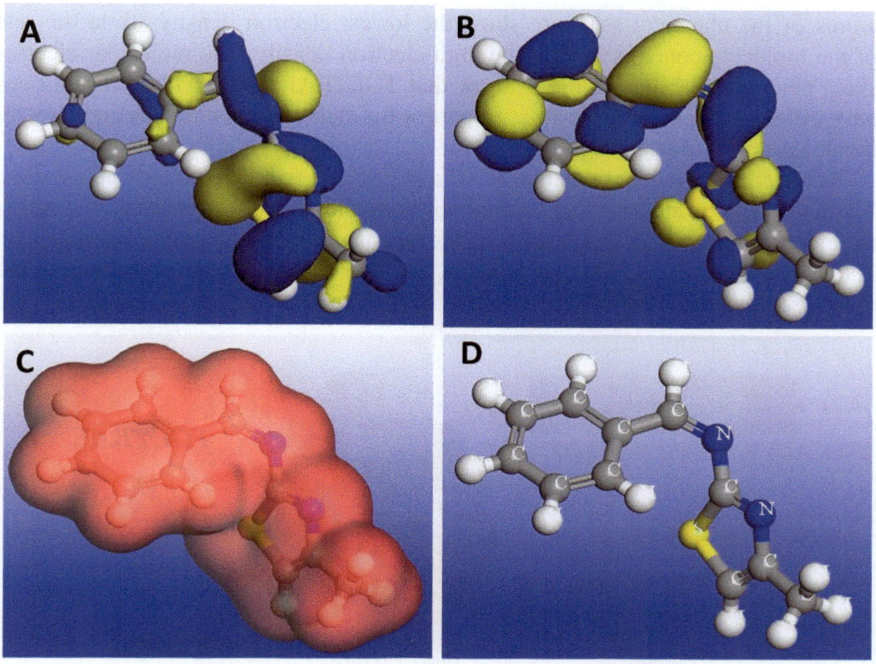

Fig. 4 A computational study based on DFT methods produced the HOMO, LUMO and electrostatic potential map and optimized structure of the Schiff base in its gas phase state [51]

inhibitor's electron-rich regions form donor–acceptor bonds with empty d orbitals on Fe atoms during corrosion inhibition processes [55].

2.2 *Molecular Electrostatic Potential*

Researchers utilize molecular electrostatic potential because it reflects the electrostatic interaction energy of evaluating fictitious charges to analyze interactions in macromolecular matrices [56]. Negative electrostatic potential regions promote electrophilic attacks, while positive electrostatic potential regions are more vulnerable to nucleophilic attacks [33]. The two polymers TGETBAE and HGTMDAE received molecular electrostatic potential analysis leading to Fig. 5 showing electron density distributions. Strong electron density appears as red while low electronic density shows up as blue in the representations [57]. The intensity of electron density follows this pattern: red stands higher than orange and then yellow and after that green before ending at blue. In TGETBAE and HGTMDAE polymers the oxygen and nitrogen atoms, which are involved in the interaction with neighbor polymers can be seen to have regions of high electron density depicted in yellow to red color. A few carbon atoms of the chemical structure show the lowest electron density while the rest of the molecule contains medium to high electron density. The molecular electrostatic potential accounts in Fig. 5 show that TGETBAE and HGTMDAE matrices have different electrostatic surface properties for their ethylenic radical and directly bonded aromatic nucleains [58].

Researchers have identified that MAC and ESP together help identify active sites. The active sites located on this Fe-surface perform the adsorption function. According to scientific research, it has been possible to conclude that favorable binding with Fe(110) surface and the DGEDDS-MDA inhibitor molecule tends to happen in the atoms bearing the most negative charge [59]. The structure of the regulator with

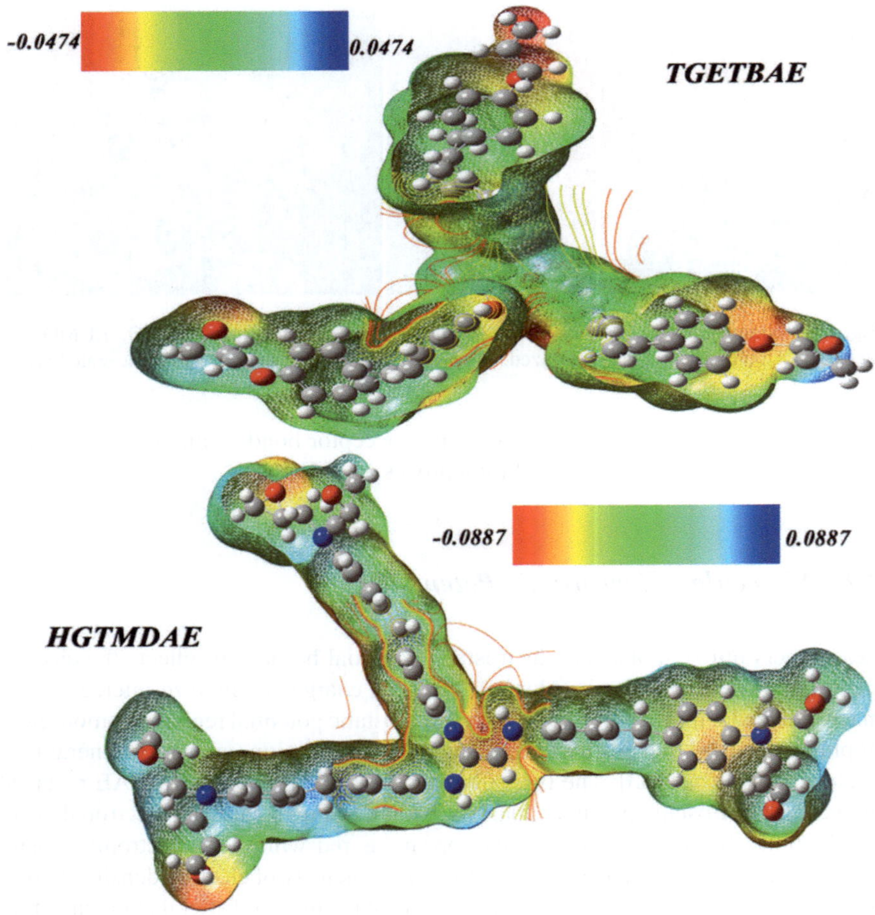

Fig. 5 Electrostatic potential analysis of the studied polymers was conducted at the molecular level; permission from Elsevier

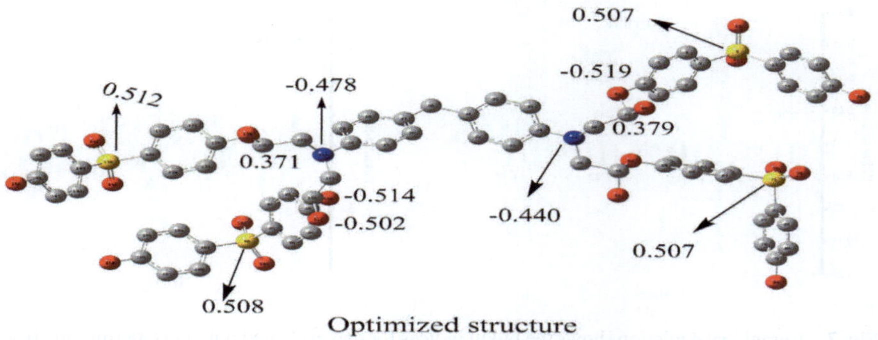

Fig. 6 The ultimate form of the DGEDDS-MDA optimized molecular structure exists. The optimized structure contains the located maximum negative and positive Mulliken atomic charges (MACs); permission from Elsevier

optimal shape opens at-sequence the points of the highest negative and positive charges within the molecule (Fig. 6) [60].

2.3 *Fukui Functions*

Fukui indices (IF) were determined using the DFT method (with the B3LYP/6-311G(d,p) basis set) of the gas phase (G). Atomic charges were approximated through natural population (PNA) analysis of localized electron pairs [61]. Electronic density acts as a key component for determining the parameters involved in chemical reactivity calculations. The PNA analysis required application of these Eqs. (7, 8) for its execution.

$$f_k^+ = P_k(N+1) - P_k(N) \tag{7}$$

$$f_k^- = P_k(N) - P_k(N-1) \tag{8}$$

The calculation uses P_k (N), P_k (N + 1) and P_k (N − 1) to represent electron populations in the k atom across N, (N + 1) and (N − 1) electron systems while f_k^+ and f_k^- indicate K atom nucleophilic and electrophilic capabilities. The maximum values of f_k^+ and f_k^- show nucleophilic and electrophilic activities at the same time [62]. The high value of f_k^+ suggests strong variations in the electron density during addition to the molecule an additional electron [63]. The f_k^- value represents the alteration experienced in the electron density in the molecule on electron reduction. The calculation results for Fukui indices of epoxy polymers TGETBAE and HGTMDAE display in Fig. 7 [17]. In Fig. 7 the C7 atom on trifunctional epoxy polymer TGETBAE displays maximum value of the f_k^+ which indicates its capacity for back-donation.

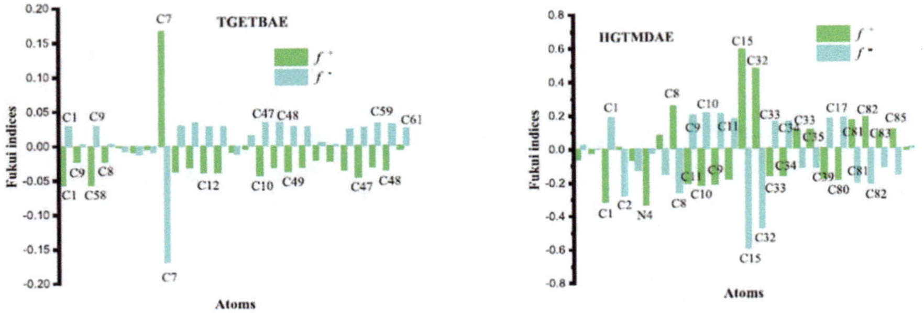

Fig. 7 A graphical depiction shows the Fukui indices for two evaluated polymers; permission from Elsevier

The main activation sites on which electron transfer occurs to the metal surface are determined from the f_k^- values at the carbon atoms C1, C9, C47, C48, C59, and C61, and others. Based on the values f_k^+ the electrophilic sites of attack on the hexa-functional epoxy polymer HGTMDAE at C8, C15, C32, C33, C35, C81, C83 and C85 are the main points. In addition, the f_k^- values, with the highest values, which are the regions susceptible to nucleophilic attack, are observed at C1, C9, C10, C11, C17, C33, C34, and C39 (Fig. 7). The results from electrostatic potential match the findings of the previous analysis.

2.4 Dual Local Descriptors

The computation of local softness measures $\sigma_k^{\pm}$ as well as electrophilicity values $\omega_k^{\pm}$ defines both electrophilic and nucleophilic attack properties [64]. The formulae for calculating $\sigma_k^{\pm}$ and $\omega_k^{\pm}$ are available in Eqs. 9 and 10, which are frequently reported in literature.

$$\sigma_k^{\pm} = \sigma f_k^{\pm} \tag{9}$$

$$\omega_k^{\pm} = \omega f_k^{\pm} \tag{10}$$

Local softness concepts are described by σ_k^+ and σ_k^- that refer to the nucleophilic and electrophilic attacks [65]. The local electrophilicity for nucleophilic and electrophilic attacks appeared as ω_k^+ and ω_k^- respectively [66]. The research team verified the dual local descriptors including dual Fukui f_k^2 (difference between nucleophilic and electrophilic Fukui functions) and dual local softness $\Delta\sigma_k$ (nucleophilic and electrophilic local softness difference) along with philicity $\Delta\omega_k$ (nucleophilic and electrophilic philicity function difference) [67]. The philicity functions' difference between their nucleophilic and electrophilic philicity functions prove to be more trustworthy than the previously mentioned local reactivity indices [68]. The calculation of f_k^2, $\Delta\sigma_k$ and $\Delta\omega_k$ follows these mathematical definitions in Eqs. (11–13):

$$f_k^2 = f_k^+ - f_k^- \tag{11}$$

$$\Delta\sigma_k = \sigma_k^+ - \sigma_k^- \tag{12}$$

$$\Delta\omega_k = \omega_k^+ - \omega_k^- \tag{13}$$

The analysis showed that a process involving electrophilic attacks will occur when the dual local descriptors (f_k^2, $\Delta\sigma_k$ and $\Delta\omega_k$) remain less than zero but a nucleophilic attack process will take place when the (f_k^2, $\Delta\sigma_k$ and $\Delta\omega_k$) measures exceed zero [69]. The investigation of electron donor and acceptor sites in the examined epoxy polymer structures utilized calculations from Table 1 that included the Fukui functions (f_k^+ and f_k^-) and local softness ($\sigma_k^{\pm}$) and local electrophilicity ($\omega_k^{\pm}$) besides dual Fukui function (f_k^2) and dual local softness ($\Delta\sigma_k$) and dual local philicity ($\Delta\omega_k$). The f_k^+ function measures the reactivity of nucleiophilic attack centers which correspond to accepting sites however f_k^- represents electrophilic attack reactivity at donation sites [70]. The paper explores dual Fukui function (f_k^2) together with dual local softness and dual local philicity results for analyzing active sites responsible for electron donation or electron acceptance in epoxy polymers. Figure 8 shows dual local descriptor (f_k^2, $\Delta\sigma_k$ and $\Delta\omega_k$) graphical data for the polymers' most crucial sites [17].

Table 2 reveals two polymers demonstrate electron acceptor properties because their three active parameter values including f_k^2, $\Delta\sigma_k$ and $\Delta\omega_k$ point to zero or less. The sites with f_k^2, $\Delta\sigma_k$ and $\Delta\omega_k$ values above zero indicate electrophilic characteristics [71]. The most active electron accepting centers, for the TGETBAE polymers (as well as for HGTMDAE), are found in the following concentration: For TGETBAE, C7 > C14; for HGTMDAE, C8 > C82 > C81 > C2 > N5 > C35 > C37 > C83 > N3.

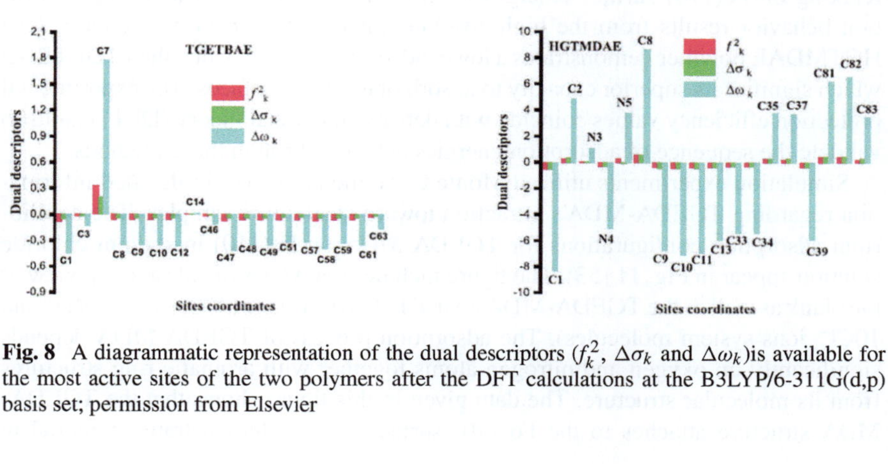

Fig. 8 A diagrammatic representation of the dual descriptors (f_k^2, $\Delta\sigma_k$ and $\Delta\omega_k$)is available for the most active sites of the two polymers after the DFT calculations at the B3LYP/6-311G(d,p) basis set; permission from Elsevier

In both polymers, the major electron-donating centres are at: C1 > C47 > C58 > C48 > C49 > C8 > C9 > C10 > C61 > C59 > C57 > C48 > C51 > C62 > C3 > C46 in TGETBAE, and in HGTMDAE at, C1 > C10 > C11 > C9 > C13 > C39 > C33 > C34 > N4 respectively [17].

2.5 Monte Carlo Simulation

For computational analysis, the researchers made use of the Condensed-phase Optimized Molecular Potentials for Atomistic Simulation Studies (COMPASS) force field [72]. The simulation MC studied the system with the studied polymer and the iron(110) surface since this surface showed maximum stability among Fe(100), Fe(111) and all other surfaces [73]. The interaction between new resin and Fe(110) surface took place within a simulation box measuring (20.12 × 20.12 × 35.8 Å). A periodic boundary condition was applied to expand the Fe(110) plane into a super-cell that measures 6.6. The surface of Fe(110) received a 30 Å vacuum plate installation. Another simulation method known as Monte Carlo was used to analyze polymer surface attachment on the Fe(110) substrate. Results from Monte Carlo simulation predictions show the ideal polymer adsorption configuration of surfaces with low adsorption energy. The most stable adsorption geometries of the studied polymers on the Fe(110) surface assumed from our simulations are shown in Figs. 9 and 10 [74].

The observation in Figs. 9 and 10 demonstrates that every studied polymer binds to carbon steel through parallel mode while effectively covering fundamental steel surface areas which indicates sufficient corrosion protection capacity. Table 3 shows all Monte Carlo derived energies and include the total energy plus the adsorption energy and the rigid adsorption energy and deformation energy and the essential value (dEads/dE (polymer)) represents the energy required to remove metal surface adsorbates while the low deformation energy that occurs as a result of adsorbates relaxing on Fe(110) surface emerges as most important [75]. The strong adsorption behavior results from the high absolute value of the adsorption energy. The HGTMDAE polymer demonstrates a lower adsorption energy value than TGETBAE which signifies its superior capacity to absorb onto metal surfaces. The experimental protection efficiency values coincide with density functional theory (DFT) results to validate the sequence of adsorption energies detected through these methods.

Simulation experiments utilized Monte Carlo methods to offer detailed information regarding TGEDA-MDA's attraction toward steel cover samples. The equilibrium adsorption configurations for TGEDA-MDA on Fe(110) in sodium chloride solution appear in Fig. 11 [53]. The figure includes two views of side and top view of the blank as well as the TGEDA-MDA coated Fe(110) surface ($400H_2O + 10Na^+$ and $10Cl^-$ ions system molecules). The adsorption process of TGEDA-MDA depends significantly on oxygen and nitrogen atoms together with aromatic ring structures from its molecular structure. The data given in this figure shows that the TGEDA-MDA structure attaches to the Fe(110) surface, which demonstrates potential in

Table 2 Fukui functions, local softness, electrophilicity, dual Fukui functions, dual softness and dual philicity of two polymers TGETBAE and HGTMDAE; permission from Elsevier

	Sites	f_k^+	f_k^-	f_k^2	σ_k^+	σ_k^-	$\Delta\sigma_k$	ω_k^+	ω_k^-	$\Delta\omega_k$
TGETBAE	C1	−0.0556	0.0288	−0.0844	−0.0332	0.0172	−0.0505	−0.2947	0.1526	−0.4474
	C3	−0.0223	0.0024	−0.0248	−0.0133	0.0014	−0.0148	−0.1182	0.0131	−0.1314
	C7	0.1668	−0.1685	0.3353	0.0997	−0.1007	0.2005	0.8834	−0.8924	1.7759
	C8	−0.0369	0.0296	−0.0665	−0.0221	0.0177	−0.0398	−0.1958	0.1567	−0.3526
	C9	−0.0318	0.0338	−0.0656	−0.0190	0.0202	−0.0392	−0.1684	0.1792	−0.3476
	C10	−0.0387	0.0277	−0.0667	−0.0231	0.0166	−0.0398	−0.2053	0.1471	−0.3525
	C12	−0.0382	0.0285	−0.0667	−0.0228	0.0170	−0.0399	0.2023	0.1512	−0.3535
	C14	−0.0088	−0.0117	0.0028	−0.0053	−0.0069	0.0016	−0.0470	−0.0619	0.0149
	C46	−0.0050	0.0147	−0.0198	−0.0030	0.0088	−0.0118	−0.0268	0.7806	−0.1049
	C47	−0.0478	0.0344	−0.0773	−0.0256	0.0205	−0.0462	−0.2270	0.1823	−0.4094
	C48	−0.0317	0.0339	−0.0656	−0.0189	0.0203	−0.0392	−0.1679	0.1798	−0.3477
	C49	−0.0374	0.0281	−0.0656	−0.0223	0.0168	−0.0392	−0.1982	0.1492	−0.3475
	C51	−0.0307	0.0284	−0.0591	−0.0183	0.0169	−0.0353	−0.1629	0.1505	−0.3134
	C57	−0.0360	0.0239	−0.0599	−0.0215	0.0143	−0.0358	−0.1908	0.1267	−0.3175
	C58	−0.0459	0.0272	−0.0732	−0.0275	0.0163	−0.0438	−0.2435	0.1445	−0.3880
	C59	−0.0305	0.0320	−0.0625	−0.0182	0.0191	−0.0374	−0.1619	0.1694	−0.3314
	C61	−0.0355	0.0309	−0.0664	−0.0212	0.0184	−0.0397	−0.1881	0.1636	−0.3518
	C63	−0.0054	0.0252	−0.0306	−0.0032	0.0150	−0.0183	−0.0287	0.1335	−0.1622
HGTMDAE	C1	−0.3116	0.1898	−0.5015	−0.3324	0.2026	−0.5351	−5.1635	3.1464	−8.3100
	C2	0.0122	−0.2784	0.2906	0.0130	−0.2971	0.3101	0.2021	−4.6139	4.8160
	N3	−0.0643	−0.1271	0.0627	−0.0686	−0.1356	0.0669	−1.0666	−2.1070	1.0404

(continued)

Table 2 (continued)

Sites	f_k^+	f_k^-	f_k^2	σ_k^+	σ_k^-	$\Delta\sigma_k$	ω_k^+	ω_k^-	$\Delta\omega_k$
N4	−0.3296	−0.0230	−0.3065	−0.3517	−0.0246	−0.3271	−5.4621	−0.3821	−5.0800
N5	0.0785	−0.1477	0.2263	0.0838	−0.1576	−0.0738	1.3015	−2.4487	3.7502
C8	0.2606	−0.2601	0.5207	0.2780	−0.2776	0.5556	4.3181	−4.3111	8.6293
C9	−0.2049	0.2006	−0.4056	−0.2186	0.2141	−0.4328	−3.3961	3.3225	−6.7212
C10	−0.2153	0.2124	−0.4278	−0.2297	0.2267	−0.4565	−3.5685	3.5209	−7.0894
C11	−0.2093	0.2073	−0.4166	−0.2233	0.2212	−0.4445	−3.4681	3.4354	−6.9035
C13	−0.1807	0.1771	−0.3578	−0.1928	0.1890	−0.3818	−2.9941	2.9360	−5.9302
C33	−0.1646	0.1628	−0.3274	−0.1785	0.1737	−0.3522	−2.7289	2.6979	−5.4268
C34	−0.1631	0.1611	−0.3242	−0.1740	0.1719	−0.3459	−2.7025	2.6695	−5.3721
C35	0.1124	−0.1127	0.2251	0.1199	−0.1202	0.2402	1.8634	−1.8676	3.7310
C37	0.1103	−0.1106	0.2210	0.1177	−0.1180	0.2358	1.8284	−1.8339	3.6624
C39	−0.1786	0.1786	−0.3573	−0.1906	0.1906	−0.3812	−2.9605	2.9605	−5.9211
C81	0.1662	−0.2059	0.3722	0.1774	−0.2197	0.3972	2.7554	−3.4132	6.1686
C82	0.1836	−0.2080	0.3916	0.1959	−0.2219	0.4191	3.0429	−3.4472	6.4901
C83	0.0957	−0.1136	0.2094	0.1021	−0.1212	0.2234	1.5867	−1.8835	3.4702

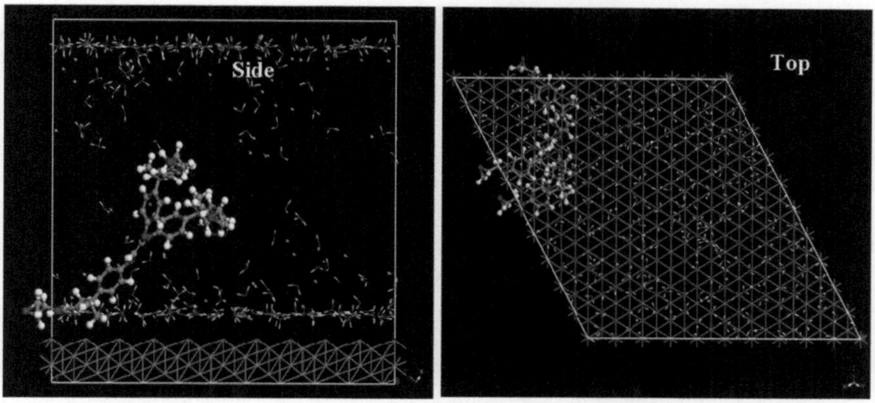

Fig. 9 The most energetically favorable adsorption arrangement of TGETBAE; permission from Elsevier

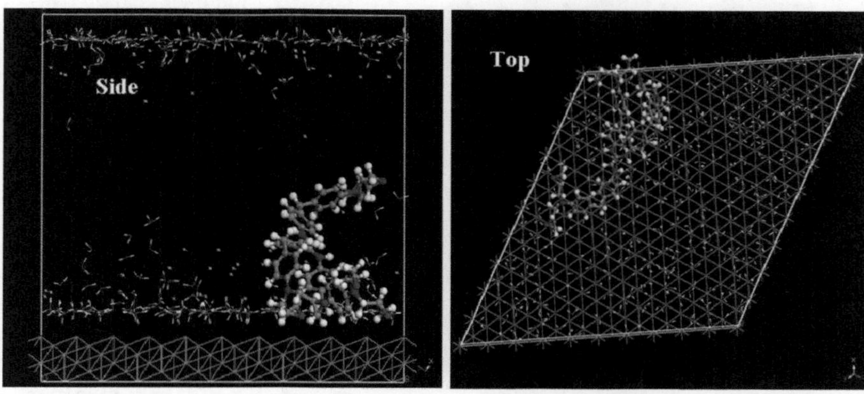

Fig. 10 The most energetically favorable adsorption arrangement of HGTMDAE; permission from Elsevier

Table 3 The Monte Carlo simulation generated outputs and descriptors as measurements for the adsorption energy of studied polymers on Fe(110) in kcal/mol; permission from Elsevier

Polymers	Total energy	Adsorption energy	Rigid adsorption energy	Deformation energy	dE_{ads}/dE (polymer)(Polymer)
Fe(110)/ HGTMDAE	−213.34	−604.70	−152.28	−452.42	−604.70
Fe(110)/ FGETBAE	−331.46	−291.77	−239.57	−52.20	−291.77

creating protective films for the Fe(110) based surfaces. The side view of Fe(110) surface with TGEDA-MDA structure shows that these molecules are spread in the upper portion of the surface, thus preventing corrosion agents [76].

The authors, Yousefi-Nasab et al. [77], performed Monte Carlo simulations of alanine gas phase adsorption in protonated and non-protonated states on iron, copper and aluminium surfaces. They used the Metropolis algorithm in unison with Adsorption Locator module of Material Studio for their analysis [78]. Information processing reached 100 K steps at each temperature level during ten temperature cycles for simulation of annealing [79]. Energies together with equilibrium configuration calculations relied on the COMPASS force field for their implementation [80]. The Ewald

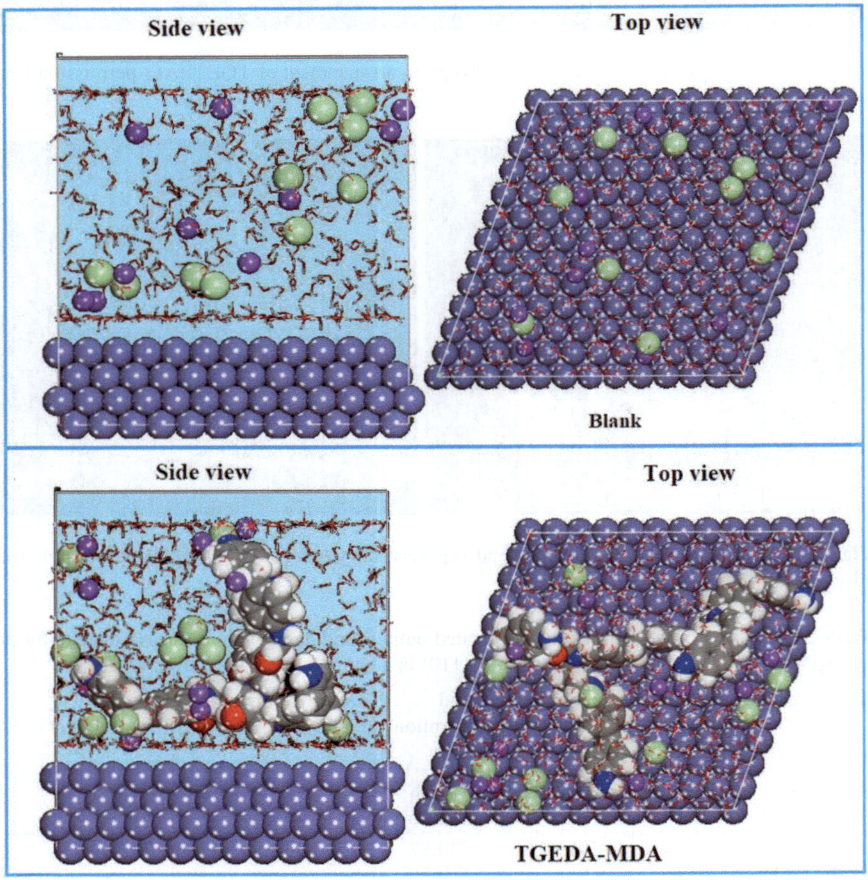

Fig. 11 The simulation of MC revealed the maximum stable low energy configurations for adsorption between Fe(110)/without (Blank) and the TGEDA-MDA + $400H_2O$ + 10 Na^+ and 10 Cl^- ions system from the top and side views; permission from Elsevier

method delivered an accuracy of 1×10^{-5} kcal/mol and Atom-based method determined Talapatra and Datta [81] interaction energies. To calculate non-bond interactions the software adopted an 18.5 A° cutoff distance employing cubic spline truncation methods. As shown in Fig. 12 [82] the periodic slab presents side-view pictures of Fe(110), Cu(111), and Al(111) surfaces which serve as input for Monte Carlo simulations [83]. Monte Carlo simulation calculation results on the adsorption of alanine, alanine dipeptide, and alanine tripeptide in their neutral forms on Fe(110) demonstrated low-energy configurations (Fig. 13) [84]. The analysis in Fig. 13 presented the favorable orientations which allowed the computation of the alanine molecules' interaction energy with metal surfaces.

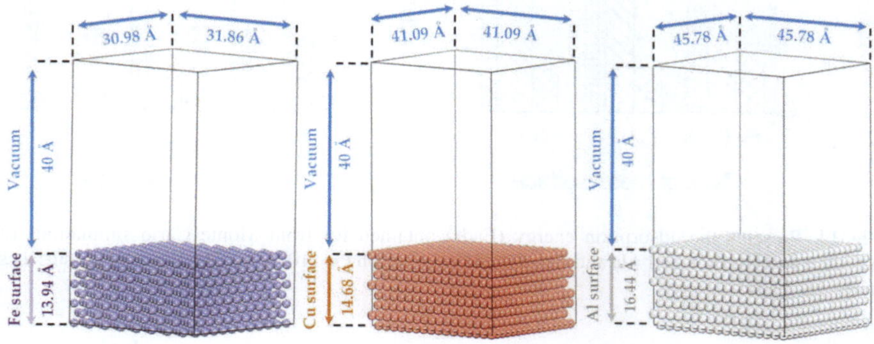

Fig. 12 The side views of periodic slabs used for Fe(110), Cu(111), and Al(111) surfaces in Monte Carlo simulations

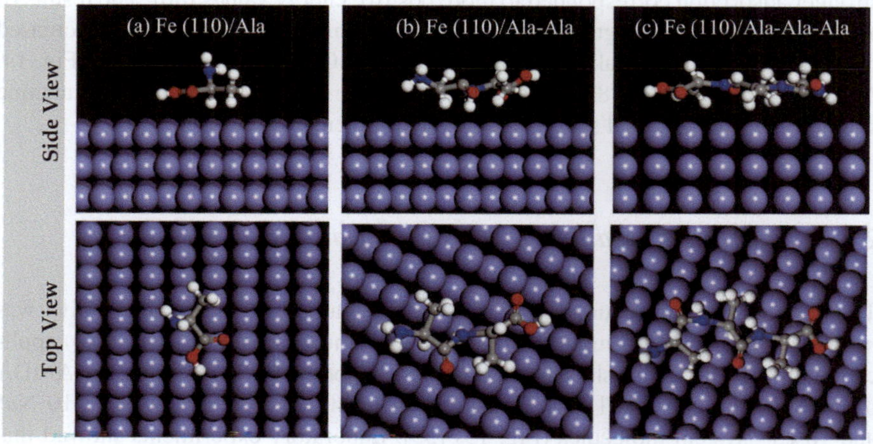

Fig. 13 Presents low-energy configurations obtained from the results of Monte Carlo simulations of adsorption of alanine, alanine dipeptide and alanine tripeptide in their neutral forms on Fe(110)

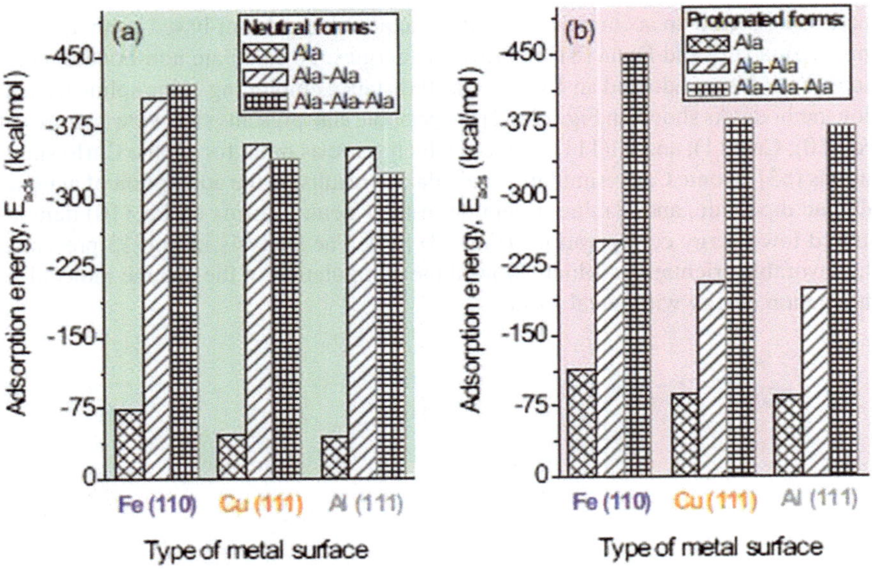

Fig. 14 Presents the adsorption energy (Eads) obtained for from Monte Carlo simulations of **a** neutral and **b** protonated Ala and Ala-derived peptides on Fe(110), Cu(111), and Al(111) surfaces in the gas phase

Figure 14 [85] presents the outcome of calculated adsorption energy measurements used for system evaluation during the study. The values of calculated adsorption energies helped establish an order of molecular adsorption strength because stronger adsorption corresponded to more positive energy values [86]. The results of absolute adsorption strength tests revealed that alanine derivatives showed increased binding affinity with metals in the order of Fe then Cu then Al according to Fig. 14. According to Yao et al. [87] experimental data the inhibition efficiency of alanine solution reached 80.2 and 56.3% in acid environment.

2.6 Molecular Dynamic (MD) Simulation

The surface bonding nature of TGEDA-MDA coating adherence to Fe(110) was investigated using MD simulation for both NaCl medium and blank Fe(110) applications [88]. The figure shows equilibrium adsorption depiction of TGEDAMDA on Fe(110) through NaCl solution surface with 400 H_2O molecules and 10 Na^+ and 10 Cl^- system molecules (Fig. 15) [53]. Laboratory experiments showed that the adsorption of TGEDA-MDA occurred on Fe(110). The visual evidence shows TGEDA-MDA coating forms an adhesive bond with the steel sample film. The molecular structure TGEDA-MDA attaches parallel to Fe(110) surfaces with this orientation optimizing the coverage capacity of TGEDA-MDA. Useful structural information from the side view of the TGEDA-MDA shows that the surface-facing oxygen and nitrogen heteroatoms (CeO and CeN atoms) on the surface are situated inside the film [89].

The final orientations of DGEDDS-MDA, with and without TiO_2 exposure on the metal surface, are presented in Fig. 16a. Adsorption behavior of DGEDDS-MDA through its heteroatoms and aromatic rings yield to a flat alignment with the surface [68]. The overall strength of adsorption of the entire system increases with TiO_2 (Fig. 16b) owing to TiO_2 powder particles that are close to the DGEDDS-MDA inhibitor molecule at the contact region. The adsorption process leads to the development of an effective barrier between materials [53].

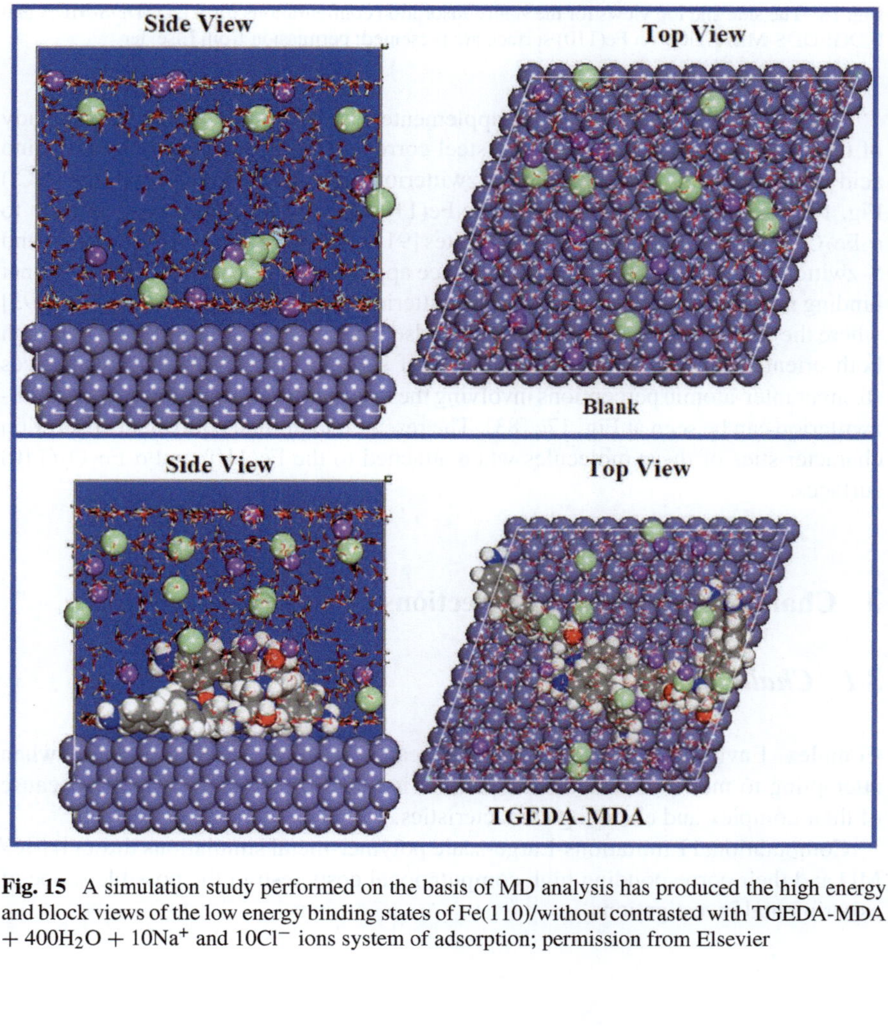

Fig. 15 A simulation study performed on the basis of MD analysis has produced the high energy and block views of the low energy binding states of Fe(110)/without contrasted with TGEDA-MDA + $400H_2O$ + $10Na^+$ and $10Cl^-$ ions system of adsorption; permission from Elsevier

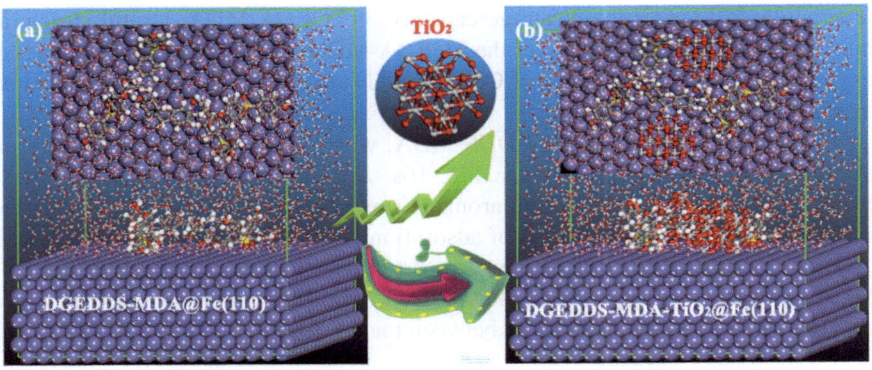

Fig. 16 The side and top views for the stable adsorption configurations for **a** DGEDDS-MDA and **b** DGEDDS-MDA-TiO_2 on Fe(110) surface are presented; permission from Elsevier

Sun et al.'s [90] studies were supplemented by MD simulations in the study of the inhibition mechanism of mild steel corrosion by the two synthesized amino acid derivatives, S-zwitterion and H-zwitterion, in 1 M hydrochloric acid (HCl) Fig. 17. For the model calculations, the Fe(110) surface was utilized in addition to α-Fe_2O_3(110) surface as the adsorption sites [91]. The side views of H-zwitterion and S-zwitterion bound to α-Fe_2O_3(110) surface appear in Fig. 16a, b [92]. Two distinct binding mechanisms appeared between zwitterions and their respective surfaces [93] where the S-zwitterion displayed superior adsorption than the H-zwitterion although both oriented planarly [94]. The presence of sulfur inside the molecule produces stronger inter-atomic perceptions involving the iron atoms. The adsorption mode of S-zwitterion can be seen at Fig. 17c [83]. The research identified equivalent absorption characteristics of these molecules when attached to the Fe(110) and α-Fe_2O_3(110) surfaces.

3 Challenges and Future Directions

3.1 Challenges

Complex Environmental Interactions-Researchers have faced obstacles when attempting to model environmental factors' impact on corrosion correctly because of their complex and changing characteristics.

Computational Limitations-Large-scale polymer-metal simulations through MC/MD and their corresponding high computational costs restrict the possible scope of experimental investigations.

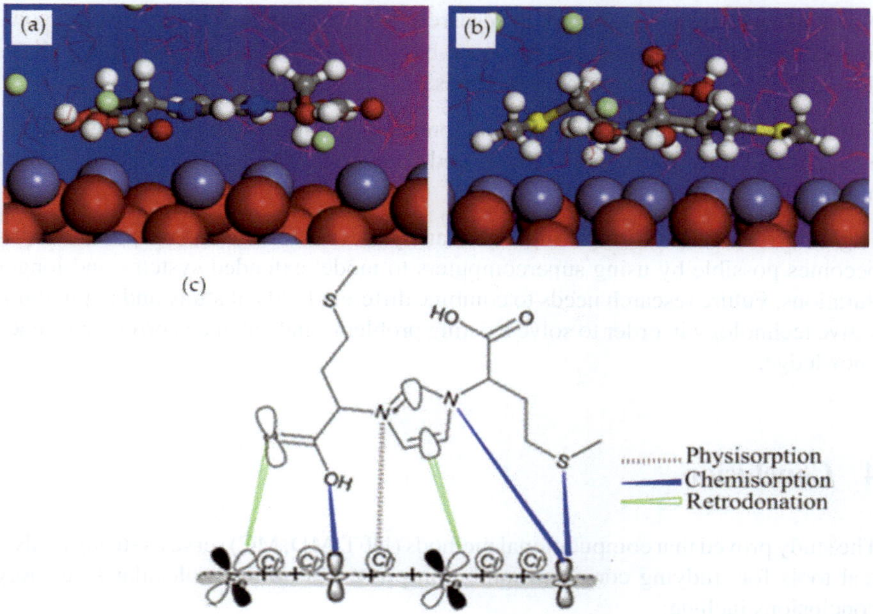

Fig. 17 **a** H-zwitterion and **b** S-zwitterion adsorbed on α-Fe_2O_3(110) surface with **c** representation of model for adsorption of S-zwitterion

Interdisciplinary Integration-People find it challenging to establish joint frameworks between material science, chemistry and engineering to create complete corrosion solutions.

Validation with Experimentation-Simulated experimental data may differ because simulation models use overly simple concepts while simulation conditions follow perfect standards.

3.2 *Future Scope*

Advanced Computational Techniques: The combination of machine learning and AI with DFT/MD would strengthen predictions used for inhibitor design and corrosion mechanism research.

Multiscale Modeling: The combination of atomic simulations with continuum models helps solve complex corrosion issues in systems like marine environments and biofoulings.

Novel Materials Development: A research investigation detects self-healing coatings alongside investigations of graphene-based composites and high-entropy alloys through computational screening methods.

Environmental Impact Studies: Development of green prevention methods requires evaluation of microbiologically influenced corrosion (MIC) and biofouling through multiple strains simulation.

High-Performance Computing: The analysis of realistic corrosion scenarios becomes possible by using supercomputers to model extended systems and longer durations. Future research needs to combine different fields of study and apply innovative technology in order to solve existing problems and enhance corrosion science knowledge.

4 Conclusion

The study proved that computational methods (DFT, MD, MC) serve as strong analytical tools for studying corrosion inhibition processes on the molecular level. Key conclusions include:

- The data analysis indicates that HGTMDAE showed better chemical softness and lower energy gap and stronger electron-donating ability than TGETBAE which predicts superior corrosion inhibiting properties.
- The parallel polymer adsorption onto Fe(110) surface was confirmed by MC and MD simulations because HGTMDAE exhibited better adsorption due to its lower adsorption energy.
- Electron transfer sites on the material surface were located through Fukui indices and MEP maps and dual local descriptors which matched experimental protection rates.
- MD simulation results demonstrated TGEDA-MDA forms stronger bonds with Fe(110) through the presence of heteroatoms (O and N) since these promote effective corrosive agent resistance barriers.

Computational approaches demonstrate their value for developing efficient corrosion inhibitors by helping to evaluate their performance characteristics in practical applications.

References

1. Beech, I.B., Sunner, J., Galperin, M.Y., Baker, A.J.: Biocorrosion: towards understanding interactions between biofilms and metals. Curr. Opin. Biotechnol. **15**, 181–186 (2004). https://doi.org/10.1016/j.copbio.2004.05.001

2. Bhat, A., Budholiya, S., Raj, S.A., Sultan, M.T.H., Hui, D., Shah, A.U.M., Safri, S.N.A.: Review on nanocomposites based on aerospace applications. Nanotechnol. Rev. **10**, 237–253 (2021). https://doi.org/10.1515/NTREV-2021-0018
3. Xavier, J.R.: Superior barrier, hydrophobic, and mechanical properties of the epoxy nanocomposite containing mixed metal oxides. J. Adhes. Sci. Technol. **37**, 1394–1418 (2023). https://doi.org/10.1080/01694243.2022.2073945
4. Jain, J., Nagar, M.: Synthesis and characterization of some acetoxime-modified hetero-metal alkoxide precursors of zinc(II) and aluminum(III) for sol-gel transformation to nano-sized $ZnAl_2O_4$. J. Sol–Gel Sci. Technol. **71**, 447–457 (2014). https://doi.org/10.1007/S10971-014-3399-0/METRICS
5. Jeevanandam, J., Kiew, S.F., Boakye-Ansah, S., Lau, S.Y., Barhoum, A., Danquah, M.K., Rodrigues, J.: Green approaches for the synthesis of metal and metal oxide nanoparticles using microbial and plant extracts. Nanoscale **14**, 2534–2571 (2022). https://doi.org/10.1039/D1NR08144F
6. Ma, I.A.W., Ammar, S., Kumar, S.S.A., Ramesh, K., Ramesh, S.: A concise review on corrosion inhibitors: types, mechanisms and electrochemical evaluation studies. J. Coat. Technol. Res. **191**, 241–268 (2021). https://doi.org/10.1007/S11998-021-00547-0
7. González-García, Y., González, S., Souto, R.M.: Electrochemical and structural properties of a polyurethane coating on steel substrates for corrosion protection. Corros. Sci. **49**, 3514–3526 (2007). https://doi.org/10.1016/J.CORSCI.2007.03.018
8. Deshpande, P.P., Jadhav, N.G., Gelling, V.J., Sazou, D.: Conducting polymers for corrosion protection: a review. J. Coat. Technol. Res. **11**, 473–494 (2014). https://doi.org/10.1007/S11998-014-9586-7/METRICS
9. McMahon, M.E., Santucci, R.J., Glover, C.F., Kannan, B., Walsh, Z.R., Scully, J.R.: A review of modern assessment methods for metal and metal-oxide based primers for substrate corrosion protection. Front. Mater. **6**, 468271 (2019). https://doi.org/10.3389/FMATS.2019.00190/BIBTEX
10. Makhmetova, A., Negim, E.-S., Ainakulova, D., Yeligbayeva, G., Khatib, J.: An overview of epoxy resins as coating to protect metals from corrosion. Kompleks. Ispolz. Miner. Syra Compl. Use Miner. Resour. **328**, 20–32 (2024)
11. Zaghloul, M.M.Y., Zaghloul, M.M.Y., Fuseini, M.: Recent progress in epoxy nanocomposites: corrosion, structural, flame retardancy and applications: a comprehensive review. Polym. Adv. Technol. **34**, 3438–3472 (2023). https://doi.org/10.1002/PAT.6144
12. Fontana, M., Greene, N.: Corros. Eng. **18**, 2025 (1967). https://www.osti.gov/biblio/4768541
13. Ali, H.M.: Titanium Dioxide: Advances and Applications (2022)
14. Aljibori, H.S., Alamiery, A., Kadhum, A.A.H.: Advances in corrosion protection coatings: a comprehensive review. Int. J. Corros. Scale Inhib. **12**, 1476–1520 (2023)
15. Kim, S., Praveen, B.M., Devendra, B.K.: A review on corrosion inhibitors: types, mechanisms, electrochemical analysis, corrosion rate and efficiency of corrosion inhibitors on mild steel in an acidic environment. Results Surf. Interf. **16**, 100258 (2024). https://doi.org/10.1016/J.RSURFI.2024.100258
16. Cao, L., Jones, A., Sikka, V., Wu, J., Gao, D.: Anti-icing superhydrophobic coatings. Langmuir **25**, 12444–12448 (2009). https://doi.org/10.1021/la902882b
17. Hsissou, R.: Review on epoxy polymers and its composites as a potential anticorrosive coatings for carbon steel in 3.5% NaCl solution: computational approaches. J. Mol. Liq. **336**, 116307 (2021). https://doi.org/10.1016/J.MOLLIQ.2021.116307
18. Dagdag, O., Berisha, A., Safi, Z., Dagdag, S., Berrani, M., Jodeh, S., Verma, C., Ebenso, E.E., Wazzan, N., El-Harfi, A.: Highly durable macromolecular epoxy resin as anticorrosive coating material for carbon steel in 3% NaCl: computational supported experimental studies. J. Appl. Polym. Sci. **137**, 49003 (2020). https://doi.org/10.1002/APP.49003
19. Hsissou, R., Benhiba, F., Echihi, S., Benzidia, B., Cherrouf, S., Haldhar, R., Ahmad-Alvi, P., Kaya, S., Serdaroğlu, G., Zarrouk, A.: Performance of curing epoxy resin as potential anticorrosive coating for carbon steel in 3.5% NaCl medium: combining experimental and computational approaches. Chem. Phys. Lett. **783**, 139081 (2021)

20. Hsissou, R., Dagdag, O., Abbout, S., Benhiba, F., Berradi, M., El-Bouchti, M., Berisha, A., Hajjaji, N., Elharfi, A.: Novel derivative epoxy resin TGETET as a corrosion inhibition of E24 carbon steel in 1.0 M HCl solution: experimental and computational (DFT and MD simulations) methods. J. Mol. Liq. **284**, 182–192 (2019)
21. Khalife, H., Rezaei, M., Amjad-Iranagh, S.: Durable superhydrophobic PTFE-rich/acrylic rGO loaded self-lubricating coating to advance wear and long-time corrosion: experimental studies and molecular dynamics simulation of friction and corrosive agents penetration. Colloids Surf A Physicochem Eng Asp **713**, 136518 (2025). https://doi.org/10.1016/J.COLSURFA.2025.136518
22. Garmer, D., Gresh, N., Roques, B.P.: Modeling of inhibitor–metalloenzyme interactions and selectivity using molecular mechanics grounded in quantum chemistry. Proteins Struct. Funct. Bioinform. **18**, 134 (2025)
23. Mobin, M., Aslam, R.: Experimental and theoretical study on corrosion inhibition performance of environmentally benign non-ionic surfactants for mild steel in 3.5% NaCl solution. Process. Saf. Environ. Prot. **114**, 279–295 (2018). https://doi.org/10.1016/J.PSEP.2018.01.001
24. Luo, X., Chen, B., Li, J., Zhou, C., Guo, M., Peng, K., Dai, H., Lan, B., Xiong, W., Liu, Y.: Zwitterion modified chitosan as a high-performance corrosion inhibitor for mild steel in hydrochloric acid solution. Int. J. Biol. Macromol. **267**, 131429 (2024). https://doi.org/10.1016/J.IJBIOMAC.2024.131429
25. Moradi, M., Rezaei, M.: Corrosion protection mechanisms of graphene oxide-reinforced polypropylene coating for mild carbon steel by advanced molecular dynamics simulations with emphasis on resistance to water and chloride ion penetration. J. Mol. Liq. **367**, 120389 (2022). https://doi.org/10.1016/J.MOLLIQ.2022.120389
26. Hsissou, R., Benhiba, F., Echihi, S., Benkhaya, S., Hilali, M., Berisha, A., Briche, S., Zarrouk, A., Nouneh, K., Elharfi, A.: New epoxy composite polymers as a potential anticorrosive coatings for carbon steel in 3.5% NaCl solution: experimental and computational approaches. Chem. Data Collect. **31**, 100619 (2021)
27. Moradi, M., Rezaei, M.: Long-term experimental evaluation and molecular dynamics simulation of polypropylene/graphene oxide nanocomposite coating in 3.5 wt% NaCl solution. Prog. Org. Coat. **164**, 106718 (2022). https://doi.org/10.1016/J.PORGCOAT.2022.106718
28. Dagdag, O., Safi, Z., Hsissou, R., Erramli, H., El-Bouchti, M., Wazzan, N., Guo, L., Verma, C., Ebenso, E.E., El-Harfi, A.: Epoxy pre-polymers as new and effective materials for corrosion inhibition of carbon steel in acidic medium: computational and experimental studies. Sci. Rep. **91**, 1–14 (2019). https://doi.org/10.1038/s41598-019-48284-0
29. Huang, L., Chen, W., Hao, W., Wang, J., et al.: Self-healing anti-corrosion coatings: a mechanism study using computational materials science. Electr. Mater. Appl. **1**, e12006 (2024). https://doi.org/10.1049/EMA3.12006
30. Batchelor, C., Martin-Martinez, F.J., Griffiths, C., Mabbett, I., Smith, J., Jewell, E.: Computationally-accelerated prediction of polyester-melamine coatings degradation to design sustainable organically-coated steels for outdoor applications. RSC Adv. **14**, 18343–18354 (2024). https://doi.org/10.1039/D3RA06744K
31. El-Amri, A., Azogagh, M., Sayed, R., El-Amri, A., Er-rahmany, N., Hamouche, F., Hsissou, R., Dkhireche, N.: Synthesis characterisation and analysis of the corrosion resistance of a new epoxy resin onto mild steel in the presence of hydrochloric acid: practical and theoretical points of view. Can. Metall. Q. **44**, 306 (2025). https://doi.org/10.1080/00084433.2025.2457306
32. Moustafa, A.H.E., Abdel-Rahman, H.H., Barakat, A., Mohamed, H.A., El-Kholany, A.S.: An exploratory experimental analysis backed by quantum mechanical modeling, spectroscopic, and surface study for C-steel surface in the presence of hydrazone-based schiff bases to fix corrosion defects in acidic media. ACS Omega **9**, 16469–16485 (2024). https://doi.org/10.1021/ACSOMEGA.4C00199
33. Baharin, N.A.N.K., Ghazali, S.A.I.S.M., Razak, F.I.A., Sirat, S.S., Tajuddin, A.M., Kamarudin, S.R.M., Dzulkifli, N.N.: Hirshfeld surface analysis, density functional theory and corrosion inhibition mechanism proposed of vanillin 4-ethylthiosemicarbazone on the mild steel in 1M HCl. Moroccan J. Chem. **13**, 80–105 (2025)

34. Bairagi, H., Vashishth, P., Narang, R., Shukla, S.K., Mangla, B.: Protection of mild steel from corrosion in an aggressive chloride media through chitosan-CeO_2 nanocomposite: experimental and computational studies. Corros. Commun. **14**, 58–71 (2024). https://doi.org/10.1016/J.CORCOM.2023.08.003
35. Matine, A., El-Maraghi, A., El-Abdallaoui, H.E.A., Zeroual, A.: Quantum, dynamic, and complexation studies of imidazole derivatives as corrosion inhibitors for mild steel in 1 M HCl. Euro-Mediter. J. Environ. Integr. **14**, 1–13 (2024). https://doi.org/10.1007/S41207-024-00690-1
36. Jha, H., Saxena, A., Kaur, J., Katin, K.P., Berdimurodov, E., Verma, D.K., Ataya, F.S.: Corrosion inhibition assessment of the Curcumis melon extract on low-carbon steel in 0.5 M H_2SO_4 using potentiodynamic polarization, electrochemical impedance spectroscopy, scanning electron microscope, DFT and MD simulation studies. Eur. Phys. J. Plus **13912**, 1–19 (2024)
37. Sun, E., Zhao, Y., Wang, G.: Influence of chloride salt erosion on the adhesion of asphalt-aggregate interfaces considering mineral anisotropy: insights from molecular dynamics. J. Mol. Model. **30**, 1–20 (2024). https://doi.org/10.1007/S00894-024-05852-0
38. Xu, Y., Li, J., Liang, H., Xiong, Y., Sun, W., Cao, X., Cao, H., Kong, F., Gao, H.: The effects of amino-modified graphene oxide on the water molecules diffusion and seawater barrier properties of DGEBA/3,3′-DDS epoxy resin: a molecular dynamics simulation. Chem. Pap. **78**, 1469–1479 (2024). https://doi.org/10.1007/S11696-023-03174-8
39. Thakran, M., Lata, S.: Polybenzopyrrole/nano-alumina composite blend with zirconium silicate reinforced epoxy as protective coating to subside corrosion of carbon steel within a dilute NaCl solution. J. Mol. Struct. **1298**, 137068 (2024). https://doi.org/10.1016/J.MOLSTRUC.2023.137068
40. Qi, Y., Wu, P., Huo, T., Li, Z., Li, X., Zhang, Y.: Enhanced corrosion resistance of epoxy coatings through the incorporation of modified TiO_2 and reduced graphene oxide. Prog. Org. Coat. **197**, 108875 (2024). https://doi.org/10.1016/J.PORGCOAT.2024.108875
41. Ashwathareddy, A., Rao, S., Shree-Subramaniyam, S., Gopala-Krishna, P., Monnppa-Rama, K., Kodange, S.: Corrosion mitigation of mild steel in acidic medium using liquid crystals: a comprehensive review. Inorg. Chem. Commun. **169**, 113071 (2024). https://doi.org/10.1016/J.INOCHE.2024.113071
42. Yao, Y., Tang, F., Yang, S., Li, B.: Improving the impermeability of cement/epoxy coating interface by g-C3N5: a molecular dynamics study. Mater. Today Commun. **41**, 110763 (2024). https://doi.org/10.1016/J.MTCOMM.2024.110763
43. Al-Jahdaly, B.A.: Electrochemical and DFT insights into 2-amino-4-(4-hydroxy-3-methoxyphenyl)-7-methyl-4 H -chromene-3-carbonitrile: an innovative strategy for antibacterial activity and corrosion protection of carbon steel. RSC Adv. **14**, 24039–24054 (2024)
44. Edoziuno, F.O., Odoni, B.U., Adediran, A.A., Byadi, S., Barhoumi, A., Nwaeju, C.C., Modebe, L.U., Okeniyi, J.O.: Surface corrosion inhibition of mild steel in an acidic environment by an anthelmintic drug: experimental, RSM, DFT and MD simulation studies. Port. Electrochim. Acta **42**, 1–29 (2024). https://doi.org/10.4152/pea.2024420101
45. El-Aouni, N., Dagdag, O., El Amri, A., Kim, H., Dkhireche, N., Elbachiri, A., Berdimurodov, E., Berisha, A., Rafik, M.: Hybrid epoxy/Br inhibitor in corrosion protection of steel: experimental and theoretical investigations. Environ. Sci. Pollut. Res. Int. **31**, 1033–1049 (2024). https://doi.org/10.1007/S11356-023-31171-7
46. Sattari, R., Khayati, G.R., Darezereshki, E.: Study of anti-corrosion properties of ellagic acid and curcumin derived from plants in saline medium: experimental, quantum chemical, Monte Carlo and molecular dynamic simulation. Colloids Surf A Physicochem Eng Asp **711**, 136349 (2025). https://doi.org/10.1016/J.COLSURFA.2025.136349
47. Galai, M., Dahmani, K., Ebn-Touhami, M., Hsissou, R., Benhiba, F., Rbaa, M., Ouakki, M., Lachhab, R., Alharbi, S.M.: Corrosion inhibition of brass electrode in 3% NaCl medium using novel quinoline derivatives based on D-glucose: synthesis, characterization, and mechanistic insights through experimental and computational approaches. Mater. Sci. Eng. B **298**, 116843 (2023). https://doi.org/10.1016/J.MSEB.2023.116843

48. Shen, P., Wen, J., Dong, B., Wang, M.: Anticorrosion mechanism of ethylene-chlorotrifluoroeethylene coatings reinforced with hydroxylated carbon nanotubes: an experimental and molecular dynamics simulation study. Prog. Org. Coat. **186**, 107991 (2024). https://doi.org/10.1016/J.PORGCOAT.2023.107991
49. El Magri, A., Hsissou, R., Hmada, A., Berisha, A., Dkhireche, N., Vaudreuil, S.: Development of new polymer composites formulated by glass as a potential protective coating for 3D printed H13 steel in acidic medium: DFT, MC and MD computational. J. Mol. Liq. **387**, 122690 (2023). https://doi.org/10.1016/J.MOLLIQ.2023.122690
50. El-Haitout, B., Sardjono, R.E., Es-Sounni, B., Chafiq, M., Salghi, R., Bakhouch, M., Al-Moubaraki, A.H., Al-Ahmari, J.M., Al-Ghamdi, A.A., Fahim, M., Hammouti, B., Chaouiki, A., Ko, Y.G.: Electrochemical and quantum chemical investigation on the adsorption behavior of a schiff base and its metal complex for corrosion protection of mild steel in 15 wt% HCl solution. Heliyon **10**, 662 (2024). https://doi.org/10.1016/J.HELIYON.2024.E40662
51. Bedir, A.G., Abd-El-Raouf, M., Abdel-Mawgoud, S., Negm, N.A., El-Basiony, N.M.: Corrosion inhibition of carbon steel in hydrochloric acid solution using ethoxylated nonionic surfactants based on schiff base: electrochemical and computational investigations. ACS Omega **6**, 4300–4312 (2021). https://doi.org/10.1021/ACSOMEGA.0C05476
52. Jafari, H., Ameri, E., Rezaeivala, M., Berisha, A.: Experimental and theoretical studies on protecting steel against 0.5 M H_2SO_4 corrosion by new schiff base. J. Indian Chem. Soc. **99**, 100665 (2022). https://doi.org/10.1016/J.JICS.2022.100665
53. Dagdag, O., Hsissou, R., El-Harfi, A., Berisha, A., et al.: Fabrication of polymer based epoxy resin as effective anti-corrosive coating for steel: computational modeling reinforced experimental studies. Surf. Interf. **18**, 100454 (2020)
54. Dagdag, O., Hsissou, R., El-Harfi, A., Safi, Z., Berisha, A., Verma, C., Ebenso, E.E., Quraishi, M., Wazzan, N., Jodeh, S., El-Gouri, M., Quraishi, M.A.: Epoxy resins and their zinc composites as novel anti-corrosive materials for copper in 3% sodium chloride solution: experimental and computational studies. J. Mol. Liq. **16**, 113757 (2020). https://doi.org/10.1016/j.molliq.2020.113757
55. Yang, X., Dissertation, A.: Structure-property relationships of corrosion and mechanical properties in polymer-coated additively manufactured stainless steel and titanium alloys (2022). https://doi.org/10.25394/PGS.21657173.V1
56. Suresh, C.H., Remya, G.S., Anjalikrishna, P.K.: Molecular electrostatic potential analysis: a powerful tool to interpret and predict chemical reactivity. Wiley Interdiscip. Rev. Comput. Mol. Sci. **12**, e1601 (2022). https://doi.org/10.1002/WCMS.1601
57. Kaya, S., Kaya, C., Guo, L., Kandemirli, F., et al.: Quantum chemical and molecular dynamics simulation studies on inhibition performances of some thiazole and thiadiazole derivatives against corrosion of iron. J. Mol. Liq. **77**, 1254 (2016)
58. Damej, M., Hsissou, R., Berisha, A., Azgaou, K., Sadiku, M., Benmessaoud, M., Labjar, N., El-Hajjaji, S.: New epoxy resin as a corrosion inhibitor for the protection of carbon steel C38 in 1M HCl experimental and theoretical studies (DFT, MC, and MD). J. Mol. Struct. **1254**, 132425 (2022). https://doi.org/10.1016/J.MOLSTRUC.2022.132425
59. Dagdag, O., Safi, Z., Hamed, O., Jodeh, S., Wazzan, N., Haldhar, R., Safi, S.K., Berisha, A., El-Gouri, M.: Comparative study of some epoxy polymers based on bisphenolic and aromatic diamines: synthesis, viscosity, thermal properties computational and statistical approaches. J. Polym. Res. **28**, 1–16 (2021). https://doi.org/10.1007/S10965-021-02530-0
60. Dagdag, O., Guo, L., Safi, Z., Verma, C., Ebenso, E.E., Wazzan, N., Masroor, S., Haldhar, R., Jodeh, S., El-Gouri, M.: Epoxy resin and TiO_2 composite as anticorrosive material for carbon steel in 3% NaCl medium: experimental and computational studies. J. Mol. Liq. **317**, 114249 (2020). https://doi.org/10.1016/J.MOLLIQ.2020.114249
61. Tabbi, H., Abbaz, T., Bendjeddou, A., Hanane-Tabbi, C., Villemin, D.: Theoretical investigations on the molecular structure, HOMO–LUMO, Fukui function, NBO analysis and NLO of amino methyl tetrathiafulvalenes compounds by DFT method. Int. J. Chem. Stud. **5**, 1368 (2017)

62. Chebli, A., Djafri, A., Boukabcha, N., Megrouss, Y., Drissi, M., Belhachemi, M.H.M., Yahiaoui, S., Guerroudj, A.R., Chouaih, A., Djafri, A.: Synthesis, crystal structure, DFT calculations, NBO, Fukui function, NCI-RDG, Hirshfeld surface analysis, NLO properties and molecular docking analysis on (E)-N'-(3-methoxybenzylidene)-2-(quinolin-8-yloxy) acetohydrazide. J. Mol. Struct. **1310**, 138287 (2024). https://doi.org/10.1016/J.MOLSTRUC.2024.138287
63. Demirel, S., Fen, Ü., Dergisi, B.E., Gören, K., Yildiko, Ü., Kelimeler, A., Türevi, S.: Aldose reductase evaluation against diabetic complications using ADME and molecular docking studies and DFT calculations of spiroindoline derivative molecule. Suleyman Demirel Univ. J. Nat. Appl. Sci. **28**, 281–292 (2024)
64. Alavanthar, A., Bhaumik, S., Paleu, V.: Investigating the failure mechanisms during the epoxy-coated bearing steel–steel and uncoated bearing steel–steel interactions (crack formation, surface tears, schallamach-like waves) and their corrosion resistance in NaCl environment. J. Bio-Tribo-Corr. **10**, 1–21 (2024). https://doi.org/10.1007/S40735-024-00887-5
65. Rahali, E., Alves, M.M., El-Bassi, L., Bousselmi, L., Montemor, M.F., Akrout, H.: Enhanced protection of hybrid polyetherimide-ZnO or CuO bilayer composite coatings against mild steel corrosion in chloride media. Prog. Org. Coat. **163**, 106602 (2022). https://doi.org/10.1016/J.PORGCOAT.2021.106602
66. Zhang, X.F., Wang, N., Li, X.D., Li, X., Wang, C.X.: Molecular dynamics study of the corrosion protection improvement of superhydrophobic dodecyltrimethoxysilane film on mild steel. J. Mol. Graph. Model. **126**, 108626 (2024). https://doi.org/10.1016/J.JMGM.2023.108626
67. Dagdag, O., Haldhar, R., Ebenso, E.E., Verma, C., El-Harfi, A., El-Gouri, M.: Epoxy resins and their nanocomposites as anticorrosive materials. Org. Corros. Inhib. Synth. Charact. Mech. Appl. **16**, 451–482 (2021). https://doi.org/10.1002/9781119794516.CH20
68. Anwar, S., Li, X.: A review of high-quality epoxy resins for corrosion-resistant applications. J. Coat. Technol. Res. **212**, 461–480 (2024). https://doi.org/10.1007/S11998-023-00865-5
69. Ou, B., Wang, Y., Lu, Y.: A review on fundamentals and strategy of epoxy-resin-based anticorrosive coating materials. Polym. Technol. Mater. **60**, 601–625 (2021). https://doi.org/10.1080/25740881.2020.1819317
70. El-Aouni, N., Berradi, M., Eddaoukhi, A., Rissouli, L., Berradi, O., El-Yacoubi, A., Rafik, M., El-Bachiri, A.: Synthesis of new epoxy polymers and their comparative viscosimetric, rheological, and thermal characterization. Interaction **246**, 1–19 (2025). https://doi.org/10.1007/S10751-024-02238-9
71. Abouchane, M., Hsissou, R., Chraka, A., Molhi, A., Damej, M., Tassaoui, K., Berisha, A., Seydou, M., Elemine, B.O., Benmessaoud, M.: Synthesis and characterization of new macromolecular epoxy resin as an effective corrosion inhibitor for C38 steel in 1 M HCl medium: electrochemical insights, surface morphological and computational approaches. J. Bio-Tribo-Corros. **10**, 1–21 (2024). https://doi.org/10.1007/S40735-024-00824-6
72. Hsissou, R., Abbout, S., Seghiri, R., Rehioui, M., Berisha, A., Erramli, H., Assouag, M., Elharfi, A.: Evaluation of corrosion inhibition performance of phosphorus polymer for carbon steel in [1 M] HCl: computational studies (DFT, MC and MD simulations). J. Mater. Res. Technol. **9**(3), 2691–2703 (2020). https://doi.org/10.1016/j.jmrt.2020.01.002
73. Dagdag, O., El-Harfi, A., El-Gouri, M., Safi, Z., Jalgham, R.T.T., Wazzan, N., Verma, C., Ebenso, E.E., Pramod-Kumar, U.: Anticorrosive properties of Hexa (3-methoxy propan-1,2-diol) cyclotri-phosphazene compound for carbon steel in 3% NaCl medium: gravimetric, electrochemical, DFT and Monte Carlo simulation studies. Heliyon **5**, 1340 (2019). https://doi.org/10.1016/J.HELIYON.2019.E01340
74. Hsissou, R., Abbout, S., Seghiri, R., et al.: HCl: computational studies (DFT, MC and MD simulations). J. Mater. Res. Technol. **12**, 19581 (2020)
75. Hsissou, R., Benzidia, B., Rehioui, M., Berradi, M., Berisha, A., Assouag, M., Hajjaji, N., Elharfi, A.: Anticorrosive property of hexafunctional epoxy polymer HGTMDAE for E24 carbon steel corrosion in 1.0 M HCl: gravimetric, electrochemical, surface morphology and molecular dynamic simulations. Polym. Bull. **77**, 3577–3601 (2020). https://doi.org/10.1007/S00289-019-02934-5

76. Aygün, B., Şakar, E., Singh, V.P., Sayyed, M.I., Korkut, T., Karabulut, A.: Experimental and Monte Carlo simulation study on potential new composite materials to moderate neutron-gamma radiation. Prog. Nucl. Energy **130**, 103538 (2020). https://doi.org/10.1016/J.PNUCENE.2020.103538
77. Yousefi-Nasab, S., Safdari, J., Karimi-Sabet, J., Norouzi, A., Amini, E.: Determination of momentum accommodation coefficients and velocity distribution function for noble gas-polymeric surface interactions using molecular dynamics simulation. Appl. Surf. Sci. **493**, 766–778 (2019). https://doi.org/10.1016/J.APSUSC.2019.07.033
78. Mashuga, M.E., Olasunkanmi, L.O., Lgaz, H., Sherif, E.S.M., Ebenso, E.E.: Aminomethylpyridazine isomers as corrosion inhibitors for mild steel in 1 M HCl: electrochemical, DFT and Monte Carlo simulation studies. J. Mol. Liq. **344**, 117882 (2021). https://doi.org/10.1016/J.MOLLIQ.2021.117882
79. Hikita, M., Yamada, K., Nakamura, A., Mizutani, T., Oohasi, A., Leda, M.: Measurements of partial discharges by computer and analysis of partial discharge distribution by the Monte Carlo method. IEEE Trans. Electr. Insul. **25**, 453–468 (1990). https://doi.org/10.1109/14.55716
80. Oya, Y., Nakazawa, M., Shirasu, K., Hino, Y., Inuyama, K., Kikugawa, G., Li, J., Kuwahara, R., Kishimoto, N., Waizumi, H., Nishikawa, M., Waas, A., Odagiri, N., Koyanagi, A., Salviato, M., Okabe, T.: Molecular dynamics simulation of cross-linking processes and material properties for epoxy resins using first-principle calculation combined with global reaction route mapping algorithms. Chem. Phys. Lett. **762**, 138104 (2021). https://doi.org/10.1016/J.CPLETT.2020.138104
81. Talapatra, A., Datta, D.: Molecular dynamics-based simulations to study structures and properties of graphene/polyurethane nanocomposites. Funct. Nanocomp. Appl. **34**, 75–135 (2024). https://doi.org/10.1201/9781003412748-4
82. Kasprzhitskii, A., et al.: Corrosion inhibition properties of small peptides: DFT and Monte Carlo simulation studies. J. Mol. Liq. **13**, 5079 (2021)
83. Ikeuba, A.I., Ita, B.I., Okonkwo, C.P., Ekuri, P.E., et al.: Review of computational methods used in the evaluation corrosion inhibition of metallic materials. Discov. Chem. Eng. **2024**, 28 (2024). https://doi.org/10.1007/s43938-024-00065-6
84. Mubarak, G., Verma, C., Barsoum, I., et al.: Internal corrosion in oil and gas wells during casings and tubing: challenges and opportunities of corrosion inhibitors. J. Taiwan Inst. Chem. Eng. **18**, 76 (2023)
85. Yang, F., Liu, Y., Nie, S., Zheng, H.: Molecular insights into the control mechanism of the solid-liquid interface interaction on shale oil occurrence: grand canonical Monte Carlo and molecular dynamics simulations investigation. Langmuir **4**, 3561 (2025). https://doi.org/10.1021/ACS.LANGMUIR.4C03561
86. Kurtulbaş, E., Ciğeroğlu, Z., Şahin, S., El-Messaoudi, N., Mehmeti, V.: Monte Carlo, molecular dynamic, and experimental studies of the removal of malachite green using g-C_3N_4/ZnO/Chitosan nanocomposite in the presence of a deep eutectic solvent. Int. J. Biol. Macromol. **274**, 133378 (2024). https://doi.org/10.1016/J.IJBIOMAC.2024.133378
87. Yao, Y., Tang, F., Yang, S.: Permeation behavior of salt solution through g-C3N5/epoxy coatings: insights from molecular dynamics simulation. Colloids Surf A Physicochem Eng Asp **690**, 133725 (2024). https://doi.org/10.1016/J.COLSURFA.2024.133725
88. Li, C., Zhou, Z., Lin, S., Meng, X., Liu, J., Zhou, C.: Studying how changes in the structure caused by hydrogen bonding affect the corrosion resistance of polybenzoxazine-based coatings using molecular dynamics simulations. Prog. Org. Coat. **192**, 108475 (2024). https://doi.org/10.1016/J.PORGCOAT.2024.108475
89. Pourhashem, S., Hadizadeh, M.H., Ji, X., Zhou, Z., Duan, J., Ci, X., Lu, H., Hou, B.: Recognizing the function of different silane coupling agents on MXene adsorption/barrier behavior in solvent-borne epoxy coatings: experimental studies, density functional theory, and molecular dynamics simulations. Prog. Org. Coat. **192**, 108453 (2024). https://doi.org/10.1016/J.PORGCOAT.2024.108453
90. Sun, J., Sun, W., Wang, L., Xu, K., Yang, Z., Ma, Y., Zhao, L., Ma, S., Xing, W., Liu, G.: Boosting the acid penetration barrier of epoxy-silica organic-inorganic hybrid coating via adjusting its

molecular structures: experimental and molecular dynamics simulation study. Colloids Surf A Physicochem Eng Asp **688**, 133634 (2024). https://doi.org/10.1016/J.COLSURFA.2024.133634

91. Ahmed, M.A., Amin, S., Mohamed, A.A.: Current and emerging trends of inorganic, organic and eco-friendly corrosion inhibitors. RSC Adv. **14**, 31877–31920 (2024). https://doi.org/10.1039/D4RA05662K
92. Agrawal, N., Munjal, S., Ansari, M., et al.: Superhydrophobic palmitic acid modified ZnO nanoparticles. Ceram. Int. **31**, 16267 (2017)
93. Hussain, Z., Huang, Y., Riaz, I., Wang, X., Nie, W., et al.: Structural protective coating based on nano-clay reinforced epoxy: a comprehensive study integrated with molecular dynamics insights (2024). https://doi.org/10.2139/SSRN.5043016
94. Chen, P., Wang, G., Li, J., Zhang, J.: Anticorrosion mechanism of FACs-GO hybrids in ER coatings by EIS and MD simulation. Prog. Org. Coat. **186**, 107996 (2024). https://doi.org/10.1016/J.PORGCOAT.2023.107996

Textile Surface Modification and Nano-finishing: A Concise Review of Techniques and Applications

Nikita Rawat, Mumal Singh, Mona Vajpayee, and Lalita Ledwani

Abstract This chapter provides a concise overview of the surface functionalization and nano-finishing processes applied to textiles. The surface properties of a material are modified to improve functionality, without affecting its bulk characteristics, a process known as surface modification. It can be achieved by different approaches such as chemical, physical and biological methods. Chemical methods proposed a wide variety of reagents for fabric surface modification; however, they require rigorous control and involve the use of many hazardous chemicals. In contrast, biological methods are environmentally friendly and do not cause hazardous waste, but they require significantly more time and patience to yield optimal results. Thus, among many surface modification procedures, the physical method is the most exact and convenient because it does not require any dangerous chemicals and takes less time than the other ways. This chapter provides a brief overview of the classification, synthesis, and applications of nanoparticles and it gives a synergetic approach of surface modification and nano-finishing in textile industry. Growing awareness of health and hygiene has increased the demand of bioactive, antibacterial and UV-protected textiles. Nanotechnology has real marketable potential for the textile industry because conventional methods have no permanent effects and lose their functions after washing. On the other hand, treating textiles with nanomaterials is a method to improve their properties, which makes them more durable, anti-static, anti-odour etc.

Keywords Textiles · Surface modification · Nanoparticles · Synergetic approach · Antimicrobial · UV-protection

---

N. Rawat · M. Vajpayee
Department of Biosciences, Manipal University Jaipur, Jaipur, India

M. Singh · L. Ledwani (✉)
Department of Chemistry, Manipal University Jaipur, Jaipur, India
e-mail: lalita.ledwani@jaipur.manipal.edu

L. Ledwani and S. Srivastava (eds.), *Experimental and Computational Advances in Materials*, Engineering Materials, https://doi.org/10.1007/978-981-95-4180-5_12

1 Introduction

Textile fibres, the fundamental building blocks of all textile materials, are composed of natural or synthetic polymers—long chain-like macromolecules formed by repeating monomeric units. The structural arrangement and chemical composition of these polymers confer specific properties that make them suitable for diverse textile applications. Broadly, textile fibres are categorized into traditional textiles, which include garments and household fabrics, and technical textiles, which serve specialized functions in sectors such as automotive, healthcare, agriculture, construction, and personal protection [21]. The increasing complexity and functionality required in modern textile applications have driven significant advancements in surface modification techniques to enhance fibre performance. Surface properties such as wettability, dyeability, adhesion, and UV resistance play a pivotal role in determining the end-use quality of textile products [25]. To meet these demands, various surface modification approaches have been developed, including physical, chemical, mechanical, and biological methods. Among them, physical and enzymatic techniques are gaining attention due to their environmentally friendly nature, precision, and reduced fibre damage. These methods eliminate the need for harsh chemicals and complex waste treatment, aligning with the global shift toward sustainable manufacturing practices [24]. In recent years, nanotechnology has emerged as a transformative approach in textile surface engineering. The incorporation of nanoparticles owing to their high surface area-to-volume ratio and unique physicochemical properties allows for the development of textiles with enhanced and long-lasting functionalities, such as antibacterial effects, anti-static behavior, UV resistance, and odor control [24]. Unlike conventional treatments, which tend to diminish after repeated washing, nanotechnology-based finishes offer durable performance and are particularly valuable in high-demand sectors such as healthcare, sportswear, military uniforms, and protective garments [60].

2 Surface Modification

Surface modification is the method of changing material surface by different approaches to impart certain features without affecting its bulk properties which distinguish it from the untreated material [24]. Conventional methods of surface modification of textiles is a major consumer of electricity, fuel and water, resulting in emissions of greenhouse gases and contaminated wastewater causing other environmental challenges [52]. To overcome the above -mentioned challenges, there are different surface modification methods which are sustainable and environment-friendly. Surface modification methods are classified into different types, such as chemical, physical, and biological methods (Fig. 1) [54].

Types of Surface Modification Methods

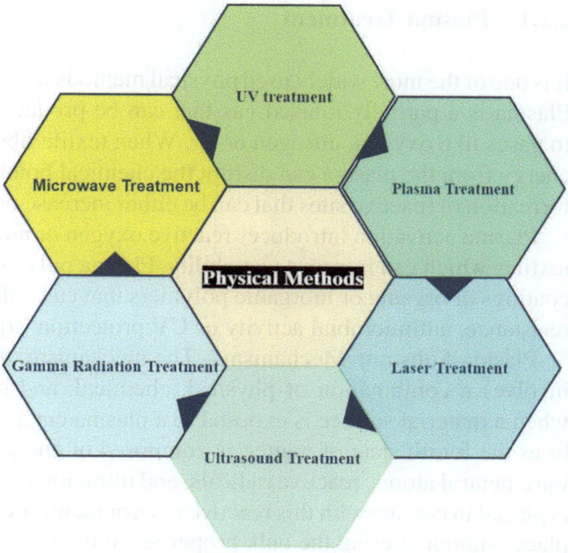

Fig. 1 Different physical surface modification method [50]

Surface modification is a valuable technique for enhancing functionality without altering the fundamental properties of fibres. By adjusting the surface energy of textile fibres, their hydrophilicity, wettability, and dyeability can be improved, or functional attributes such as water and oil repellency, better adhesion, soil release, and antistatic properties can be enhanced [50].

- Physical surface modification includes plasma, ozone gas therapy, microwave, ultrasound, gamma radiation, neutron and electron beam irradiations [26].
- Chemical surface modification involves changing fibre composition through surface grafting or enzymatic alteration using various chemicals, such as aminolysis, hydrolysis, acetylation, and isocyanate treatment [13].
- Biological surface modification includes enzymatic treatment, antimicrobial treatment, coating with plant-based compounds, fungal treatment, bio-functional coating and grafting [32].

2.1 Physical Methods

Physical treatments involve the use of reactive gases or radiation to modify textile surfaces. Molecular deposition from these gases can occur during plasma or flame spray processes. Furthermore, accelerated ions generated through ion beam techniques, radiofrequency plasma, or gamma irradiation can induce surface changes such as etching and oxidation. Cross-linking of surface molecules by free radical formation during UV treatment, corona discharge etc. [10].

Some physical methods are described as follows.

2.1.1 Plasma Treatment

It is one of the most widely used physical methods for surface modification of textiles. Plasma is a partially ionised gas that can be produced by applying a high voltage to gases like oxygen, nitrogen or air. When textile fibres are exposed to plasma, the energy from the plasma can disrupt the chemical bonds at the surface, leading to the formation of reactive sites that can be either increase or decrease surface energy [55].

Plasma activation introduces reactive oxygen or nitrogen species to the surface of textiles which can improve wettability. Plasma polymerization creates thin, uniform coatings of organic or inorganic polymers that can enhance properties such as water resistance, antimicrobial activity or UV protection [9].

Plasma-Substrate Mechanism—The mechanism of plasma–substrate interaction involves a combination of physical, chemical, and photonic processes that occur when a material surface is exposed to a plasma environment. Plasma, often referred to as the fourth state of matter, is composed of energetic species such as electrons, ions, neutral atoms, reactive radicals, and ultraviolet (UV) photons. When a substrate is placed in contact with this reactive environment, several surface-level changes take place without altering the bulk properties of the material [35].

Physically, energetic ions and electrons bombard the surface, leading to sputtering or etching, which removes weakly bound contaminants and alters the surface topography. This bombardment can also cause surface roughening, which increases the surface area and enhances the adhesion of coatings or other materials. Chemically, highly reactive species such as oxygen or nitrogen radicals interact with the substrate, breaking existing bonds and introducing new functional groups like hydroxyl (–OH), carboxyl (–COOH), or amine ($-NH_2$) groups. These modifications improve surface energy, wettability, and chemical reactivity [44].

Simultaneously, UV photons generated within the plasma can break molecular bonds in a process called photolysis, creating reactive sites on the surface. In polymeric materials, plasma exposure can induce either chain scission, where long molecular chains are broken into smaller segments, or crosslinking, where new bonds form between chains, altering mechanical and chemical properties. Overall, plasma–substrate interaction results in surface cleaning, activation, functionalization, and modification of morphology, all of which are critical for improving the performance of materials in various industrial and scientific applications. The interaction of gas plasma with the substrate typically results in different processes (Fig. 2) [40] which are as follows:

- Plasma Generation-A working gas (e.g., O_2, N_2, Ar, CF_4) is ionized under the influence of an electrical discharge, typically in low- or atmospheric-pressure systems. This ionization process yields a chemically reactive plasma phase enriched with ions, electrons, radicals, metastable species, and UV photons, which collectively constitute the active medium for surface modification [17]. Radical Formation and Surface Activation-The generation of free radicals at the outermost layers results in surface activation, providing chemically active sites that can readily participate in further reactions. These sites are transient in

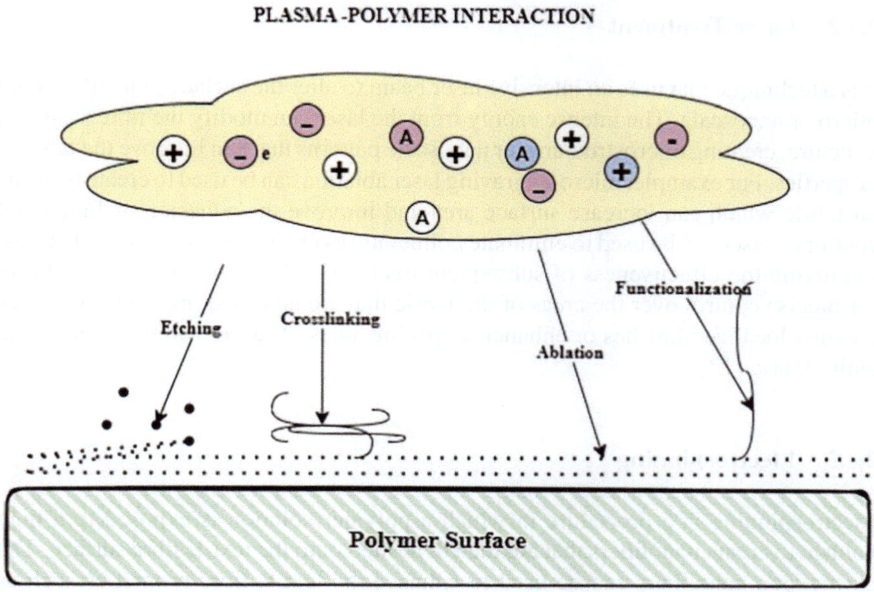

Fig. 2 Depicts the plasma-polymer interaction [40]

nature and must rapidly stabilize through interactions with surrounding plasma species [34]. Chemical Functionalization (Grafting of New Functional Groups)-Plasma-derived radicals and ions interact with the activated polymer surface, introducing new functionalities whose nature depends on the plasma gas composition. Oxygen plasma promotes the formation of hydroxyl (–OH), carbonyl (–C=O), and carboxyl (–COOH) groups, enhancing surface hydrophilicity. Nitrogen plasma leads to the incorporation of amine ($-NH_2$), cyano (–CN), and nitroso (–NO) groups, improving adhesion and bioactivity. Conversely, fluorocarbon plasma results in the grafting of –CF, $-CF_2$, and $-CF_3$ groups, lowering surface energy and imparting hydrophobicity [6]. Crosslinking and Stabilization-In parallel, plasma exposure can promote crosslinking of adjacent polymer chains at the interface. This structural reinforcement stabilizes newly introduced functionalities, preventing their desorption and improving the durability of surface modifications under subsequent processing or environmental exposure [9].

Plasma technology is considered a clean and sustainable alternative to conventional surface treatment methods, as it operates without the use of water or chemicals. It has been extensively utilised for textile surface modification owing to its surface-selective action, leaving the fibre's core properties unaffected [37]. Since most textile fibres are heat-sensitive, non-thermal plasma is especially advantageous in textile processing. Atmospheric pressure plasma can initiate surface reactions without excessive thermal degradation of the substrate [37].

2.1.2 Laser Treatment

It is a technique that uses an intensive laser beam to alter the surface of textiles at the micro or nanoscale. The intense energy from the laser can modify the fibre's surface structure, creating microstructures or nanoscale patterns that can improve the fabric's properties. For example, micro-engraving laser ablation can be used to create patterns on fabric which can increase surface area and improve the adhesion of functional coatings. Laser can be used to eliminate pollutants or oils from the exterior of textiles, improving the effectiveness of subsequent treatments. Laser treatment also allows for precise control over the areas of the fabric that are modified, making it ideal for creating localized finishes or enhancing specific areas of fabric without altering the entire fabric [12].

2.1.3 Electrospinning

Electrospinning is a procedure in which a polymer solution is exposed to a high voltage to create nanofibers that are then deposited onto the textile fibre surface. The electrospun nanofibers can be used to enhance the textile fibre properties such as filtration efficiency, water repellence or thermal insulation. This technique is particularly effective for creating functional coatings that can impart properties such as antimicrobial activity, UV protection or self-healing abilities. The fibres created by electrospinning can be very thin (often in the nano-meter range) allowing for a high surface area to volume ratio and providing textiles with enhanced functional properties.

2.1.4 UV Treatment

Surface modification by using UV (ultraviolet radiation) is well known as one of the physical surface modification techniques. Ultraviolet (UV) rays, occupying the wavelength range of 100–400 nm within the electromagnetic spectrum, were first identified by Johann Ritter in 1801. These rays are classified into UV-A (320–400 nm), UV-B (280–320 nm), and UV-C (200–280 nm) based on their respective wavelength ranges. These rays are a form of electromagnetic radiation emitted by both natural sources, such as the sun, and artificial sources, such as UV lamps. According to previous research, UV radiation increases the surface's hydrophilicity and dye uptake by forming reactive groups such as carbonyl, hydroxyl, aldehyde and carboxyl. The fact that the modification doesn't significantly alter the fabric's strength during this application is another benefit. These benefits are further enhanced by the following advantages in terms of environmental sustainability, easily accessibility, dry process and decrease in the amount of chemicals used to carry out the necessary activities [45, 46].

2.1.5 Electron Beam Radiation

Gamma radiation and electron beam irradiation are among the most widely employed techniques in surface treatment processes. Electron beam technology, in particular, offers numerous advantages such as rapid processing, shallow penetration depth, and high energy efficiency. As it operates without the use of solvents, this method significantly minimises the release of volatile organic compounds during the drying phase. When compared to gamma radiation, electron beams exhibit superior performance in terms of energy efficiency, safety, maintenance requirements, processing speed, operational costs, and achievable dosage rates [31].

In surface modification via electron beam treatment, electron–electron interactions generate active sites on the substrate surface. The electron beam is produced at high voltages, typically ranging from 300 keV to 12 MeV. For surface coating treatments that require low penetration, electron beam accelerators operating at energies between 150 and 300 keV are commonly used. Upon exposure to accelerated electrons, polymers undergo a series of interactions classified as physical, physico-chemical, and chemical. In the physical phase, transient reactive species are formed. These are subsequently converted into polymer radicals during the physico-chemical stage, which then initiate various chemical reactions at the polymer surface [13].

2.1.6 Microwave Radiation

Microwave radiation is a clean, eco-friendly, and efficient heating method, that transforms microwave energy into heat within dielectric materials. Substances with high dielectric coefficients, such as water, salt, and ethanol, can generate heat through dipole oscillations when exposed to microwave radiation. In most microwave dyeing processes, this radiation enhances the dyeing rate and improves dyestuff absorption. Furthermore, microwave irradiation induces a mass transfer effect from the substrate to the solvent, resulting in cell wall rupture. The reduced dyeing time demonstrates that this technique is cost-effective and time-efficient, contributing to its eco-friendly nature. Consequently, this radiation method can modify various dyeing techniques for both natural and synthetic fabrics [23].

Depending on the desired functionality such as improved wettability, adhesion, printability, hydrophilicity, or hydrophobicity various functional groups can be introduced onto the fabric surface by using different gases during physical surface treatment methods [25].

2.2 *Chemical Methods*

In chemical surface modification, textile fibres are exposed to liquid reagents to introduce reactive functional groups onto their surfaces (Fig. 3). One commonly used chemical agent is chromic acid, which introduces oxygen-containing functionalities

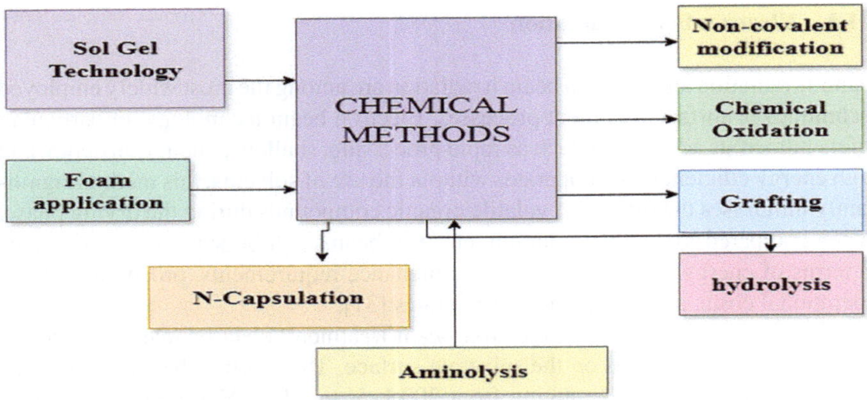

Fig. 3 Different chemical surface modification methods [4]

on synthetic fibres such as polyester and nylon. Another widely applied technique is grafting, where the process begins with the generation of free radicals on the fibre surface to initiate copolymerization with selected monomers. Additionally, sol–gel technology represents an effective chemical treatment method that enables the incorporation of functional coatings onto fibre surfaces.

2.2.1 Sol–Gel Technique

Many types of polymers including textile fibres can be coated with sol–gel techniques. This method is used to enhance many properties of polymers like water repellence, ultraviolet protection and electrical conductivity. Solvents used in this technique could be recycled or reused [1].

The sol–gel method involves several key terms. A sol is a colloidal suspension where solid particles are uniformly dispersed within a liquid. When these particles form a three-dimensional interconnected network, the system is called a gel, which exists as a structure between solid and liquid phases. The sol–gel process refers to the transformation of the sol into this two-phase gel system. Additionally, a nano sol is a nano-structured form of the sol–gel process, where the particles or molecules in the sol are at the nanoscale, enabling advanced material properties.

The sol–gel process typically involves three primary components: initiators such as metal salts or metal oxides, solvents—which may include water or organic solvents like methanol, ethanol, propanol, or butanol—and catalysts, which can be either acidic or basic in nature. Stable colloidal solutions are formed by dispersing metal oxides into aqueous or nonaqueous media, with colloid particle sizes generally below 100 nm. This highly adaptable technique is utilized in a wide range of applications, including the fabrication of composites, porous membranes and gels, fibres, glass–ceramics, and thin-film coatings [11].

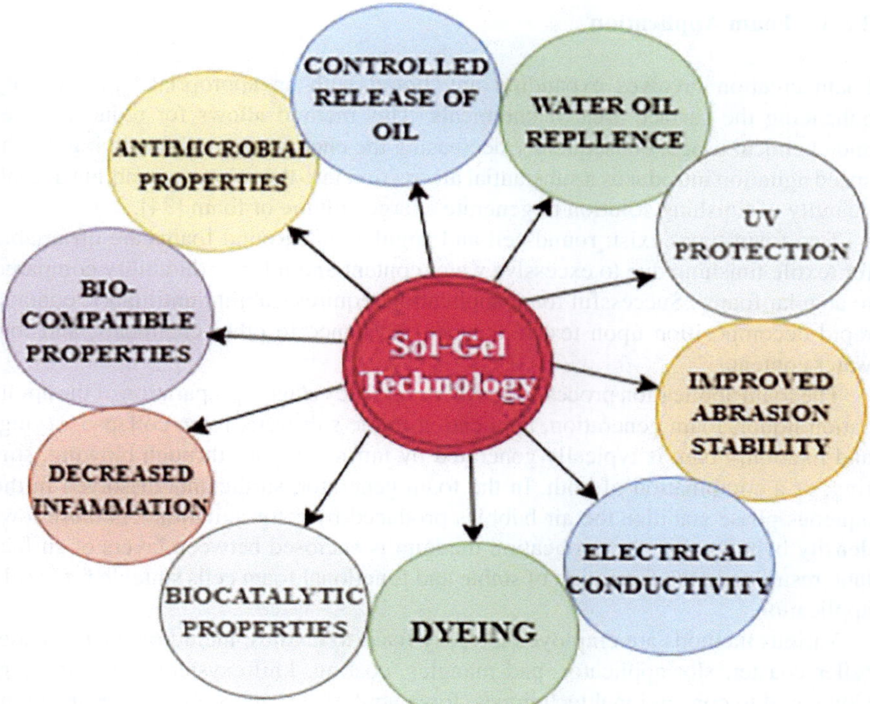

Fig. 4 Different application of sol–gel technology [11]

Some advantages of sol–gel methods are easy production processes, high efficiency, and production in demanded shapes such as aerosols and fibres. Production in low-temperature conditions, low particle size distribution, and functional coating materials for textiles [11] (Fig. 4).

2.2.2 Hydrolysis

Textile fibres such as polylactic acid (PLA) and polyester can undergo enzymatic hydrolysis using hydrolases, cutinases, or lipases to increase the number of hydroxyl and carboxylic acid groups on the polymer surface. This surface modification has been evaluated through contact angle measurements, reactions involving radiolabelled butyric acid (^{14}C), and X-ray photoelectron spectroscopy (XPS). The enzymatic hydrolysis results in a higher surface density of anionic functional groups, which significantly promotes nucleation. Consequently, the increase in surface –COOH and –OH groups lead to enhanced polymer wettability [19].

2.2.3 Foam Application

Foam creation involves expanding any liquid with an appropriate gas, thereby enhancing the surface area of chemicals. This method allows for reduced water and chemical usage, consequently decreasing the energy required for drying. High-speed agitation introduces a substantial air fraction into the solution, enabling a small quantity of finishing solution to generate a large volume of foam [34].

Two foam types exist: round-cell and angular-cell. Round foams are unsuitable for textile finishing due to excessive water content and inferior durability compared to angular foams. Successful foam application requires stability until fabric contact, rapid decomposition upon textile contact, resistance to other chemicals, and low water content.

The foam application process involves several key stages: preparation of the application liquor, foam generation, application to the substrate, foam collapse, drying, and fixation. Foam is typically generated by introducing air through blowing, stirring, or a combination of both. In the foam generator, surfactants dissolved in the aqueous phase stabilize the air bubbles produced by rotor agitation. As these low-density bubbles rise, the application medium is enclosed between layers of surfactant, resulting in the formation of stable and functional foam cells suitable for textile application.

Various methods are employed to apply foam to textiles, including doctor blade, roller coaster, slot applicator, pad mangles, coating, knife systems, and printing. Compared to conventional techniques, foam application offers advantages in chemical cost, working time, and flexibility. It allows for significantly reduced water consumption by increasing foam volume, eliminates the need for defoaming chemicals, and enhances working speed and energy efficiency through rapid wetting and high penetration [41].

The foam application method provides lower water, chemical, and energy consumption, as well as reduced working time compared to classical approaches. It enables coating on both fabric sides and offers several benefits to the textile production process, including: applicability to diverse surfaces, independent of surface composition and structure, homogeneous coating provision, increased efficiency through controlled chemical application, reduced thickener usage, uniform and repeatable foam production, low wet pick-up rate, improved first-quality production, capability for application on both fabric sides.

2.2.4 N-Capsulation

The N-capsulation technique enables the creation of capsules with multiple functional properties, including antimicrobial, insect-repelling, and anti-inflammatory effects. The core, which contains the functional elements, is enveloped by a shell material. The selection of wall material is influenced by factors such as the core material's reactive groups and the textile surface. Capsules are typically categorized by size: nanocapsules ($< 1\ \mu m$), microcapsules ($1–1000\ \mu m$), and macrocapsules ($> 1000\ \mu m$).

To prevent coating disruption during the textile finishing process, capsule sizes below 40 μm are generally advised.

Microcapsules can be fabricated in multiple configurations, including different shapes, core numbers, and shell or matrix structures. In matrix-type capsules, the active material is uniformly embedded within the wall. Diverse methods are employed for capsule synthesis, integration into textiles, shell degradation, and controlled release of encapsulated agents [58].

Microencapsulation is a versatile technique with broad applications across various industries, largely determined by the properties of the encapsulated core materials. In textiles, it is employed to confer functional attributes such as controlled fragrance release, malodour absorption, antimicrobial and antioxidant effects, as well as the delivery of vitamins to support wound healing [7].

2.2.5 Grafting

Grafting techniques can enhance polymers with various desirable characteristics. When applied to textile surfaces, this method can produce fabrics with specialized properties such as antimicrobial, hydrophobic, or fire-resistant qualities. Grafting, when employed as a preliminary surface modification technique, can significantly improve the efficiency of subsequent textile processing steps while minimizing water, energy, and chemical usage. The strategy involves forming physical, chemical, or ionic bonds between functional groups and the polymeric chains of the textile substrate. This is typically initiated by generating free radical sites on the polymer backbone. Four principal methods are commonly used to achieve grafting on textile surfaces: chemical grafting, radiation-induced grafting, plasma-assisted grafting, and UV-induced grafting. The free radical sites generated by these methods act as initiators for copolymerization [5].

The type of monomer used determines the specific properties that can be imparted, including Water repellency: for outerwear, outdoor furniture, tents, and awnings; Water, oil, and stain resistance: for interior and automotive textiles, Antibacterial properties: for hospital textiles, public transport fabrics, and hotel linens, Antistatic qualities: for gas station uniforms, Flame retardant features: for firefighter suits [53].

2.2.6 Aminolysis Oxidation

Textile fibres like PLA are aminolyzed to introduce amino functionalities, allowing glycosaminoglycans to bind electrostatically to the positively charged ammonium-modified surface. These surface-bonded amines lead to a positively charged surface. IR ascertains this, modification in particular through the amides' bonds at 1655 and 1537 cm^{-1} [60].

2.2.7 Chemical Oxidation

Numerous synthetic polymers, such as polyethene and polypropylene, come in various forms based on density and branching. A concentrated chromic acid solution was initially used at 25–75 °C, followed by oxidation with 70% aqueous nitric acid at 50 °C. These severe treatment conditions resulted in surface etching and the generation of carbonyl derivatives, approximately 60% of which were carboxylic acids, with the remaining 40% comprising ketones and aldehydes. Ozone is utilised for the surface oxidation of polyethene and polypropylene, both with and without UV irradiation. The ozonation process produces surface peroxides that break down into radicals. This ozonation method was applied to enhance both the adhesion to bone and the biocompatibility of these polymers. Moreover, ozonation enhanced the antimicrobial properties of the grafted polymers [23].

2.2.8 Non-covalent Surface Modification

Non-woven fabric can be treated with polycationic chitosan via impregnation followed by squeezing, and subsequently coated with polyanionic sodium alginate. Ionic interactions occur between the $-NH_3^+$ groups of chitosan and the $-COO^-$ groups of the PET substrate [4].

2.3 *Biological Methods*

The escalating global population and economic growth in developing nations are intensify the strain on worldwide consumption and environmental resources. Implementing alternative technologies that require fewer natural resources to fulfil human needs is a crucial step towards sustainability. Industrial biotechnology, which involves the use of bio-based materials and natural processes in manufacturing, presents a cleaner production alternative to conventional methods. The textile and apparel industry are recognized as a major contributor to water pollution due to its high water and energy usage, its dependence on non-biodegradable artificial chemicals. With the rise of environmentally conscious societies, the need for chemical-free and eco-friendly processes has been recognized by textile and garment manufacturers. This chapter categorises the biological surface modification techniques employed in the textile industry as follows (Fig. 5) [57]:

- Enzymes
- Chitosan
- Plants-based compounds
- β-Cyclodextrin
- Sericin.

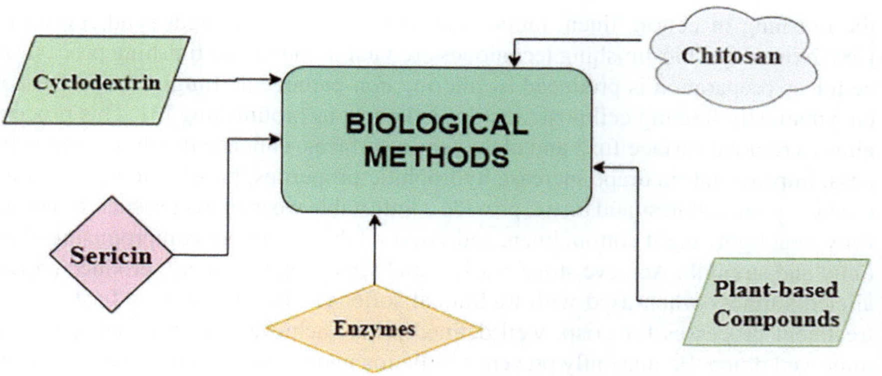

Fig. 5 Different biological methods of surface modifications [57]

2.3.1 Enzyme Treatment

Enzymes, commonly sourced from bacterial metabolic processes, are classified as biocatalysts and consist of or include protein structures. Catalysts facilitate reactions without being consumed in the process. Enzymes are named by adding "-ase" to the substrate or reaction they catalyse. The molecule acted upon is referred to as the substrate, and the enzyme's turnover number quantifies how many substrate molecules are converted per second. In cellular systems, enzymatic reactions often occur in succession, with the product of one enzyme serving as the substrate for the following enzyme. For example, the enzyme amylase transforms starch into two-chain maltose, while the enzyme maltase converts one-chain maltose into glucose [55].

Enzymes have been widely utilized for many years in various fields, including medicine, analysis, food chemistry, the beverage industry, and household detergents.

Textile surface modification through enzymatic processes involves treating fibres or adding functional groups to alter their surface properties, both physically and chemically. Various enzymes can be utilized in textile manufacturing, offering several advantages: enhancing appearance and reducing surface fuzziness through biotreatment, creating a fashionable worn look via bio-sanding, removing H_2O_2 residue before dyeing through bleaching, eliminating cotton contaminants (in de-sizing, washing, and bleaching processes [6].

Enzymes, which are sourced from numerous origins, play a crucial role in promoting sustainability within the textile industry. Amylases, used for starch decomposition in the textile industry, have been known since the 1910s. In recent years, interesting applications have been discovered for wool finishing. These methods employ enzymes to remove the flake layer on the surface, modifying the wool fabric surface. Additionally, this process imparts anti-felting properties and enhances shininess and texture. Producers of textile enzymes have framed cellulases to improve the characteristics of fabrics made from cellulose and have introduced new enzymes for the finishing of cotton, linen, ramie, and their combinations with synthetic fibres [38]. Nowadays, bio-finishing techniques are vital in the textile finishing process. A cellulose preparation is produced by altering non-pathogenic fungi. The process of enzymatically treating cellulosic textiles is known as biofinishing [3]. This process aims to remove surface fuzz and clean fabric surfaces, enhance flexibility and softness, improve fabric drape increase hydrophilic properties, boost colour attraction, efficiency, smoothness and lustre, provide comfortable wear characteristics. Successfully treat lightweight cotton, linen, and viscose fabrics without compromising elasticity and strength. Achieve stone-washed and vintage appearances, produce unparalleled softness when used with traditional softeners. Reduce surface fuzz in pretreatment processes for crisp, well-defined prints, achieve a natural, soft feel and improved drape. Permanently prevent fibrillation and pilling. enhance water absorption, especially in towels and bath textiles. Create innovative finishing effects, ensure compatibility with all ecological processes.

The effects of bio-finishing are long-lasting, even after multiple washes. In contrast, traditional methods leave chemicals in the fabric, degrading its appearance with each wash. This approach competes with many textile chemicals, offering a modern, biodegradable alternative that doesn't cause pollution. Enzymes help conserve water and energy, reduce waste, and minimize environmental impact in the industry [6].

2.3.2 Chitosan

Chitin, the world's second most predominant natural polymer, is typically extracted from fungal and crustacean exoskeletons. When exposed to potassium hydroxide boiling, it becomes soluble in organic acids. This process contributes to the Roadmap to Sustainable Textiles. The chain structure of chitosan, containing reactive amino and hydroxyl functional groups, enables efficient chemical bonding to inorganic strengthening networks and improves processability through various changes.

As a biodegradable, odourless, and non-toxic substance, chitosan serves as a versatile green material that promotes environmental sustainability. Its unique properties make it one of the most auspicious bio-based polymers, with applications across biotechnology, pharmaceuticals, food, medicine, textiles, environmental science, paper production, and agriculture [28].

The solubility of chitosan enhances its applicability and increases the reactivity of its functional groups. Nevertheless, its performance is constrained by several factors. Physical characteristics such as particle size, porosity, and crystallinity significantly influence its adsorption efficiency. Common limitations include low adsorption capacity, limited surface area, pronounced hydrophilicity, high crystallinity, reduced porosity, and resistance to mass transfer. Despite these challenges, the polycationic nature of chitosan allows it to form strong ionic interactions with textile substrates, supporting its extensive use in dyeing and finishing applications.

Chitosan's nanofiber structure makes it especially valuable in medical textiles, particularly for wound dressings. Polysaccharide fibres coated with chitosan film or chemically modified chitosan can significantly enhance textile properties, imparting features such as antistatic behavior, stain resistance, and antibacterial activity [14].

2.3.3 Plant-Based Compounds

A variety of active compounds, including diketones, flavonoids, and chalcones, are found in plants and are associated with antibacterial, antifungal, and antioxidant effects. In recent years, the textile industry has faced growing environmental scrutiny, leading to the imposition of strict environmental and economic regulations—and, in some cases, complete bans—on certain chemical substances. As a result, there is increasing interest in using environmentally friendly, biocompatible, low-toxicity, and sustainable plant-based compounds in textile finishing process.

Natural and plant-derived materials are being increasingly investigated for textile finishing processes due to their sustainability, affordability, and environmental compatibility. Their application in producing eco-friendly textiles is regarded as a key advancement in the industry, particularly with the growing demand for antibacterial fabrics in response to rising infectious disease rates worldwide. Assuming their sustainable and environmentally friendly nature, plant-based compounds are widely used to provide cleaning, aromatising, insect-repelling, antioxidant, flame-retardant, UV protective, antifungal, and antimicrobial characteristics to textile products [27].

The use of plant-based compounds in the textile industry is both eco-friendly and sustainable. Since harmful chemicals are not involved, there is minimal or no need for wastewater treatment, which in turn reduces overall production costs [47].

2.3.4 β-Cyclodextrin

In the textile sector, there is a growing change towards the production of functional textiles made from eco-friendly materials and methods. β-Cyclodextrin (β-CD) is non-toxic, biodegradable, and environmentally safe. Although various auxiliaries are used alongside fundamental chemicals in textile finishing, cyclodextrins (CD) are among one of them. The present requirements of the industry for textile include reduced fabric usage, biodegradability, minimal waste production, and multifunctional properties. Cyclodextrin molecules, being biopolymers are compatible with sustainable, and environmentally friendly biological systems, and have been thoroughly studied [49].

The most widely studied cyclodextrins include α-cyclodextrin (α-CD), β-cyclodextrin (β-CD), and γ-cyclodextrin (γ-CD), which comprise 6, 7, and 8 glucose units, respectively. These cyclic oligosaccharides possess a hydrophilic outer surface and a hydrophobic inner cavity, and are enzymatically synthesized. Cyclodextrins are water-soluble, although they exhibit poor solubility in organic solvents, with β-CD showing the lowest solubility in water among the three. Owing to its affordability, processability, non-toxic nature, and absence of skin irritation or mutagenic effects, β-cyclodextrin and its derivatives are extensively utilized. Additionally, the hydroxyl groups on β-CD allow it to bind with metal oxides and adsorb contaminants from aqueous systems.

In the textile industry, cyclodextrins are primarily employed in printing, dyeing, and finishing processes. They offer multiple functional advantages, including UV protection, antifungal and antibacterial effects, anti-odour properties, insect repellency, and improved dyeing performance. When used in dyeing, β-cyclodextrins enhance dye uptake, reduce dye loss in wastewater, improve colour uniformity, and prevent fading during washing. As a biocompatible biopolymer, cyclodextrin has strong potential for use in finishing treatments aimed at developing high-value, functional textile products [12].

2.3.5 Sericin

Silk, a naturally occurring polymer produced by silkworms and spiders, is predominantly derived from the *Bombyx mori* silkworm and select spider species, including *Araneus diadematus* and *Nephila clavipes.* Its favourable characteristics, including ease of processing, compatibility with biological systems, ability to decompose naturally, and excellent mechanical properties in both its natural and restored states, make it an ideal choice for crafting functional products. Silk comprises two primary proteins: fibroin, a fibrous protein, and sericin, a soft, globular protein. Fibroin, which accounts for 70–80% of silk, is crystalline and water-insoluble, whereas sericin, representing 20–30%, is amorphous and dissolves in water. Sericin functions as an adhesive, holding the fibroin filaments together [29].

Although silk fibroin is traditionally valued for its role in textiles, it has garnered increasing interest as a multifunctional biomaterial in areas such as tissue engineering, drug delivery, electronics, and environmental remediation. In contrast, sericin—once overlooked in conventional applications—has recently attracted attention due to its biocompatibility, immunological tolerance, biodegradability, and its range of advantageous properties, including anti-inflammatory, antibacterial, antioxidant, and photoprotective activities. In the textile industry, sericin is used to enhance moisture absorption, antibacterial activity, nanofiber production, self-cleaning capabilities, and UV protection. Furthermore, sericin contributes to improving key textile properties such as tear resistance, anti-wrinkle performance, dyeability, fibre strength, tensile strength, and fabric handle [36].

Prior to processing, it is necessary to separate silk fibroin and sericin due to their marked differences in abundance, solubility, amino acid profiles, and reactive functional groups. In textile manufacturing, this separation is carried out through degumming—a crucial step to obtain fibroin filaments with desirable properties such as high tensile strength, pronounced lustre, and enhanced capillarity. Conventional degumming techniques, typically involving soap, alkali, or both, are associated with environmental concerns, high water and energy demands, and potential fibre degradation. In response, more sustainable alternatives have been introduced, including enzymatic degumming, supercritical CO_2 (CSCF) treatment, acid hydrolysis, steam treatment, and ultrasonic methods. Simultaneously, there is a growing interest in valorizing sericin—once considered waste—for its bioactive potential in pharmaceutical and cosmetic applications. By repurposing waste sericin as part of environmentally conscious and sustainable production efforts, textile manufacturers can reduce wastewater treatment costs and generate additional revenue through value-added products (Table 1).

Table 1 Overview of different surface modification methods

Method	Type	Effect	Advantage	Disadvantage
Physical method	Plasma treatment, microwave, electron beam, UV treatment etc [2]	Efficient fibre processing entails the elimination of non-fibre constituents, enhancement of fibre–matrix interfacial adhesion, and surface cleaning to promote compatibility with composite materials	There is no requirement for dangerous chemicals or solvents, reduced environmental impact, quick treatment duration, low operational expenses, and increased versatility [15, 22, 30, 33, 51]	Low-pressure plasma needs a carefully designed plasma reactor system accompanied by costly vacuum systems, and only batch processing can be performed [56]
Chemical method	Chemical oxidation, hydrolysis, aminolysis, sol–gel technology, isocyanate treatment, polymer grafting, etc [23]	The fibre surface was cleaned, and the surface was chemically altered to prevent moisture absorption. This resulted in enhanced interfacial adhesion, increased fibre strength, and a rougher surface texture	The process is up-front, practical, and suitable for various applications employed on an industrial scale, typically employing the most frequently used methods	Utilization of dangerous substances, disposal of solvent waste, appropriate management, and higher expenses for the end product [12]
Biological method	Enzymes like hemicellulose, laccase, pectinase, cellulase, plant-based compounds, β-Cyclodextrin etc [17]	Fibre separation, surface cleaning, and improved interfacial bonding	Environment-friendly high-quality fibres, low fermentation	Expensive and limited to experimental scale [18]

3 Nanotechnology

In the current age of nanotechnology, nanoparticles have demonstrated extensive utility across various domains, including optical polarizability, catalysis, electronics, and medicine. Metals such as copper, silver, and gold possess distinctive characteristics that enable their application in audiophile technology, photonics, photography, and biological labelling, as well as in scientific research. Research has documented the use of various biological sources, including plant extracts, algae, diatoms, fungi, and bacteria, for nanoparticle synthesis. However, the biogenic production of these nanoparticles using plant biomass has proven more feasible due to its straightforward methodology and easily scalable processes. This expands the potential for studying and controlling nanoparticles at the cellular level, with applications in drug delivery, hepatitis B treatment, cancer detection and therapy, environmental pollution management, ceramic coatings, HIV/AIDS treatment, and the textile industry, among others [59].

3.1 Green Nanotechnology

Environmentally friendly and advanced, green technology aims to preserve natural resources for sustainable development. Its main objectives include minimizing the use of non-renewable resources, reducing energy consumption, and improving the quality of human life without exhausting the Earth's natural resources. These goals are achieved through strategies such as environmental sustainability, source reduction, energy conservation, and the adoption of green chemistry principles [16].

Green nanotechnology, which is a branch of green technology that includes green energy systems and green information technology, follows the principles of green chemistry and engineering. Its primary goal is to develop safer and more sustainable nanoparticles [48]. To promote the sustainable development of these nanoparticles, biogenic processes are now preferred because of their eco-friendliness, cost-efficiency, and low energy demands. These methods avoid damaging environmental effects and produce safer products and by-products [43].

3.2 Synthesis of Nanoparticles

To effectively utilise nanoparticles in practical applications, several requirements must be fulfilled. These include maintaining a consistent and controlled particle size by managing their high surface energy, preventing particle agglomeration, and ensuring uniformity in morphology, chemical composition, microstructure, and crystallinity. Nanostructures are typically created using two main approaches. The bottom-up approach builds nanoparticles by assembling atoms and molecules, allowing for precise control over shape and size. Common bottom-up techniques include sol–gel processing, chemical vapor deposition (CVD), and laser pyrolysis. Conversely, the top-down approach works from larger to smaller objects and is widely used in photolithography, electron beam treatment, and plasma etching (Fig. 4) [42].

Nanoparticle synthesis methods are categorised as Physical, Chemical, and Biological and operate based on one of the two above-mentioned approaches in Fig. 4. Physical methods employ high-energy radiation, heat, electrical energy, or mechanical force to induce processes such as etching, melting, or evaporation of materials. These methods pose minimal contamination risk due to the absence of solvents. In contrast to chemical synthesis, which frequently employs toxic reagents and produces non-biodegradable, hazardous by-products, biological synthesis methods adhere to green chemistry principles, offering an eco-friendly, safe, and energy- and cost-efficient alternative [8]. Microorganisms such as bacteria, fungi, and algae are capable of withstanding harsh conditions, including high solvent bath temperatures and elevated metal ion concentrations. Additionally, plant-mediated synthesis integrates the therapeutic properties of plants into the synthesized nanostructures, enhancing the antibacterial functionality of materials such as textiles, polymers, and medical devices [12, 52] (Fig. 6).

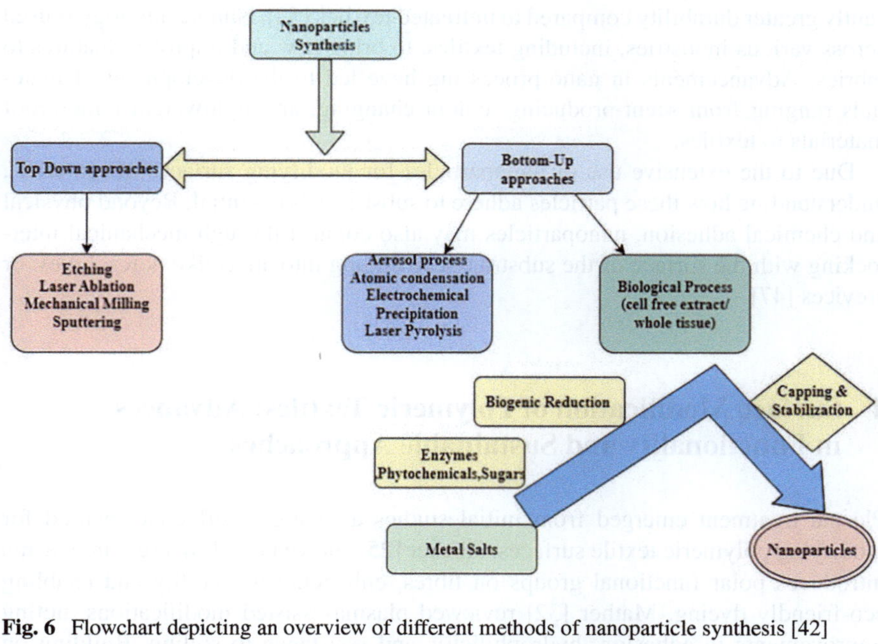

Fig. 6 Flowchart depicting an overview of different methods of nanoparticle synthesis [42]

3.3 Surface Functionalization of Textiles Through Nanoparticles

Nanoparticles have numerous uses in the textile industry, whether incorporated as nano-sols, nano-emulsions, or applied directly to fabrics. They greatly improve the finishing process by making it more consistent, efficient, and accurate. Their high surface area makes them more effective than traditional micron-sized particles, contributing to the production of high-performance and functional textiles with improved efficiency. At the nanoscale, nano-functionalization modifies garments to be resistant to stains and dirt, which in turn reduces laundering frequency, time, and associated costs [48].

Nano-finishing techniques ensure optimal fabric breathability and softness, thereby minimising the occurrence of skin-related issues such as allergies and dermatitis, particularly in individuals with heightened skin sensitivity. The microscopic protective layer formed by nano-finishing imparts a variety of desirable attributes to the fabric, such as heat resistance, antibacterial activity, oil and water repellence, wrinkle resistance, UV protection, and enhanced dyeability—all without compromising the fabric's colour or brightness. Furthermore, nano-treated fabrics are environmentally safe, retain their brightness and freshness, and exhibit significantly greater durability compared to untreated textiles [52]. Nanotechnology is used across various industries, including textiles, to bring new and improved features to fabrics. Advancements in nano-processing have led to the development of materials ranging from scent-producing, colour-changing, and lightweight bulletproof materials to textiles.

Due to the extensive use of nanoparticles for modifying surfaces of materials, understanding how these particles adhere to substrates is essential. Beyond physical and chemical adhesion, nanoparticles may also connect through mechanical interlocking with the surface of the substrate, embedding into areas like pores, holes, or crevices [47].

4 Surface Modification of Polymeric Textiles: Advances in Functionality and Sustainable Approaches

Plasma treatment emerged from initial studies as a highly effective method for modifying polymeric textile surfaces. Höcker [25] showed that low-pressure plasma introduces polar functional groups on fibres, enhancing wettability and enabling eco-friendly dyeing. Mather [32] reviewed plasma-assisted modifications, noting improvements in adhesion, hydrophilicity, and resistance to soiling. Building on these findings, Gorjanc et al. [17] combined CF_4 plasma with silver nanoparticles on cotton, achieving durable antibacterial activity, while Gorjanc et al. [18] integrated plasma pre-treatment with nanoparticle finishing to develop multifunctional fabrics for protective use. Vertommen et al. [57] demonstrated enzymatic modification of

PET, improving hydrophilicity and biocompatibility. Zhu et al. [60] used aminolysis to enhance cell adhesion on polyesters, supporting their use in implants and scaffolds. Morshed et al. [35] advanced the field by employing a plasma–dendrimer system to immobilize glucose oxidase on polyester, creating bioactive smart textiles for diagnostic and therapeutic applications. These studies highlight the progression of modified polymers from textiles to biomedical-grade materials. For protective and environmental uses, plasma–nanotechnology approaches have enabled multifunctional textiles. Raslan et al. [44] reported UV protection, flame retardancy, and antibacterial properties in plasma-treated polyester, suitable for military and outdoor applications. Haji and Naebe [20] reviewed plasma modification with natural dyes, emphasizing cleaner production and reduced chemical waste, while Naebe et al. [37] focused on plasma-assisted antimicrobial finishing, reinforcing its role in infection control. Nanotechnology integration has expanded textile functionalities.. Yetisen et al. [59] reviewed nanotechnology in textiles, covering self-cleaning, UV resistance, flame retardancy, and stimuli-responsive behavior. Shah et al. [48] emphasized applications in wearable electronics, sensors, and energy-harvesting fabrics. Irfan et al. [26] demonstrated practical advances by loading ZnO nanoparticles onto plasma-treated cotton via ultrasonics, yielding UV shielding, antibacterial activity, and photocatalytic self-cleaning. Sustainability has become a central focus of plasma-based treatments. Dave et al. [9] used ammonia dielectric barrier discharge plasma on polyester to enhance dye uptake with reduced chemical use, an approach later extended with natural dyes [10]. Samanta et al. [45, 46] showed the synergy of plasma and UV in eco-friendly processing, while Shabbir et al. [47] positioned such surface modification technologies as essential to sustainable textiles and aligned with the UN Sustainable Development Goals. Together, these studies demonstrate the potential of plasma-based polymer modification as a green, multifunctional, and future-oriented technology.

5 Conclusion

Recent advances and development in textile polymer industries promoted the acceptance of various surface modification techniques. Such techniques have the capability to increase the durability of textile polymers, supporting sustainability initiatives through reduction, reuse, and recycling. Certain alternative methods offer more effective reduction capabilities compared to traditional approaches. Moreover, some of these approaches indirectly contribute to sustainability. For instance, textiles treated with eco-friendly, non-toxic, and biodegradable antimicrobial agents not only inhibit microbial growth and odour formation but also reduce the need for frequent laundering, thereby conserving water, energy, and detergent resources.

As we continue to deplete our resources, it is essential to integrate the principles of reduction, reuse, and recycling into our everyday activities. To ensure a sustainable future, we suggest the following: adopting eco-conscious green production methods, encouraging research on the development of surface modification techniques, and incorporating these sustainability-enhancing production methods as new standard practices once they achieve sufficient development.

In modern times, plasma technology has surfaced as a fascinating and efficient approach for the physicochemical surface modification of textiles, potentially serving as a substitute or enhancement for traditional wet chemical methods. The properties of selected materials can be changed by employing different plasma gases and modifying treatment parameters such as duration, voltage, and power. Furthermore, nanotechnology offers a promising path for altering and improving fabrics. Current studies have introduced a novel strategy for the fabrication and application of nanoparticles (NPs) in textile materials, employing sustainable and non-toxic biopolymers such as chitosan, alginate, and starch. The unique nanoscale dimensions and extensive surface area of these particles allow for the tailored enhancement of surface properties that are typically difficult to attain using traditional methods. Beyond attributes like antimicrobial activity, UV protection, stain resistance, and improved dye retention, plasma-treated polymers coated with nanoparticles can also enhance anti-felting properties, reduce static buildup, increase electrical conductivity, and improve abrasion resistance. This process additionally reinforces the mechanical strength of fabrics by increasing surface roughness and optical reflectivity—critical for applications like photocatalysis.

Despite these promising benefits, industrial-scale adoption of plasma and nanoparticle-treated textiles remains limited, with progress toward consumer-ready garments still in the early research phase. Key challenges include concerns over the long-term durability of the treatments and the need for varied reactors and gas compositions depending on fabric type. Nevertheless, growing interdisciplinary collaboration among researchers, textile engineers, and designers is helping to address these barriers. These emerging technologies, if fully optimized, could contribute significantly to the advancement of multifunctional textiles designed for use in critical areas such as medicine, healthcare, military operations, and intelligent textile systems.

References

1. Ahlawat, D. S., Kumari, R., Rachna, Yadav, I.: Synthesis and characterization of sol-gel prepared silver nanoparticles. Int. J. Nanosci. **13**(1), 1450004 (2014)
2. Alimohammadi, F., Gashti, M.P., Shamei, A.: A novel method for coating of carbon nanotube on cellulose fiber using 1, 2, 3, 4-butanetetracarboxylic acid as a cross-linking agent. Prog. Org. Coat. **74**(3), 470–478 (2012)
3. Araujo, R., Casal, M., Cavaco-Paulo, A.: Application of enzymes for textile fibres processing. Biocatal. Biotransform. **26**(5), 332–349 (2008)
4. Başyiğit, Z.Ö.: Effects of chemical and surface modification on mechanical and chemical properties of polyester fabrics. Duzce Univ. J. Sci. Technol. **6**(4), 1344–1353 (2018)
5. Bech, L., Meylheuc, T., Lepoittevin, B., Roger, P.: Chemical surface modification of poly (ethylene terephthalate) fibers by aminolysis and grafting of carbohydrates. J. Polym. Sci., Part A: Polym. Chem. **45**(11), 2172–2183 (2007)
6. Bhat, M.: Cellulases and related enzymes in biotechnology. Biotechnol. Adv. **18**(5), 355–383 (2000)

7. Bhaumik, N.S., Konar, A., Roy, A.N., Samanta, A.K.: comparative studies on dyeability with direct, acid and reactive dyes after chemical modification of jute with mixed amino acids obtained from extract of waste soya bean seeds. J. Inst. Eng. (India) Series E **98**(2), 121–133 (2017)
8. Datta, A., Patra, C., Bharadwaj, H., Kaur, S., Dimri, N., Khajuria, R.: Green synthesis of zinc oxide nanoparticles using *Parthenium hysterophorus* leaf extract and evaluation of their antibacterial properties. J. Biotechnol. Biomater. **07**(03), 1000271 (2017)
9. Dave, H., Ledwani, L., Chandwani, N., Desai, B., Nema, S.K.: Surface activation of polyester fabric using ammonia dielectric barrier discharge and improvement in colour depth. Ind. J. Fibre Textile Res. **39**(3), 274–281 (2014)
10. Dave, H., Ledwani, L., Nema, S.K.: Surface modification by atmospheric pressure air plasma treatment to improve dyeing with natural dyes: an environment friendly approach for leather processing. Plasma Chem. Plasma Process. **36**(2), 599–613 (2016)
11. Elkashouty, M., Elsyaed, H., Twaffiek, S., Salem, T., Elhadad, S.S.M.: An overview: textile surface modification by using sol-gel technology. Egypt. J. Chem. **63**(9), 3301–3311 (2020)
12. Essa, D.: 04. Recent techniques of surface modification for textile fabrics. J. Meas. Sci. Appl. (JMSA) **4**(1), 66–87 (2024)
13. Fabbri, P., Messori, M.: Surface modification of polymers: chemical, physical, and biological routes. In: Modification of Polymer Properties (pp. 109–130) (2017)
14. Farouk, A., Moussa, S., Ulbricht, M., Textor, T.: ZnO nanoparticles-chitosan composite as antibacterial finish for textiles. Int. J. Carbohydrate Chem. **2012**(1), 693629 (2012)
15. Gashti, M.P., Hajiraissi, R., Gashti, M.P.: Morphological, optical and electromagnetic characterization of polybutylene terephthalate/silica nanocomposites. Fibers Polymers **14**(8), 1324–1331 (2013)
16. Ghosh, J., Rupanty, N.S., Noor, T., Asif, T.R., Islam, T., Reukov, V.: Functional coatings for textiles: advancements in flame resistance, antimicrobial defense, and self-cleaning performance. RSC Adv. **15**(14) (2025)
17. Gorjanc, M., Bukošek, V., Gorenšek, M., Mozetič, M.: CF4 plasma and silver functionalized cotton. Text. Res. J. **80**(20), 2204–2213 (2010)
18. Gorjanc, M., Gorensek, M., Jovancic, P., Mozetic, M.: Eco-friendly textile dyeing and finishing. In: Multifunctional Textiles—Modification by Plasma, Dyeing and Nanoparticles (pp. 3–33) (2013)
19. Gupta, D., Chaudhary, H., Gupta, C.: Topographical changes in polyester after chemical, physical and enzymatic hydrolysis. J. Text. Inst. **106**(7), 690–698 (2015)
20. Haji, A., Naebe, M.: Cleaner dyeing of textiles using plasma treatment and natural dyes: a review. J. Clean. Prod. **265**, 121866 (2020)
21. Haq, I.U., Naz, M.Y., Shukrullah, S., Anwar, M., Ullah, S., Al-Sehemi, A.G., Irshad, M.I.: Dielectric barrier plasma effect on surface functionality and coating properties of ultrasonically coated cotton with ZnO nanoparticles and *Aloe vera* extraction. Chem. Pap. **76**(2), 889–900 (2022)
22. Hoffman, A.S.: Surface modification of polymers: Physical, chemical, mechanical and biological methods. Macromol. Symposia **101**(1), 443–454 (1996)
23. Hofmann, A.I., Cloutet, E., Hadziioannou, G.: Materials for transparent electrodes: from metal oxides to organic alternatives. Adv. Electron. Mater. **4**(10), 1700412 (2018)
24. Hussain, I., Singh, N.B., Singh, A., Singh, H., Singh, S.C.: Green synthesis of nanoparticles and their potential application. Biotech. Lett. **38**(4), 545–560 (2016)
25. Höcker, H.: Plasma treatment of textile fibers. Pure Appl. Chem. **74**(3), 423–427 (2002)
26. Irfan, M., Hussain, H., Saleem, B., Saleem, M., Shukrullah, S., Legutko, S., Petrů, J., Naz, M.Y., Pagáč, M., Rahman, S., Khan, R.: Evaluation of utrasonically ZnO loading effect on photocatalytic self-cleaning, UV protection and antibacterial activity of plasma/citric acid-activated cotton fabric. Nanomaterials **12**(12), 2122 (2022)
27. Islam, S., Hasan, B.: An overview of the effects of water and moisture absorption on the performance of hemp fiber and its composites. SPE Polymers **6**(1), e10167 (2024)

28. Islam, M.H., Islam, M.R., Dulal, M., Afroj, S., Karim, N.: The effect of surface treatments and graphene-based modifications on mechanical properties of natural jute fiber composites: a review. Iscience **25**(1) (2022)
29. Joshi, M., Bhattacharyya, A.: Nanotechnology–a new route to high-performance functional textiles. Text. Prog. **43**(3), 155–233 (2011)
30. Kingshott, P., Andersson, G., McArthur, S.L., Griesser, H.J.: Surface modification and chemical surface analysis of biomaterials. Curr. Opin. Chem. Biol. **15**(5), 667–676 (2011)
31. Korkmaz, G., Kılınç, M., Kılınç, N., Kut, Y.D.: The role of surface modification methods for sustainable textiles. In Roadmap to Sustainable Textiles. IntechOpen (2023)
32. Mather, R.R.: Surface Modification of Textiles by Plasma Treatments: Surface Modification of Textiles (pp. 296–317) (2009)
33. Mohammad, F.: Emerging green technologies and environment friendly products for sustainable textiles. In: Roadmap to Sustainable Textiles and Clothing: Environmental and Social Aspects of Textiles and Clothing Supply Chain (pp. 63–82) (2014)
34. Moloney, M.G.: Functionalized polymers by chemical surface modification. J. Phys. D Appl. Phys. **41**(17), 174006 (2008)
35. Morshed, M.N., Behary, N., Bouazizi, N., Guan, J., Chen, G., Nierstrasz, V.: Surface modification of polyester fabric using plasma-dendrimer for robust immobilisation of glucose oxidase enzyme. Sci. Rep. **9**(1), 15730 (2019)
36. Nadi, A., Boukhriss, A., Bentis, A., Jabrane, E., Gmouh, S.: Evolution in the surface modification of textiles: a review. Text. Prog. **50**(2), 67–108 (2018)
37. Naebe, M., Haque, A.N.M.A., Haji, A.: Plasma-assisted antimicrobial finishing of textiles: a review. Engineering **12**, 145–163 (2022)
38. Nierstrasz, V.A.: Enzyme Surface Modification of Textiles: Surface Modification of Textiles (pp. 139–163) (2009)
39. Novák, I., Valentin, M., Špitalský, Z., Popelka, A., Sestak, J., Krupa, I.: Superhydrophobic polyester/cotton fabrics modified by barrier discharge plasma and organosilanes. Polym. Plast. Technol. Eng. **57**(5), 440–448 (2018)
40. Oehrlein, G.S., Phaneuf, R.J., Graves, D.B.: Plasma-polymer interactions: a review of progress in understanding polymer resist mask durability during plasma etching for nanoscale fabrication. J. Vacuum Sci. Technol. B Nanotechnol. Microelectron. Mater. Process. Measur. Phenomena **29**(1) (2011)
41. Ozdemir, M., Yurteri, C.U., Sadikoglu, H.: Physical polymer surface modification methods and applications in food packaging polymers. Crit. Rev. Food Sci. Nutr. **39**(5), 457–477 (1999)
42. Patra, S., Mukherjee, S., Barui, A.K., Ganguly, A., Sreedhar, B., Patra, C.R.: Green synthesis, characterisation of gold and silver nanoparticles and their potential application for cancer therapeutics. Mater. Sci. Eng. C **53**, 298–309 (2015)
43. Patra, J.K., Baek, K.H.: Green nanobiotechnology: factors affecting synthesis and characterization techniques. J. Nanomater. **2014**(1), 417305 (2014)
44. Raslan, W.M., Rashed, U.S., El-Sayad, H., El-Halwagy, A.A.: Ultraviolet protection, flame retardancy and antibacterial properties of treated polyester fabric using plasma-nano technology. Mater. Sci. Appl. **02**(10), 1432–1442 (2011)
45. Samanta, K.K., Basak, S., Chattopadhyay, S.K.: Eco-friendly coloration and functionalization of textile using plant extracts. In: Roadmap to Sustainable Textiles and Clothing: Environmental and Social Aspects of Textiles and Clothing Supply Chain (pp. 263–287) (2014)
46. Samanta, K.K., Basak, S., Chattopadhyay, S.K.: Environment-friendly textile processing using plasma and UV treatment. In: Roadmap to Sustainable Textiles and Clothing: Eco-Friendly Raw Materials, Technologies, and Processing Methods (pp. 161–201) (2014)
47. Shabbir, M., Rather, L.J., Bukhari, M.N., Ul-Islam, S., Shahid, M., Khan, M.A., Mohammad, F.: Light fastness and shade variability of tannin colorant dyed wool with the effect of mordanting methods. J. Nat. Fibers **16**(1), 100–113 (2019)
48. Shah, M.A., Pirzada, B.M., Price, G., Shibiru, A.L., Qurashi, A.: Applications of nanotechnology in the smart textile industry: a critical review. J. Adv. Res. **38**, 55–75 (2022)

49. Shahzad, A.: Effects of fibre surface treatments on mechanical properties of hemp fibre composites. Composite Interfaces **18**(9) (2011)
50. Singh, M., Vajpayee, M., Ledwani, L.: Eco-friendly surface modification of natural fibres to improve dye uptake using natural dyes and application of natural dyes in fabric finishing: a review. Mater. Today Proc. **43**, 2868–2871 (2021)
51. Singh, A., Singh, N.B., Hussain, I., Singh, H., Singh, S.C.: Plant-nanoparticle interaction: an approach to improve agricultural practices and plant productivity. Int. J. Pharmaceutical Sci. Invent. **4** (2015)
52. Singh, M., Vajpayee, M., Ledwani, L.: Eco-friendly surface modification and nanofinishing of textile polymers to enhance functionalisation. Nanotechnol. Energy Environ. Eng. 529–559 (2020)
53. Uyama, Y., Kato, K., Ikada, Y.: Surface modification of polymers by grafting. In: Grafting/Characterisation Techniques/Kinetic Modelling, pp. 1–39 (1999)
54. Vajpayee, M., Singh, M., Ledwani, L.: Non-thermal plasma treatment of cellulosic biopolymer to enhance its surface property for various applications: a review. Mater. Today Proc. **43**, 3250–3255 (2021)
55. Vajpayee, M., Dave, H., Singh, M., Ledwani, L.: Cellulase enzyme-based wet-pretreatment of lotus fabric to improve antimicrobial finishing with *A. indica* extract and enhance natural dyeing: sustainable approach for textile finishing. ChemistrySelect **7**(25), e202200382 (2022)
56. Vajpayee, M., Singh, M., Ledwani, L., Prakash, R., Nema, S.K.: Investigation of antimicrobial activity of DBD air plasma-treated banana fabric coated with natural leaf extracts. ACS Omega **5**(30), 19034–19049 (2020)
57. Vertommen, M.A.M.E., Nierstrasz, V.A., Veer, M. Van Der Warmoeskerken, M.M.C.G.: Enzymatic surface modification of poly (ethylene terephthalate). J. Biotechnol. **120**(4), 376–386 (2005)
58. Xu, C., Cao, L., Zhao, P., Zhou, Z., Cao, C., Li, F., Huang, Q.: Emulsion-based synchronous pesticide encapsulation and surface modification of mesoporous silica nanoparticles with carboxymethyl chitosan for controlled azoxystrobin release. Chem. Eng. J. **348**, 244–254 (2018)
59. Yetisen, A.K., Qu, H., Manbachi, A., Butt, H., Dokmeci, M.R., Hinestroza, J.P., Skorobogatiy, M., Khademhosseini, A., Yun, S.H.: Nanotechnology in textiles. ACS Nano **10**(3), 3042–3068 (2016)
60. Zhu, Y., Mao, Z., Gao, C.: Aminolysis-based surface modification of polyesters for biomedical applications. RSC Adv. **3**(8), 2509–2519 (2013)

Sensors and Actuators

Bioengineered Endophyte-Integrated Biosensors: A New Frontier in Environmental Monitoring and Remediation

Irra Dhar, Anchal Sharma, Ankita Kumawat, Abhishek Dadhich, Himanshu Priyadarshi, Lakshika Sharma, and Madan Mohan Sharma

Abstract Environmental pollution remains a global concern, driven by the widespread release of heavy metals, persistent organic pollutants, microplastics, and endocrine-disrupting chemicals. Traditional detection and remediation technologies often fall short due to high costs, slow response times, and limitations in real-time field applicability. Recent progress in materials science has enabled the development of biosensor-based systems and green remediation strategies, offering more sustainable, efficient, and field-deployable solutions. This chapter explores integrated approaches that combine biological materials—such as plant endophytes and enzyme systems—with nanomaterial-based biosensors for real-time environmental monitoring. Endophytes, known for their metal chelation, enzymatic activity, and stress tolerance, serve dual roles in pollutant detection and remediation. Coupled with green synthesis of nanoparticles and phytoremediation techniques, these biological systems contribute to low-impact, resource-efficient cleanup processes. Advances in material characterization, nanostructure design, and surface engineering have improved biosensor sensitivity and selectivity across complex environmental matrices. Additionally, computational tools such as life cycle assessment (LCA), predictive modeling, and machine learning aid in optimizing biosensor performance and assessing long-term environmental impacts. The chapter highlights how the convergence of experimental innovations and computational methodologies in materials science enables the design of smart, multifunctional systems for pollution management. Emphasis is placed on sustainability, adaptability, and real-world applicability, aligning with the broader goals of circular economy and green engineering. This interdisciplinary perspective

Irra Dhar and Anchal Sharma equally contributed as a first author to this work

I. Dhar · A. Sharma · A. Kumawat · A. Dadhich · M. M. Sharma (✉)
Department of Biosciences, Manipal University Jaipur, Jaipur, Rajasthan, India
e-mail: madanmohan.sharma@jaipur.manipal.edu

H. Priyadarshi
Department of Electrical Engineering, Manipal University Jaipur, Jaipur, Rajasthan, India

L. Sharma
S.S. Jain Subodh P.G. College, Jaipur, India

L. Ledwani and S. Srivastava (eds.), *Experimental and Computational Advances in Materials*, Engineering Materials, https://doi.org/10.1007/978-981-95-4180-5_13

supports the ongoing transition toward environmentally responsible technologies within the framework of materials science and engineering.

Keywords Environmental biosensing · Green remediation technologies · Endophyte-assisted detoxification · Sustainable functional materials · Computational Modeling in Remediation

1 Introduction

1.1 Environmental Pollution and Its Complexity

Environmental pollution has emerged as a defining challenge of the Anthropocene, threatening planetary health, ecosystem functionality and human well-being. Rapid urbanisation, intensive industrialisation and resource-hungry economic activity have pushed waste streams beyond safe operating limits, unleashing complex cocktails of heavy metals, industrial chemicals, persistent organic pollutants (POPs) and emerging contaminants such as micro- and nanoplastics into soils, waters and the atmosphere [90, 140]. The World Health Organization now estimates that 99% of the global population breathes air that exceeds its clean-air guidelines, contributing to 8.1 million premature deaths in 2021 alone and siphoning an estimated 4.98% of global GDP-losses felt most acutely by low- and middle-income nations [139, 161]. Parallel crises are unfolding in aquatic and terrestrial realms: 19–23 million tonnes of plastic waste enter aquatic systems each year, while indiscriminate discharge of agro-industrial effluents has left vast tracts of farmland and food chains burdened with toxic metals and xenobiotics [149, 150].

Compounding the scale of contamination is an escalating chemical complexity. Polychlorinated biphenyls, dioxins and other POPs defy natural degradation and bio-accumulate across trophic levels [105]. Trace-level exposures to lead, mercury or cadmium can elicit neurological, renal and developmental disorders even decades after deposition [131]. New-age threatsfrom bisphenol-A and endocrine-disrupting pesticides to microplastic fibres detected in human placentademand analytical sensitivity far finer than that offered by conventional laboratory assays [122, 155].

1.2 Need for Biosensors in Contaminant Detection and Remediation

In response to these pollution challenges, policy frameworks increasingly prioritise integrative strategies that (i) sense contaminants in real time, (ii) couple monitoring with adaptive, low-impact clean-up, and (iii) close resource loops through recovery and reuse [59, 125]. Yet legacy techniques: atomic-absorption spectrometry, high-performance liquid chromatography or ICP-MS—remain expensive, lab-bound and often blind to mixtures, making rapid decision-making impractical in dynamic field conditions [14, 76].

Biosensor technology offers a transformative alternative. By fusing highly selective biorecognition modules (enzymes, antibodies, DNA aptamers or whole microbial cells) with electrochemical, optical, piezoelectric or thermal transducers, modern biosensors deliver on-site, sub-nanomolar detection of metals, pesticides, pharmaceuticals and even microplastic fragments [50, 55]. Sensor miniaturisation, wireless telemetry and IoT integration further enable continuous air–water–soil observatories that feed real-time data to urban planners, farmers and regulatory agencies [67, 136].

Concurrently, green remediation approaches- bioremediation, phytoremediation, green-synthesised nanoparticles and life-cycle-assessed engineeringseek to restore contaminated sites without transferring risks elsewhere [64, 119]. Microbial consortia degrade hydrocarbons; hyperaccumulator plants stabilise or phytoextract metals; renewable-energy-powered pump-and-treat systems curb carbon footprints; and plant- or microbe-derived nanomaterials catalyse pollutant breakdown while minimising secondary toxicity [69, 158, 159] (Table 1).

1.3 The Emerging Role of Endophytes in Smart Biosensing and Remediation

Recent advances reveal a potent synergy when plant endophyte, bacteria or fungi residing asymptomatically within plant tissues, are engineered or harnessed within biosensor platforms. Endophytes naturally sequester metals, degrade organics and modulate host stress responses via siderophores, laccases, ACC-deaminase and antioxidant enzymes [28, 129]. CRISPR/Cas-mediated metabolic engineering now allows precision boosts in pollutant-targeted pathways, while their biomolecules (e.g., laccase, lipase, metallothioneins) can be immobilised onto conductive nanomaterials to create hybrid bio-nano interfaces with femtomolar sensitivity [42, 166]. Endophyte-mediated biosynthesis of Ag, ZnO or Fe_3O_4 nanoparticles further couples detection with in situ degradation or phyto-extraction, realising the ideal of smart, self-reporting remediation [86, 138].

This chapter illustrates the potential of endophyte-driven biosensing systems to close the crucial gap between environmental monitoring and repair by combining cutting-edge biotechnology, materials science, and ecological engineering. By providing targeted mitigation of Anthropogenic impacts—the defining influence of human activity on Earth's geology, climate, and ecosystems—these low-carbon, adaptable solutions are in line with planetary health goals and the concepts of the circular economy.

Emerging contaminants (such microplastics and pharmaceuticals) and xenobiotics (like synthetic pesticides and industrial chemicals) that are difficult to detect but pose serious threats to human health and the environment are highlighted in the

Table 1 Overview of principal green remediation strategies, their mechanisms, and environmental applications

Technique	Principle	Target pollutants	Key benefits	Limitations	Example applications	References
Bioremediation	Use of microorganisms to degrade or detoxify contaminants	Hydrocarbons, heavy metals, solvents	Cost-effective, eco-friendly, minimal disturbance	Slow process, may need bioaugmentation	Oil spill cleanups, landfill leachates	[69]
Phytoremediation	Use of plants to absorb, accumulate, or degrade contaminants	Metals, pesticides, organics	In situ, enhances soil health, biomass recovery possible	Depth limitation, seasonal dependency	Arsenic-contaminated soils, wastewater treatment	[4]
Life cycle and footprint assessment	LCA tools to evaluate long-term impact of remediation choices	Not applicable	Avoids long-term harm, ensures net environmental gain	Requires data, may not guide operational methods	Decision-making in remediation planning	[60]
Resource efficiency and renewables	Use of solar, wind, or energy recovery systems in remediation	General remediation processes	Reduces emissions, lowers operational cost	Initial setup cost	Solar-powered groundwater pumping systems	[64]
Green nanotechnology	Eco-friendly synthesis of nanoparticles using plant extracts, microbes, or biopolymers	Organic dyes, metals, industrial waste	Low toxicity, reduced chemical use, safer synthesis	May have scalability issues, newer field	AgNPs for dye degradation, ZnO NPs for heavy metal removal	[158, 159, 174]

conversation. To address heavy metal stress, microbial remediation techniques like metallothionein-based detoxification or siderophore-mediated metal chelation can be guided by real-time sensor data via IoT-enabled networks. These collaborations highlight how microbial biotechnology can provide scalable, long-lasting treatments in the Anthropocene.

2 Materials and Methods

This chapter employed a multidisciplinary experimental framework integrating microbiological, nanotechnological, and material science approaches to develop endophyte-assisted biosensing and remediation platforms for environmental applications. Endophytic microorganisms were isolated from surface-sterilized tissues of selected hyperaccumulator and medicinal plants using culture-based methods on nutrient agar and potato dextrose agar. Isolates were purified and subjected to molecular identification through 16S rRNA and ITS sequencing. Functional screening of endophytes focused on traits essential for nanoparticle biosynthesis and pollutant mitigation, including siderophore production, phosphate solubilization, ACC deaminase activity, and secretion of antioxidant enzymes such as catalase and laccase.

Following functional characterization, selected bacterial and fungal strains were employed for the green synthesis of nanoparticles. Endophyte culture supernatants containing bioactive metabolites (e.g., phenolics, flavonoids, nitrate reductase) were incubated with aqueous solutions of metal precursors such as silver nitrate, zinc acetate, ferric chloride, and sodium selenite. Biosynthesis was carried out under optimized conditions of pH, temperature, and time. The resulting nanoparticles were harvested by centrifugation, washed, and lyophilized. Structural and physicochemical characterization was conducted using UV–Vis spectroscopy (to monitor surface plasmon resonance), Fourier-transform infrared spectroscopy (FTIR, for functional group analysis), X-ray diffraction (XRD), scanning electron microscopy (SEM), and dynamic light scattering (DLS) to determine particle size distribution, morphology, surface charge, and crystallinity.

To explore the sensing capabilities of endophyte-derived materials, biosensor platforms were fabricated using nanomaterials as conductive scaffolds. These included carbon nanotubes (CNTs), graphene oxide (GO), gold nanoparticles (AuNPs), conducting polymers (e.g., polyaniline, polypyrrole), and metal–organic frameworks (MOFs). Enzymes or whole cells from endophytic strains were immobilized onto these substrates using cross-linkers such as glutaraldehyde to ensure surface stability and biocompatibility. The biosensors were designed to detect key environmental pollutants, including heavy metals (Cd, Cr, Pb, As), organic compounds (phenols, dyes, pesticides), and microbial contaminants. Electrochemical techniques (amperometry and electrochemical impedance spectroscopy) and optical assays (colorimetric and fluorescence-based detection) were used to assess biosensor performance. Analytical parameters such as sensitivity, linear dynamic range, detection limit,

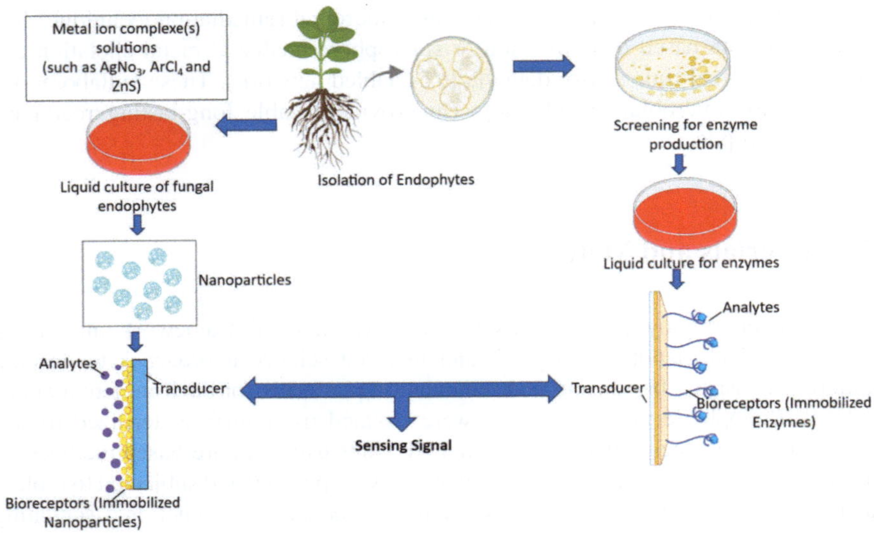

Fig. 1 Endophyte isolation and screening and biosensor fabrication

and response time were established using standard pollutant solutions, followed by validation with real samples such as wastewater and soil leachates (Fig. 1).

In parallel, the interaction between endophytes, nanoparticles, and host plants was studied in greenhouse microcosms to assess bio-delivery, pollutant mitigation, and plant health effects. Plants were inoculated with endophytic strains and treated with biosynthesized nanoparticles to evaluate root colonization, nanoparticle transport, and pollutant uptake. Tracer studies using fluorescent or magnetic nanoparticles confirmed targeted delivery and internal translocation. Metal accumulation in plant tissues was quantified using inductively coupled plasma optical emission spectroscopy (ICP-OES), while oxidative stress markers and antioxidant enzyme activity (SOD, CAT, POD) were also measured. These integrated experiments confirmed the dual role of endophytes as nanoparticle biofactories and biocarriers, enhancing both pollutant detection and sustainable remediation.

3 Biosensors in Environmental Monitoring

Biosensors are analytical instruments that combine a biological recognition element with a transducer to detect specific analytes and generate measurable signals. Depending on the nature of the biological component or the type of transduction mechanism employed, biosensors can be classified into various types. The biological recognition elements include enzymes, antibodies, whole cells, DNA, aptamers, tissues, and molecularly imprinted polymers (MIPs), which are selected based on their specificity to the target analyte [55]. The interaction between the analyte and the bio-recognition element leads to a physicochemical change that is picked up by the transducer. These transducers may be optical, electrochemical, piezoelectric, or thermal in nature, each converting the biological interaction into a quantifiable signal [121, 165].

The core working principle of a biosensor revolves around three main components functioning in a coordinated manner. First is the bio-recognition element, which is the sensor's biological component and may comprise enzymes, antibodies, DNA sequences, microbial cells, or other biomolecules. These components selectively interact with the target pollutant, ensuring specificity and reliability in detection [48, 142]. Second is the transducer, which translates the biological interaction into a measurable signal. Common types of transducers include electrochemical (such as amperometric, potentiometric, and conductometric), optical (fluorescence or absorbance-based), piezoelectric (mass change sensing), and thermal (heat release measurement) systems. These transducers enhance sensitivity and allow for miniaturized and portable devices that can operate in situ under varying environmental conditions [48, 142]. Finally, the signal processor and display unit amplify and process the raw signal generated by the transducer, presenting the final output in a readable format such as digital display or wireless transmission for remote monitoring.

The versatility in biosensor design has led to the development of devices tailored for specific environmental matrices and contaminants. Electrochemical biosensors are particularly valued for their simplicity, affordability, and potential for miniaturization, making them ideal for field applications. Optical biosensors offer high sensitivity and rapid detection, while piezoelectric sensors are useful for detecting changes in mass or acoustic properties associated with molecular interactions. The integration of biosensors with wireless technologies, IoT systems, and data processing algorithms further extends their utility for large-scale environmental monitoring programs.

3.1 Functional Types of Biosensors in Environmental Applications

Biosensor is described as the tool of analysis, which is defined relative to its biorecognition component and the method of signalling transduction. This is because its design and its operative concept has direct effect on its performance in sensing different pollutants in different environmental conditions e.g. in air, water, and soil. Another of the most common biosensors would be electrochemical biosensors which measure biochemical interactions as electrical signals, which can be in the form of voltage, current or impedance. Their applicability, flexibility and low-cost to detect pollutant, predominantly nitrates, heavy metals and pesticides make them extremely appealing to be utilized in real-time environmental monitoring [27, 51, 52, 79, 121]. In parallel with them, optical biosensors, sensing optical property changes, refractive index, absorbance, fluorescence etc., are used. The devices are sensitive and are more appreciated because they are immune to electromagnetic interferences. They are being facilitated by recent inventions, even smartphone-based systems available, which employ internal cameras and light emitters to quantify environmental pollutants using heavy metals and pathogens in even remote areas [151, 164].

Other piezoelectric materials may be used or quartz crystals in piezoelectric biosensors to sense the small mass variations caused by binding of the analyte. The model of analysis is frequency shifts caused by such interactions. They have good stability in aqueous phase and have label-free detection sensors and are suited for detecting airborne toxins and pesticides but their surface fouling and modest selectivity occur still [98, 110]. At the same time, microbial or whole-cell biosensors deal with microorganisms that are metabolically active that undergo a measurable physiological or biochemical response to certain pollutants. One application that these biosenseors find particular value is in the detection of trace contaminants in a complicated sample such as municipal wastewater or industrial effluents. Their cost-effective nature and immediate response and importantly the biological value makes them irreplaceable by businesses that wish to achieve long-term sustainable waste management processes, although their reliability can be influenced by microbial stability and environmental fluctuations [77, 120] (Table 2).

Even driving the detection capabilities to the limits, nanomaterial-based biosensors that modify the behaviour of engineered nanoparticles to enhance the mechanism of signal transduction. The sensitivity of these novel biosensors reaches at the femtomolar range and allows one to detect ultra-trace environmental toxins. A well-known encapsulation is that of a nanosensor built on a gold electrode by the cells of the bacteria Cyanobacteria of extracellular polymeric substances (EPSs), able to detect microplastic using electrochemical impedance spectroscopy. Such designs indicate an increasing relationship between biosensors and nanotechnology on environmental monitoring [41, 68, 128]. Accumulatively, they offer highly relevant yet sensitive biosensor-based platforms to counter the emerging concerns of environmental pollution by utilizing selective, quick, and proximity-based indicators.

These diverse classes of biosensors, despite differing in materials and mechanisms, all operate through a common functional architecture comprising a biological recognition component, a transduction unit, and a signal processing system. Understanding this general working principle is essential to appreciate how biosensors convert biochemical interactions into measurable analytical signals for environmental applications. The schematic below (Fig. 2) illustrates the fundamental components and signal flow in a typical electrochemical biosensor, one of the most widely used formats in real-time environmental monitoring.

3.2 Materials Used in Biosensors

The performance of biosensors heavily depends on the materials used for signal transduction, bio-recognition, and analyte interaction. Advances in nanotechnology,

Table 2 Comparative analysis of functional types of biosensors in environmental applications

Biosensor type	Detection mechanism	Sensitivity	Selectivity	Target pollutants	Advantages	Limitations	References
Electrochemical biosensor	Redox reaction; electrical signal output (current/voltage/impedance)	pM–nM	High	Heavy metals, toxins, nitrates, pesticides	Low-cost, portable, suitable for in situ applications	Affected by fouling, requires calibration	[27, 50, 51, 79, 121]
Optical biosensor	Change in light intensity, absorbance, fluorescence or refractive index	nM–pM	Very high	Pesticides, heavy metals, pathogens	High accuracy, non-invasive, real-time detection using smartphone-based optics	Can be expensive, sensitive to ambient light conditions	[151, 164]
Piezoelectric biosensor	Mass change detection via frequency shift on quartz crystal	pg–ng (ppb levels for gases)	Medium	Pesticides, toxic gases, heavy metals	Label-free detection, stable at elevated temperatures	Lower selectivity, surface contamination may affect output	[98, 110]
Microbial (whole-cell) biosensor	Biochemical/electrical response from microbial metabolism	~ 10^{-9}–10^{-6} M	Medium	Heavy metals, pesticides, organics	Cost-effective, high sensitivity for complex wastewater samples, real-time monitoring	Slower response time, microbial stability and reproducibility issues	[77, 120]
Nanomaterial-integrated Biosensor	Nanomaterial-assisted signal amplification (e.g., fluorescence, impedance)	~ 10^{-14} M	High	Microplastics, heavy metals (Cr^{3+}, Pb^{2+}), toxins	Ultra-sensitive, precise detection at trace levels	Potential environmental toxicity, scalability concerns	[41, 68, 128]

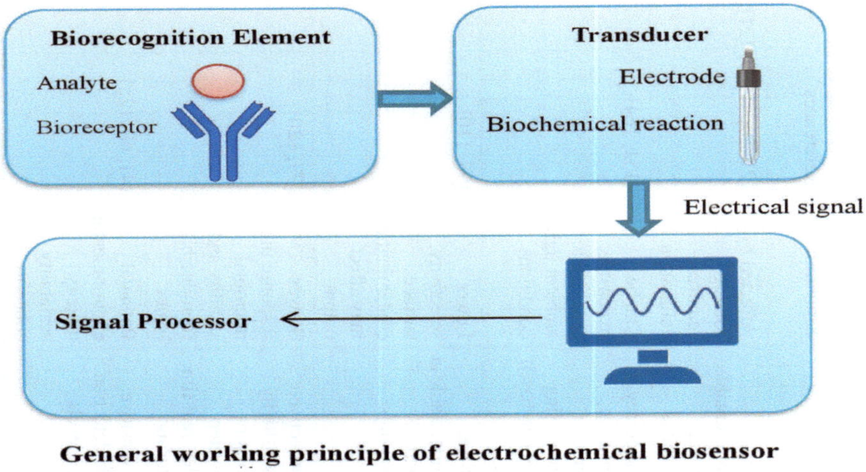

Fig. 2 Schematic illustration of the biosensor mechanism depicting target-specific biorecognition, physicochemical signal transduction, and quantitative output generation

polymer science, and materials engineering have significantly enhanced the sensitivity, specificity, and adaptability of biosensors across diverse environmental applications. Materials such as nanomaterials, biopolymers, hydrogels, conducting polymers, and metal–organic frameworks (MOFs) play distinct and synergistic roles in biosensor platforms, enabling efficient immobilization of biomolecules, signal amplification, and real-time detection under challenging environmental conditions (Tables 3 and 4).

Table 3 Link between each material and why it matters for biosensors

Material	Why it matters
Graphene/C nanotubes	Exceptional conductivity, high surface area, fast electron-transfer, versatile functionalization
Au, Ag, Fe_3O_4 nanoparticles	Catalytic activity, large reactive surface, improved signal amplification
ITO (Indium tin oxide)	Transparent conductive surface, functionalizable, affordable
Silica nanoparticles	Tunable porosity, stable, customizable surfaces
Cellulose/nanocellulose	Biocompatible, low-cost, good immobilization support, fast response
MOFs (metal–organic frameworks)	High chemical stability, signal enhancement
Metal oxide nanostructures (e.g. ZnO, TiO_2)	High sensitivity, biocompatibility, catalytic and immobilization capability

Table 4 Functional materials in biosensing applications

Material type	Specific example	Functional role in biosensing	Advantages	Limitations	References
Nanomaterial	Gold Nanoparticles (AuNPs)	Surface plasmon resonance (SPR) enhancement; signal amplification	High conductivity, strong light-matter interaction, chemical stability	Aggregation in biological matrices; cost of synthesis	[44, 109]
Nanomaterial	Carbon Nanotubes (CNTs)	Electron mediation and electrode surface enhancement	High surface area, excellent mechanical/electrical properties	Cytotoxicity concerns, dispersion issues	[87, 134]
Nanomaterial	Graphene and Graphene Oxide (GO)	Electron transport, bioreceptor immobilization platform	High conductivity, functionalization flexibility	Oxidation sensitivity, hydrophobic nature	[137]
Nanomaterial	Quantum Dots (QDs)	Fluorescence resonance energy transfer (FRET); signal transduction	High sensitivity, wide detection range	Photobleaching, potential toxicity	[114]
Polymeric Material	Biopolymers	Matrix for enzyme/microbial immobilization	Renewable, eco-friendly, biocompatible	Mechanical fragility, limited electrical properties	[1]
Polymeric Material	Hydrogels	Analyte diffusion control; 3D receptor hosting	Stimuli-responsive, high loading capacity	Drying sensitivity, poor mechanical strength	[36, 47, 148]

(continued)

Table 4 (continued)

Material type	Specific example	Functional role in biosensing	Advantages	Limitations	References
Synthetic Polymer	Conducting Polymers (e.g., polyaniline, polypyrrole)	Signal transduction via electrical conductivity	Tunable chemical structure, flexible fabrication	Low selectivity, environmental instability	[5, 54, 108]
Porous Framework	Metal–Organic Frameworks (MOFs)	Selective analyte binding; high porosity scaffold	High surface area, modularity, chemical/ thermal stability	Water sensitivity, complex synthesis	[49, 97, 175]

3.2.1 Nanomaterials

Gold nanoparticles (AuNPs):Gold nanoparticles (AuNPs) have been at the centre stage of new research in the last few years because of its special properties of optics and electricity sensing. This is why their high stability, low resistance, and high conductivity, as well as high interaction with light, make them highly useful in biological sensing. The high extinction coefficient and the ability to change in optical behaviour due to their size and spacing are common elements in colorimetric sensing methods as AuNPs are quickly becoming one of the most generally used methods of creating sensitive and sustainable methods of detection [44, 109].Carbon nanotubes (CNTs):Carbon nanotubes (CNTs) are one special kind of one-dimensional nanomaterial that have a broad potential deployment in biosensing, tissue engineering, drug delivery and even medical diagnostics and cell tracking or labelling. The most remarkable examples of properties are such features of CNTs as an extreme strength and stability, good electrical conductivity, and dirt resistance along with a very large surface area given their size and thus high performance in biological applications [134]. In specific, multi-walled CNT sensors are able spy real-time biomolecules with high accuracy and the best thing is that, they can spy the biomolecules without the need to be labelled with any extra labelling which makes it simpler [87].

Graphene and graphene oxide (GO):Graphene oxide (GO), a single-atomic-layer derivative of graphite whose surface is functionalized with oxygen-containing groups, has become a very promising candidate in biosensor development owing to its inherent electrical stability, exquisite conductivity and high level of chemical functionality. These characteristics render GO to be very compatible in terms of functionalization and inculcation in diverse sensing platforms. As an example, the synthesis of a paper-based electrochemical sensor which represented a detection limit of 30 nM of mercury (Hg 2+) in aqueous samples was achieved by adding carbon nanotubes (CNTs), graphene oxide and gold nanoparticles (AuNPs) [137].

Quantum dots (QDs):Quantum dots (QDs) are nanoparticles produced by the disintegration of bigger carbon-based products e.g. graphene oxide, carbon nanotubes, and carbon fibers. These microscale devices find a lot of application in optical biosensors due to their signifying sensitivity and large detection range. The types of targets that QDs can identify are wide-ranged, such as organic compounds, pharmaceutical-related analytes, biomolecules (including proteins, amino acids, and nucleic acids), and neurotransmitters as well. The sensor construction leaves a QD with a recognized element (as in functionalization) bound to a sheet of graphene. Upon the introduction of a pollutant or a target substance, a measurable change usually through the disruption or alteration of the interaction between the QD and the layer of graphene is induced that can then be translated into a readable signal by the sensor [114].

3.2.2 Biopolymers and Hydrogels

The three-dimensional morphology of hydrogels is a specific addition to biosensing and bioanalytical systems over conventional two-dimensional surfaces, or polymer brush-modified planar substrates. The ability to incorporate multiple recognition elements, in addition to the fact that such an environment supports biological activity, is a great advantage, as it makes the sensing platform highly functional and sensitive [36, 47]. One of the biggest advantages of hydrogels is the ease of manipulation of their physical properties, thus their ability to respond to a large variety of external stimuli like light, sound, pH, temperature, and ionic strength [148]. Due to this adaptability, they are a perfect candidate to come up with intelligent and responsive biosensors. Simultaneously, there has been massive attention given to the use of biopolymers that are renewable and sustainably-based into the development of sensors. Such materials are eco-friendly, in addition to possessing the functionality of being highly selective, sensitive and biocompatible. Since the market pressures increase to use eco-friendly and efficient sensing techniques, biopolymer sensing provides a potential future direction. Scientists are currently studying their potential in such directions as environmental, medical diagnostics, and other tasks where high accuracy in the detection and description of physical properties is needed [1].

3.2.3 Conducting Polymers

Gas sensors frequently use CPs because they provide a great deal of design flexibility. They can create discrete layers where the interaction between the conducting matrix and the analyte gas alters the transduction mechanism's key physical parameter [24, 37, 54, 91, 106]. Conductive polymers can be manufactured artificially or purchased commercially to improve their sensing capabilities. It is also possible to alter their chemical structure. Once the sensor architecture and conductive polymer have been selected, the conductive polymer is coated onto the substrate and patterned using the selected patterning technique to build the sensor structure. The next step is to integrate the sensor into the sensing system, which includes adding electrodes and any necessary readout or data collection components [22, 70, 108].

3.2.4 Metal–Organic Frameworks (MOFs)

A new type of hybrid material includes metal–organic frameworks (MOFs) which have been demonstrated to have great potential in the development of biosensors. They have such properties due to which they are advantageous: they are well-designed crystalline structures, with high surface area and porosity, chemical, and thermal stability, as well as their structures can be easily modified into tunable ones according to the targeted sensing sensibilities [97]. Such properties of MOFs have made them very effective in the capture and detection of a broad group of target analytes. Out of all the applications of theirs, chemical and biological sensing is one of the most effective ones. The selective binding of desired analytes in liquid and solid phases, as well as a modular structure, have made MOFs effective solutions in discovering analytes of interest in both the liquid and solid phases [49, 175, 176]. This has consequently led to MOFs being incorporated in the next-generation biosensors that intend to enhance sensitivity, selectivity, and flexibility of operation.

3.3 Biosensors for Detection of Pollutants

3.3.1 Detection of Heavy Metals

Due to industrial processes, agriculture, and inadequate waste disposal, especially during the untreated wastewater disposal and household waste, things like mercury, lead, copper, cadmium are becoming increasingly prevalent in our environment [130]. These metals are able to accumulate in soil and water thus posing both health and environmental hazards. Scientists have prepared some of the intelligent and sensitive devices to easily identify these toxic metals. An example is graphene oxide and silver nanoparticle-based sensor that can sense mercury ions (Hg^{2+}) by sensing changes in DNA structure wrought by the metal [163].

The other ingenious item is a paper-based sensor which links to a smartphone. It detects presence of mercury in the well and river water using copper nanoclusters, which are formed using sound waves (known as sonication). It is easy, portable, and able to sense mercury even in very low amounts, that is, 1.7 nM [89].

Feng et al., in another report developed a hollow carbon and MoS 2 nanotubes-based colour-changing biosensor. Such sensor is able to detect mercury in wastewater in nanomolar (2.7 nM) concentrations and therefore, can be used as an effective measure of pollution.

3.3.2 Detection of Pesticides and Herbicides

Plant protection agents and pesticides are good because they help keep the crops healthy by averting the diseases and other pests that may destroy the crops. Such substances comprise an enormous selection of chemicals such as plant growth regulators, repellents, acaricides (intended to attack mites and ticks), and herbicides, insecticides, fungicides, etc. [167]. Pesticides can be however dangerous to the environment and human health in as much as they are beneficial in farming. When used in fields, they may leak into soil and get into groundwater, thus possibly propagating into the drinking water supplies [170].

In an attempt to address this, scientists have come up with a number of applications, when it comes to pesticides. Together with classical chromatographic methods, the use of electrochemical biosensors has had increasing success, being widely considered a faster, more portable, and more economical system to track pesticide concentrations in environment [9, 71, 178].

Table 5 Detection capabilities of modern biosensors for environmental pollutants, including their sensitivity (detection limits), materials used, and detection methods

Pollutant type	Example of pollutant	Biosensor (limit of detection)	Biosensor type/material	References
Heavy metal ion	Pb^{2+}	≈ 0.07 μM (~ 70 nM)	Electrochemical with polypyrrole/CNT composite	[153]
Heavy metal ion	Hg^{2+}	0.1 ppb optical (~ 0.5 nM); 0.035 nM electrochemical	Chitosan/AuNP or polyaniline sensors	[153]
Heavy metal (multi-ion)	Pb, Cd, Cu, Hg etc	~ 0.0001–300 nM (broad range)	DNAzyme / electrochemical biosensors	[50]
Pesticides	Atrazine (by algae sensor)	0.11 μmol/L	Algal photosynthesis-inhibition sensor	[7]
Pesticides	organophosphates, chlorpyrifos	~ 3.75×10^{-11} M (≈ 0.04 nM)	Electrochemical enzyme/ aptamer sensors	[45, 93]
Herbicides	Glyphosphate	≈ 3.03 μM (~3×10^{-6} M)	Enzymatic-Electrochemical	[104]
Organic pollutants	Bisphenol A	0.01 nM (~1×10^{-11} M)	Enzymatic-Electrochemical	[50]
Microbial detection	*Salmonella typhimurium*	46 CFU/mL	Aptamer-Elactrochemical	[172]

3.3.3 Detection of Organic Pollutants

Biosensors turned out to be great devices to recognize a substance used to pollute wastewater and dangerous pathogens, which significantly contribute to environmental and population health [55]. Such sensors may not only note the presence of contaminants, but they are also able to measure the available, as well as toxic, parts of toxic substances, such as cadmium (Cd), and toluene, in groundwater and river water [51, 52, 154, 156]. Actually, biosensors have been effectively adopted in determining the extent of availability of some pollutants like mercury, phenanthrene to enter living systems. This helps scientists get a better idea about the practical environmental hazards and proves how application of biosensors can prove beneficial when determining bioavailability of contaminants in natural water lines [23, 46] (Table 5; Fig. 3).

3.3.4 Pathogen and Microbial Detection

Microbial sensors are proving to be very promising in the identification of specific pollutants. They operate based on a mixture of microorganisms such as bacteria with a transducer so as to detect harmful substances in a variety of sources with

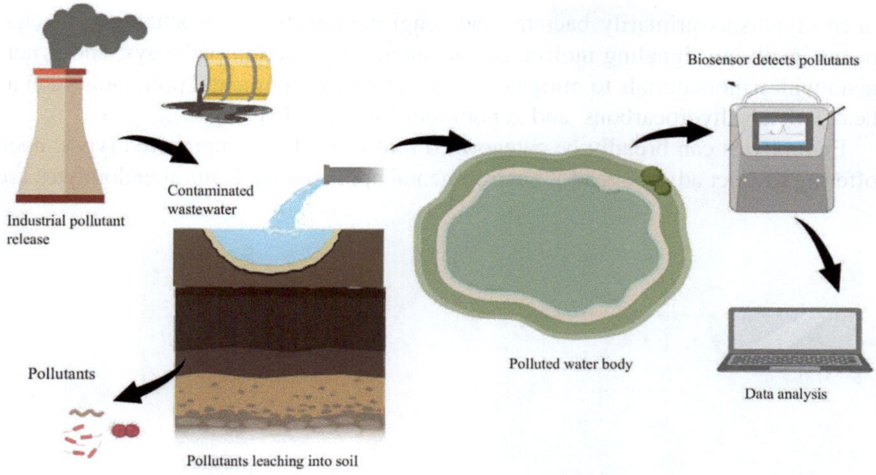

Fig. 3 Biosensor application for water and soil pollutant detection

ease, sensitivity, and with a lot of accuracy as well as being very fast. Bacteria may interact with a variety of chemical substances owing to their very high metabolic dexterity, the flexibility of their genes, and their resilience to various environmental settings. That is why microbial biosensors are particularly useful in the monitoring of the pollutants like fungicides [11]. Traditional lab-based techniques can be used in detecting bacterial contamination on food, water and environmental samples too, but these techniques can be augmented with these biosensors. Nonetheless, there are significant benefits involved in biosensors, such as, they are faster, sensitive, highly specific and usually save on time and few efforts. Due to such advantages, they are increasingly being favoured in surveillance of bacterial infection in both public health and environmental safety [107].

4 Role of Endophytes in Environmental Applications

Endophytic microorganisms possess the ability to sequester metals through a variety of methods, including metal biosorption. Beneficial characteristics of endophytic organisms, such as bacteria and fungi, aid plants in enhancing their well-being, lowering stress levels, and detoxifying metals [129]. Pollutants like metals and dyes are becoming more prevalent in our surroundings. Urbanization and industrialization (mining, electroplating, and wielding) release metals into the environment, including chromium (Cr), copper (Cu), lead (Pb), zinc (Zn), cadmium (Cd), and aluminium (Al) [26, 96, 133, 157]. They engage in mutualistic associations that enhance plant fitness and play vital roles in bioremediation, detoxification of pollutants, biosynthesis of nanoparticles, and environmental resilience of host plants [129, 143]. These microorganisms primarily bacteria and fungi deploy diverse biochemical mechanisms involving signaling molecules, enzymatic degradation pathways, and synergism with nanomaterials to mitigate the impact of environmental pollutants such as heavy metals, hydrocarbons, and xenobiotics [88, 95, 113] (Fig. 4).

Endophytes can broadly be categorized into natural and engineered types, each offering distinct advantages for environmental applications. Natural endophytes are

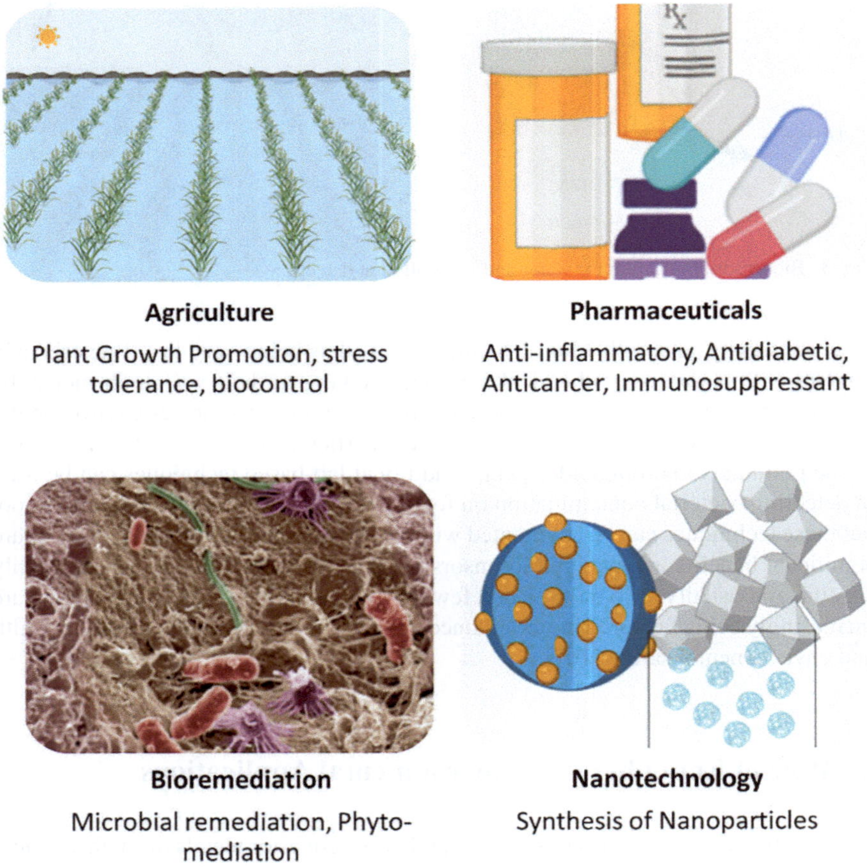

Fig. 4 Diverse applications of microorganisms in biotechnology: Microorganisms play pivotal roles across key sectors, including agriculture (enhancing plant growth, stress resistance, and pest control), pharmaceuticals (producing bioactive compounds like anticancer and antidiabetic drugs), bioremediation (degrading pollutants through microbial or phytoremediation strategies), and nanotechnology (enabling eco-friendly synthesis of nanoparticles). This integration underscores their potential in driving sustainable solutions for agriculture, medicine, environmental conservation, and advanced materials. Harnessing these microbial capabilities is critical to advancing green technologies, biosafety, and global health initiatives

wild-type strains that have coevolved with plants and naturally have functional characteristics like metal chelation, phytohormone production, enzymatic degradation of pollutants, and antioxidant activity. They are ecologically resilient and well-suited to colonize and clean up contaminated areas because of these traits [28]. Engineered endophytes, on the other hand, are genetically altered to improve or add particular metabolic processes utilizing cutting-edge technologies like CRISPR/Cas9. These alterations frequently entail the overexpression of genes that code for detoxifying enzymes, metal-binding proteins, or biosensing reporter elements, allowing for precise, targeted roles in stress reduction, pollutant detection, and biosensor system integration [42, 166]. Natural and artificial endophytes work together to provide a flexible toolkit for environmentally friendly monitoring and cleanup.

4.1 *Endophytes as Bioremediators*

By mobilizing, degrading, or immobilizing contaminants in the soil, encouraging plant growth, reducing phytotoxicity, and enhancing plants' metal tolerance, among other aspects, endophytes have been shown to be important in helping host plants adapt to polluted environments and to improve phytoremediation [25, 38, 74, 160, 177].

Phytoremediation—the use of plants to remove, transfer, or neutralize environmental contaminants—is significantly enhanced when supported by endophytic microbes. These microbes not only assist in pollutant solubilization and mobilization but also alleviate plant stress, enhance biomass production, and contribute to detoxification mechanisms [10] (Fig. 5).

4.2 *Chelation and Solubilization of Metals*

Endophytes produce siderophores, organic acids, and metallothioneins that chelate and solubilize toxic metals such as cadmium (Cd), lead (Pb), chromium (Cr), and arsenic (As), thereby facilitating their uptake and translocation within host plants [28]. In the context of metal mobilization, endophytes facilitate the bioavailability of heavy metals through the secretion of siderophores and biosurfactants. These molecules chelate metal ions and increase their solubility in the rhizosphere and endosphere, enhancing uptake by plant roots. For instance, biosurfactant-producing strains of *Bacillus subtilis* improved arsenic and chromium mobilization in rice (*Oryza sativa*), thereby accelerating phytoextraction [102]. Similarly, siderophore-producing endophytic bacteria from *Miscanthus* species have been shown to assist in mobilizing lead and cadmium in mining soils [154, 156].

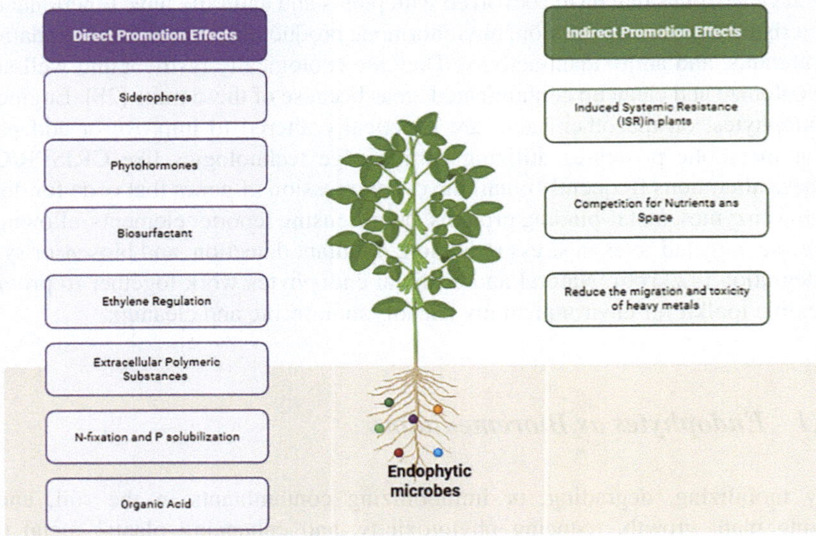

Fig. 5 Endophyte-assisted phytoremediation of heavy metals

4.3 Detoxification Enzymes

Enzymes such as catalase, peroxidase, superoxide dismutase (SOD), and glutathione Stransferase (GST) neutralize reactive oxygen species (ROS) generated under heavy metal stress [171]. The synthesis of antioxidant enzymes (such as catalase, superoxide dismutase, and glutathione S-transferase), the sequestration of toxic metals in vacuoles, and the modification of stress-responsive genes are some of the methods used to reduce phytotoxicity. Endophytes preserve cellular homeostasis by reducing the buildup of reactive oxygen species (ROS) that usually follows exposure to heavy metals [127, 171]. Additionally, the bioavailability of harmful ions in their most reactive forms is decreased by the synthesis of metal-chelating molecules such as metallothioneins, siderophores, and organic acids (such as citric, oxalic, and malic acids), which lessens the toxicity of these ions to host plants [28]. Certain fungal endophytes like *Trichoderma harzianum* and *Piriformospora indica* secrete organic acids (e.g., oxalic and citric acids) and metallothioneins that bind and neutralize toxic ions, reducing their bioavailability and toxicity within plant tissues [63].

4.4 Phytohormone Modulation

The production of phytohormones such indole-3-acetic acid (IAA), gibberellins, cytokinins, and the activity of 1-aminocyclopropane-1-carboxylate (ACC) deaminase are the main mechanisms that promote plant development in contaminated environments. Even in the face of metal stress, these biomolecules support the maintenance of normal biomass production and root architecture [32, 39]. It has been demonstrated that endophytes such as *Bacillus subtilis, Azospirillum brasilense*, and *Pseudomonas fluorescens* create IAA and solubilize phosphate, which further facilitates root growth and nutrient uptake [61, 78]. This improves the surface area available for metal uptake and translocation in addition to improving plant vigor. Recent studies show that inoculation with *Pseudomonas koreensis* significantly improved biomass and cadmium tolerance in *Brassica juncea* by regulating auxin biosynthesis and metal transporter expression [75].

4.5 Ethylene Regulation

ACC deaminase-producing endophytes help plants tolerate stress by lowering levels of stress-induced ethylene, a hormone that can inhibit root elongation [40].

4.6 Endophytic Signaling Molecules and Enzymes

Numerous signaling molecules and enzymes that facilitate plant–microbe interactions and pollution detoxification are produced by endophytes. For protection, adaption, and communication in contaminated environments, these biochemical compounds are essential (Table 6).

4.7 Synergistic Action with Nanomaterials

Nanotechnological methods have been used in industry, health, and agriculture [12, 33, 179]. Sizes between 1 and 100 nm in diameter, as well as a variety of physical and chemical characteristics, biological activity, and enhanced reactivity, are characteristics of nanoparticles (NPs) [3, 30]. Gold (Au), silver (Ag), nickel (Ni), platinum (Pt), titanium (Ti), zinc (Zn), palladium (Pd), and other NPs are synthesized in a variety of shapes and colors for use in gas sensing [84, 147, 173], chemical sensing, bioimaging [18, 29, 53, 169], bioimaging, gas sensing, CO_2 capturing [35, 117], and other related applications [92] (Fig. 6).

Table 6 Enzymes and signaling molecules produced by endophytes involved in bioremediation

Molecule/ enzyme	Function	Endophyte source	Target pollutants	References
Siderophores	Metal chelation and solubilization	*Bacillus subtilis*	Fe, Cr, As	[28]
ACC Deaminase	Reduces ethylene stress to enhance root growth	*Pseudomonas putida*	Cd, Pb	[39]
Laccases	Oxidation of phenolic and aromatic pollutants	*Aspergillus niger*	PAHs, phenols	[123]
Glutathione S-transferase (GST)	Detoxification of xenobiotics by glutathione conjugation	*Burkholderia cepacia*	Organic pollutants	[66]
Biosurfactants	Solubilize hydrophobic pollutants	*Pseudomonas aeruginosa*	Hydrocarbons, PAHs	[15]
Catalase	Degrades hydrogen peroxide (ROS scavenging)	*Fusarium oxysporum*	Heavy metal-induced ROS	[168]
Superoxide dismutase (SOD)	Converts superoxide radicals to hydrogen peroxide	*Penicillium chrysogenum*	Cd, Cr, Pb	[6]
Peroxidase	Breaks down peroxides, reduces oxidative stress	*Colletotrichum gloeosporioides*	Ni, Zn, Cu	[78]
Phenol oxidase	Degrades phenolic compounds	*Phoma sp.*	Phenols, industrial dyes	[135]
Organic acids (e.g., oxalic, citric)	Metal solubilization and chelation	*Trichoderma harzianum*	Pb, Cd, Zn	[115]
Indole-3-acetic acid (IAA)	Enhances root development and pollutant uptake	*Enterobacter cloacae*	Indirectly supports remediation	[124]
Nitric oxide (NO)	Signaling molecule regulating stress responses	*Piriformospora indica*	Oxidative stress due to metals	[73]

(continued)

Table 6 (continued)

Molecule/ enzyme	Function	Endophyte source	Target pollutants	References
Hydrogen cyanide (HCN)	Suppresses metal-tolerant pathogens and aids metal mobility	*Pseudomonas fluorescens*	Ni, Zn, Cu	[21]
Azoreductase	Degrades azo dyes through reductive cleavage	*Lysinibacillus sphaericus*	Synthetic dyes	[138]
Nitroreductase	Reduces nitroaromatic pollutants	*Bacillus amyloliquefaciens*	TNT, DNT	[99]

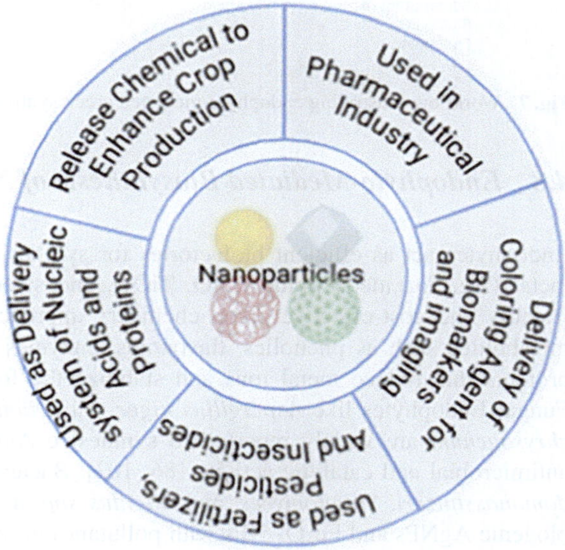

Fig. 6 Schematic representation of the versatile applications of nanoparticles

Endophytes are gaining attention for their ability to adapt and produce useful compounds. They can help create, modify, and work with nanomaterials (NMs). These plant-associated microbes support the green synthesis of nanoparticles (NPs), improve their stability and availability, and assist in breaking down pollutants or accumulating metals.

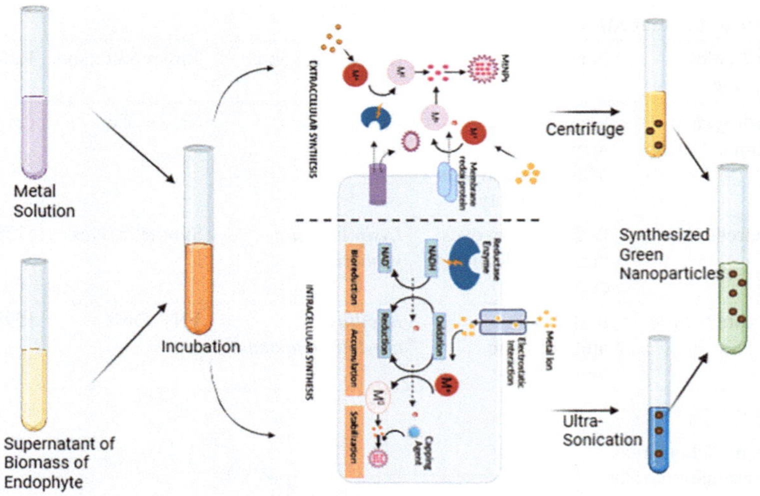

Fig. 7 Workflow illustrating endophyte-mediated green synthesis of nanoparticles

4.8 Endophyte-Mediated Biosynthesis of Nanoparticles

Endophytes act as efficient biofactories for synthesizing nanoparticles, including metal (Ag, Au),metal oxide (ZnO, TiO_2), and selenium nanoparticles, via eco-friendly and cost-effective green chemistry approaches. These microbes secrete metabolites such as phenolics, flavonoids, enzymes (e.g., nitrate reductase), and proteins that reduce metal ions and stabilize the formed nanoparticles (Fig. 7). Fungal Endophytes like *Aspergillus niger*, *Fusarium oxysporum*, and *Penicillium chrysogenum* are widely reported to synthesize AgNPs and ZnONPs with high antimicrobial and catalytic activity [86, 103]. Bacterial endophytes such as *Pseudomonasstutzeri*, *Streptomyces*, and *Bacillus spp.* have been shown to synthesize biogenic AgNPs and Fe_3O_4 NPs with pollutant-removing capabilities [138]. These nanoparticles often show enhanced pollutant degradation, owing to high surface area ancatalytic potential, while maintaining lower toxicity due to the biocompatible synthesis process.

4.9 Enhancement of Pollutant Degradation and Metal Remediation

In contaminated environments, endophyte-assisted nanoparticles have synergistic effects in the mobilization of heavy metals and the breakdown of xenobiotic substances. A few of the mechanisms are: increased metal solubilization and chelation: Metals such as Cr, Pb, Cd, and as are mobilized more effectively by biogenic

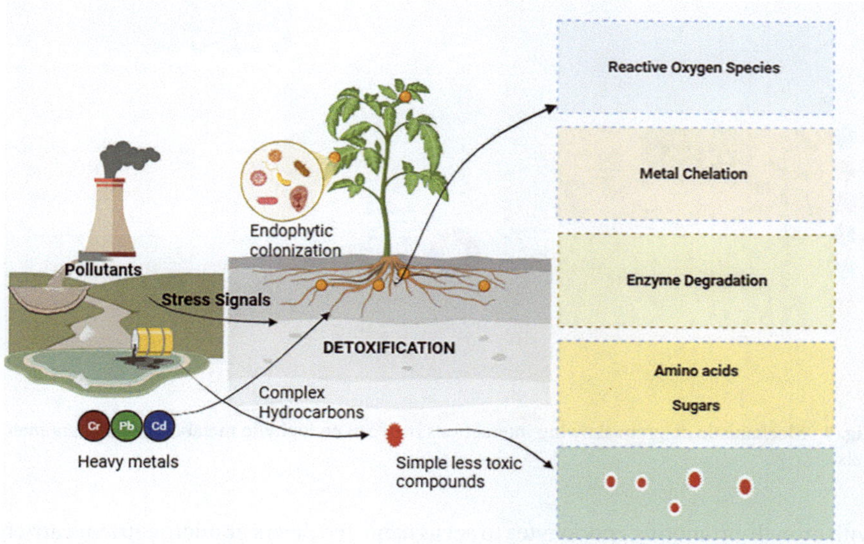

Fig. 8 Plant endophytes' contribution to environmental cleanup. Heavy metals, oil, and pollutants from industrial sources cause plants to experience stress, which activates endophytes in the plants. By regulating reactive oxygen species, chelating metals, and producing degradation enzymes, these endophytes aid in detoxification and make it possible to effectively clean up contaminated soil

NPs functionalized with siderophores and biosurfactants generatedby endophytes [5]. Catalytic degradation of organic pollutants: Endophyte-produced ZnONPs and FeO_4 NPshave potent redox and photocatalytic capabilities that break down dyes, phenolic chemicals, andinsecticides [101]. Oxidative stress mitigation: When plants are exposed to heavy metals, ROS-induced damage is lessened by nanoparticles functionalized with antioxidant-rich fungal extracts [85]. In one study, ZnONPs produced from *Trichoderma harzianum* improved cadmium phytoextraction and plant tolerance in *Brassica* species [65]. AgNPs produced from *Streptomyces* also enhanced rice development under arsenic stress and broke down methylene blue [138] (Fig. 8).

4.9.1 Delivery of Nanoparticles Through Endophytic Carriers

Endophytes can serve as nanomaterial carriers and delivery systems by infiltrating internal plant tissues and enabling the movement of nanoparticles to particular locations (e.g., roots, vascular tissues).Targeted pollutant removal in contaminated plant tissues; sustained release of nanoparticles for long-term remediation; minimized toxicity and environmental accumulation, due to biological encapsulation and controlled release. The distribution of nanoparticles through endophytes is gaining attention, particularly in nano-enableagriculture, where biosynthesized NPs work in concert

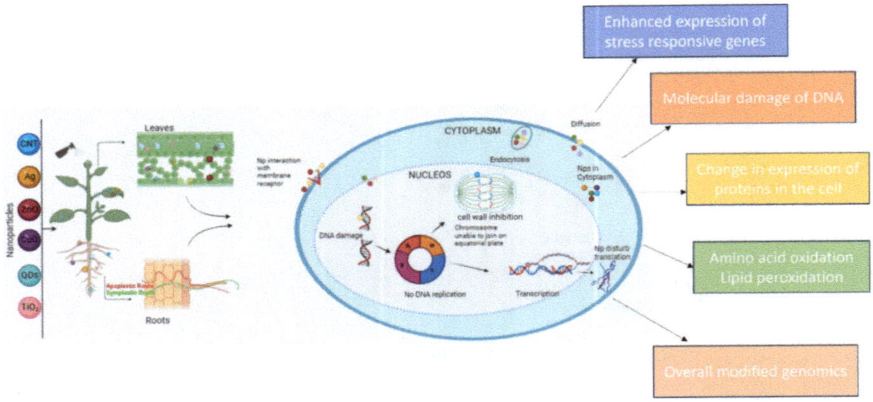

Fig. 9 Mechanistic diagram showing interactions between endophytic metabolites and nanomaterials

with growth-promoting endophytes to act as nano-fertilizers or micronutrient carriers [144].

4.9.2 Applications in Nano-phytoremediation

The use of nanoparticles to improve plant-based remediation is being referred to as "nano-phytoremediation" Endophytes play a major role in this by: Improving the absorption and dispersion of nanoparticles in host plants; modulating plant responses to metal stress through hormone synthesis and stress-responsive gene activation; strengthening detoxification and antioxidant defense systems. For example, in *Spinacia oleracea* (spinach) cultivated in contaminated soils, endophyticconsortia seeded with Fe_3O_4 nanoparticles enhanced chromium accumulation and decreased phytotoxicity [116] (Fig. 9).

5 Integration of Endophytes in Biosensor Design

Biosensors and endophyte-aided phytoremediation are transforming traditional phytoremediation into a smart, sustainable solution for environmental monitoring and clean-up [16]. phytoremediation techniques do not offer real time information about the level of pollutants and the condition of the plant hence their low efficacy. The biosensors eliminate this gap and allow one to monitor the contaminants, the expected conditions of the soil, and the activities of the beneficial microbes [19]. New types of biosensors are microbial whole-cell sensors, genetically designed biosensors, enzyme-based sensors, and nanomaterial-amplified devices detecting extremely low quantities of heavy metals, hydrocarbons, or different toxins [94, 132]. This is

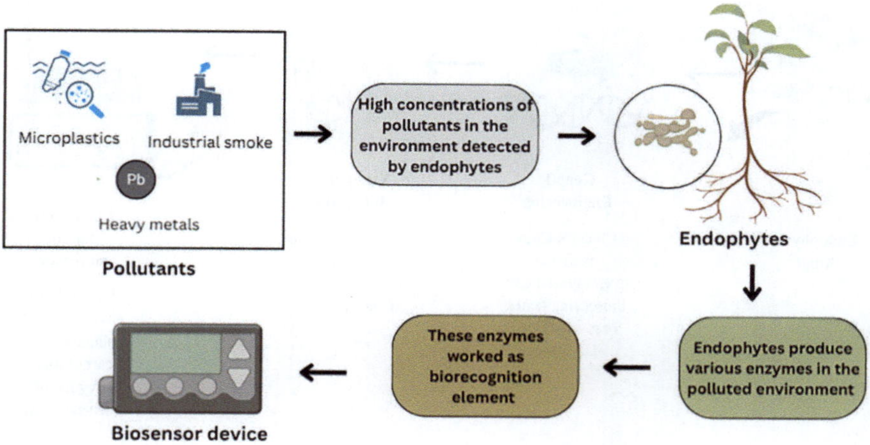

Fig. 10 Schematic overview showing the synergy between endophytes and biosensors in environmental monitoring

because miniaturization of the biosensors and incorporation of wireless technology can enable site managers to monitor field data by remote means thus they can make site management decisions promptly. Nanotechnology and genetic engineering have recently advanced sensor stability and specificity, thus being more appropriate in the field [62] (Fig. 10).

In the meantime, endophytes are beneficial bacteria and fungi residing within the plant tissues and essential factors that support plant growth during the period of stress, clean up or transform pollutants and enhance the effectiveness of phytoremediation [2, 162]. The most recent developments indicate that engineered endophytes may be more effective in breaking down resistant contaminants and even be able to emit warning signals capable of being registered by biosensors, forming an entirely self-sustaining circle of monitoring and control [100]. New work indicates that it is possible to use microbial biosensors to track the redox conditions in soils, directly controlling the metabolic processes of endophytes and pollutant degradation. The combination has the potential to make traditional phytoremediation more flexible, low impact and in line with green engineering and circular economy objectives [19, 132]. The synergy appears when the biosensors allow one to detect monitoring the activity and health of endophytes within the host plant, making sure that these so-called hidden helpers are working well [94]. Also, systems of integrated biosensors-endophytes can be integrated within a wider context of intelligent environmental management. As an example, microbial biosensors might monitor the redox status of soils, to show when conditions are most favorable to stimulate endophytes to degrade hydrocarbons [16]. These are promising but still challenging to implement in practice: the calibration of sensors under the changing conditions of the field, consistency of biological elements of the sensory system, the long-term viability of settled (introduced or engineered) endophytes [62].

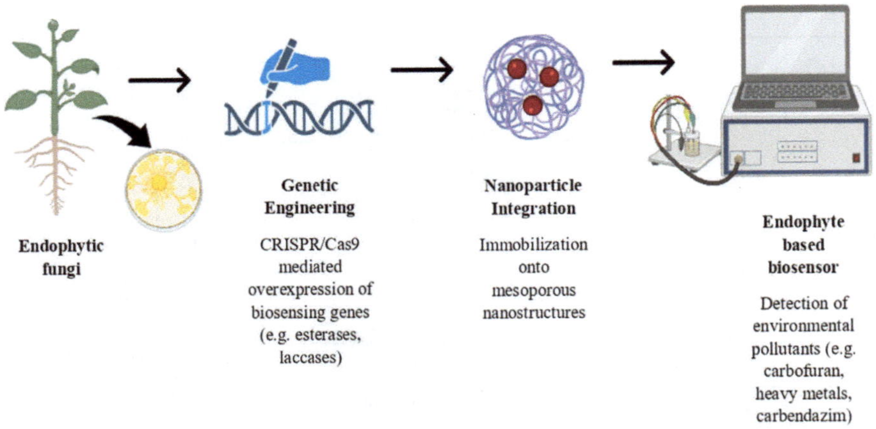

Fig. 11 Bioengineering endophytes for biosensor applications

5.1 Genetic and Metabolic Engineering of Endophytes

By examining genetic advances and discovering their control mechanisms through genetic engineering, especially with next-generation sequencing, we can gain a deeper understanding of the enigmatic nature of endophytic fungi [118]. The CRISPR/Cas9 system allows detailed and efficient modifications to a cell's DNA, making it a tool for genome editing. One of the application of endophytic fungi is that; they are essential to agriculture because they grow inside plants and have advantages for their host. Endophytic fungi will gain from overexpressing genes that produce secondary metabolites using CRISPR/CAS9. For genetically altering fungal endophytes, CRISPR/Cas9 is an advanced gene-editing technique that is essential [152]. A study found that the endophytic fungus *P. fici* effectively created a CRISPR/Cas9-dependent genome editing system. To the best of our knowledge, this is the first report of successfully using CRISPR/Cas9 to induce genetic modifications in endophytic fungi. With this system, we achieved single-locus genome editing, including site-specific gene insertion and deletions of long DNA fragments (5–10 kb), with accuracy rates as high as 27.2% and 48.0%, and simultaneous dual-locus genome editing of multiple loci with only one selection marker integration via a single transformation [166] (Fig. 11).

5.2 Functionalization of Sensor Platforms with Endophytic Biomolecules

According to recent studies, fungal endophytes' exceptionally have high stability, biocompatibility, and utility make them useful biosensing elements [17, 42]. The

Table 7 Examples of endophyte derived biomolecules used in sensor interfaces

Biomolecule	Endophyte source	Function in sensor	Sensor application
Lipase	Fungal endophyte	Enzyme immobilized on Zn layered NP_s	Carbendazim detection
Esterases	*Eupenicillium shearii* FREI-39	Biocatalyst for hydrolysis	Carbofuran pesticide detection
Silver nanoparticles ($AgNP_{s)}$)	*Rothia endophytica*	Conductive interface + antimicrobial	Colorimetric sensor paltform
ZnS nanoparticles	*Aspergillus flavus*	Optical probe/ fluorescence	Metal ion detection
Fe_3O_4 nanocomposites	*Aspergillus flavus*	Magnetic + biocatalytic surface	Dye/pollutant sensing
Laccase/tyrosinase	Endophytic fungi	Oxidoreductase enzyme for biosensing	Phenol/pesticide biosensors

utility of microorganisms in biotechnology is increased when endophytic fungi are used to produce biosensors. Biomolecules that are derived from different endophytic fungi having different sensor applications shown in Table 7.

5.3 *Endophyte-Nanomaterial Hybrids: Bio-Nano Interfaces*

Uddandarao and Balakrishnan [145] state that the fungal endophyte-derived nanoparticles are promising candidates for the development of nano-biosensors due to their notable physicochemical properties as well as their electronic, fluorescence, and surface plasmon resonance characteristics. Although the use of microorganisms to produce metal nanoparticles is growing in popularity, the potential of endophytes in nanoparticle synthesis is still largely untapped. Endophyte intervention in the creation of nanoparticles opens up new possibilities for novel nanomaterials with a variety of applications [13]. Combining endophytes and nanoparticles may lead to the development of synergistic systems that more efficiently target and degrade water contaminants (Fig. 10). Nanoparticles (NPs) like AgNPs, ZnONPs, and FeO_4 NPs can be biosynthesised and functionalized by endophytes, which are known to produce laccases, siderophores, and antioxidant enzymes. In complex environmental matrices, these biogenic nanoparticles interface with endophytic enzymes or whole-cell structures to produce synergistic configurations that enhance binding specificity, reduce background interference, and amplify detection signals [86, 138]. By removing the requirement for dangerous chemicals commonly employed in the creation of nanoparticles, these technologies also promote environmentally friendly fabrication methods.

The self-regulating properties of these hybrids are a noteworthy benefit. During pollutant contact, endophyte-mediated redox enzymes, including laccase or glutathione-S-transferase, are adsorbed onto NP substrates, which improves electron transfer. For instance, sub-nanomolar sensitivity in phenol detection using electrochemical impedance spectroscopy has been demonstrated using fungal endophyte-derived AgNPs in combination with tyrosinase enzymes [17, 42]. According to Uddandarao and Balakrishnan [145], FeO_4 nanoparticles that are biosynthesized using *Aspergillus flavus* have also been included into magnetic biosensors to detect industrial dyes in wastewater with great stability and reusability.

Both signal transduction and real-time remediation are functionally supported by these hybrid systems. By reacting to environmental stimuli like pH changes, the presence of metal ions, or organic pollutant fingerprints, the embedded biological elements can activate localized detoxification processes and send detectable optical or electrochemical signals. They are good candidates for creating self-sufficient, field-deployable biosensors that can perform feedback-based remediation and continuous monitoring because of their versatility. Essentially, Endophyte–Nanoparticle Hybrid Systems represent the fusion of environmental technology, materials science, and microbial symbiosis, advancing the development of intelligent biosensors that are both active environmental cleaners and passive detectors. As they continue to advance, they could have revolutionary uses in industrial pollutant management, precision agriculture, and water quality monitoring (Fig. 12).

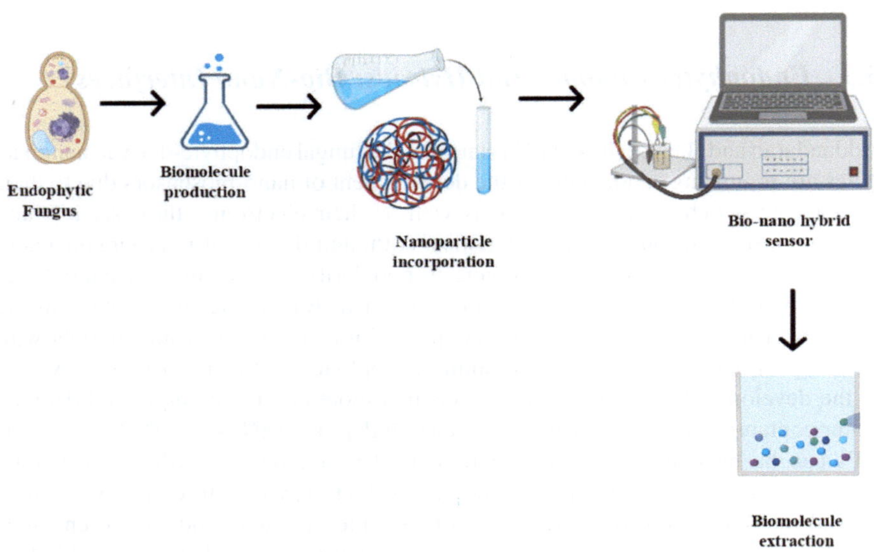

Fig. 12 Conceptual figure showing bio-nano hybrid sensor platform integrating endophyte molecules

6 Case Studies

6.1 Endophyte-Based Biosensors for Arsenic and Mercury

Heavy metals such as arsenic (As) and mercury (Hg) are among the most toxic environmental pollutants, with significant implications for soil quality, groundwater safety, and food chain contamination [8]. Recent advances have demonstrated that endophyte-based biosensors offer sensitive and selective detection of these metals, especially when integrated with nanomaterials for signal enhancement. In one study, *Bacillus cereus* endophytes isolated from rice roots were engineered to express metallothionein and phytochelatin synthase genes, enabling arsenic biorecognition [56, 57]. Immobilization of these engineered strains onto a gold nanoparticle (AuNP)-modified electrode produced a biosensor capable of detecting arsenite (As^{3+}) at nanomolar concentrations through differential pulse voltammetry [81].

Similarly, mercury biosensors have been developed using fungal endophytes such as *Penicillium chrysogenum*, which naturally produce mercuric reductase enzymes [126]. When interfaced with carbon nanotube (CNT)-based electrodes, these bioelements demonstrated high specificity for Hg^{2+} ions with detection limits below 1 μg/L [111]. The incorporation of green-synthesized silver nanoparticles further improved electron transfer and signal stability [34]. Table 5 summarizes selected examples of endophyte-based biosensors developed for metal pollutants, highlighting their recognition elements, sensor configurations, detection ranges, and environmental applications.

6.2 Real-Time Monitoring of Soil and Water Pollutants

The transition from laboratory-based detection to real-time field monitoring is essential for effective environmental risk assessment and decision-making [112]. Field-deployable biosensors leveraging endophyte-derived enzymes and cells have been tested in various contaminated ecosystems [43]. In one case, a portable electrochemical biosensor was constructed using *Streptomyces*-derived lipase immobilized on a graphene oxide (GO) matrix [20, 146]. This device was capable of detecting phenolic compounds in irrigation runoff with real-time readout via a mobile interface. The biosensor showed high repeatability and accuracy over 48-h continuous operation in a semi-urban field site.

Another study utilized an endophyte-based colorimetric biosensor to detect nitrate and pesticide residues in agricultural drainage systems. Here, nitrate reductase-expressing *Pseudomonas aeruginosa* strains were encapsulated in a hydrogel matrix integrated into a paper-based lateral flow device. The system provided visible color change within minutes of sample application, with potential for deployment in remote farming regions [72].

Figure 8 presents an image of a functional biosensor deployed at a contaminated wetland site, demonstrating the potential for integrating microfluidic platforms, sensor arrays, and wireless telemetry for continuous environmental monitoring.

6.3 Field-Deployable Devices in Agriculture and Industry

The application of endophyte-based biosensing systems is rapidly expanding into agriculture and industrial sectors, where real-time diagnostics can inform resource use, contamination control, and environmental compliance. In agriculture, biosensors developed using endophytic *Trichoderma harzianum* have been adapted for monitoring cadmium levels in soil and evaluating the bioavailability of micronutrients [58]. These systems, when integrated with GPS-enabled mobile devices, allow spatial mapping of soil health and contaminant distribution, aiding precision farming practices.

In the industrial context, wastewater treatment plants have piloted the use of endophyte-based electrochemical sensors for detecting residual chromium and phenol concentrations. A case study involving a tannery effluent site utilized *Aspergillus niger*-derived biosensors for Cr^{6+} detection and feedback-based optimization of treatment conditions [80]. These biosensors demonstrated high durability under harsh pH and temperature variations, making them suitable for continuous monitoring applications [82]. Overall, the case studies illustrate the versatility, scalability, and reliability of endophyte-based biosensors in diverse environmental settings. Their adaptability to different transducer formats, pollutant classes, and deployment conditions underscores their potential as next-generation tools for sustainable monitoring and remediation systems.

6.4 Sector-Specific Applications of Endophyte-Driven Biosensors

Agriculture: The current agricultural strategies can change the way pollutants are detected and plant protection through endophyte-integrated biosensors. Agrochemical overloading causes problems to croplands, and these include pesticides and metal-based fertilizers. Besides the capacities of increasing the plant growth, endophytic microbes also ensure biotransformation of pollutants as biological filtering, due to the action of specific enzymes, such as laccase, and peroxidase. To illustrate, siderophore- and ACC deaminase producing endophytes could increase the plant resilience and at the same time rhizosphere detoxification [141]. This incorporation into sensor ion may provide real-time measurements of soil heavy metals, pesticide presence, which may be useful in precision farming and food safety.

Industrial Management: In manufacturing, mainly when related to industries such as mining, tanning, and chemical production, the pollution may prove complex and cannot be controlled by the conventional approach. In this case, both the detecting and detoxification of pollutants could be done through endophyte-mediated biosynthesis of nanoparticles (including Ag, ZnO, Fe_3O_4). These industrial toxin-seeking nanomaterials can be involved as an active component of biosensors that provide real-time feedback to them [83]. These endophytes can be modified to render the management of waste a proactive and sustainable activity when implemented using pollutant specific genetic circuits.

Urban Areas: Biosensing with endophytes is also advantageous in urban habitats whose pollution sources, like particles in the air, car exhaust, plastic waste, constitute a wide range of pollution. Such sensors implanted in green infrastructure such as biofilters, rooftop gardens or bioswales systems would be able to continuously monitor urban pollution. Their nature of connectivity through IoT allows the instant supply of data to the city planners or environmental agencies, offering the possibility of intervention with the quick reaction to specific areas as well as local action [31]. By integrating into a single device detection and degradation, the bioengineered endophytes will allow the cities to switch to the more intelligent and green environmental management systems.

7 Challenges and Limitations: A Balanced Critique

1. ***Field Deployment Instability***

Endophytes engineered in the laboratory have displayed promising potential in a regulated setting, however, but their performance and stability tend to decrease in open fields environment. This has been seen to be due to the uncontrollable factors like the variability in composition of soil, competition by microbes as well as climatic changes.

Recommendations: There is a need to conduct multi location field trials to test efficacy and uniformity of various agro climatic conditions.

2. ***Regulatory and Biosafety Barriers***

The outbreak of genetically engineered microbes into the environment has regulatory and biosafety issues. The approvals are usually excessively slow and highly restrictive, with a view to environmental security and chances of horizontal gene transfer.

Future Direction: Research should be more oriented towards the building of risk assessment models and biosafety verification that allows safe and ethical deployment.

3. ***Genetic material; stability and function***

Genetically modified strains can also end up becoming dysfunctional due to the development of genetic mutations, metabolic load or due to the conditions in an environment. This limits their long term use and applicability.

Ways to escape: Forming the genetic stability affected by nature in particular conditions and functional stability and desirable in real conditions may be done with the help of modern technologies, specifically, genome editing using CRISPR/Cas or adaptive laboratory evolution.

8 Conclusion and Future Perspective

The convergence of biosensor technology and endophyte-assisted green remediation presents a transformative approach to address the growing complexity of environmental pollution. Traditional methods often lack the sensitivity, sustainability, or adaptability required for real-time monitoring and field-scale application. Endophytes, with their innate ability to tolerate and detoxify pollutants, not only enhance phytoremediation but also serve as platforms for biosynthesis and delivery of functional nanomaterials. These microbes modulate stress responses, secrete chelating agents and antioxidant enzymes, and can be genetically engineered to augment pollutant-specific pathways. Biosensors integrated with nanomaterials and endophytic biomolecules provide rapid, sensitive, and field-deployable systems for detecting heavy metals, organic pollutants, and microbial contaminants. Additionally, the functionalization of sensor interfaces with endophytic enzymes and metabolites improves specificity and environmental stability. Emerging concepts such as nano-phytoremediation, endophyte-mediated nanoparticle delivery, and bio-nano hybrid sensor platforms offer scalable and low-impact alternatives to conventional remediation. Together, these technologies enable adaptive pollution management aligned with circular economy principles and planetary health goals. Advancing interdisciplinary collaborations across materials science, biotechnology, and ecological engineering will be essential to transition these integrative systems from lab-scale innovation to real-world environmental applications.

Figure and Data Permissions

All figures and data presented in this chapter are the original work of the authors. Therefore, no external permission was required for their inclusion.

Acknowledgements The authors would like to express their gratitude for the financial assistance provided by the DST-FIST(DST/2022/1012) and DST PURSE Project to Manipal University Jaipur from the Government of India. The Dr. Ram Das Pai Fellowship, awarded by Manipal University Jaipur, is highly acknowledged.

Figure and Data Permissions All figures and data presented in this chapter are the original work of the authors. Therefore, no external permission was required for their inclusion.

References

1. Abady, M.M., Shawky, A.M., Sakr, F.A., Mohammed, D.M., Goda, E.S.: Recent advancements in biosensors using biopolymers. In: Bio-Based Polymers: Farm to Industry, vol. 2, pp. 81–11 (2024).
2. Abou-Shanab, R.A.I., El-Sheekh, M.M., Sadowsky, M.J.: Role of rhizobacteria in phytoremediation of metal-impacted sites. In: Emerging and Eco-Friendly Approaches for Waste Management, pp. 299–328. Springer, Singapore (2018). https://doi.org/10.1007/978-981-10-8669-4_14
3. Adeleke, B.S., Akinola, S.A., Adedayo, A.A., Glick, B.R., Babalola, O.O.: Synergistic relationship of endophyte-nanomaterials to alleviate abiotic stress in plants. Front. Environ. Sci. **10** (2022). https://doi.org/10.3389/fenvs.2022.1015897
4. Aghili, S., Golzary, A.: Greening the earth, healing the soil: a comprehensive life cycle assessment of phytoremediation for heavy metal contamination. Environ. Technol. Innov. **32**, 103241 (2023). https://doi.org/10.1016/J.ETI.2023.103241
5. Ahmed, T., Masood, H.A., Noman, M., AL-Huqail, A.A., Alghanem, S.M.S., Khan, M.M., Muhammad, S., Manzoor, N., Rizwan, M., Qi, X., Abeed, A.H.A., Li, B.: Biogenic silicon nanoparticles mitigate cadmium (Cd) toxicity in rapeseed (*Brassica napus* L.) by modulating the cellular oxidative stress metabolism and reducing Cd translocation. J. Hazard. Mater. **459**, 132070 (2023). https://doi.org/10.1016/j.jhazmat.2023.132070
6. Ali, E.A.M., Sayed, M.A., Abdel-Rahman, T.M.A., Hussein, R.: Fungal remediation of Cd (ii) from wastewater using immobilization techniques. RSC Adv. **11**(8), 4853–4863 (2021)
7. Amaral, D.S.D., Tholozan, L.V., Bonemann, D.H., Jansen-Alves, C., Boschetti, W., Novo, D.L.R., Carreno, N.L.V., Pereira, C.M.P.: Algal Biosensors for detection of potentially toxic pollutants and validation by advanced methods: a brief review. Chemosensors **12**(11), 235 (2024)
8. Angon, P.B., Islam, M.S., Shreejana, K.C., Das, A., Anjum, N., Poudel, A., Suchi, S.A.: Sources, effects and present perspectives of heavy metals contamination: soil, plants and human food chain. Heliyon **10**(7), e28357 (2024). https://doi.org/10.1016/j.heliyon.2024.e28357
9. Arduini, F., Cinti, S., Caratelli, V., Amendola, L., Palleschi, G., Moscone, D.: Origami multiple paper-based electrochemical biosensors for pesticide detection. Biosens. Bioelectron. **126**, 346–354 (2019)
10. Aryal, M.: Phytoremediation strategies for mitigating environmental toxicants. Heliyon **10**(19), e38683 (2024). https://doi.org/10.1016/j.heliyon.2024.e38683
11. Asghar, U., Malik, M.F., Javed, A.: Pesticide exposure and human health: review. J. Ecosyst. Ecogr. **5**, 1–2 (2016)
12. Audah, K.A.: Drug discovery: a biodiversity perspective. In: Siddiquee, S., Melvin, G.J.H., Rahman, M.M. (eds.) Nanotechnology: Applications in Energy, Drug and Food, pp. 249–265. Springer International Publishing (2019). https://doi.org/10.1007/978-3-319-99602-8_12
13. Baker, S., Santhosh, P., Rakshith, D., Satish, S.: Screening of bacterial endophytes inhabiting *Mimosa pudica* L. Sci. J. Microbiol. **1**(5), 136–140 (2012)
14. Balakrishnan, R.M., Uddandarao, P., Raval, K., Raval, R.: A perspective of advanced biosensors for environmental monitoring. In: Tools, Techniques and protocols for monitoring Environmental Contaminants, pp. 19–51 (2019)
15. Banat, I.M., De Rienzo, M.A.D., Quinn, G.A.: Microbial biofilms: biosurfactants as antibiofilm agents. Appl. Microbiol. Biotechnol. **98**(24), 9915–9929 (2014). https://doi.org/10.1007/s00253-014-6169-6
16. Barathi, S., Lee, J., Venkatesan, R., Vetcher, A.A.: Current status of biotechnological approaches to enhance the phytoremediation of heavy metals in India—a review. Plants **12**(22), 3816 (2023). https://doi.org/10.3390/plants12223816
17. Barboza, L.G.A., Cózar, A., Gimenez, B.C.G., Barros, T.L., Kershaw, P.J., Guilhermino, L.: Chapter 17 - Macroplastics pollution in the marine environment. In: Sheppard, E. (ed.) World

Seas: An Environmental Evaluation, 2nd edn., pp. 305–328. Academic Press (2019). https://doi.org/10.1016/B978-0-12-805052-1.00019-X

18. Bareket, L., Inzelberg, L., Rand, D., David-Pur, M., Rabinovich, D., Brandes, B., Hanein, Y.: Temporary-tattoo for long-term high fidelity biopotential recordings. Sci. Rep. **6**(1), 25727 (2016). https://doi.org/10.1038/srep25727
19. Basu, S.K., Kovalchuk, I.: Biosensing with plants: plant receptors for sensing environmental pollution. In: Zourob, M. (eds.) Recognition Receptors in Biosensors. Springer, New York (2010). https://doi.org/10.1007/978-1-4419-0919-0_10
20. Beroigui, O., Errachidi, F.: Streptomyces at the Heart of several sectors to support practical and sustainable applications: a review. Progr. Microb. Mol. Biol. **6**(1 SE-) (2023). https://doi.org/10.36877/pmmb.a0000345
21. Bharti, A., Maheshwari, H.S., Sharma, M.P., Prakash, A.: Microbial-mediated abiotic stress tolerance in soybean plants. In: Microbial Mitigation of Stress Response of Food Legumes, pp. 209–228. CRC Press (2020)
22. Bhattacharyya, A.S.: Conducting polymers in biosensing: a review. Chem. Phys. Impact **8**, 100642 (2024). https://doi.org/10.1016/j.chphi.2024.100642
23. Brányiková, I., Lucáková, S., Kuncová, G., et al.: Estimation of Hg (II) in soil samples by bioluminescent bacterial bioreporter *E. coli* ARL1, and the effect of humic acids and metal ions on the biosensor performance. Sensors **20**(11), 3138 (2020)
24. Chen, C.H., Luo, S.C.: ACS Appl. Mater. Interfaces **7**, 21931–21938 (2015)
25. Chen, L., Luo, S., Xiao, X., Guo, H., Chen, J., Wan, Y., Li, B., Xu, T., Xi, Q., Rao, C., Liu, C., Zeng, G.: Application of plant growth-promoting endophytes (PGPE) isolated from *Solanum nigrum* L. for phytoextraction of Cd-polluted soils. Appl. Soil Ecol. **46**(3), 383–389. https://doi.org/10.1016/j.apsoil.2010.10.003
26. Chowdhary, P., Yadav, A., Kaithwas, G., Bharagava, R.: Distillery wastewater: a major source of environmental pollution and its biological treatment for environmental safety. In: Green Technologies and Environmental Sustainability, pp. 409–435 (2017). https://doi.org/10.1007/978-3-319-50654-8_18
27. Díaz-González, M., Gutiérrez-Capitán, M., Niu, P., Baldi, A., Jiménez-Jorquera, C., Fernandez-Sanchez, C.: Electrochemical devices for the detection of priority pollutants listed in the EU Water Framework Directive. Trends Anal. Chem. **77**, 186–202 (2016)
28. Dimkpa, C.O., Merten, D., Svatoš, A., Büchel, G., Kothe, E.: Metal-induced oxidative stress impacting plant growth in contaminated soil is alleviated by microbial siderophores. Soil Biol. Biochem. **41**(1), 154–162 (2009). https://doi.org/10.1016/j.soilbio.2008.10.010
29. Dreaden, E.C., Alkilany, A.M., Huang, X., Murphy, C.J., El-Sayed, M.A.: The golden age: gold nanoparticles for biomedicine. Chem. Soc. Rev. **41**(7), 2740–2779 (2012). https://doi.org/10.1039/C1CS15237H
30. Dutta, G., Sugumaran, A.: Bioengineered zinc oxide nanoparticles: Chemical, green, biological fabrication methods and its potential biomedical applications. J. Drug Delivery Sci. Technol. **66**, 102853 (2021). https://doi.org/10.1016/j.jddst.2021.102853
31. Dwibedi, V., Rath, S.K., Joshi, M., Kaur, R., Kaur, G.: Microbial endophytes: application towards sustainable agriculture and food security. Appl. Microbiol. Biotechnol. **106**, 7801–7818 (2022). https://doi.org/10.1007/s00253-022-12078-8
32. Egamberdieva, D., Zoghi, Z., Nazarov, K., Wirth, S., Bellingrath-Kimura, S.D.: Plant growth response of broad bean (*Vicia faba* L.) to biochar amendment of loamy sand soil under irrigated and drought conditions. Environ. Sustain. **3**(3), 319–324 (2020). https://doi.org/10.1007/s42398-020-00116-y
33. Elemike, E.E., Uzoh, I.M., Onwudiwe, D.C., Babalola, O.O.: The role of nanotechnology in the fortification of plant nutrients and improvement of crop production. Appl. Sci. **9**(3) (2019). https://doi.org/10.3390/app9030499
34. Fahim, M., Shahzaib, A., Nishat, N., Jahan, A., Bhat, T.A., Inam, A.: Green synthesis of silver nanoparticles: a comprehensive review of methods, influencing factors, and applications. JCIS Open **16**, 100125 (2024). https://doi.org/10.1016/j.jciso.2024.100125

35. Ganesh, M., Hemalatha, P., Peng, M.M., Jang, H.T.: One pot synthesized Li, Zr doped porous silica nanoparticle for low temperature CO_2 adsorption. Arab. J. Chem. **10**, S1501–S1505 (2017). https://doi.org/10.1016/j.arabjc.2013.04.031
36. Gehrke, S.H., Uhden, L.H., McBride, J.F.: Hydrogel performance in controlled release. J. Control. Release **55**, 21–33 (1998)
37. Gerard, M., Chaubey, A., Malhotra, B.D.: Biosensors. Biosens. Bioelectron. **17**, 345 (2002)
38. Germaine, K.J., Keogh, E., Ryan, D., Dowling, D.N.: Bacterial endophyte-mediated naphthalene phytoprotection and phytoremediation. FEMS Microbiol. Lett. **296**(2), 226–234 (2009). https://doi.org/10.1111/j.1574-6968.2009.01637.x
39. Glick, B.R.: Using soil bacteria to facilitate phytoremediation. Biotechnol. Adv. **28**(3), 367–374 (2010). https://doi.org/10.1016/j.biotechadv.2010.02.001
40. Glick, B.R., Cheng, Z., Czarny, J., Duan, J.: Promotion of plant growth by ACC deaminase-producing soil bacteria. In: Bakker, P.A.H.M., Raaijmakers, J.M., Bloemberg, G., Höfte, M., Lemanceau, P., Cooke, B.M. (eds.) New Perspectives and Approaches in Plant Growth-Promoting Rhizobacteria Research, pp. 329–339. Springer, Netherlands (2007). https://doi.org/10.1007/978-1-4020-6776-1_8
41. Gongi, W., Touzi, H., Sadly, I., Ben Ouada, H., Tamarin, O., Ben Ouada, H.: A novel impedimetric sensor based on cyanobacterial extracellular polymeric substances for microplastics detection. J. Polym. Environ. **30**(11), 4738–4748 (2022)
42. Grawe, G., Oliveira, T., Moccelini, S., Terezo, A., Soares, M., Castilho, M.: Electrochemical biosensor for carbofuran pesticide based on esterases from *Eupenicillium shearii FREI*-39 endophytic fungi. Biosens. Bioelectron. **63**, 407–413 (2014). https://doi.org/10.1016/j.bios.2014.07.069
43. Gui, Q., Lawson, T., Shan, S., Yan, L., Liu, Y.: The application of whole cell-based biosensors for use in environmental analysis and in medical diagnostics. Sensors **17**(7) (2017). https://doi.org/10.3390/s17071623
44. Hakimian, F., Ghourchian, H., Hashemi, A.S., Arastoo, M.R., Rad, M.B.: Ultrasensitive optical biosensor for detection of miRNA-155 using positively charged Au nanoparticles. Sci. Rep. **8**, 1–9 (2018)
45. Hara, T.O., Singh, B.: Electrochemical biosensors for detection of pesticides and heavy metal toxicants in water: recent trends and progress. ACS ES&T Water **1**(3), 462–478 (2021)
46. He, W., Hu, Z.H., Yuan, S., et al.: Bacterial bioreporter-based mercury and phenanthrene assessment in Yangtze River delta soils of China. J. Environ. Qual. **47**(3), 562–570 (2018)
47. Herrmann, A.: Ph.D. thesis. Freie Universität Berlin (2020)
48. Hossain, S.M.Z., Mansour, N.: Biosensors for on-line water quality monitoring—a review. Arab. J. Basic Appl. Sci. **26**(1), 502–518 (2019)
49. Hu, M., et al.: Label-free electrochemical immunosensor based on AuNPs/Zn/Ni-ZIF-8-800@graphene composites for sensitive detection of monensin in milk. Sens. Actuators B: Chem. (2019)
50. Huang, C.W., Lin, C., Nguyen, M.K., Hussain, A., Bui, X.T., Ngo, H.H.: A review of biosensor for environmental monitoring: principle, application, and corresponding achievement of sustainable development goals. Bioengineered **14**(1), 58–80. https://doi.org/10.1080/21655979.2022.2095089
51. Hui, C.-Y., Guo, Y., Gao, C.-X., et al.: A tailored indigoidine-based whole-cell biosensor for detecting toxic cadmium in environmental water samples. Environ. Technol. Innov. **27**, 102511 (2022)
52. Hui, Y., Huang, Z., Alahi, M.E.E., Nag, A., Feng, S., Mukhopadhyay, S.C.: Recent advancements in electrochemical biosensors for monitoring the water quality. Biosensors **12**(7), 551 (2022)
53. Islam, M.T., Saenz-Arana, R., Wang, H., Bernal, R., Noveron, J.C.: Green synthesis of gold, silver, platinum, and palladium nanoparticles reduced and stabilized by sodium rhodizonate and their catalytic reduction of 4-nitrophenol and methyl orange. New J. Chem. **42**(8), 6472–6478 (2018). https://doi.org/10.1039/C8NJ01223G

54. Janata, J., Josowicz, M.: Conducting polymers in electronic chemical sensors. Nat. Mater. **2**, 19–24 (2003)
55. Justino, C.I.L., Duarte, A.C., Rocha-Santos, T.A.P.: Recent progress in biosensors for environmental monitoring: a review. Sensors. **17**, 2918 (2017). https://doi.org/10.3390/s17122918
56. Kalaimurugan, D., Balamuralikrishnan, B., Durairaj, K., Vasudhevan, P., Shivakumar, M.S., Kaul, T., Chang, S.W., Ravindran, B., Venkatesan, S.: Isolation and characterization of heavy-metal-resistant bacteria and their applications in environmental bioremediation. Int. J. Environ. Sci. Technol. **17**(3), 1455–1462 (2020). https://doi.org/10.1007/s13762-019-02563-5
57. Karnwal, A., Kumar, G., El Din, A., Mahmoud, J.D., Singh, R., Al-Tawaha, A.R.M.S., Malik, T.: Eco-engineered remediation: microbial and rhizosphere-based strategies for heavy metal detoxification. Curr. Res. Biotechnol. **9**, 100297 (2025). https://doi.org/10.1016/j.crbiot.2025.100297
58. Karoń, S., Porycka, K., Jr Lapitan Lorico, D.S., Drozd, M., Pietrzak, M., Malinowska, E.: A versatile approach to quality control of protein-based receptor layers by reversible, nonspecific staining for multiplex SPRi Immunosensing. Sens. Actuators, B Chem. **421**, 136512 (2024). https://doi.org/10.1016/j.snb.2024.136512
59. Khalid, S., Shahid, M., Alothman, Z.A., Al-Kahtani, A.A., Murtaza, B., Dumat, C.: Predicting chemical speciation of metals in soil using Visual Minteq. Soil Ecol Lett **5**, 220162 (2023)
60. Khan, M.A.A., Qadir, Z., Asad, M., Kouzani, A.Z., Parvez Mahmud, M.A.: Environmental footprint assessment of a cleanup at hypothetical contaminated site. Appl. Sci. **11**(11), 4907 (2021). https://doi.org/10.3390/app11114907
61. Khan, M.M.A., Khatun, A., Islam, M.T.: Promotion of plant growth by phytohormone producing bacteria. Microbes in action, pp. 1–43. Nova Science Publishers, New York (2016)
62. Khanam, Z., Gupta, S., Verma, A.: Endophytic fungi-based biosensors for environmental contaminants—a perspective. S. Afr. J. Bot. **134**, 401–406 (2020). https://doi.org/10.1016/j.sajb.2020.08.007
63. Kidwai, M.K., Malik, A., Dhull, S.B., Rose, P.K., Garg, V.K.: Bioremediation potential of Trichoderma species for metal (loid) s. In: Bioremediation of Toxic Metal (loid)s, pp. 137–152. CRC Press (2022)
64. Kowalska, A., Biczak, R.: Phytoremediation and environmental law: harnessing biomass and microbes to restore soils and advance biofuel innovation. Energies **18**(7), 1860 (2025). https://doi.org/10.3390/en18071860
65. Kumar, K., Shyamlal, B.R.K., Gupta, A., Mathur, M., Swami, A.K., Chaudhary, S.: Efficacious fungicidal potential of composite derived from nano-aggregates of Cu-Diclofenac complexes and ZnO nanoparticles. Compos. Commun. **10**, 81–88 (2018). https://doi.org/10.1016/j.coco.2018.07.004
66. Kumar, D.: Biodegradation of γ-Hexachlorocyclohexane by Burkholderia sp. IPL04. Biocatal. Agric. Biotechnol. **16**, 331–339 (2018). https://doi.org/10.1016/j.bcab.2018.09.001
67. Kumar, P., Singh, R.: IoT applications in environmental control systems: a review of recent developments. J. Emerg. Technol. **20**(4), 567–580 (2023)
68. Kumar, R., Singh, D. P., Muñoz, R., Amami, M., Singh, R. K., Singh, S., Kumar, V.: Graphene-based materials for biotechnological and biomedical applications: drug delivery, bioimaging and biosensing. Mater. Today Chem. **33**, Article 101750 (2023)
69. Kuppan, N., Padman, M., Mahadeva, M., Srinivasan, S., Devarajan, R.: A comprehensive review of sustainable bioremediation techniques: Eco friendly solutions for waste and pollution management. Waste Manage. Bull. **2**(3), 154–171 (2024). https://doi.org/10.1016/J.WMB.2024.07.005
70. Kushwaha, C.S., et al.: Mater. Sci. Eng. B, **284**, Article 115856 (2022)
71. Lang, Q., Han, L., Hou, C., Wang, F., Liu, A.: A sensitive acetylcholinesterase biosensor based on gold nanorods modified electrode for detection of organophosphate pesticide. Talanta, 156–157 (2016).
72. Lapitan Jr, L.D.S., Felisilda, B.M.B., Tiangco, C.E., Rosin Jose, A.: Advances in bioreceptor layer engineering in nanomaterial-based sensing of *Pseudomonas aeruginosa* and its

metabolites. Chem. Asian J. **19**(18), e202400090 (2024). https://doi.org/10.1002/asia.202400090

73. Lata, R., Chowdhury, S., Gond, S.K., White, J.F., Jr.: Induction of abiotic stress tolerance in plants by endophytic microbes. Lett. Appl. Microbiol. **66**(4), 268–276 (2018). https://doi.org/10.1111/lam.12855
74. Li, H.-Y., Wei, D.-Q., Shen, M., Zhou, Z.-P.: Endophytes and their role in phytoremediation. Fungal Diversity **54**(1), 11–18 (2012). https://doi.org/10.1007/s13225-012-0165-x
75. Li, L., Wang, S., Wu, S., Rao, S., Li, L., Cheng, S., Cheng, H.: Morphological and physiological indicators and transcriptome analyses reveal the mechanism of selenium multilevel mitigation of cadmium damage in *Brassica juncea*. Plants **12**(8) (2023). https://doi.org/10.3390/plants12081583
76. Li, Y., Chen, B., Yang, S., Jiao, Z., Zhang, M., Yang, Y., Gao, Y.: Advances in environmental pollutant detection techniques: enhancing public health monitoring and risk assessment. Environ. Int. **197**, 109365 (2025). https://doi.org/10.1016/J.ENVINT.2025.109365
77. Lim, J.W., Ha, D., Lee, J., Lee, S.K., Kim, T.: Review of micro/nanotechnologies for microbial biosensors. Front. Bioeng. Biotechnol. **3**, 61 (2015)
78. Ma, Y., Prasad, M.N.V, Rajkumar, M., Freitas, H.: Plant growth promoting rhizobacteria and endophytes accelerate phytoremediation of metalliferous soils. Biotechnol. Adv. **29**(2), 248–258 (2011). https://doi.org/10.1016/j.biotechadv.2010.12.001
79. Maduraiveeran, G., Jin, W.: Nanomaterials based electrochemical sensor and biosensor platforms for environmental applications. Trends Environ. Anal. Chem. **13**, 10–23 (2017)
80. Magesh, R., Ramasamy, D., Jayaprakash, K., Kandhasamy, S.: Biodegradation of tannery effluent hexavalent chromium by *Aspergillus niger* using egg shell packed bed reactor. Pollut. Res. **36**, 43–47 (2017)
81. Male, K.B., Hrapovic, S., Santini, J.M., Luong, J.H.T.: Biosensor for arsenite using arsenite oxidase and multiwalled carbon nanotube modified electrodes. Anal. Chem. **79**(20), 7831–7837 (2007). https://doi.org/10.1021/ac070766i
82. Malik, S., Singh, J., Goyat, R., Saharan, Y., Chaudhry, V., Umar, A., Ibrahim, A.A., Akbar, S., Ameen, S., Baskoutas, S.: Nanomaterials-based biosensor and their applications: a review. Heliyon **9**(9), e19929 (2023). https://doi.org/10.1016/j.heliyon.2023.e19929
83. Manganyi, M.C., Dikobe, T.B., Maseme, M.R.: Exploring the potential of endophytic microorganisms and nanoparticles for enhanced water remediation. Molecules **29**(12), 2858 (2024). https://www.mdpi.com/1420-3049/29/12/2858
84. Mansha, M., Qurashi, A., Ullah, N., Bakare, F.O., Khan, I., Yamani, Z.H.: Synthesis of In2O3/graphene heterostructure and their hydrogen gas sensing properties. Ceramics Int. **42**(9), 11490–11495 (2016). https://doi.org/10.1016/j.ceramint.2016.04.035
85. Mansoor, S., Ali, A., Kour, N., Bornhorst, J., AlHarbi, K., Rinklebe, J., Abd El Moneim, D., Ahmad, P., Chung, Y.S.: Heavy metal induced oxidative stress mitigation and ros scavenging in plants. Plants (Basel, Switzerland) **12**(16) (2023). https://doi.org/10.3390/plants12163003
86. Mansoor, S., Zahoor, I., Baba, T.R., Padder, S.A., Bhat, Z.A., Koul, A.M., Jiang, L.: Fabrication of silver nanoparticles against fungal pathogens. Front. Nanotechnol. **3**(October), 1–12 (2021). https://doi.org/10.3389/fnano.2021.679358
87. Mao, Y., Cui, S., Li, W., Fan, X., Liu, Y., Xu, S., Luo, X.: A lab-on-a-carbon nanodot sensor array for simultaneous pattern recognition of multiple antibiotics. Sens. Actuators B: Chem. **296**, Article 126694 (2019)
88. Maqsood, Q., Hussain, N., Sumrin, A., Ali, S.W., Tariq, M.R., Mahnoor, M.: Monitoring and abatement of synthetic pollutants using engineered microbial systems. Discover Life **54**(1), 9 (2024). https://doi.org/10.1007/s11084-024-09652-7
89. Maruthupandi, M., Thiruppathi, D., Vasimalai, N.: One minute synthesis of green fluorescent copper nanocluster: the preparation of smartphone-aided paper-based kit for on-site monitoring of nanomolar level mercury and sulfide ions in environmental samples. J. Hazard. Mater. **392**, 122451 (2020)
90. Matlin, S.A., Cornell, S.E., Krief, A., et al.: Chemistry must respond to the crisis of transgression of planetary boundaries. Chem. Sci. **13**(11710–11720), 3 (2022)

91. McQuade, D.T., Pullen, A.E., Swager, T.M.: Conjugated polymer-based chemical sensors. Chem. Rev. **100**(7), 2537–2574 (2000)
92. Meena, M., Zehra, A., Swapnil, P., Harish, Marwal, A., Yadav, G., Sonigra, P.: Endophytic nanotechnology: an approach to study scope and potential applications. Front. Chem. **9** (2021). https://doi.org/10.3389/fchem.2021.613343
93. Mendes, F., Machado, B.O., Castro, B.B., et al.: Harnessing the power of biosensors for environmental monitoring of pesticides in water. Appl. Microbiol. Biotechnol. **109**, 92 (2025)
94. Meyer, L., Guyot, S., Chalot, M., Capelli, N.: The potential of microorganisms as biomonitoring and bioremediation tools for mercury-contaminated soils. Ecotoxicol. Environ. Saf. **262**, 115185 (2023). https://doi.org/10.1016/J.ECOENV.2023.115185
95. Miglani, R., Parveen, N., Kumar, A., Ansari, M. A., Khanna, S., Rawat, G., Panda, A. K., Bisht, S.S., Upadhyay, J., Ansari, M.N.: Degradation of xenobiotic pollutants: an environmentally sustainable approach. Metabolites, **12**(9) (2022). https://doi.org/10.3390/metabo12090818
96. Mishra, S., Bharagava, R.N.: Toxic and genotoxic effects of hexavalent chromium in environment and its bioremediation strategies. J. Environ. Sci. Health. Part C Environ. Carcinogen. Ecotoxicol. Rev. **34**(1), 1–32 (2016). https://doi.org/10.1080/10590501.2015.1096883
97. Mohanty, B., Kumari, S., Yadav, P., Kanoo, P., Chakraborty, A.: Metal-organic frameworks (MOFs) and MOF composites-based biosensors. Coord. Chem. Rev. **519**, 216102 (2024)
98. Monosik, R., Stredansky, M., Sturdik, E.: Biosensors—Classification, characterization and new trends. Acta Chimica Slovaca **5**(1), 109–120 (2012)
99. Morcillo, R.J.L., Manzanera, M.: The Effects of plant-associated bacterial exopolysaccharides on plant abiotic stress tolerance. Metabolites **11**(6) (2021). https://doi.org/10.3390/metabo110 60337
100. Mukherjee, P., Dutta, J., Roy, M., et al.: Plant growth-promoting rhizobacterial secondary metabolites in augmenting heavy metal(loid) phytoremediation: an integrated green in situ ecorestorative technology. Environ. Sci. Pollut. Res. **31**, 55851–55894 (2024). https://doi.org/ 10.1007/s11356-024-34706-8
101. Nadeem, N., Habib, A., Hussain, S., Sufian, A., Ahmad, I., Noreen, F., Mehmood, A., Ali, F., Batoo, K.M., Ijaz, M.F.: Ecofriendly synthesis of silver nanoparticle for phytochemical screening, photocatalytic and biological applications. J. Inorg. Organomet. Polym. Mater. **35**(2), 1036–1051 (2025). https://doi.org/10.1007/s10904-024-03326-7
102. Naik, H., Manoharadas, S., Bommayasamy, N., John, T., Gobi, M., Dewala, S., Amaresan, N.: Green synthesis of iron oxide nanoparticles using Bacillus subtilis mitigate salinity stress in rice (*Oryza sativa* L.) plants and enhance physiological activities. Environ. Sci. Nano **12** (2025). https://doi.org/10.1039/D4EN01184H
103. Narayanan, Z., Glick, B.R.: Secondary metabolites produced by plant growth-promoting bacterial endophytes. Microorganisms **10**(10) (2022). https://doi.org/10.3390/microorganis ms10102008
104. Octobre, G., Delprat, N., Doumèche, B., Leca-Bouvier, B.: Herbicide detection: a review of enzyme-and cell-based biosensors. Environ. Res. **249**, 118330 (2024)
105. Othman, N., Ismail, Z., Selamat, M.I., Sheikh Abdul Kadir, S.H., Shibraumalisi, N.A.: A review of polychlorinated biphenyls (PCBs) pollution in the air: where and how much are we exposed to? Int. J. Environ. Res. Public Health. **19**(21), 13923 (2022). https://doi.org/10. 3390/ijerph192113923
106. Park, S.J., Lee, S.H., Yang, H., Park, C.S., Lee, C.S., Kwon, O.S., et al.: ACS Appl. Mater. Interfaces **8**, 28897–28903 (2016)
107. Paul, R., Saville, A.C., Hansel, J.C., Ye, Y., Ball, C., Williams, A., Chang, X., Chen, G., Gu, Z., Ristaino, J.B., Wei, Q.: Extraction of plant DNA by microneedle patch for rapid detection of plant diseases. ACS Nano **13**, 6540–6549 (2019)
108. Pavel, I.A., Lakard, S., Lakard, B.: Chemosensors **10**(3), 97 (2022)
109. Peixoto de Almeida, M., Pereira, E., Baptista, P., Gomes, I., Figueiredo, S., Soares, L., Franco, R.: Gold nanoparticles as (bio)chemical sensors. Mater. Today: Proc. (2014)
110. Pohanka, M.: The piezoelectric biosensors: principles and applications, a review. Int. J. Electrochem. Sci. **12**, 496–506 (2017)

111. Pokhrel, L.R., Ettore, N., Jacobs, Z.L., Zarr, A., Weir, M.H., Scheuerman, P.R., Kanel, S.R., Dubey, B.: Novel carbon nanotube (CNT)-based ultrasensitive sensors for trace mercury(II) detection in water: a review. Sci. Total. Environ. **574**, 1379–1388 (2017). https://doi.org/10.1016/j.scitotenv.2016.08.055
112. Popescu, S.M., Mansoor, S., Wani, O.A., Kumar, S.S., Sharma, V., Sharma, A., Arya, V.M., Kirkham, M.B., Hou, D., Bolan, N., Chung, Y.S.: Artificial intelligence and IoT driven technologies for environmental pollution monitoring and management. Front. Environ. Sci. **12** (2024). https://doi.org/10.3389/fenvs.2024.1336088
113. Qattan, S.Y.A.: Harnessing bacterial consortia for effective bioremediation: targeted removal of heavy metals, hydrocarbons, and persistent pollutants. Environ. Sci. Eur. **37**(1), 85 (2025). https://doi.org/10.1186/s12302-025-01103-y
114. Raeyani, D., Shojaei, S., Ahmadi-Kandjani, S.: Optical graphene quantum dots gas sensors: experimental study. Mater. Res. Express **7**(1), Article 015608 (2020)
115. Rajkumar, M., Ae, N., Prasad, M.N.V., Freitas, H.: Potential of siderophore-producing bacteria for improving heavy metal phytoextraction. Trends Biotechnol. **28**(3), 142–149 (2010). https://doi.org/10.1016/j.tibtech.2009.12.002
116. Rajput, V., Minkina, T., Mazarji, M., Shende, S., Sushkova, S., Mandzhieva, S., Burachevskaya, M., Chaplygin, V., Singh, A., Jatav, H.: Accumulation of nanoparticles in the soil-plant systems and their effects on human health. Ann. Agric. Sci. **65**(2), 137–143 (2020). https://doi.org/10.1016/j.aoas.2020.08.001
117. Ramacharyulu, P.V.R.K., Muhammad, R., Praveen Kumar, J., Prasad, G.K., Mohanty, P.: Iron phthalocyanine modified mesoporous titania nanoparticles for photocatalytic activity and CO_2 capture applications. Phys. Chem. Chem. Phys. **17**(39), 26456–26462 (2015). https://doi.org/10.1039/C5CP03576G
118. Rautela, I., Oraon, R.K.K, Kalia, S., Sharma, M.D.: Chapter 20 - Genetic engineering of endophytic fungi for enhanced bioactivity. In: Dwibedi, V., Kapoor, N., Gambhir, L., Kajale, S. (eds.) Navigating Endophytic Research for Next-Generation Therapeutics, pp. 403–417 (2025). https://doi.org/10.1016/B978-0-443-31484-1.00020-7
119. Reddy, K., Adams, J.: Sustainable Remediation of Contaminated Sites. Momentum Press (2015)
120. Rojas-Villacorta, W., Rojas-Flores, S., De La Cruz-Noriega, M., Chinchay Espino, H., Diaz, F., Gallozzo Cardenas, M.: Microbial biosensors for wastewater monitoring: mini-review. Processes **10**(10), 2002 (2022)
121. Ronkainen, N.J., Halsall, H.B., Heineman, W.R.: Electrochemical biosensors. Chem. Soc. Rev. **39**(5), 1747–1763 (2010)
122. Rykowska, I., Wasiak, W.: Research trends on emerging environment pollutants—a review. Open Chem. **13**(1), 1353–1370 (2015). https://doi.org/10.1515/chem-2015-0151
123. Sahoo, A., Meher, I., Chhotaray, S.P., Meena, M., Swapnil, P., Arora, J., Zehra, A.: Fungal endophytes: a resource for producing extracellular enzymes. In: Meena, M., Zehra, A., Swapnil, P., Seth, C.S. (eds.) Metabolic Sustainability of Endophytes: Current Status, Challenges and Potential, pp. 105–145. Springer, Singapore (2025). https://doi.org/10.1007/978-981-96-4004-1_6
124. Santoyo, G., Moreno-Hagelsieb, G., del Carmen Orozco-Mosqueda, M., Glick, B.R.: Plant growth-promoting bacterial endophytes. Microbiol. Res. **183**, 92–99 (2016). https://doi.org/10.1016/j.micres.2015.11.008
125. Shahid, M., Bakhat, H.F., Shah, G.M., Murtaza, B.: Recent trends in environmental sustainability. Environ. Sci. Pollution Res. **30**(44), 99198–99201 (2023). https://doi.org/10.1007/s11356-023-29348-1
126. Shaifali, C.K.G.: Microbial remediation of heavy metals. Int J Chem Stud **9**(2), 92–107 (2021)
127. Shankar, A., Sharma, K.K.: Fungal secondary metabolites in food and pharmaceuticals in the era of multi-omics. Appl. Microbiol. Biotechnol. **106**(9), 3465–3488 (2022). https://doi.org/10.1007/s00253-022-11945-8
128. Sharma, M., Mahajan, P., Alsubaie, A.S., Khanna, V., Chahal, S., Thakur, A., Yadav, A., Arya, A., Singh, A., Singh, G.: Next-generation nanomaterials-based biosensors: real-time

biosensing devices for detecting emerging environmental pollutants. Mater. Today Sustain. **29**, Article 101068 (2025)
129. Sharma, P., Kumar, S.: Bioremediation of heavy metals from industrial effluents by endophytes and their metabolic activity: recent advances. Bioresour. Technol. **339**, 125589 (2021). https://doi.org/10.1016/j.biortech.2021.125589
130. Sharma, P., Pandey, V., Sharma, M.M.M., Patra, A., Singh, B., Mehta, S., Husen, A.: A review on biosensors and nanosensors application in agroecosystems. Nanoscale Res. Lett. **16**, 136 (2021)
131. Shetty, S. S., Deepthi, D., Harshitha, S., Sonkusare, S., Naik, P. B., Kumari N.S., Madhyastha, H.: Environmental pollutants and their effects on human health. Heliyon **9**(9), e19496 (2023). https://doi.org/10.1016/J.HELIYON.2023.E19496
132. Shukla, A., Srivastava, S., D'Souza, S.F.: An integrative approach toward biosensing and bioremediation of metals and metalloids. Int. J. Environ. Sci. Technol. **15**, 2701–2712 (2018). https://doi.org/10.1007/s13762-018-1766-z
133. Sim, C.S.F., Chen, S.H., Ting, A.S.Y.: Endophytes: emerging tools for the bioremediation of pollutant. In: Bharagava, R.N., Chowdhary, P. (eds.) Emerging and Eco-Friendly Approaches for Waste Management, pp. 189–217. Springer, Singapore (2019). https://doi.org/10.1007/978-981-10-8669-4_10
134. Simon, J., Flahaut, E., Golzio, M.: Overview of carbon nanotubes for biomedical applications. Materials **12**(4), 624 (2019)
135. Singh, G., Dwivedi, S.K., Mishra, J.: Role of fungal enzymes in the removal of azo dyes. In: Arora NK, Mishra J, Mishra V (eds) Microbial Enzymes: Roles and Applications in Industries, pp. 231–257. Springer, Singapore (2020). https://doi.org/10.1007/978-981-15-1710-5_9
136. Smith, J., Brown, P.: IoT-enabled environmental monitoring: challenges and future directions. Environ. Sci. Technol. **54**(3), 1234–1245 (2020)
137. Sridhar, V., Jung, K.H., Park, H.: Vitamin derived nitrogen doped carbon nanotubes for efficient oxygen reduction reaction and arsenic removal from contaminated water. Materials 13(7) (2020). https://doi.org/10.3390/ma13071686
138. Srinivasan, S., Bankole, P.O., Sadasivam, S.K.: Biodecolorization and degradation of textile azo dyes using *Lysinibacillus sphaericus* MTCC 9523. Front. Environ. Sci. **10** (2022). https://doi.org/10.3389/fenvs.2022.990855
139. State of Global Air: State of Global Air Report 2024. Health Effects Institute and UNICEF (2024). https://www.stateofglobalair.org/resources/report/state-global-air-report-2024
140. Steffen, W., Richardson, K., Rockström, J., et al.: Planetary boundaries: guiding human development on a changing planet. Science **347**, 1259855 (2015)
141. Suman, A., Yadav, A.N., Verma, P.: Endophytic microbes in crops: diversity and beneficial impact for sustainable agriculture. In: New and Future Developments in Microbial Biotechnology and Bioengineering. Springer (2016). https://doi.org/10.1007/978-81-322-2647-5_7
142. Tetyana, P., Shumbula, P.M., Njengele-Tetyana, Z.: Biosensors: design, development and applications. Nanopores. IntechOpen (2021)
143. Tiwari, P., Kang, S., Bae, H.: Plant-endophyte associations: rich yet under-explored sources of novel bioactive molecules and applications. Microbiol. Res. **266** (2023), 127241. https://doi.org/10.1016/j.micres.2022.127241
144. Tripathi, S., Sharma, S., Rai, P., Mahra, S., Tripathi, D. K., Sharma, S.: Synergy of plant growth promoting rhizobacteria and silicon in regulation of AgNPs induced stress of rice seedlings. Plant Physiol. Biochem.: PPB **213**, 108720 (2024). https://doi.org/10.1016/j.plaphy.2024.108720
145. Uddandarao, P., Balakrishnan, R.M.: Thermal and optical characterization of biologically synthesized ZnS nanoparticles synthesized from an endophytic fungus *Aspergillus flavus*: a colorimetric probe in metal detection. Spectrochim. Acta Part A: Mol. Biomol. Spectrosc. **175**, 200–207 (2017). https://doi.org//doi.org/10.1016/j.saa.2016.12.021
146. Ukirade, N., Choudhari, U., Bhapkar, S., Jadhav, U., Jagtap, S., Rane, S.: Enzyme-immobilized graphene oxide-based electrochemical biosensor for glutathione detection. RSC Adv. **15**(17), 12987–12996 (2025). https://doi.org/10.1039/D4RA09033K

147. Ullah, A.K.M.A., Kibria, A.K.M.F., Akter, M., Khan, M.N.I., Maksud, M.A., Jahan, R.A., Firoz, S.H.: Synthesis of Mn_3O_4 nanoparticles via a facile gel formation route and study of their phase and structural transformation with distinct surface morphology upon heat treatment. J. Saudi Chem. Soc. **21**(7), 830–836 (2017). https://doi.org/10.1016/j.jscs.2017.03.008
148. Ullah, F., Othman, M.B.H., Javed, F., Ahmad, Z., Akil, H.M.: Classification, processing and application of hydrogels: a review. Mater. Sci. Eng. C **57**, 414–433 (2015)
149. United Nations Environment Programme: Plastic Pollution (2022). https://www.unep.org/plastic-pollution
150. United Nations Environment Programme: Emissions Gap Report 2024 (2024). https://www.unep.org/resources/emissions-gap-report-2024
151. Velasco-Garcia, M.N.: Optical biosensors for probing at the cellular level: a review of recent progress and future prospects. Semin. Cell Dev. Biol. **20**(1), 27–33 (2009)
152. Verma, V., Batta, A., Singh, H.B., Srivastava, A., Garg, S.K., Singh, V.P., Arora, P.K.: Bioengineering of fungal endophytes through the CRISPR/Cas9 system. Front. Microbiol. **14** (2023). https://doi.org/10.3389/fmicb.2023.1146650
153. Wang, B., Huang, D., Weng, Z.: Polymer-Based biosensors for detection of heavy metals. Encyclopedia (2023). https://encyclopedia.pub/entry/47774
154. Wang, C., Kong, Y., Hu, R., Zhou, G.: Miscanthus: A fast-growing crop for environmental remediation and biofuel production. GCB Bioenergy **13**(1), 58–69 (2021). https://doi.org/10.1111/gcbb.12761
155. Wang, F., Xiang, L., Sze-Yin Leung, K., Elsner, M., Zhang, Y., Guo, Y., Pan, B., Sun, H., An, T., Ying, G., Brooks, B. W., Hou, D., Helbling, D. E., Sun, J., Qiu, H., Vogel, T. M., Zhang, W., Gao, Y., Simpson, M.J., et al.: Emerging contaminants: a one health perspective. Innovation **5**(4) (2024). https://doi.org/10.1016/j.xinn.2024.100612
156. Wang, G.-H., Tsai, T.-H., Kui, C.-C., et al.: Analysis of bioavailable toluene by using recombinant luminescent bacterial biosensors with different promoters. J. Biol. Eng. **15**(1), 2 (2021)
157. Wang, J., Chen, C.: Biosorbents for heavy metals removal and their future. Biotechnol. Adv. **27**(2), 195–226 (2009). https://doi.org/10.1016/j.biotechadv.2008.11.002
158. Wang, Y., Li, Q., Zhang, P., O'Connor, D., Varma, R.S., Yu, M., Hou, D.: One-pot green synthesis of bimetallic hollow palladium-platinum nanotubes for enhanced catalytic reduction of p-nitrophenol. J. Colloid and Interface Sci. **539**, 161e167 (2019)
159. Wang, Y., O'Connor, D., Shen, Z., Lo, I.M.C., Tsang, D.C.W., Pehkonen, S., et al.: Green synthesis of nanoparticles for the remediation of contaminated waters and soils: constituents, synthesizing methods, and influencing factors. J. Clean. Prod. **226**, 540e549 (20190)
160. Weyens, N., Truyens, S., Dupae, J., Newman, L., Taghavi, S., van der Lelie, D., Carleer, R., Vangronsveld, J.: Potential of the TCE-degrading endophyte Pseudomonas putida W619-TCE to improve plant growth and reduce TCE phytotoxicity and evapotranspiration in poplar cuttings. Environment. Pollution **158**(9), 2915–2919 (2010). https://doi.org/10.1016/j.envpol.2010.06.004
161. World Health Organization: Seventy-eighth World Health Assembly—Daily update: 26 May 2025 (2025). Retrieved from https://www.who.int/news/item/26-05-2025-seventy-eighth-world-health-assembly---daily-update--26-may-2025
162. Wu, J., Kamal, N., Hao, H., Qian, C., Liu, Z., Shao, Y., Zhong, X., Xu, B.: Endophytic *Bacillus megaterium* BM18-2 mutated for cadmium accumulation and improving plant growth in Hybrid Pennisetum. Biotechnol. Rep. **24**, e00374 (2019). https://doi.org/10.1016/J.BTRE.2019.E00374
163. Xia, N., Feng, F., Liu, C., Li, R., Xiang, W., Shi, H., Gao, L.: The detection of mercury ion using DNA as sensors based on fluorescence resonance energy transfer. Talanta **192**, 500–507 (2019)
164. Xia, Y., Hu, J., Zhao, S., Tao, L., Li, Z., Yue, T., Kong, J.: Build-in sensors and analysis algorithms aided smartphone-based sensors for point-of-care tests. Biosens. Bioelectron. X **11**, 100195 (2022)

165. Xiao, Y., De Araujo, C., Sze, C.C., Stuckey, D.C.: Toxicity measurement in biological wastewater treatment processes: a review. J. Hazard. Mater. **286**, 15–29 (2015)
166. Xu, X., Huang, R., Yin, W.-B.: An optimized and efficient CRISPR/Cas9 system for the endophytic fungus *Pestalotiopsis fici*. J. Fungi **7**(10), 809 (2021). https://doi.org/10.3390/jof7100809
167. Yadav, I.C., Devi, N.L.: Pesticides classification and its impact on environment. Environ. Sci. Eng. **6**, 140–158 (2017)
168. Yadav, N., Saini, O., Debnath, N., Singh, S., Thakur, T.K., Rajendra, K., Meena, R., Thakur, I.S., Srivastava, S.: Optimizing enhanced heavy metal detoxification by novel hybrid fungal hyphae-nano-biocomposite functionalized with graphene oxide: unravelling process parameters & adsorption modelling. Process Safety Environ. Protect. **188**, 917–928 (2024). https://doi.org/10.1016/j.psep.2024.05.086
169. Yew, Y.P., Shameli, K., Miyake, M., Ahmad Khairudin, N.B.B., Mohamad, S.E.B., Naiki, T., Lee, K.X.: Green biosynthesis of superparamagnetic magnetite Fe_3O_4 nanoparticles and biomedical applications in targeted anticancer drug delivery system: a review. Arab. J. Chem. **13**(1), 2287–2308 (2020). https://doi.org/10.1016/j.arabjc.2018.04.013
170. Zambito Marsala, R., Capri, E., Russo, E., Bisagni, M., Colla, R., Lucini, L., Gallo, A., Suciu, N.A.: First evaluation of pesticides occurrence in groundwater of Tidone Valley, an area with intensive viticulture. Sci. Total. Environ. **736**, 139730 (2020)
171. Zandi, P., Schnug, E.: Reactive Oxygen species, antioxidant responses and implications from a microbial modulation perspective. Biology **11**(2) (2022). https://doi.org/10.3390/biology11020155
172. Zhang, W., Bu, S., Bai, H., Ma, C., Ma, L., Wei, H., Liu, X., Li, Z., Wan, J.: A sensitive biosensor for determination of pathogenic bacteria using aldehyde dehydrogenase signaling system. Anal. Bioanal. Chem. **412**(28), 7955–7962 (2020)
173. Zhang, H., Wang, Y., Zhu, X., Li, Y., Cai, W.: Bilayer Au nanoparticle-decorated WO3 porous thin films: on-chip fabrication and enhanced NO_2 gas sensing performances with high selectivity. Sens. Actuators B: Chem. **280**, 192–200 (2019). https://doi.org/10.1016/j.snb.2018.10.065
174. Zhang, P., Hou, D., O'Connor, D., Li, X., Pehkonen, S.O., Varma, R.S., et al.: Green and size-specific synthesis of stable Fe-Cu oxides as earth-abundant adsorbents for malachite green removal. ACS Sustain. Chem. Eng. 9229–9236 (2018)
175. Zhang, P., Lo, I., O'Connor, D., Pehkonen, S., Cheng, H., Hou, D.: High efficiency removal of methylene blue using SDS surface-modified ZnFe2O4 nanoparticles. J. Colloid Interface Sci. **508**, 39–48 (2017)
176. Zhang, W., et al.: Novel nanostructured MIL-101(Cr)/XC-72 modified electrode sensor: a highly sensitive and selective determination of chloramphenicol. Sens. Actuators B: Chem (2017)
177. Zhang, Y., He, L., Chen, Z., Zhang, W., Wang, Q., Qian, M., Sheng, X.: Characterization of lead-resistant and ACC deaminase-producing endophytic bacteria and their potential in promoting lead accumulation of rape. J. Hazard. Mater. **186**(2), 1720–1725 (2011). https://doi.org/10.1016/j.jhazmat.2010.12.069
178. Zhao, H., Ji, X., Wang, B., Wang, N., Li, X., Ni, R., Ren, J.: An ultra-sensitive acetylcholinesterase biosensor based on reduced graphene oxide-Au nanoparticles-β-cyclodextrin/Prussian Blue-chitosan nanocomposites for organophosphorus pesticides detection. Biosens. Bioelectron. **65**, 23–30 (2015)
179. Zulfiqar, F., Navarro, M., Ashraf, M., Akram, N.A., Munné-Bosch, S.: Nanofertilizer use for sustainable agriculture: advantages and limitations. Plant Sci. **289**, 110270 (2019). https://doi.org/10.1016/j.plantsci.2019.110270

Composite Materials

Material Matters: Unlocking the Future of Technology Through Innovative Composites

Rashi Nathawat, Ashish Somwanshi, and Satyapal S. Rathore

Abstract This book chapter delves into the transformative role of composite materials in modern technology, emphasizing their design, properties, and applications to achieve superior performance over traditional materials. It provides a comprehensive overview of four key types: polymer matrix composites (PMCs), metal matrix composites (MMCs), ceramic matrix composites (CMCs), and carbon-carbon composites (CCCs). PMCs, featuring thermoset or thermoplastic matrices reinforced with fibers like carbon, glass, or aramid, are discussed for their high strength-to-weight ratios, corrosion resistance, and uses in aerospace, automotive, and biomedical fields, enhanced by nano-fillers such as graphene. MMCs, with metallic matrices like aluminum or titanium embedded with ceramic reinforcements, highlight improved mechanical and thermal properties, synthesized via liquid- or solid-state processes, and applied in superconducting wires and structural components. CMCs address ceramic brittleness through fiber integration, covering synthesis methods like sol-gel, hot pressing, and chemical vapor deposition, with applications in high-temperature environments such as jet engines and cutting tools. CCCs, composed entirely of carbon, excel in extreme heat resistance, detailing high-pressure processing and uses in spacecraft heat shields, brakes, and medical implants. Drawing on historical inspirations from nature and recent advancements, the chapter illustrates how composites enable lighter, more durable, and multifunctional solutions, supported by examples and references to ongoing research.

R. Nathawat (✉) · A. Somwanshi
Functional Ceramics and Smart Materials Lab, Department of Physics, Manipal University Jaipur, Jaipur, India
e-mail: rashi.nathawat@gmail.com; rashi.nathawat@jaipur.manipal.edu

A. Somwanshi
e-mail: ashish.2410190002@muj.manipal.edu

S. S. Rathore
Department of Physics, Cluster University of Jammu, Jammu, India

L. Ledwani and S. Srivastava (eds.), *Experimental and Computational Advances in Materials*, Engineering Materials, https://doi.org/10.1007/978-981-95-4180-5_14

1 Introduction

Advancements in technology are deeply tied to innovations in material science, as the effectiveness of any cutting-edge design–whether for a turbine, aircraft, or other system–relies heavily on the capabilities of the materials used. In the modern pursuit of creating products that are more efficient, lighter, and more durable, composite materials have become a game-changer, reflecting the ongoing effort to improve and refine the substances that shape our world.

While the study of composite materials may appear to be a contemporary field, the concept itself is far from novel. Nature provides numerous examples of composite structures: the coconut palm leaf, for instance, functions as a cantilever thanks to its naturally reinforced fibers, while wood achieves a balance of resilience and rigidity by combining pliable cellulose fibers with a stiff lignin framework. Similarly, human bones act as a biological composite, blending soft collagen with hard apatite minerals to provide both support and protection. Beyond nature, human-engineered solutions like rubber strengthened with carbon black, cement mixed with sand, or resin embedded with glass fibers illustrate that the practice of merging different materials to enhance performance has been a longstanding aspect of innovation.

The1960srepresented a pivotal moment for composite materials, as systematic research and development began to unlock their vast potential. Growing needs in sectors such as aerospace, energy, and infrastructure highlighted the limitations of single-material solutions, prompting a surge in efforts to develop custom-engineered materials. These materials were crafted to offer exceptional qualities like high strength, rigidity, and reduced weight. By strategically combining various elements, engineers can now create materials that surpass the performance of their individual parts, a principle that has also transformed the design process itself. As early experts in the field have pointed out, composites encourage designers to think creatively, essentially developing a bespoke material for each specific use case to maximize efficiency and reduce costs.

A key benefit of composite materials lies in their adaptability. They can be tailored not only to deliver outstanding mechanical properties but also to fulfill specialized non-mechanical roles, such as improved resistance to wear, corrosion, or even unique electrical characteristics for advanced technologies like superconductors. While fiber-reinforced composites excel at providing strength ratios along the fiber direction, achieving consistent performance in all directions often necessitates alternative methods, such as using laminates or particle-reinforced configurations.

For the purposes of this book, we define a composite material as an engineered product consisting of distinct phases–each with unique physical or chemical properties–that are intentionally combined to yield attributes unattainable by any single component such as a second order phase transition. This definition excludes naturally occurring composites like wood, focusing instead on the purposeful design of modern composite systems.

In this chapter, we will delve into the multifaceted realm of composites, with a primary emphasis on fiber-reinforced systems, examining their creation, manufacturing processes, and groundbreaking applications that are continually expanding the limits of material performance in today's technological landscape.

Polymer Matrix Composites (PMCs) consist of a polymer-based matrix reinforced with fibers or particles, providing high strength-to-weight ratios, excellent corrosion resistance, and ease of fabrication. PMCs can be classified into two main categories: thermosetting PMCs and thermoplastic PMCs. Thermosetting polymers, such as epoxy, polyester, and phenolic resins, are more commonly used than thermoplastics due to their superior mechanical properties and high thermal stability. Thermoplastics, including polypropylene (PP), polyetheretherketone (PEEK), and polyimide (PI), offer advantages like recyclability and improved toughness. Common reinforcements in PMCs include carbon fibers (CFRP), glass fibers (GFRP), Kevlar (AFRP), and natural fibers. Recent advancements in PMCs have incorporated nano-fillers such as graphene oxide (GO) and carbon nanotubes (CNTs) to improve mechanical and electrical properties. For instance, recent studies [1] have demonstrated that graphene-reinforced epoxy composites exhibit enhanced thermal conductivity and mechanical strength, making them suitable for aerospace applications.

PMCs are extensively used in the automotive industry to reduce vehicle weight, thereby improving fuel efficiency and reducing emissions. For example, BMW's i3 electric vehicle features CFRP components to achieve weight reduction. In the aerospace sector, Boeing's 787 Dreamliner utilizes GFRP and CFRP to enhance structural performance while reducing overall weight. In biomedical applications, PMCs reinforced with bioactive ceramics are being developed for bone implants and prosthetics due to their biocompatibility and strength [2].

Metal Matrix Composites (MMCs) are composed of a metallic matrix reinforced with ceramic or metallic phases to improve mechanical, thermal, and wear resistance properties. The matrix material is typically aluminum (Al), magnesium (Mg), titanium (Ti), or copper (Cu), while reinforcements include silicon carbide (SiC), aluminum oxide (Al_2O_3), tungsten (W), and boron carbide (B4C). Aluminum-based MMCs (Al-MMCs) are among the most widely studied materials due to their lightweight nature, high thermal conductivity, and excellent mechanical properties. Recent research [3] has focused on developing SiC-reinforced Al composites for aerospace and automotive applications. Al-MMCs have been used in engine components, pistons, and brake discs, where they provide superior wear resistance and reduced thermal expansion. Titanium-based MMCs (Ti-MMCs) are employed in biomedical implants and aerospace components due to their excellent corrosion resistance and biocompatibility. A recent study [4] reported the development of Ti-MMCs reinforced with boron fibers for use in orthopedic applications, demonstrating high strength and fatigue resistance.

Mg-based MMCs are attracting attention for lightweight structural applications in the aerospace and defense sectors. Recent studies [5] have explored the incorporation of graphene nanoplatelets (GNPs) into Mg matrices, significantly improving the tensile strength and thermal stability of the composite.

Due to the high tool wear associated with MMCs, advanced machining techniques such as electrical discharge machining (EDM) and laser-assisted machining are commonly used [6]. These techniques ensure precision cutting while minimizing damage to the reinforcement phase.

Ceramic Matrix Composites (CMCs) are a class of composites in which a ceramic matrix is reinforced with ceramic fibers to overcome the brittleness of monolithic ceramics. The most commonly used matrix materials include silicon carbide (SiC), aluminum oxide (Al_2O_3), and silicon nitride (Si_3N_4), while reinforcement materials include carbon fibers and SiC whiskers.

CMCs are primarily used in high-temperature applications due to their excellent thermal shock resistance and oxidation stability. Recent advancements in SiC/SiC CMCs have made them suitable for jet engine components, gas turbine blades, and heat shields in aerospace applications. A study by NASA demonstrated that SiC-reinforced CMCs exhibit higher strength retention at elevated temperatures compared to traditional nickel-based superalloys.

In the biomedical field, Al_2O_3-based CMCs are being explored for load-bearing implants and dental restorations, offering improved wear resistance and biocompatibility. CMCs are also widely used in nuclear applications, where their high radiation resistance makes them suitable for structural components in nuclear reactors.

Carbon-Carbon Composites (CCCs) consist of a carbon matrix reinforced with carbon fibers, offering exceptional high-temperature stability, low density, and excellent strength-to-weight ratio. CCCs are primarily used in applications where materials must withstand extreme environments, such as hypersonic aircraft, spacecraft, and thermal protection systems.

Recent advancements [7] have focused on improving the oxidation resistance of CCCs by incorporating protective coatings such as SiC, HfC, and ZrC. Studies have shown that SiC-coated CCCs exhibit enhanced durability under oxidative environments, making them ideal for reusable spacecraft applications.

CCCs are extensively used in brake discs for high-performance cars and aircraft due to their superior friction properties and resistance to thermal degradation. The medical field has also explored CCCs for use in prosthetics and bone implants, leveraging their bio-inert properties and high strength [8].

2 Polymer Matrix Composites (PMCs)

2.1 Structure of PMCs

PMCs are composed of two kinds of polymer matrices. One is linear chains of hydrocarbon molecules and other is branched chains of hydrocarbons. The one with linear arrangement of carbons at the microlevel is called thermoplastic polymer and the other is called thermoset polymer. They can be comined with metal organic frameworks as dispersed phase to make novel polymer matrix composite materials [9]. Thermosetting polymers are having branched structure as shown in the Figs. 1 and 2.

Metal–organic frameworks (MOFs) have emerged as innovative nano-fillers for polymer matrix composites (PMCs) due to their crystalline, highly porous structures and tunable chemical functionalities, which enable not only enhanced mechanical reinforcement but also the integration of smart properties such as selective gas adsorption, catalysis, and sensing. By offering exceptionally high surface areas, MOFs promote strong interfacial adhesion and efficient load transfer when incorporated into polymer matrices, and their surfaces can be chemically tailored or polymer-grafted to improve dispersion and compatibility. Fabrication techniques such as melt blending, and solution causting are used to create these composites, though challenges remain in maintaining the structural integrity of MOFs during processing. Overall, MOF-based PMCs represent a promising class of multifunctional materials that combine lightweight, high-strength characteristics with additional functionalities applicable to energy storage, environmental remediation, and advanced sensor technologies.

Some notable examples of MOF constructed by researchers are: ZIF-8 (Zeolitic Imidazolate Framework-8): Structure: A zinc-based MOF with a unique, highly porous structure. Applications: Gas storage (e.g., CO_2 capture), catalysis, and drug delivery. MIL-53 (Materials Institute Lavoisier-53): Structure: An aluminum-based MOF with a "breathing" framework. Applications: Gas storage, catalysis, and drug delivery. UiO-66: Structure: A zirconium-based MOF known for its high stability and tunable pore size. Applications: Gas storage, catalysis, and drug delivery.

During recent times, organometallic compounds, that is, chemical compounds have become popular in several industries and pharmaceutical products. These compounds can be incorporated into a polymer matrix to synthesize innovative layers of polymer matrix composites. Some notable examples of organometallic compounds include: Ferrocene, Cobaltocene, etc. (Fig. 3).

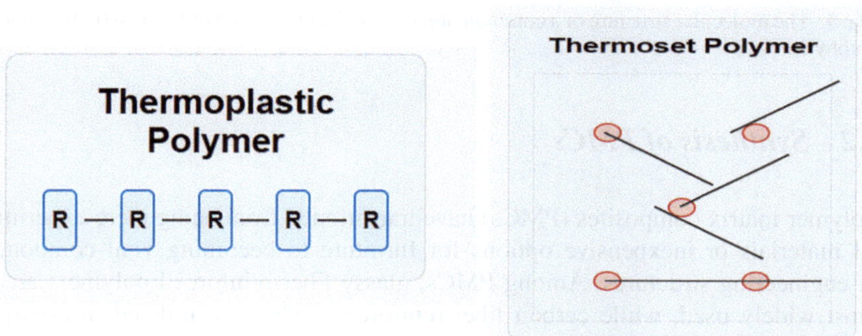

Fig. 1 The structure of (**a**) thermoplastic polymer with linear chains and (**b**) thermoset polymer with zigzag branched chains at the microlevel

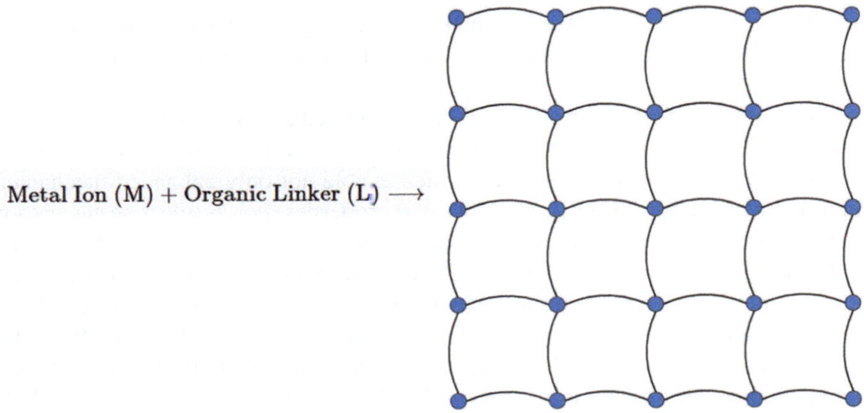

Fig. 2 The structure of a metal organic framework as a 3D network of metal ions shown as blue spheres and organic linkers represented by the curly lines

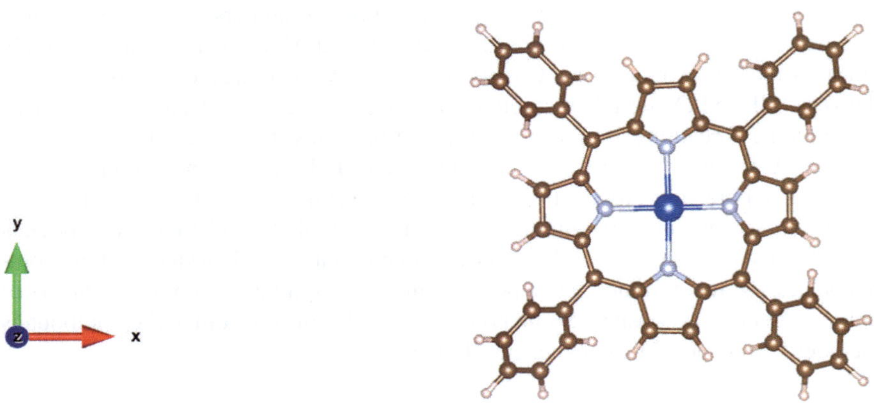

Fig. 3 The molecular structure of a common organometallic compound of Copper(II) Tetraphenylporphyrin

2.2 *Synthesis of PMCs*

Polymer matrix composites (PMCs) have transitioned from being mere experimental materials or inexpensive options for furniture to becoming vital components in engineering structures. Among PMCs, glassy fiber-reinforced polymers are the most widely used, while carbon fiber-reinforced PMCs are utilized in aerospace applications due to their superior strength and lightweight properties.

Many of the manufacturing processes initially developed for glass fiber-reinforced PMCs have been adapted for other fiber-based composites as well. Polymer matrices can be categorized into thermosets and thermoplastics. Thermosets, once cured, undergo a permanent chemical transformation through cross-linking, which is typically achieved using heat, pressure, or chemical agents. Traditional thermal curing, often requiring temperatures above 200 °C, can introduce issues such as residual stresses, thermal gradients, and long processing times. In thick or asymmetrical laminates, these stresses may lead to warping, fiber misalignment, microcracking, or layer separation. To address these challenges, electron beam curing has emerged as an alternative method, offering a more efficient process that minimizes these drawbacks.

2.2.1 Manual Layup and Spray Techniques

Manual layup and spray methods are among the most straightforward techniques for processing polymer composites. In the manual layup process, fibers are manually located into a modular form, followed by the application of resin, such as unsaturated polyester, which is either brushed or sprayed on. Alternatively, both chopped fibers and resin can be sprayed together onto the mold surface. To ensure proper compaction and uniformity, rollers are used to press and densify the layers. Curing is typically carried out at room temperature or with moderate heating in an oven, often with the help of accelerators and catalysts to speed up the process [10].

2.2.2 Filament Winding

Filament winding is a highly flexible technique used to manufacture composite structures. In this process, continuous fiber are firstly submerged in resin heat bath for impregnation and then wound onto a rotating or stationary structure. A roving consists of thousands of fine filaments bundled together.

This technique is commonly used to fabricate large cylindrical structures such as pipes, as well as spherical tanks for chemical storage. Filament winding is compatible with a variety of reinforcement fibers, including glass, carbon, and aramid, in combination with resin systems such as epoxy, polyester, and vinyl ester. There are two main filament winding methods: wet winding and prepreg winding. In wet winding, a low-viscosity resin is applied to the fibers during the winding process, with commonly used polyesters and epoxies having viscosities below 2 Pas (2,000 cP). Potential voids in filament-wound structures are most likely to form at fiber crossovers or in regions where layers with different orientations meet [11].

2.2.3 Resin Transfer Molding

Resin transfer molding (RTM) is a widely used technique for fabricating polymer matrix composites (PMCs). The process carried out at low pressures involves placing the resin into a modular form [12].

For effective impregnation, the resin must have a sufficiently low viscosity (less than 1 Pas) to ensure thorough fiber wetting. Various additives can be incorporated into the resin to improve surface finish, flame resistance, weather durability, and curing speed. The flow behavior of the resin within the fiber preform is governed a mathematical law, which describes fluid movement through a porous medium. In RTM, this principle is crucial for optimizing resin distribution and ensuring uniform composite properties. According to this physical law for single-phase fluid flow, the volume current density J of a fluid is mathematically expressed as:

$$J = -\frac{k}{\eta} gradP \tag{1}$$

where, k represents the permeability of the porous medium, η is the viscosity of the fluid, and gradP denotes the pressure gradient driving the flow. Notably, Darcy's law closely resembles Ohm's law in electrical conduction, where fluid flow corresponds to electric current, pressure gradient to voltage, and permeability to electrical conductivity. It is important to recognize that permeability k is a characteristic of the porous medium, determined by its microstructure, and is independent of the fluid's properties. Darcy's law is generally not applicable to gravel because it assumes laminar flow, which requires low fluid velocity and small pore sizes. Gravel has large voids, which cause water to drain quickly due to gravity, leading to turbulent flow rather than the laminar flow required for Darcy's law.

1. The closed-mold system ensures uniform resin distribution, leading to improved dimensional accuracy and consistent mechanical properties.
2. The controlled injection process minimizes the formation of air pockets, enhancing the composite's strength and durability.
3. Compared to other molding techniques, RTM operates at relatively low pressures, reducing equipment costs and energy consumption.

2.3 *Properties of PMCs*

Continuous fiber-reinforced polymer composites exhibit highly anisotropic properties, meaning their mechanical behavior varies with direction. The overall performance of these materials are influenced by various factors described below [13].

2.3.1 Structural Defects in PMCs

The final stage of PMC fabrication, known as debulking, is crucial for minimizing voids, one of the most common defects. However, other structural defects can also occur, including:

Resin-rich (fiber-poor) regions: Areas with an excess of resin and a deficiency of reinforcing fibers. Voids: These are spaces in the material devoid of any substance. Debonded and delaminated regions: Poor adhesion between layers can cause weakening of the composite structure [14]. Variations in fiber alignment: Misalignment can lead to inconsistent mechanical properties.

2.3.2 Mechanical Properties

Polymer matrix composites (PMCs) are fascinating materials whose mechanical traits–like strength and stiffness–shift depending on how their fibers are arranged. This variation, known as anisotropy, sets them apart from uniform materials like metals and stems from the way fibers are embedded within the polymer base.

1. Take a composite where continuous glass fibers run in a single direction within an unsaturated polyester matrix. If you slice it crosswise, you'll see the fibers spread out evenly. This consistent layout means the material behaves predictably–its strength and rigidity are steady and dependable along the fiber path. It's a reliable choice when you need performance you can count on in a specific direction.
2. Now, consider a different setup: short glass fibers mixed into a polyethylene terephthalate (PET) matrix and shaped through injection molding. Here, the structure gets more complex. The material forms three distinct layers. The outer layers have more fibers lined up with the flow of the mold, while the middle layer has fewer aligned fibers. Because of this uneven alignment, the composite's properties aren't the same everywhere–its strength depends on which direction you test it in, showcasing its anisotropic nature.
3. Then there are laminated composites, built by stacking sheets (called plies) with fibers angled in chosen directions. This layering creates a material that's much stronger along the fibers than across them, again highlighting anisotropy. Data on these composites often focuses on their top performance along the fiber lines, but it's worth remembering they don't match that strength in other directions. This makes them quite different from metals, which perform consistently no matter how you measure them.

Most thermoset matrix composites fail through fracture rather than yielding, as they lack plastic deformation before failure. Their strain-to-failure values are generally low (typically below 0.5%), limiting their ability to absorb energy during fracture. This characteristic poses challenges in structural design, as engineers cannot rely on localized yielding to relieve stress concentrations. The stiffness and strength of continuous fiber-reinforced composites are highest along the fiber axis but decrease significantly at off-axis orientations. Fiber volume fractions in PMCs typically vary as follows:

1. Unidirectional composites: Up to 65% fiber content
2. Bidirectional composites: Around 50% fiber content
3. Randomly oriented in-plane fibers: Below 30% fiber content

The Young's modulus of a fiber-reinforced material composite can be estimated using the rule-of-mixtures equation:

$$E_c = E_f V_f + E_m V_m \tag{2}$$

where E and V represent the Young's modulus and volume fraction of each component, respectively, and the subscripts c, f, and m denote composite, fiber, and matrix. From previously discussed fiber properties, it can be observed that:

1. Glass fiber/polymer composites provide a modest modulus increase.
2. Aramid, polyethylene, and carbon fiber composites offer substantial stiffness and strength improvements.
3. Aramid fiber composites have higher creep rates than glass fiber composites but provide superior impact resistance and ballistic protection.
4. Carbon fiber composites offer the highest stiffness and strength enhancements.
5. Polyethylene fiber composites provide high strength at a low density.

Since fiber properties significantly influence the mechanical characteristics of PMCs, material selection plays a crucial role in composite design.

2.3.3 Damping Characteristics

One of the crucial characteristics PMCs have is damping – their ability for soaking up vibrational energy. This is a big deal in places where you want to keep noise low, things steady, and structures from shaking apart. This is crucial as when the resonant frequency of the structure equals the frequency of driving force, the amplitude of oscillations diverges to infinity so break up of structure is a real physical possibility due these oscillations. In rocket engines, these oscillations are called pogo oscillations. Think about machines that rev up and slow down all the time; good damping stops them from rattling themselves to pieces. Or picture a tennis racket or golf club–damping makes them feel smoother in your hands and perform better by keeping unwanted buzz in check.

In composite materials, damping behavior is influenced by several factors, including the reinforcement orientation relative to the loading axis, the viscoelastic properties of both the matrix and reinforcement, and the strength of interfacial bonding. Slippage at the matrix-reinforcement interface can alter energy dissipation, while structural damage—such as microcracks, fiber breakage, or environmental degradation–may further affect damping performance. Among different composite systems, aramid fiber/polymer composites exhibit superior damping characteristics due to the inherently high energy absorption capacity of aramid fibers [15].

2.4 Applications of PMCs

Polymer matrix composites find widespread use across various industries due to their tailored properties: in aerospace, they form lightweight structural components for aircraft and spacecraft; in automotive, they contribute to lighter and more fuel-efficient vehicles; in marine, they create durable and corrosion-resistant boats and offshore structures; in construction, they offer strong and long-lasting infrastructure solutions; in sports, they enhance the performance of equipment; and in medicine, they provide lightweight and strong prosthetics. Essentially, PMCs are implemented in diverse fields where a combination of high strength, stiffness, and reduced weight is advantageous [16] (Fig. 4).

2.4.1 Ballistic Protection

In the realm of ballistic protection, polymer matrix composites, particularly those incorporating high-performance fibers like aramids (e.g., Kevlar, Twaron) or high-strength polyethylene, are fundamental. These woven fabrics, often layered to create ballistic panels, serve as the core of concealable body armor, effectively mitigating the impact of low- to medium-energy handgun rounds by absorbing and dissipating projectile energy. The intricate engineering of these composite structures within the ballistic panel prioritizes not only maximum protection but also flexibility and comfort for everyday wear, highlighting the critical role of advanced polymer composites in safeguarding individuals in potentially dangerous situations [17].

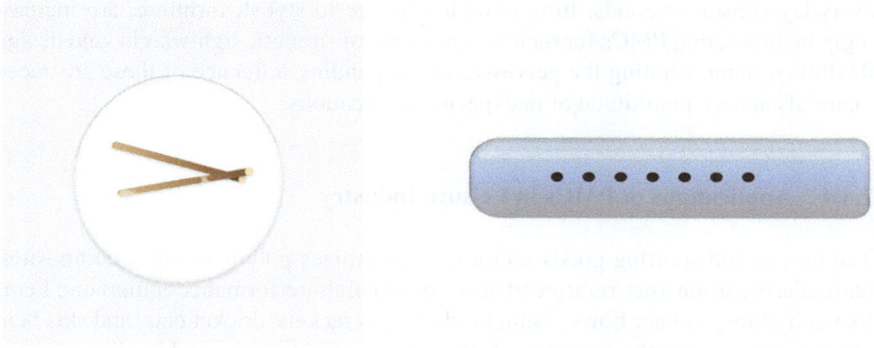

Fig. 4 The drumsticks used in musical instruments and fuselage of an aeroplane are made up of polymer matrix composite

2.4.2 Applications of PMCs in Aerospace

The Boeing 787 exemplifies a significant advancement in aerospace applications of polymer matrix composites, demonstrating their potential to revolutionize aircraft design and performance. By utilizing carbon fiber-reinforced epoxy for major structural components like the fuselage and wings, and other composites for smaller parts, the 787 achieves substantial weight savings, leading to improved fuel efficiency and longer flight ranges. Beyond weight reduction, the superior strength and stiffness of these composites enable enhanced passenger comfort through higher cabin pressurization and humidity, as well as design innovations like larger windows. This case study underscores the critical role of advanced polymer composites in creating lighter, more efficient, and more comfortable aircraft, while also driving the development of innovative manufacturing techniques like automated tape laying and co-curing processes within the aerospace sector [18].

2.4.3 Miscellaneous Aerospace Applications of PMCs

Beyond the prominent roles in aerospace, automotive, marine, and construction, polymer matrix composites are quietly revolutionizing a diverse array of other sectors. In the realm of sports and recreation, PMCs enable the creation of lighter yet stronger equipment like golf clubs, bicycles, and tennis rackets, enhancing athletic performance. The medical field benefits from PMCs in the development of lightweight and durable prosthetics and orthotics, improving patient comfort and mobility. Furthermore, specialized PMCs find utility in chemical processing plants due to their corrosion resistance, and in electronics as lightweight, insulating housings. Even everyday consumer goods, from durable luggage to stylish furniture, are increasingly incorporating PMCs for their unique blend of strength, lightweight, and design flexibility, demonstrating the pervasive and expanding influence of these advanced materials across a multitude of unexpected applications.

2.4.4 Applications of PMCs in Leisure Industry

The leisure and sporting goods sector heavily utilizes polymer matrix composites, particularly carbon fiber-reinforced ones, to craft high-performance equipment. Items like golf clubs, archery bows, fishing rods, tennis rackets, cricket bats, and skis benefit immensely from the exceptional strength-to-weight ratio and durability of these composites, leading to enhanced athletic performance. Beyond structural advantages, certain carbon fiber-reinforced thermoplastic composites offer excellent electrical conductivity, a crucial property for dissipating static charge in sensitive applications like high-speed computer components and musical instruments. This inherent conductivity eliminates the risk of electrostatic buildup and discharge, which can damage electronics or cause discomfort. Moreover, carbon fibers dispersed in thermoplastic resins not only provide electrical conductivity but also simultaneously boost the material's strength and wear resistance, often negating the need for separate conductive elements. This unique combination of properties has also led to the increasing use of carbon fiber composites in manufacturing musical instruments such as guitars, violins, violas, and bows, where their lightweight nature, durability, and favorable acoustic qualities are highly valued [19].

3 Metal Matrix Composites (MMCs)

3.1 Structure of MMCs

The matrix material of MMCs are made up of metallic substances and the reinforcing phase can be ceramics, oxides particles, fibers of continuous materials. Metal Matrix Composites (MMCs) are advanced materials engineered by embedding reinforcing elements into a metal matrix to combine the best properties of both components. The reinforcement–often ceramic materials like silicon carbide (SiC), alumina (Al_2O_3), or carbon fibers–comes in forms such as particles, fibers, or whiskers [20]. In particle-reinforced MMCs, the particles are uniformly dispersed for consistent properties.

3.2 Synthesis of MMCs

Metal matrix composites (MMCs) are engineered materials where a metal or a metal alloy forms the main, continuous part, which is strengthened by the addition of other materials in the form of tiny particles, short fibers or whiskers, or long, continuous fibers. The creation of these MMCs involves various key manufacturing methods, which are followed by an examination of their resulting characteristics and a review of their wide-ranging uses in different industries. Broadly, MMCs can be categorized based on the type of reinforcement they contain: those strengthened by particles, those with short fibers or whiskers, and those with continuous fibers or sheets. Particle-reinforced MMCs have gained significant importance due to their lower cost compared to those with continuous fibers and their more uniform properties in all directions, making them a versatile choice. The fabrication of MMCs primarily involves two main approaches: processes that start with the metal in a liquid state and those that begin with the metal in a solid state.

3.2.1 Liquid State Process

Creating metal matrix composites using liquid-state methods, particularly with lower-melting-point metals like aluminum, involves melting the base metal and then introducing it to a pre-arranged framework of strengthening materials, be they fibers, whiskers, or tiny particles. A key challenge lies in ensuring the molten metal properly adheres to and surrounds these reinforcements, often requiring surface treatments on the strengthening components. Think of it like trying to mix oil and water—you need a way to make them interact effectively. One successful approach involves carefully dispersing treated strengthening particles into the molten metal, stirring the mixture at a precise temperature just above melting before shaping it. The size of these strengthening particles is a delicate balance; too small makes the liquid metal sluggish, while too large might not provide enough overall reinforcement. To overcome the challenges of wetting and ensure thorough filling, techniques employing pressure are also used, essentially forcing the liquid metal into the reinforcement structure. Scientists use principles of fluid flow to understand and optimize these processes, and maintaining the correct temperature is crucial to ensure smooth flow and prevent the formation of weak or brittle areas within the composite material [21].

3.2.2 Solid State Process

Solid-state methods offer an alternative approach to creating metal matrix composites by sidestepping the melting of the primary metal. Instead, these techniques rely on the movement of atoms and the application of mechanical force to fuse the composite together. Imagine pressing and heating layers of different materials so that their surfaces intermingle and bond at an atomic level, this is the essence of diffusion bonding, a key solid-state process. Variations like the foil-fiber-foil method allow for precise arrangement of reinforcing fibers and strong connections between the layers. Another set of solid-state techniques uses brute force to shape and combine the materials. Processes like squeezing, pushing through a die, pulling, and rolling can deform a mixed block of materials, causing one phase to stretch into strengthening fibers within the other. Think of repeatedly rolling out dough with nuts inside, where the nuts become more aligned. Finally, deposition techniques build the composite layer by layer, directly applying the matrix material onto the reinforcement. This can involve dipping fibers in a metallic solution, using electricity to plate metal onto fiber bundles, or even using chemical vapor to create thin coatings. While these methods can take longer, they provide exceptional control over how the different materials meet, allowing for tailored interfaces that can improve bonding or prevent unwanted chemical reactions. Ultimately, while melting offers an easy way to fill complex shapes, solid-state methods provide a finer degree of control over the internal structure and the connections between the metal and the reinforcement [22].

3.3 Properties of MMCs

Metal Matrix Composites (MMCs) are advanced materials that integrate the strength and durability of metals with the enhanced mechanical, thermal, and wear-resistant properties of reinforcing phases, such as ceramics or carbon-based materials. This combination results in superior performance.

3.3.1 Mechanical Properties

Compared to the original metals they are made from, metal matrix composites generally boast greater strength and rigidity. By adding tough ceramic materials like silicon carbide or aluminum oxide, these composites can handle much heavier loads. How strong an MMC becomes depends on how much reinforcing material is added, the size of its particles, and how evenly these particles are spread throughout the metal. While these hard reinforcements make MMCs more resistant to scratching and denting, they often become less flexible and more prone to breaking. The ability of an MMC to resist fracturing is mainly determined by how easily the base metal can deform without breaking and how well the reinforcing material is bonded to the metal. Because the added materials are usually brittle, MMCs tend to be less ductile than pure metals, meaning they can crack more easily under extreme stress. However, by carefully choosing the type of metal, the shape and arrangement of the reinforcement, and the manufacturing process, it's possible to create MMCs that have a good combination of both strength and flexibility [23].

3.3.2 Thermal Properties

Consider a highway system designed for heat, not cars. In a pure metal, heat travels somewhat directly. Now, picture adding tiny, highly conductive "lane" (the reinforcements) throughout this system. These lanes, even if they aren't the primary material, can significantly speed up the flow of thermal energy, leading to a composite that dissipates heat more efficiently than the original metal alone. The effectiveness of this "heat highway" depends on how many extra lanes are added, how well they're connected, and the overall layout.

Furthermore, we can think of how different materials react to heat – some expand a lot, like a poorly designed bridge on a hot day. By embedding stiff, less expansive ceramic materials within a metal, we're essentially creating internal "anchors" that resist overall expansion. This results in a composite that maintains its shape much better across temperature swings, a crucial benefit for delicate applications where even slight changes in size can cause major problems, like in the intricate workings of spacecraft or precision electronics [24].

3.4 Applications of MMCs

The world of metal matrix composites finds its primary playgrounds in two major arenas: the skies and the ground. When it comes to things that fly, like aircraft and spacecraft, the name of the game is peak performance, often without much concern for the price tag. Here, MMCs shine because they are lightweight yet incredibly strong and stiff, and engineers can even fine-tune how they respond to heat and conduct electricity. For critical flying components where strength and efficiency are paramount, MMCs with continuous reinforcing fibers are the top contenders, offering a significant performance edge.

Down on Earth, however, the story shifts to a balance between how well something works and how much it costs. This is where MMCs reinforced with particles become more common, finding their place in a wide range of structural and non-structural uses across various industries. They offer a good mix of strength and affordability. Interestingly, the single biggest commercial use of MMCs isn't in aerospace or your car; it's in the realm of superconducting materials. These specialized composites, with their unique electrical and thermal behaviors, are the unsung heroes in advanced technologies that rely on the amazing properties of superconductivity.

3.4.1 Multifilamentary Superconducting Composites

Consider certain special materials that, when chilled to the extreme cold of just a few degrees above absolute zero, suddenly become perfect electrical highways with absolutely no resistance. This bizarre ability, called superconductivity, allows them to carry immense amounts of electricity without losing any energy, creating incredibly powerful magnetic fields. This unlocks amazing technologies like MRI machines that peer inside the human body, massive experiments that smash atoms to understand the universe, and even futuristic trains that float above their tracks.

The story of this strange phenomenon began over a century ago when scientists noticed this vanishing electrical resistance in super-cooled mercury. Since then, a surprising number of elements and mixtures have been found to share this ability at very low temperatures. To make these superconductors practical, engineers figured out how to embed them as incredibly thin threads within a normal metal like copper, making them more robust for real-world use.

For decades, the most common superconductors relied on being cooled with super-cold liquid helium, which is expensive and not very abundant. These "conventional" superconductors had a temperature limit to their magic. Then, a groundbreaking discovery revealed a new class of "high-temperature" superconductors, strange ceramic materials that could become superconducting at much warmer temperatures, above the boiling point of the more readily available liquid nitrogen. This was a game-changer, potentially making the technology much more accessible and cheaper to operate.

However, these high-temperature superconductors have their quirks. They tend to be brittle, like delicate ceramics, making it hard to form them into the thin, flexible wires needed for many applications. They also struggle to carry as much electrical current as desired and can lose their superconducting ability in the presence of strong magnetic fields. Scientists and engineers are working hard to overcome these challenges, trying to make these materials stronger, carry more current, and

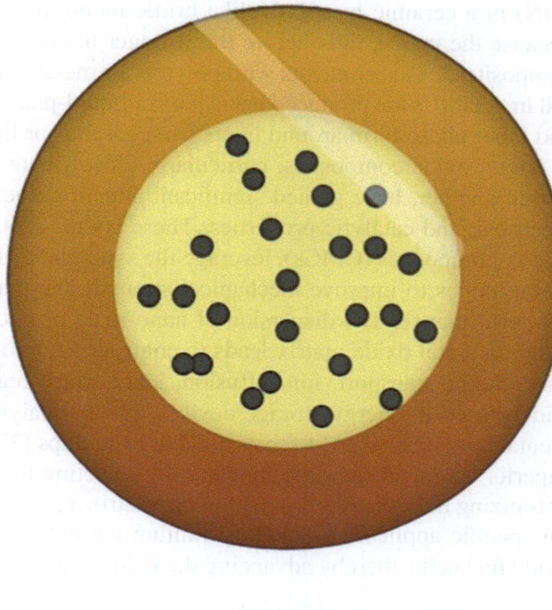

Fig. 5 The cross-section of a high temperature superconducting wire showing the dark superconducting filaments embedded in a copper matrix

be less sensitive to magnetic fields. The ultimate goal is to create superconducting wires that are flexible, robust, and can efficiently carry huge amounts of electricity in strong magnetic fields at relatively high temperatures, paving the way for even more revolutionary technologies (Fig. 5).

Extensive research and development efforts have focused on improving HTS wire and system designs to overcome these limitations. To be commercially viable, HTS wires must be flexible, mechanically robust, and capable of carrying high currents in strong magnetic fields. The key factors influencing the success of HTS wire systems include critical current density, operating temperature, magnetic field tolerance, and cost. Critical current (Jn) represents the maximum current a wire can carry while maintaining its superconducting state, with an ideal target range of 1,000–100,000 A/cm^2. To minimize cooling challenges, HTS materials should operate efficiently at temperatures between 4.2 K and 77 K. Additionally, superconducting devices must be resilient to surrounding magnetic fields to function reliably in real-world applications [25].

4 Ceramic Matrix Composites (CMCs)

4.1 Structure of CMCs

Ceramic Matrix Composites (CMCs) are advanced ceramics made by embedding fibers like carbon, silicon carbide (SiC), aluminum oxide (Al_2O_3), or silicon nitride

(SiN) in a ceramic matrix. Unlike brittle monolithic ceramics, CMCs are tougher because the matrix fails before the stronger fibers, earning them the term "inverse composites." This contrasts with polymer or metal matrix composites, where fibers fail first. CMCs are produced using gas- or liquid-phase techniques, where the matrix and inner phase form around fibers from gaseous or liquid precursors [26].

Metal oxide composites, particularly those where one metal oxide is dispersed within another, have gained significant attention due to their enhanced structural, electrical, and catalytic properties. These composites, often referred to as ceramic matrix composites (CMCs), leverage the synergistic effects between the constituent metal oxides to improve mechanical strength, thermal stability, and electrochemical performance. The dispersion of nano-scale or microscale metal oxide particles within a metal oxide matrix leads to controlled micro-structural features that influence charge transport, ion diffusion, and surface reactivity. Such composites are widely used in energy storage devices, photo-catalysis, gas sensors, and environmental remediation due to their tunable band-gaps [27, 28], high surface areas, and superior chemical stability. By carefully selecting the metal oxide components and optimizing their distribution within the matrix, researchers can tailor these materials for specific applications, such as lithium-ion batteries, super-capacitors, and solid oxide fuel cells, thereby advancing the field of functional materials [29–31] (Fig. 6).

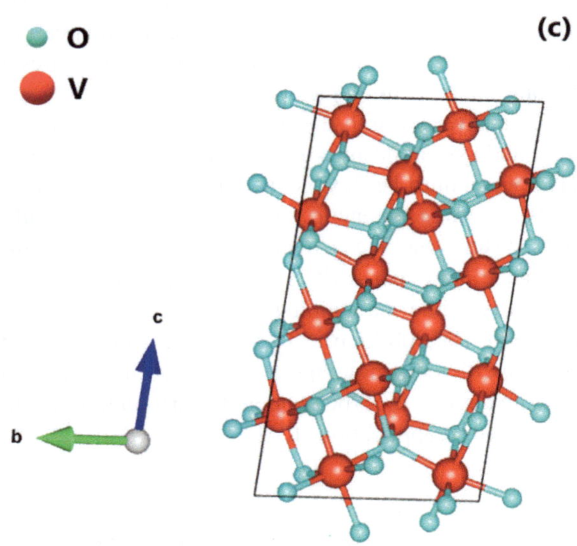

Fig. 6 The crystal structure of V4O7, one of the lesser known and metastable magneli phase of vanadium oxides

4.2 Synthesis of CMCs

Ceramic materials offer a unique combination of benefits, such as high strength and stiffness at elevated temperatures, low density, and excellent chemical resistance. However, their inherent brittleness and low toughness mean that even minor flaws can lead to sudden, catastrophic failure. To overcome these limitations, ceramic matrix composites (CMCs) are engineered by incorporating fibers into the ceramic matrix. This approach not only helps to enhance toughness but also leverages the high-temperature strength and environmental durability of ceramics. Unlike other composites–where the load is mostly carried by the fiber due to a high fiber-to-matrix modulus ratio–in CMCs, the elastic moduli of the fiber and the matrix can be quite similar. This, along with the high processing temperatures and limited ductility of ceramics, makes thermal mismatch a critical factor in CMC performance.

Several processing techniques are used to synthesize CMCs exists, each with its own merits and challenges:

4.2.1 Cold Pressing and Sintering

This traditional approach involves mixing ceramic powders with fibers, pressing the mixture at room temperature, and then sintering it at high temperatures. While this method is a direct extension of conventional ceramic processing, the significant shrinkage that occurs during sintering often leads to cracks. The presence of high-aspect-ratio reinforcements further complicates densification, as these fibers can interfere with the shrinkage process and, if there is a mismatch in thermal expansion between the fiber and matrix, can induce stresses that hinder proper sintering [32, 33].

4.2.2 Hot Pressing

To improve densification and minimize cracking, hot pressing is frequently employed. This process applies both high temperature and pressure simultaneously, promoting rapid consolidation of the material into a dense, pore-free compact. A common variant is the slurry infiltration process, where a fiber preform is first impregnated with a slurry of ceramic powder, a carrier liquid, and an organic binder. After drying and binder removal, the preform is hot pressed. This method allows for precise fiber alignment and stacking, leading to composites with low porosity and enhanced mechanical properties. However, the process is best suited for matrices with lower melting or softening points, such as glass or glass-ceramic systems [34].

4.2.3 Chemical Vapour Deposition

In these processes, the ceramic material deposits gradually onto and between the fibers. Although CVI allows for excellent control over the matrix formation and results in composites with outstanding high-temperature properties, it is often a slow process due to diffusion limitations. Modifications, such as forced gas flow and temperature gradients, can accelerate the process and improve uniformity .

Each of these synthesis and processing methods offers a different balance of advantages and challenges. The choice of technique depends largely on the specific application requirements, such as the desired density, mechanical properties, and operating temperature range of the final composite. Advances in these processes continue to enhance the performance of CMCs, making them increasingly attractive for critical applications in high-temperature environments [35].

4.2.4 Reaction Bonding Processes

Reaction bonding offers an alternative to traditional sintering by forming the matrix through chemical reactions at temperatures lower than conventional sintering. This method minimizes issues with shrinkage and fiber degradation and permits the use of high reinforcement volume fractions and multidirectional fiber preforms. The downside is that achieving a fully dense composite can be challenging, as reaction bonding often results in a higher degree of porosity compared to hot pressing. Hybrid methods that combine aspects of both reaction bonding and hot pressing have been developed to improve the final composite density [36].

4.2.5 Directed Oxidation or the Lanxide Process

Another innovative approach is the directed oxidation process, where a molten metal is oxidized in a controlled manner to form the ceramic matrix. In this method, a barrier is applied to guide the oxidation so that a continuous ceramic network develops, often alongside a small percentage of residual metal. This residual metal can sometimes contribute beneficially to fracture toughness, though its complete removal is necessary for applications requiring high-temperature stability. While this technique offers near-net-shape fabrication and cost advantages, it can be difficult to control the chemical reactions fully to achieve a pure ceramic matrix [37].

4.2.6 Sol-Gel

Consider using a special liquid solution, like a very fine paint containing the building blocks of ceramics, to create strong composite materials. This liquid, called a sol, is made by mixing and reacting certain metallic compounds. It then thickens into a gel, which, when carefully heated, transforms into a solid ceramic. A major perk of this "sol-gel" method is that this hardening happens at much lower temperatures than traditional ceramic making.

Just like with regular ceramics made this way, using sols to create composites offers benefits like gentler processing, more even mixing of ingredients in simple ceramics, and the ability to create unique combinations of different materials. Plus, because the starting liquid is so thin, it can easily seep into the spaces between reinforcing fibers. However, this method also has its downsides. As the gel dries and turns into a solid, it can shrink a lot, sometimes causing cracks in the material. To get a solid, dense ceramic around the fibers, you often have to repeat the liquid soaking process multiple times.

Interestingly, some techniques used for working with polymers can also be adapted for this sol-gel process. For example, you can use vacuum to help the liquid soak thoroughly into the fibers, or you can wind fiber strands through a tank of the sol and then around a mold to create a specific shape. Once the sol hardens into a gel, it's removed from the mold and heated to become a final ceramic or glass structure.

Another way to use the sol-gel method is to create pre-coated fibers, where the sol acts like a glue, binding fibers and ceramic particles together. A big advantage here is that you don't need a separate step to burn off the glue, as it becomes part of the final ceramic. Also, the material can often be solidified at lower temperatures. However, getting a consistent, non-porous coating without carbon-rich areas, which could weaken the final composite, remains a challenge [38] (Fig. 7).

4.3 *Properties of CMCs*

Ceramic Matrix Composites (CMCs) are advanced materials consisting of a ceramic matrix reinforced with ceramic fibers, whiskers, or particles. Unlike conventional ceramics, which are inherently brittle and susceptible to catastrophic failure, CMCs

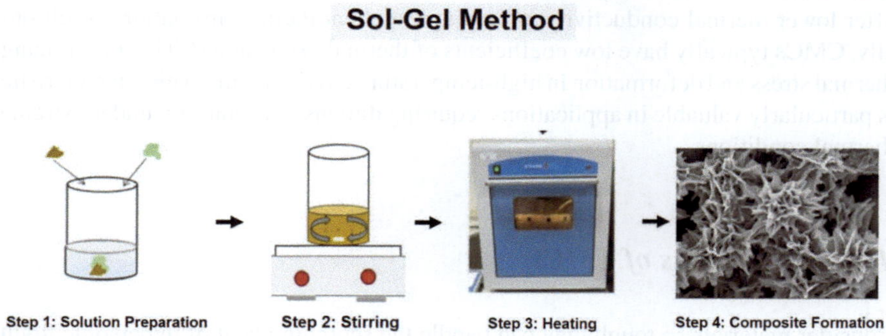

Fig. 7 The process flow of a traditional sol-gel process to synthesize the three phase composite system of vanadium oxides and zinc oxide

exhibit enhanced toughness, thermal stability, and damage tolerance. These properties make them well-suited for high-temperature applications in aerospace, defense, and energy sectors. The overall performance of CMCs is influenced by factors such as matrix composition, reinforcement type, interfacial bonding, and processing techniques.

4.3.1 Mechanical Properties

CMCs possess high strength and stiffness due to their ceramic constituents, with the reinforcement phase significantly enhancing load-bearing capacity and inhibiting crack propagation. Unlike monolithic ceramics, CMCs retain structural integrity under mechanical stress, making them ideal for high-performance applications. One of the key advantages of CMCs over conventional ceramics is their superior fracture toughness. The presence of reinforcing fibers or whiskers enables crack deflection and bridging, reducing the likelihood of sudden failure. This damage-tolerant behavior is particularly crucial for aerospace and turbine components. Additionally, CMCs exhibit exceptional hardness, providing excellent resistance to wear and abrasion, which is especially beneficial for applications involving sliding contact, such as cutting tools, bearings, and protective coatings.

4.3.2 Thermal Properties

A key advantage of CMCs is their ability to endure extremely high temperatures without significant degradation. Unlike metals, which lose strength at elevated temperatures, CMCs maintain their mechanical integrity, making them ideal for applications such as jet engines, hypersonic vehicles, and thermal barrier coatings. The thermal conductivity of CMCs depends on the matrix and reinforcement materials. For instance, SiC/SiC composites exhibit high thermal conductivity, making them well-suited for heat exchangers and electronic applications, while oxide-based CMCs offer lower thermal conductivity, providing excellent thermal insulation. Additionally, CMCs typically have low coefficients of thermal expansion (CTE), minimizing thermal stress and deformation in high-temperature environments. This characteristic is particularly valuable in applications requiring dimensional stability under extreme thermal conditions.

4.4 Applications of CMCs

Consider materials so tough they can handle the scorching heat of hypersonic flight while also being incredibly light and strong – that's the promise of ceramic matrix composites in the aerospace world. These materials offer a unique combination of superpowers: they are exceptionally stiff and strong for their weight, which means lighter aircraft, less fuel burned, and better overall efficiency. Surprisingly, they can also lead to lower costs over the long run because they last longer and need less upkeep compared to traditional super-hot metal alloys. Their ability to operate at much higher temperatures pushes the boundaries of what's possible for high-speed, high-altitude machines. Plus, they are incredibly resilient against wear, tear, and the harsh effects of the atmosphere, meaning parts don't need replacing as often. Interestingly, CMCs can even help stealth technology by making aircraft harder to detect by heat sensors and radar.

From airplanes and helicopters to missiles and spacecraft re-entering the atmosphere, CMCs are finding their place. Future aircraft that will travel at many times the speed of sound, where the outer skin gets incredibly hot, will heavily rely on these materials for protection. Key parts like the very front of the aircraft, the edges of wings and control surfaces, and even the protective domes for radar will face extreme heat, and CMCs are being developed to withstand these conditions. Right now, aircraft often use special heat-shielding materials that don't actually support any weight. The goal with advanced CMCs is to create materials that can both handle these extreme temperatures and be structural parts themselves, making for better performance and more reliable high-temperature aircraft.

4.4.1 Cutting Tool Inserts

Consider of the tips of high-speed machine tools that carve away at super-tough metals like Inconel 718. These aren't made of ordinary stuff; often, they're crafted from ceramic matrix composites, like alumina strengthened with tiny silicon carbide whiskers. These advanced cutting tips can outlast regular ceramic tools by threefold and even outperform cemented carbide tools by a factor of eight! What makes CMCs so exceptional for this demanding job? Firstly, they are incredibly resistant to wear and tear, meaning they stay sharp for much longer. Secondly, they can handle sudden and extreme temperature changes without cracking. Thirdly, they possess high mechanical strength, allowing them to withstand the immense forces involved in heavy cutting. Fourthly, they are less likely to chip or break under stress. And finally, they efficiently conduct heat away from the cutting edge, preventing overheating and ensuring a more stable and precise machining process. All these qualities together translate to faster production, less machine downtime for tool changes, and ultimately, better results in precision manufacturing.

4.4.2 Miscellaneous Applications of CMCs

Imagine materials so robust they could form the heart of advanced technologies, handling intense heat in engines, resisting harsh chemical attacks, or playing crucial roles in specialized electronics and power generation. Ceramic matrix composites hold this promise, even extending their potential to sophisticated military applications.

Consider, for instance, a high-tech heater using a special tube made of these ceramics. Heat can travel in three main ways: by direct touch, through moving fluids, or as invisible energy waves, like the warmth you feel from the sun. In this ceramic heating tube, the hot gases inside first transfer their energy to the tube itself. Then, the tube glows, radiating this energy outwards to heat whatever needs warming. The hotter the tube gets, the much, much more efficient this radiant heating becomes – the efficiency jumps dramatically with temperature. Because ceramic tubes can withstand much higher temperatures than metal ones, they can glow hotter and transfer heat much more effectively, leading to better performance and less wasted energy in the overall heating system.

5 Carbon-Carbon Composites (CCCs)

5.1 Structure of CCCs

Carbon-carbon (C/C) composites are a class of materials where both the reinforcing fibers and the matrix are made of pure carbon, offering high strength and resistance to extreme temperatures. They can be structured in various ways, including unidirectional, bidirectional, and multi-directional (3D, 4D, etc.) arrangements of carbon fibers [39].

5.2 Synthesis of CCCs

Carbon, in its many forms—from shapeless soot to the hardness of diamond and the intricate structures of nanotubes—offers a remarkable palette of properties. Among these, carbon-carbon composites stand out as a unique creation: carbon fibers embraced by a matrix of pure carbon. Whether these fibers are long and continuous or short and scattered, the resulting material boasts impressive strength, incredibly low weight, and the ability to withstand extreme heat.

These C/C composites truly shine in environments where temperatures soar beyond the limits of most materials, often enduring over 1000°C for extended periods, and in some cases even pushing towards a scorching 2200°C, as long as oxygen isn't present to cause trouble. Originally conceived for the demanding conditions of space travel, they've since found their niche in high-performance engineering where extreme heat resistance is a must-have. Think of the robust brakes in aircraft and high-performance cars, the molds used in intense hot-pressing processes, the reliable fasteners in high-temperature machinery, the heating elements in industrial furnaces, and the protective linings in extreme environments.

However, creating these exceptional materials comes at a cost, primarily due to the lengthy and intricate manufacturing processes. Whether it involves transforming a polymer into carbon through intense heating or depositing carbon atoms from a gas onto a carbon fiber framework, both paths require multiple cycles and can take days to complete, significantly adding to the final price.

In practice, three main techniques are employed to build these carbon-carbon marvels. One involves repeatedly soaking a carbon fiber structure in a hot, pressurized tar-like substance called pitch, then baking it to convert the pitch into carbon. This cycle is often repeated to make the material denser. A variation uses high pressure during this baking process to further compact the material. Another method starts with a carbon fiber reinforced plastic, which is then subjected to intense heat to convert the plastic itself into a carbon matrix, again often requiring multiple impregnation and heating steps. The third technique, chemical vapor deposition, involves flowing a carbon-containing gas over a hot carbon fiber structure, causing pure carbon to deposit onto and within the fibers. Adding inert gases helps this process happen more evenly. Each of these methods, while effective, contributes to the overall complexity and cost of carbon-carbon composites.

5.2.1 High Pressure Processing

Consider a pressure cooker specifically designed to bake carbon materials. When making carbon-carbon composites using a tar-like substance called pitch, applying immense pressure during the baking process dramatically improves how much of that pitch actually turns into useful carbon. Instead of losing about half the material at normal pressure, pushing down with forces equivalent to hundreds of atmospheres can preserve over 90% of it. This high-pressure baking method, known as HIPIC, involves placing the partially formed carbon structure and extra pitch inside a sealed metal can. This can is then put into a giant pressure chamber that's heated to extremely high temperatures while being squeezed with tremendous force. The thin metal can acts like a flexible bag, ensuring the pressure is evenly applied all around. This entire high-pressure cooking cycle can take a full day. Once finished, the metal can is removed, and the carbon composite usually undergoes an even hotter treatment, exceeding 2300°C, to fine-tune its internal structure and make it even stronger.The final quality of these carbon-carbon materials is greatly affected by how many tiny air pockets, mostly interconnected, are left inside. A simple way to track these pockets is by measuring how dense the material is. Typically, a starting carbon fiber structure might have a density of around 1.0. After a few standard soaking and baking cycles, this might increase to about 1.4. However, using the high-pressure baking method, the density can jump to as high as 1.8 after a similar number of cycles. For reference, pure, solid carbon would ideally be around 2.25. Even the carbon fibers themselves usually fall in the range of 1.6 to 2.2, with the really dense ones being rare and usually requiring extra high-temperature treatment. By significantly boosting how much pitch turns into carbon and how tightly packed the final material becomes, HIPIC offers a more efficient way to create high-performance carbon-carbon composites, making them a better fit for tough jobs that demand exceptional resistance to heat and stress [40].

5.3 *Properties of CCCs*

Carbon-Carbon (C/C) composites are high-performance materials consisting of carbon fiber reinforcement embedded within a carbon matrix. These composites are distinguished by their exceptional strength, high-temperature stability, and resistance to thermal shock. Unlike conventional materials, C/C composites maintain their mechanical integrity even under extreme thermal conditions, making them ideal for demanding applications in aerospace, automotive, and industrial sectors. Their unique combination of low density, high strength, and superior thermal performance has led to their widespread use in jet engines, space vehicles, and high-performance braking systems [41].

5.3.1 Mechanical Properties

C/C composites possess exceptional strength-to-weight and stiffness-to-weight ratios, outperforming many metallic and ceramic materials. Their mechanical properties are influenced by factors such as fiber orientation, fiber type, and processing techniques. The strong covalent bonding in carbon enhances the composite's load-bearing capacity, contributing to its superior performance. Unlike traditional ceramics or graphite, C/C composites exhibit high fracture toughness due to their layered fiber architecture. This reinforcement structure effectively dissipates stress, reducing the risk of catastrophic failure and enhancing damage tolerance. Additionally, C/C composites offer excellent wear resistance, particularly in high-friction applications such as aerospace braking systems and racing car disc brakes. Their ability to maintain mechanical integrity under repeated stress cycles ensures durability across a wide range of demanding applications [42].

5.3.2 Thermal Properties

Carbon/carbon (C/C) composites can exhibit high thermal conductivity, particularly when reinforced with mesophase pitch-based carbon fibers, which possess a highly oriented graphitic structure. These composites display significant anisotropy, with high thermal conductivity along the fiber direction and much lower conductivity in the transverse direction. As a result, achieving the desired thermal performance requires careful selection and arrangement of fiber orientations to optimize heat dissipation or insulation based on application needs [43].

5.4 Applications of CCCs

Consider things approaching to a very high temperatures, the fundamental limit of the material and its constituent particles and need to stay strong and modular, that's where carbon-carbon composites truly shine. Their unique ability to handle extreme temperatures and mechanical stress makes them perfect for some very demanding jobs. Think of the outer shields of spacecraft as they plunge back into Earth's atmosphere, enduring immense heat—C/C composites are often the first line of defense. They're also the material of choice for the brake systems in both high-performance aircraft and some top-tier cars, where they offer exceptional stopping power even under intense heat. Furthermore, in industrial settings that involve extreme temperatures and pressures, like the molds used in hot-pressing manufacturing, C/C composites provide the necessary stability and longevity. Even critical components like the nozzles in high-temperature systems benefit from the robust nature of these materials. In essence, wherever the heat is on and structural integrity is crucial, carbon-carbon composites are often the go-to solution [44].

5.4.1 Carbon/Carbon Composite Brakes

Braking systems rely on friction between contact surfaces to decelerate or stop a moving component. During this process, the kinetic energy of the moving object is converted into heat, which must be effectively absorbed and dissipated by the brake materials.

A typical braking system consists of two primary components: a rotating part and a stationary part that is pressed against it to generate friction. As kinetic energy is transferred into heat, the friction pad material and other brake components experience significant temperature increases. An efficient braking system must be capable of withstanding thermal shock, dissipating heat rapidly, and maintaining consistent braking performance under varying operating conditions [45].

5.4.2 Miscellaneous Applications of Carbon/Carbon Composites

The sky and beyond are key domains for carbon-carbon composites. In the realm of rocket engines, these materials form critical parts that must withstand the intense heat and stress of launch. For spacecraft returning to Earth, C/C composites act as vital heat shields, designed to gradually wear away on the surface, like a sacrificial layer, while protecting the crucial structure underneath from the extreme temperatures of reentry. To make these structures even stronger and more resilient, engineers often create 3D weaves of carbon fibers, oriented in multiple directions, providing enhanced durability.

Interestingly, the applications of these high-tech materials extend even to the delicate field of medicine. Due to their excellent compatibility with the human body,

Fig. 8 The leading edges of spacecraft are made up of carbon-carbon composites to protect them from the heat of re-entry

carbon-carbon composites are used in implants and devices that help heal broken bones inside the body. Furthermore, their robustness at high temperatures and under pressure makes them ideal for creating high-performance molds used in industrial hot pressing, outlasting traditional graphite molds and leading to longer operational life in demanding manufacturing processes [46] (Fig. 8).

5.4.3 Use in Carbon/SiC Brake Disks

C/SiC composites exhibit a relatively high strain at fracture, typically ranging from 0.1 to 0.3%, which is significantly greater than that of conventional ceramics. This results in a material that is both lightweight and highly resistant to thermal shock, allowing it to endure extreme braking conditions without becoming brittle. These characteristics make C/SiC brake disks particularly well-suited for high-performance applications, such as motor-sports and aerospace braking systems, where reliability and durability under high thermal and mechanical loads are critical [47].

References

1. Hao, C., Mehdi, Y., António, R.M.: Dynamic inter-fibre failure of unidirectional composite laminates with through-thickness reinforcement. Compos. Sci. Technol. **176**, 64–71 (2019)
2. Sharmistha, N., Asish K, P., Ashirbad, J., Subramani, K., Bikramjit, B.: UHMWPE-MWCNT-nHA based hybrid trilayer nanobiocomposite: processing approach, physical properties,

stem/bone cell functionality, and blood compatibility. J. Biomed. Mater. Res. B Appl. Biomater. **108**(5), 2320–2343 (2020)

3. Suraj, P.R.: Metal-matrix composites for space applications. JOM **53**(4), 14–17 (2001)
4. Hrudayanjali, P., Tapan, K., Mishri, S., Rani, P., Bijayalaxmi, K., Biswajit, D.: A review on processing routes, properties, applications, and challenges of titanium metal matrix composite. East Eur. J. Phys. **3**, 5–17 (2021)
5. Sangaraju, S., Siddheshwar, S., Ahmed, S.A., Pravin, P.P., Irfan, K., Rajesh, S., Lavish, K., Ankita, A.: Significance of reinforcement in mg-based MMCs for various applications: a review. Mater. Today (2023)
6. Marimuthu, S., Dunleavey, J., Smith, B.: Laser based machining of aluminum metal matrix composites. Procedia CIRP **85**, 243–248 (2019)
7. Yao, X.-Y., Li, H.-J., Zhang, Y.-L., Ren, J.-J., Yao, D.-J., Tao, J.: A SiC/ZrB2-SiC/SiC oxidation resistance multilayer coating for carbon/carbon composites. Corros. Sci. **57**, 148–153 (2012)
8. Piyush, G., Chandrasekar, M., Parthasarathy, V.: Friction and wear properties of biocomposites for dental, orthopedic, and biomedical applications (2024)
9. Soo-Jin, P., Min-Kang, S.: Types of composites. In: Interface science and composites, interface science and technology, pp 501–629. Elsevier (2011)
10. Elkington, M., Bloom, D., Ward, C., Chatzimichali, A., Potter, K.: Hand layup: understanding the manual process. Adv. Manuf. Polym. Compos. Sci. **1**(3), 138–151 (2015)
11. Minsch, N., Herrmann, F.H., Gereke, T., Nocke, A., Cherif, C.: Analysis of filament winding processes and potential equipment technologies. Procedia CIRP **66**, 125–130 (2017)
12. Shojaei, A., Ghaffarian, S.R., Karimian, S.M.H.: Modeling and simulation approaches in the resin transfer molding process: a review. Polym. Compos. **24**(4), 525–544 (2003)
13. Dipen Kumar, R., Durgesh, D.P., Ravinder, K., Catalin, I.P.: Recent progress of reinforcement materials: a comprehensive overview of composite materials. J. Mater. Res. Technol. **8**(6), 6354–6374 (2019)
14. Bleay, S.M., Loader, C.B., Hawyes, V.J., Humberstone, L., Curtis, P.T.: A smart repair system for polymer matrix composites. Compos. Part A Appl. Sci. Manuf. **32**(12), 1767–1776 (2001)
15. Tang, X., Yan, X.: A review on the damping properties of fiber reinforced polymer composites. J. Ind. Text. **49**(6), 693–721 (2020)
16. McDanels, D.L., Serafini, T.T., DiCarlo, J.A.: Polymer, metal, and ceramic matrix composites for advanced aircraft engine applications. J. Mater. Energy Syst. **8**(1), 80–91 (1986)
17. Benzait, Z., Trabzon, L.: A review of recent research on materials used in polymer-matrix composites for body armor application. J. Compos. Mater. **52**(23), 3241–3263 (2018)
18. Muhamad, H., Mat, Y., Izwan, I.: Aerospace structures and engines from polymer composites. In: Aerospace Materials, pp. 349–383. Elsevier (2025)
19. Shubham, S., Sudhakara, P., Misra, S.K., Singh, J.: A comprehensive review of current developments on the waste-reinforced polymer-matrix composites for automotive, sports goods and construction applications: materials, processes and properties. Mater. Today **33**, 1671–1679 (2020)
20. Marigoudar, R.N., Sadashivappa, K.: Comparison of tool life and surface characteristics of uncoated, coated carbide and ceramic tools during machining of SiC reinforced ZA43 alloy MMC. Mater. Sci. Technol. **30**(8), 876–887 (2014)
21. Barada Prasanna Sahoo and Diptikanta Das: Critical review on liquid state processing of aluminium based metal matrix nano-composites. Mater. Today **19**, 493–500 (2019)
22. Daulat Kumar, S., Devang, M., Gautam, U.: Manufacturing of metal matrix composites: a state of review. Mater. Today *26*, 506–519 (2020)
23. Tjong, S.: Microstructural and mechanical characteristics of in situ metal matrix composites. Mater. Sci. Eng. R. Rep. **29**(3–4), 49–113 (2000)
24. Sunil Kumar, K., Reddy, M., Kannan, R., Karthikeyan, S., Prashanth, S., Rohith Reddy, B.: A review on mechanical and thermal properties of aluminum metal matrix composites. E3S Web Conference **184**, 201033 (2020)
25. Critchlow, P.R., Gregory, E., Zeitlin, B.: Multifilamentary superconducting composites. Cryogenics (Guildf.) **11**(1), 3–10 (1971)

26. Sarin, V.K., Ruhle, M.: Microstructural studies of ceramic-matrix composites. Composites **18**(2), 129–134 (1987)
27. Ashish, K.K., Rashi, N., Satyapal, S.R., Anoop Kumar, M.: Band gap tuning possibilities in vanadium oxide. Mater. Today *43*, 2939–2943 (2021)
28. Rashi, N., Ashish, K.K., Satyapal, S.R., Anoop, K.M., Kabra, K.: Effect of heat treatment on band gap of V2O5. J. Nano- Electron. Phys. **13**(1), 01030–1–01030–5 (2021)
29. Ashish, K.K., Satyapal, S.R., Sarika, S., Rashi, N.: Structural transition and photoluminescence behavior of (V2O5)1-x (Ag0.33V2O5)x (x = 0 to 0.1) nanocomposites. Results Chem. **5**(100802), 100802 (2023)
30. Upadhyay, A.K., Ajay, K., Rashi, N.: Taguchi design of experiments based optimization and experimental investigation of mechanical performance of hybrid epoxy nanocomposites. Zast. Mater. **64**(4), 433–443 (2023)
31. Synergistic properties of (Au2V)0.01(V2O5) 0.99 composite: Synthesis, characterization, and enhanced functionalities. NanoWorld J. (2023)
32. Salvatore, G., Mattia, B., Luca, Z., Gianmarco, T., Andrew, I.D., Daoyao, K., Anna, J., Michael, J.R.: A review of cold sintering processes. Adv. Appl. Ceram. Struct. Funct. Bioceram. **119**(3), 115–143 (2020)
33. Jon-Paul, M., Xiaoyu, K., Richard, D.F., Elizabeth, C.D., Hanzheng, G., Jing, G., Amanda, B., Shuichi, F., Clive, A.R.: Cold sintering: current status and prospects. J. Mater. Res. *32*(17), 3205–3218 (2017)
34. Loh, N.L., Sia, K.Y.: An overview of hot isostatic pressing. J. Mater. Process. Technol. **30**(1), 45–65 (1992)
35. Choy, K.: Chemical vapour deposition of coatings. Prog. Mater Sci. **48**(2), 57–170 (2003)
36. Borbidge WE, Allen RV, Whelan A review of the reaction bonding technique for joining ceramics to metals. J. Phys. Colloq. *47*(C1), C1–131–C1–137 (1986)
37. Unique methods of making mmc and cmc by lanxide process; lanxide hoshiki ni yoru cmc oyobi mmc no seiho
38. Larry LH, Jon KW.: The sol-gel process. Chem. Rev. **90**(1), 33–72 (1990)
39. Windhorst, T., Blount, G.: Carbon-carbon composites: a summary of recent developments and applications. Mater. Eng. **18**(1), 11–15 (1997)
40. Alireza, B., Ying, Y., Chi, W.H., Gordon, M.: A review: synthesis of carbon-based nano and micro materials by high temperature and high pressure. Ind. Eng. Chem. Res. **5**2(36), 12689–12702 (2013)
41. Chłopek, J., Błażewicz, S., Powroźnik, A.: Mechanical properties of carbon-carbon composites. Ceram. Int. **19**(4), 251–257 (1993)
42. Tatsuo, O.: Carbon/carbon composites and their properties. Carbon Alloys 523–544 (2003)
43. Kulkarni, M.R., Brady, R.P.: A model of global thermal conductivity in laminated carbon/carbon composites. Compos. Sci. Technol. **57**(3), 277–285 (1997)
44. Nilesh, A., Aditya, R., Kathan, B., Shreyash, S.V., Aldrin, A.A.F., Utkarsh, C., Divyansh, S., Albert, E.P., Vezhavendhan, R.: An overview of carbon-carbon composite materials and their applications. Front. Mater. **11** (2024)
45. Shuai, W., Liu, Y., Ge, Y., Ran, L., Peng, K., Yi, M.: Structural transformation of carbon/carbon composites for aircraft brake pairs in the braking process. Tribol. Int. **102**, 497–506 (2016)
46. Hiroshi, H., Ken, G., Sato, T., Tanatsugu, N.: Applications of carbon-carbon composites to an engine for a future space vehicle. Adv. Compos. Mater. *12*(2-3), 237–259 (2003)
47. Krnel, K., Stadler, Z., Kosmač, T.: Carbon/carbon-silicon-carbide dual-matrix composites for brake discs. Mater. Manuf. Process. **23**(6), 587–590 (2008)

Computational Modelling for Materials

Modelling the MITF-MAX Interface: Biomaterials-Inspired Insights into Melanoma Pathways

Gurpreet Kaur Bamrah and Saurabh Srivastava

Abstract Biomaterials have emerged as promising platforms in cancer research by mimicking physiological environments and enabling targeted therapeutic delivery. While this study does not employ biomaterials experimentally, it conceptually explores how structural insights into oncogenic protein interaction can inform future biomaterials-based interventions. Focusing on transcription factors MITF and MAX complex, central to melanocyte biology and melanoma progression, we performed Co-expression and gene-survival analyses, revealing a weak but significant correlation between MITF and MAX in skin cutaneous melanoma (SKCM) and no relation in uveal melanoma (UVM). Computational modelling using GRAMMDOCK and molecular dynamics (MD) simulation in GROMACS software identified a structurally stable MITF-MAX heterodimer. RMSD analysis confirmed complex stability and key interfacial interactions, including hydrogen bonds and salt bridge between ASP17 and ARG366, reinforced interface integrity. These results provide structural insights into MITF-MAX dimerization and provide a foundational framework for guiding future design of biomaterials with computational modelling and advance therapeutic strategies targeting interactions in melanoma therapy.

1 Introduction

Biomaterials are natural or synthetic materials designed to interact with biological systems for medical or biological applications. They are often used to replace, restore, or enhance the function of tissues or organs or to facilitate healing [1]. Biomaterials are engineered to be biocompatible, which means that they should not provoke an adverse immune response in the body [2]. The Clemson University Advisory Board for Biomaterials has officially defined a Biomaterial as "a systematically and pharmacologically inert substance designed for implantation within or incorporation with living systems [3]". Biomaterials involve precisely immobilizing signaling groups

G. K. Bamrah · S. Srivastava (✉)
Department of Chemistry, Computational and Theoretical Chemistry Laboratory, Manipal University Jaipur, Jaipur, Rajasthan, India
e-mail: Saurabh.srivastava@jaipur.manipal.edu

L. Ledwani and S. Srivastava (eds.), *Experimental and Computational Advances in Materials*, Engineering Materials, https://doi.org/10.1007/978-981-95-4180-5_15

on surfaces, developing synthetic materials with regulated characteristics for drug delivery, and surface modification of materials to prevent nonspecific protein adsorption in vivo, and cell carriers, materials with biological inspiration that replicate natural processes, and the creation of complex three-dimensional architectures to generate distinct patterns for tissue engineering, biological microelectromechanical systems (bioMEMs), and diagnostics. The study of biomaterials lies at the intersection of several fields, including biology, chemistry, chemical engineering material science, mechanics, surface science, and medicine, with significant contribution from entrepreneurs' government-regulated standards bodies, and ethicists [4].

Over the past five decades, the field of biomaterials and medical devices has grown into an industry valued at $100 billion [4]. Though the main focus is on medical applications, biomaterials are also used to grow cells in culture, assay blood proteins in clinical laboratories, process biomolecules for biotechnological applications, grow implants to control cattle fertility, create diagnostic gene arrays, aquaculture, and create experimental cell-silicon "biochips" [5]. "Biomaterial science" refers to the physical and biological evaluation of materials and how they interact with the biological environment. The synthesis, optimization, characterization, testing, and biology of host-material interactions of biomaterials have historically been the focus of the most rigorous research and development. A generic biological reaction is introduced by the majority of biomaterials. In fact, "biocompatibility" is a supplementary term that is crucial to comprehending the objective of biomaterials science [5].

Based on their material class, origin and properties, biomaterials can be broadly divided into four types: metals, polymers, ceramics and composites. Metals like titanium and stainless steel are known for their strength and endurance, making them perfect for load-bearing applications like bone plates and joint replacements [6]. Conversely, polymers are flexible and malleable, and they are frequently utilized in medical equipment, from catheters to contact lenses [7]. Owning to exceptional biocompatibility, ceramics, like hydroxyapatite, are widely utilized in bone restoration [8]. Composites and other biomaterials including graphene oxide and nanocellulose are utilized in tissue engineering, bone regeneration, drug delivery etc. [9, 10]. Each of these classes have distinct advantages and disadvantages like for instance, metallic biomaterials can wear resistance but are susceptible to bio-corrosion and release of metal ions in the body, that can be detrimental to the host, ceramics have high biocompatibility and inertness but are brittle in nature., polymeric biomaterials are easy in processing due to their versatile nature, but they have lower mechanical strength compared to metals and ceramics., composite biomaterials face complexity in manufacturing and potential for interfacial issues even though it can be tailored with different materials (Fig. 1).

Recognizing the cutting-edge platforms for studying tumor microenvironment, drug delivery, and tissue engineering, biomaterials are significant in cancer research [11]. In order to accurately represent the behavior and interactions of cancer cells engineered polymers, nanomaterials and hydrogels are commonly employed to mimic the extracellular matrix of the tumor. These compounds enable highly precise drug delivery, minimizing off-target effects and enhancing efficacy [12, 13]. Thus, introducing biomaterial-based systems can be positive for studying transcription

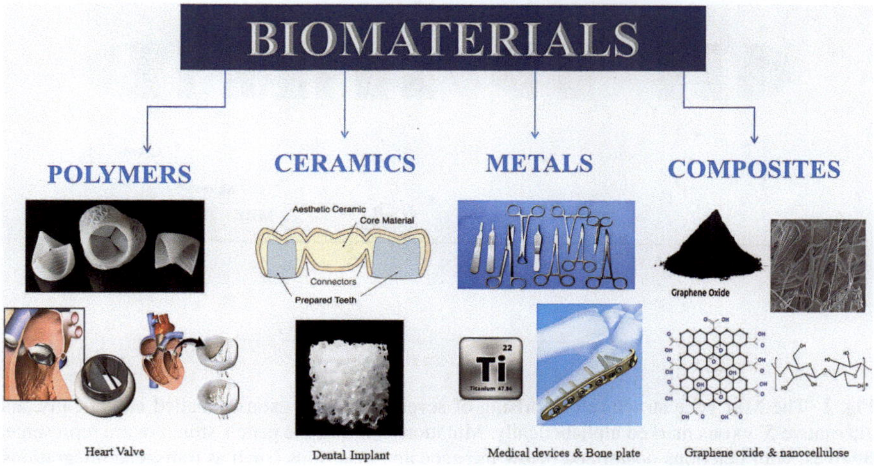

Fig. 1 Biomaterial types and their examples illustrated

factors such as MITF-MAX that regulate critical pathways in cancer progression. For example, biomaterials can serve as template for simulating protein interactions or, as vehicles for delivering inhibitors, bridging the gap between molecular insights and therapeutic development [14]. By combing biomaterial-based systems with inhibition studies, novel drug delivery mechanisms and precision targeting strategies can be explored for therapeutic interventions in melanoma and other cancers driven by MITF dysregulation.

1.1 MITF-MAX: Structure, Function and Role in Cancer Progression

MITF (Microphthalmia-associated transcription factor) is a transcription factor protein belonging to the basic helix-loop-helix zipper (bHLHZip) family. The formation, survival, and differentiation of melanocytes depend on MITF, which acts as a master regulator in these cells. Since its initial isolation 30 years ago, the gene encoding MITF has become a crucial coordinator of numerous facets of melanocyte and melanoma biology. MITF binds to particular DNA sequences and, like all transcription factors it up-regulates the genes targets [15]. It also plays critical roles in osteoclasts, mast cells, stem cells of the hair bulge and, the retinal pigment epithelium.

A. MITF mutation and their Phenotypic Impact

Mutations in MITF results in two pigmentations and deafness syndromes namely, com (WSA2) and Tietz syndrome [16] Mice has about 30 distinct alleles for MITF, the most severe of them is MITF^{mi} (R217del), linked to premature death. Less lethal

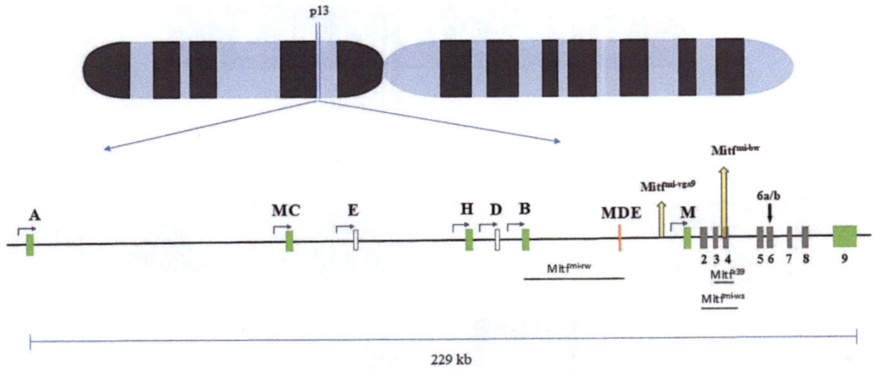

Fig. 2 The Mitf gene structure comprising of several common exons, labelled numerically, and alternative 5' exons marked alphabetically. Mutations altering the gene's structure are represented by lines, with deletions positioned below the gene and insertions (such as transgene integrations) placed above. The distal regulatory element is identified as MDE

alleles include $MITF^{mi-sp}$ and $MITF^{mi-spotted}$ lacks 18-base-pair exons (approximately 180–190 residues) and does not encode a phenotype in homozygous mice. When paired with additional mutations at the loci, an observable increase in spots or patterns due to interaction of mutations is discerned (see Fig. 2) [17]. Mutations in the MITF gene can lead to changes in pigmentation and structural abnormalities in mice, as it is critical for the development and function of the retinal pigment epithelium (RPE) and other ocular cells.

Regarding its role in retinal vasculature, studies show that $MITF^{mi-vga}$ mice exhibit a decrease in retinal vascularization compared to control mice, detailed analysis of retinal and choroidal vasculature in other MITF mutant mice remain limited. This is a loss-of-function mutation causing severe microphthalmia (a developmental disorder of the eye in which one or both eyes are abnormally small and have anatomic malformations) and a complete white coat in homozygous mice, in contrast to heterozygous $MITF^{mi-vga9/+}$ mice, eye size and pigmentation is normal with pigment mottling and few fundus lesions [18] The $MITF^{mi-rw}$ allele involves deletion of 1H-, 1D-, and 1B-exons, leading to random coat pigmentation, varying eye sizes, highlighting significance of alternative promoter usage and translation in initiation functional diversity and disease outcomes [19] $MITF^{mi-bw}$ mutant mice have an L1 element insertion that abolishes MITF-M isoform expression resulting in white coat, black eyes and normally pigmented RPE [17]. In mice and zebrafish, $MITF^{x39}$ deletion includes the exon with premature stop codon in MITF-A gene, yet a normal development of RPE can occur implying that alternative splicing or compensation may produce functional protein isoforms [20].

MITF is a conserved gene with a significant role in pigmentation across different species. In horses and dogs, changes in the promoter of the main melanocyte isoform results in white spotting [21, 22]. However, its function extends beyond pigmentation., example, in *Drosophila melanogaster*, MITF is primarily involved in gut function, and mutations in the eye disc increase the neuronal field [23]., meanwhile in mice, MITF mutations cause retinal hypoplasia [24]. Other semidominant mutations related to MITF in mice exhibit the following abnormalities: early development of deafness, loss of pigmentation, reduced eye size, failure of secondary bone resorption and decreased mast cell count [25]. Homozygous $MITF^{mi-ws}$ mutant is one such semidominant allele., these are white with pink eyes, indicating interallelic complementation at the MITF locus affecting eye and coat color. This mutation also causes a white belly spot in heterozygotes, while heteroallelic combinations with $MITF^{mi-wh}$ produce spotted mice with age-related depigmentation (refer Table 1) [26].

With 94.4% similarity to MITF-M in humans, the first melanocyte-specific isoform of Mitf is made up of 419 amino acids [27]. Human RPE cell lines and other cell types do not express MITF-M mRNA as it is only found in pigmented melanoma cells and melanocytes. It effectively transactivates melanogenesis enzyme genes like *tyrosinase* and *tyrosine-related protein* (TRP-1) in cultured cells [28]. Human MITF mutations, mainly in bHLH-LZ domain, cause pigmentary disorders, deafness, and conditions like Waardenburg syndrome IIa and COMMAD (coloboma, osteopetrosis, microphthalmia, macrocephaly, albinism, and deafness) syndrome [29]. Unlike mice, symptoms appear in heterozygous individuals. A specific mutation increases the risk of melanoma and renal cell carcinoma, highlighting MITF's role in development, evolution and cancer.

Table 1 Genetic mutation of MITF, their full forms and phenotype description

Mutation	Full form	Phenotype
$MITF^{mi-or}$	Micropthalmia-osteopetrotic	White coat, small red eyes, osteopetrosis, failure of incisors to erupt
$MITF^{mi-vit}$	Vitiligo mutation	Early pigmentation on the thorax and abdomen, progressive loss of coat and eye color, RPE-photoreceptor interaction defects
$MITF^{mi-h}$	Mutation hooded	White coat, close eyelids at birth
$MITF^{mi-rw}$	Mutation red-eyed white	The neck region is marked with color, small red eyes
$MITF^{mi-bw}$	Mutation black-eyed white	White coat with colored patches on the rump and head, small red eyes
$MITF^{mi-vga9}$	Variant allele identified in research	White coat, small red eyes
$MITF^{mi-b}$	Mutation brownish	White coat, reduced pigmentation in the eyes
$MITF^{mi}$	Mutation microphthalmia	White coat, small red eyes osteopetrosis, inner ear abnormalities, failure of incisor eruption mast cell deficiency
$MITF^{wh}$	Mutation white	White coat, small eyes with slight pigmentation

MITF is of melanocytic lineage, playing a crucial role in malignant melanoma. Its function extends across various melanoma subtypes and phenotypes, owning to distinct gene expression profiles, although the direct effect of mutation expression is found limited in a number of melanoma cases. Several transcription factors and signaling pathways such as Wnt/β catenin pathway, are commonly involved in regulating MITF expression and activity across different types of tumors, when in fact, pathways like BRAFV600E/ERK1/2 are more uniquely associated with melanoma. Primary activity of MITF is predominantly regulated through upstream activators and suppressors through transcriptional, post-transcriptional, and post translational mechanisms including epigenetic modifications and other microenvironmental influences [30].

MITF is a basic helix-loop-helix leucine zipper (b-HLH-LZ) transcription factor and is a part of MYC superfamily. It features a dimerization motif with a positively charged region for DNA binding, along with a transactivation domain (TAD). Alternative promoter usage allows the MITF gene to produce variegated isoforms, including MITF-A, MITF-B MITF-C, MITF-D MITF-E, MITF-H, MITF-J, MITF-MC and MITF-M (see Fig. 3). These isoforms differ in their N-terminal regions encoded by exon 1 and exhibit tissue-specific expression. The human MITF gene spans approximately 229,000 base pairs on the short arm of chromosome 3, extending from exon 1A to the poly(A) site in exon 9. It consists of nine upstream exons, each with its own transcriptional start site, and eight downstream exons (exons 2–9), which are typically present in all transcripts [15]. Through alternative splicing, the gene produces multiple mature transcripts and distinct protein isoforms. However, certain exons, such as 1B1b and those encoding parts of basic domain and HLH-Zip domain, are essential and cannot be spliced out without resulting in truncated proteins or mRNA decay. Each upstream exon has a unique expression pattern, regulated by tissue-specific enhancers and transcription factors (Fig. 4).

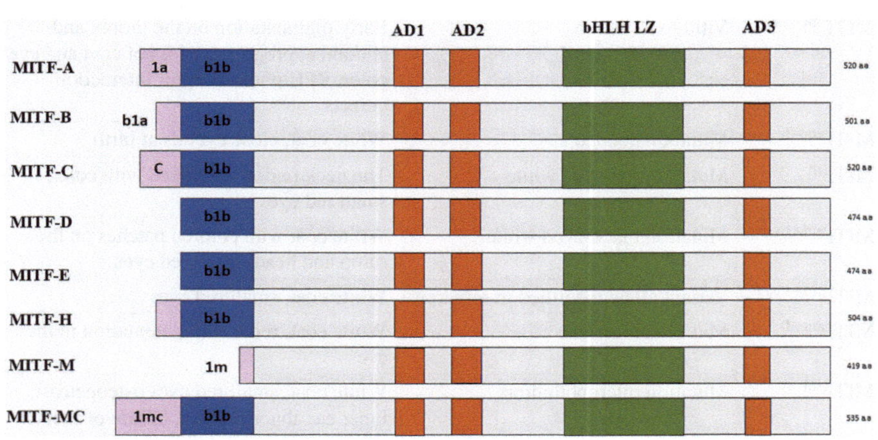

Fig. 3 Different isoforms of MITF

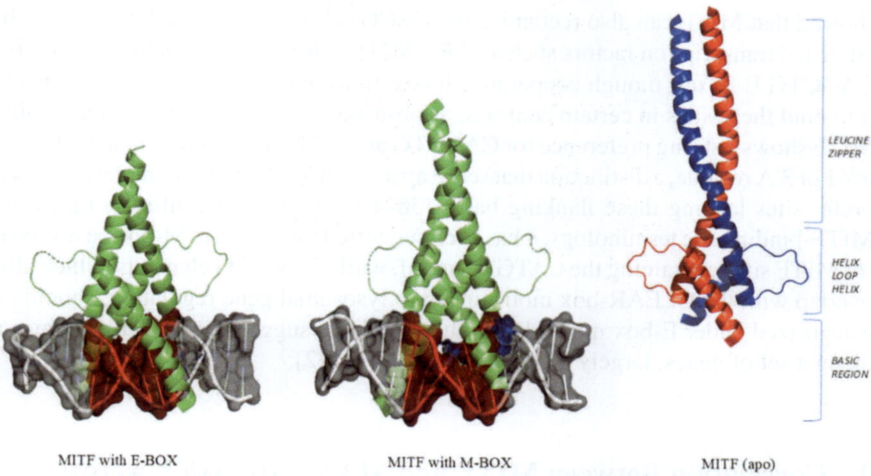

Fig. 4 Structural characterization of the MITF-DNA binding and assembly region. DNA nucleotides included sequences for E-BOX and M-BOX. Central motifs of M-BOX and E-BOX are highlighted in red, with difference in the M-BOX marked in blue. Crystal structures of MITF were analyzed in its apo form (without DNA) and when bound to 16-base pair oligonucleotides E-BOX and M-BOX motifs. MITF dimer is shown in green cartoon rep and DNA in hybrid cartoon and surface models. Missing protein loops are represented by dotted lines

This gene structure enables the production of MITF protein isoforms with varied N-terminal regions, allowing for diverse functions across various tissues. Figure 4 illustrates isoforms of the MITF gene, highlighting variations in their domain structures. Each isoform shares a conserved b-HLH-LZ domain (green), crucial for DNA binding and dimerization, along with activation domains, AD1, AD2, AD3 (brown) involved in transcriptional regulation. The key differences lie in their N-terminal regions (pink and blue), where distinct exon usage (1a, 1b, C, 1m, 1mc) leads to functional diversity. The total amino acid length varies among isoforms, influencing their expression and biological roles. These isoforms are expressed in different tissues, with MITF-M being melanocyte specific and essential for melanogenesis, while others contribute to diverse biological processes, including pigmentation, eye development, and cancer.

MITF functions as a homodimer or heterodimer or by pairing with related proteins from MiT family TFEB and TFE3, which are widely expressed [31]. Another family member, TFEC, is more selectively expressed and likely plays a regulatory role in specific tissues [32, 33] and has been implicated in controlling hematopoietic vascular niche during zebrafish development, although its functioning in mammals remains uncertain [34]. MITF specifically binds to E-box motifs (CANNTG) through its bHLH-LZ domain, with its binding preference influenced by the surrounding nucleotide sequence and its own protein structure. Initial research identified M-box motifs (CATGTG E-box variants) in pigmentation-related genes, but later findings showed that MITF can also recognize the CACGTG motif [35–37]. Unlike some b-HLH-LZ transcription factors such as AP4, MITF exhibits weak binding affinity for CAGCTG E-boxes, though cooperative interactions with other proteins may enable it to bind these sites in certain contexts, as observed in ChIP-seq data. Additionally, MITF shows a strong preference for CACGTG and CATGTG motifs when flanked by a 5′T or 3′A residue, a distinction that sets it apart from MYC-MAX complexes, which prefer sites lacking these flanking bases [38–40]. Given past confusion regarding MITF-binding site terminology, it has been suggested that the term M-box be reserved for MITF sites containing the CATGTG motif, while CACGTG elements, which also overlap with the CLEAR-box motif linked to lysosomal gene regulation, should be categorized under E-box motifs [41]. These findings suggest that MITF regulates a distinct set of genes, largely independent of MYC [42].

2 Correlation Between MITF and MAX Expression Across Different Types of Melanomas

MITF plays a pivotal role in melanoma biology, serving as a lineage-specific oncogene, which basically means that it promotes cancer development in certain cell types or tissues. Such genes are typically involved in regulating normal cellular processes within a particular lineage (group of cells from a common progenitor), such as differentiation, proliferation, or survival [43]. Together, MITF-MAX complex plays a crucial role in regulating gene expression related to melanocyte differentiation and function. Their binding forms a transcriptional activator complex that regulates expression of gene involved in pigmentation and can influence survival and proliferation [44, 45]. Understanding the correlation between MITF and MAX expression provides insights into their potential functional interplay in different types of cancer. The graphs are sourced from co-expression analysis of ENCORI database, examining this correlation across two cancer types: SKCM (Skin Cutaneous Melanoma) and UVM (Uveal Melanoma) [46].

Graph A from Fig. 5 gives the analysis of 471 SKCM (Skin Cutaneous Melanoma) samples examining the relationship between MITF and MAX expression levels, revealing a regression equation $y = 0.0430x + 3.9428$, indicating a positive but shallow slope. The correlation coefficient (r) is 0.147, suggesting a weak positive correlation, while the p-value of 1.38×10^{-3} demonstrates statistical significance. This weak positive correlation implies that as MITF expression increases, MAX expression levels show a slight but consistent increase in SKCM. The significant p-value indicates that the observed relationship is unlikely due to random chance. This correlation may suggest a functional interaction between MITF and MAX in SKCM biology, potentially contributing to melanoma progression.

Graph B in Fig. 5 communicates data from 80 UVM (Uveal Melanoma) samples analyzing the correlation between MITF and MAX expression., exhibiting a regression equation of $y = 0.0077x + 4.6090$, indicating a nearly flat slope. The correlation

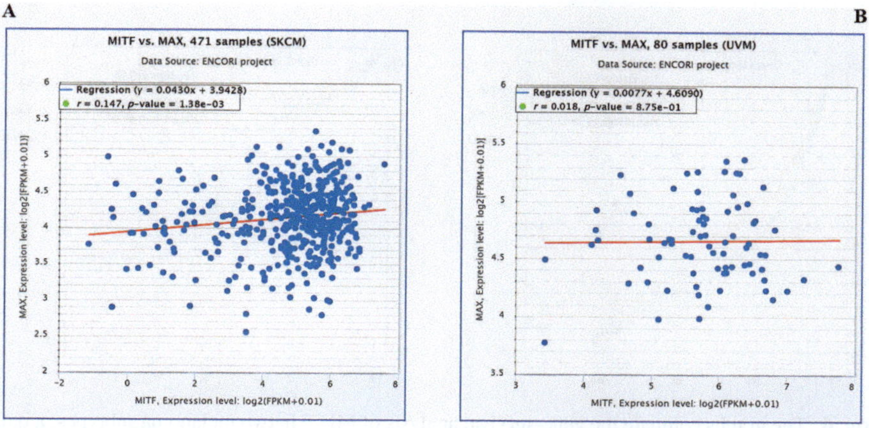

Fig. 5 The graphs compare the correlation between MITF and MAX expression levels in two melanoma subtypes: **a** data from SKCM (Skin Cutaneous Melanoma) samples. **b** Data from UVM (Uveal Melanoma) samples

coefficient (r) is 0.018, suggesting no meaningful correlation, while the *p*-value of 0.875 shows no statistical significance. These findings suggest that MITF and MAX expression levels are not functionally linked in UVM, as evidenced by the lack of correlation and non-significant *p*-value, indicating the observed trend is likely attributed to random variation rather than a true biological association. This suggests that MITF and MAX may not be functionally linked in UVM, or their interaction could be context-dependent, varying across different cancer types.

3 Gene Survival Analysis for MITF Across Different Types of Cancer

Gene-survival analysis is a critical tool in cancer research that assesses the relationship between the expression levels of a gene and patient survival outcomes. By stratifying patients into high and low expression groups and generating Kaplan–Meier survival curves, this analysis helps identifying genes that may serve as prognostic biomarkers or therapeutic targets. It provides insights into how specific genes influence disease progression, treatment response, or overall survival [47]. We performed survival analysis of MITF in SKCM and UVM subtypes of melanoma using binary, gene-survival analysis of ENCORI [46].

The Kaplan–Meier survival analysis of MITF expression in SKCM (Fig. 6, graph a) demonstrates a significant difference in overall survival between low (orange line) and high (green line) expression groups. Patients with low MITF expression exhibited poorer survival outcomes compared to those with high expression, as evidenced by a hazard ratio (HR) of 1.39 and a statistically log-rank *p*-value of 0.017. This indicates

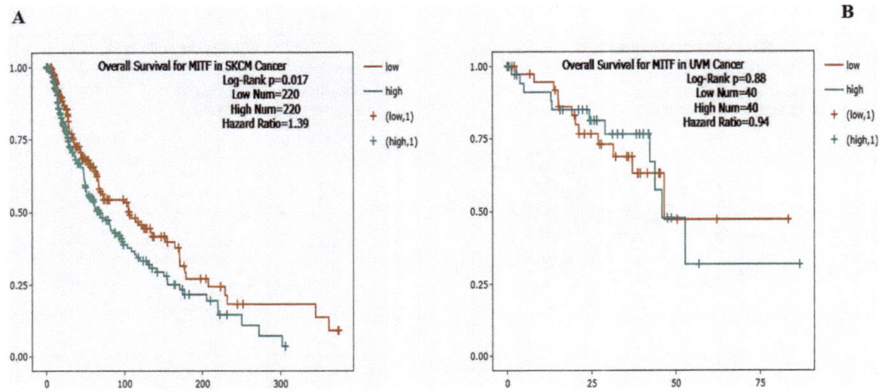

Fig. 6 The graphs compare the gene-survival analysis of MITF in two melanoma subtypes: **a** data from SKCM (Skin Cutaneous Melanoma) samples. **b** Data from UVM (Uveal Melanoma) patients. The green line shows high expression levels while the orange line shows the low MITF expression levels

that low MITF expression is associated with a 39% increased risk of mortality. The expression is calculated using formula (HR-1) $\times$ 100. The clear separation of the survival curves highlights the potential of MITF as a prognostic marker for SKCM. These findings suggest that MITF may play a critical role in the progression of melanoma and warrant further investigation into its biological and therapeutic implications.

In contrast, the Kaplan–Meier survival analysis for MITF expression in UVM (graph B of Fig. 6) reveals no significant difference in overall survival between the low and high expression groups. The survival curves largely overlap, with a hazard ratio (HR) of 0.94 and a non-significant log-rank *p*-value of 0.88, indicating that MITF expression does not significantly influence survival outcomes in UVM. These results suggest that the prognostic role of MITF observed in SKCM is not replicated in UVM, potentially reflecting biological differences between the two melanoma subtypes. The small sample size (40 patients per group) may also limit the statistical power of this analysis, emphasizing the need for further studies with larger cohorts to validate these findings.

4 Molecular Docking and Simulation Studies of MITF-MAX Complex

The 3D structure of MITF (PDB ID: 7D8T) and MAX (PDB ID: 1R05) proteins were obtained from RCSB Protein Data Bank database. To investigate the potential interaction between MITF and MAX, molecular docking was performed using GRAMMDOCK (Global RAnge Molecular Matching) [48], a robust protein–protein docking algorithm, designed to predict the binding modes of such complexes, thus providing insights into the structural orientation and interaction patterns between the two proteins. The GRAMM docking algorithm runs on FFT-based rigid-body search where the receptor protein is held in a fixed position, while the ligand protein is rotated and translated to identify the optimal binding pose. It biases the sampling towards orientations where the selected residues face each other rather than checking all orientations. The algorithm calculates the energy of the interaction in each position, based on the geometric fit and energy-based scoring, correlated with van der Waals energy of interaction, in each position and selects the ligand pose with the lowest energy. This targeted approach reduces false positives with fewer biologically implausible poses., speeds convergence as less CPU is wasted on impossible poses, and integrates low-resolution or partial data to combine small hints from different experiments, making it especially useful for complex systems like virus-host interactions.

In protein–protein docking, the terms "receptor" and "ligand" describe the two interacting proteins. Here, MITF serves as receptor and MAX as ligand. The receptor structure consisted of 1492 atoms, while ligand contained 1424 atoms. Grid-based energy calculations and matching projections were used to identify optimal binding orientations. Results will produce a list of matches with details like energy rotation, and translation values for receptor-ligand interactions. The docking analysis generated a total of 30,000 matches, out of which 10 structural models of MITF-MAX complex were selected for detailed evaluation based on energy scores. Among these, the first model and tenth model, with the highest and least binding energy respectively, have been selected for comparison-analysis study, based on the following docking parameters (Table 2).

These results indicate a strong potential interaction between MITF and MAX, providing structural foundation for further investigations. GRAMMDOCK uses arbitrary energy units (a.u.) rather than physical units like kcal/mol or KJ/mol since it is based on shape complementarity and empirical scoring. A lower or more negative energy value indicates a better docking score and a more favorable interaction. Since model 1 has a lower energy (− 487 a.u.) compared to model 10 (− 455 a.u.), it suggests that model 1 has more stable and favorable binding interactions. To refine the docking results and to simulate the behavior of the complex in a more realistic environment, input files for both the models were generated using CHARMM-GUI's solution builder binary. The setup included system solvation, mdp files for ions, equilibration and production run and other pre-simulation configurations to mimic physiologic conditions. Molecular dynamics simulations were performed on both models using GROMACS software for 500 ns. The production run employed the leap-frog

Table 2 Comparison of highest and lowest binding energy values

Parameters	Highest binding energy (a.u.) (Model 1)	Lowest binding energy (a.u.) (Model 10)
Energy	− 487	− 455
Rotation	335, 131, 343	122, 21, 191
Translation	− 196.69, − 43.99, 19.64	− 175.69, − 19.49, 33.64

integrator (md), a reliable scheme for time integration that updates atomic positions and velocities throughout the simulation with a 2 fs timestep over 250 million steps. Coordinates were recorded every 100 ps, while energies and logs were saved at 2 ps and 20 ps intervals, respectively. The simulation utilized the CHARMM36 force field in combination with the TIP3P water model, ensuring compatibility and accurate representation of protein and solvent interactions with CHARMM parameterization framework. Interactions were treated using the Verlet cutoff scheme, which ensures efficient and accurate force calculations within a specified distance. Van der Waals and Coulomb interactions were both limited to 1.2 nm, while long-range electrostatics were treated with the Particle Mesh Ewald (PME) method, allowing for precise treatment of interactions beyond the cutoff range. The system was maintained at 303.15 K using the V-rescale thermostat, with separate coupling for solute and solvent, and pressure was controlled isotropically with the C-rescale algorithm. Hydrogen bonds were constrained via the LINCS algorithm, to allow for a larger timestep, and center-of-mass motion was removed every 100 steps to stabilize the system. The simulation provides an in-depth understanding of the dynamic behavior, stability, and conformational changes of the MITF-MAX complex under near-physiologic conditions (Figs. 7 and 8).

Model 1 exhibits early stabilization, with RMSD reaching around 1 nm within the first 50 ns and remaining relatively stable between 0.8 and 1.2 nm throughout the simulation. The minimal fluctuations suggest that structure has equilibrated well and maintains conformational stability over time. In contrast, Model 10 shows a higher degree of fluctuation with RMSD increasing gradually and reaching 1.5–1.7 nm in the later stages. The continuous variation suggests possible structural flexibility or instability, indicating that the model may not have fully converged. The last 100 ns show continued fluctuation between 1.3 and 1.7 nm, suggesting incomplete convergence or structural instability. These findings highlight the MITF-MAX interaction as a structurally significant and potentially functionally relevant complex. Docking studies and MD simulation results suggest that model 1 forms a more stable complex, likely supporting stronger transcriptional regulation, while

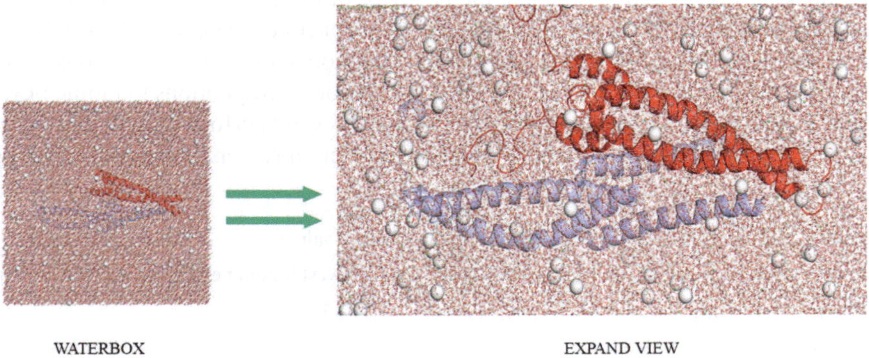

Fig. 7 Solvated MITF-MAX heterodimer within a waterbox

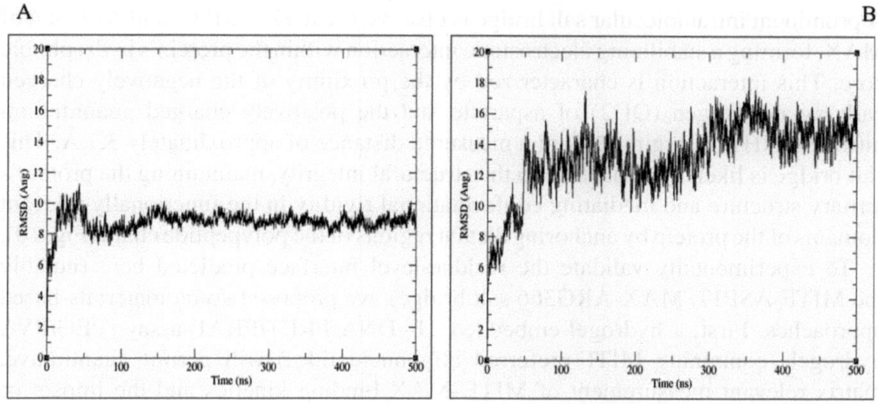

Fig. 8 RMSD analysis of **a** Model 1 and **b** Model 10 over 500 ns, shows model 1 stabilizing around 1.0 nm, while model 10 exhibits higher fluctuations, indicating greater flexibility

model 10 exhibits greater flexibility which could influence its functional efficiency. The stability of MITF-MAX complex is essential for its role in DNA binding and transcriptional activation and these structural insights lay the groundwork for future investigations into its regulatory mechanisms and potential as a therapeutic target in diseases such as melanoma.

A. Hydrogen Bond Network Mediating MITF-MAX Interface Stability

Hydrogen bonding is a fundamental non-covalent interaction that plays a pivotal role in protein–protein complex formation and stabilization. Using Visual Molecular Dynamics (VMD), we calculated and visualized hydrogen bond interaction at the interface of MITF-MAX complex within 3.0 nm distance of polar contacts, focusing on key residues between MITF (Tan) and MAX (Lilac). The analysis revealed a coordinated network of hydrogen bonds that likely contributed to interface specificity and conformational rigidity. SER341 and ASP345 forming multiple polar interactions through side-chain hydroxyl and amide groups, often anchoring flexible loop regions. ARG366 forms a strong electrostatic hydrogen bond with negatively charged ASP17, reinforcing the structural core and acting as a central anchor in the interface. ASP3, located near the N-terminal region, engages in directional hydrogen bonding that may influence initial docking recognition. The VMD-based hydrogen bond map highlights the presence, orientation, occupancy, and temporal stability of these interactions, revealing that these residues form a complementary and cooperative hydrogen bond scaffold that only reinforces dimer stabilization but also mediates directional specificity, restricts interfacial flexibility, and potential governs the kinetics of transient recognition and complex assembly (Fig. 9).

B. Intramolecular Salt Bridge Between ASP17 and ARG366

A prominent intramolecular salt bridge is observed ASP17 of MITF and ARG366 of MAX, forming a stabilizing electrostatic interaction within the protein's hydrophobic core. This interaction is characterized by the proximity of the negatively charged carboxylate oxygen (OD2) of aspartate and the positively charged guanidinium nitrogen (NH1) of arginine, with a measured distance of approximately 3.7 Å. This salt bridge is likely to contribute to the structural integrity, maintaining the protein's tertiary structure and mediating conformational rigidity in the functionally relevant domains of the protein by anchoring distant regions of the polypeptide chain (Fig. 10).

To experimentally validate the residue-level interface predicted here (notably the MITF-ASP17: MAX-ARG366 salt bridge), we propose two biomaterials-based approaches. First, a hydrogel-embedded TF–DNA FRET/FRAP assay (PEG/PVA hydrogels containing MITF-preferred oligonucleotides) will permit quantitative, matrix-relevant measurement of MITF–MAX binding kinetics and the impact of

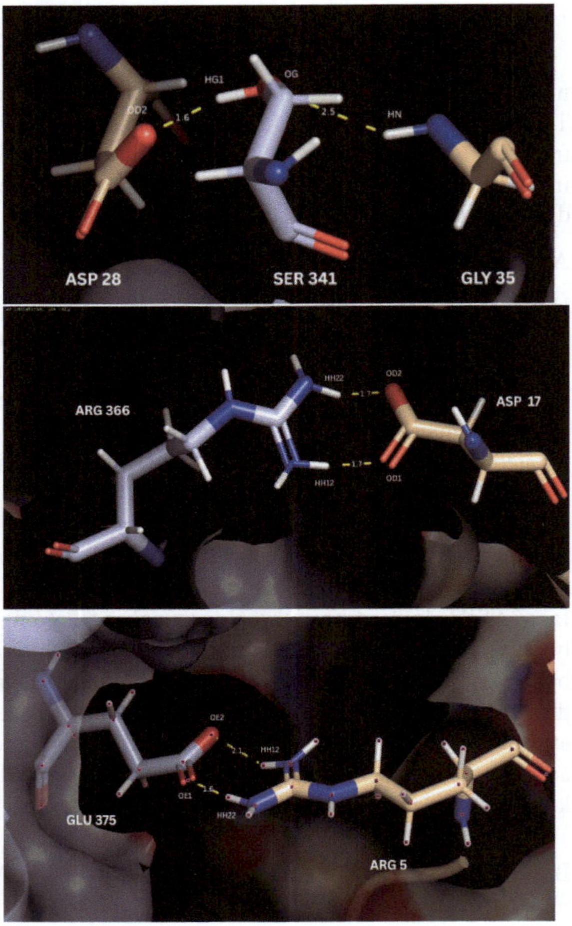

Fig. 9 Visualization of key hydrogen bond interactions at MITF (tan) and MAX (lilac) interface

Fig. 9 (continued)

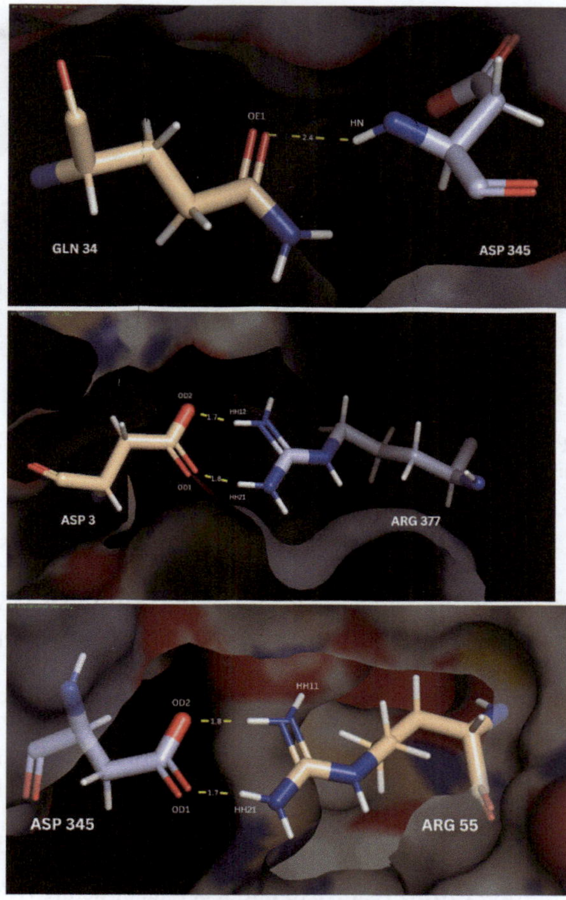

targeted point mutations. Similar hydrogel-TF sensors and FRAP approaches have been used to quantify TF–DNA binding and molecular diffusion in 3D matrices [49, 50]. Second, targeted lipid nanoparticle delivery of MITF-directed siRNAs or interface-mimetic peptides can test whether perturbing the predicted interface alters MITF transcriptional activity and melanoma growth in vitro and in vivo; nanoparticle platforms have demonstrated effective delivery of regulatory RNAs and peptides to melanoma models [51, 52].

C. How this differs from known MYC–MAX structures:

- **DNA-site and sequence preference:** MITF prefers CATGTG/CACGTG variants (M-box/E-box with specific flanking bases), whereas MYC–MAX shows distinct E-box preferences; this implies different DNA-contact geometry and flanking-base recognition that can change dimer orientation and DNA-anchoring footprints.

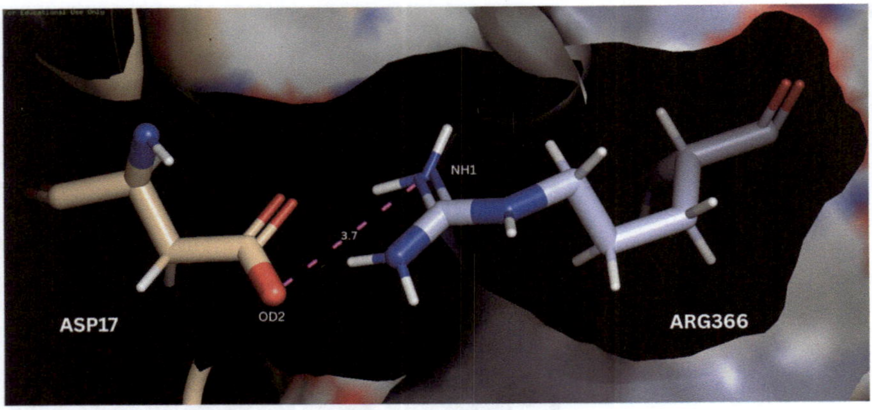

Fig. 10 Visualization of key intermolecular salt bridge between MITF (tan) and MAX (lilac) interface

- **Interface chemistry and anchoring residues:** In our models MITF–MAX dimerization is reinforced by the ASP17–ARG366 intermolecular salt bridge and a coordinated polar scaffold (SER/ASP network). MYC–MAX published structures rely on a different set of helix-loop-helix contacts and leucine-zipper packing.
- **Dynamic behaviour.** Our MD shows a relatively rigid core stabilized by the salt bridge and polar network; by contrast, MYC–MAX has been reported to exhibit different flexibility patterns in the basic region and zipper packing, implying potentially different kinetics of DNA binding and partner exchange.

D. Structural Deviation from the crystal reference and flexibility profiles of Models:

(a) RMSD relative to initial crystal structure (RMSD_XTAL)

Relative to the crystal reference, model 10 shows an initial rise within ~ 0–50 ns followed by sustained excursions that hover in ~ 12–17 Å band, indicating that once crystallographic packing/partners are removed the chain relaxes into alternative solution conformations and continues to sample them. By contrast, the heterodimer (Model 1) remains clustered in a narrow, stable band (~8–10 Å over the 300–500 ns window), i.e., it is consistently different from the crystal yet internally stable. This pattern is exactly what you'd expect if (i) some parts of the crystal pose are shaped by DNA or packing and (ii) MAX binding provides a stabilizing scaffold that restricts MITF's conformational drift. The key takeaway is not "which is closer to the crystal", but that Model 1 adopts a reproducible solution ensemble with reduced large-scale drift, suggesting a more robust and stable architecture (ref. Fig. 11).

(b) Root Mean Square Fluctuation (RMSF) analysis

Per-residue fluctuation analysis reveals a consistent pattern of alternating peaks and troughs in both heterodimer models, corresponding to flexible loops or termini and

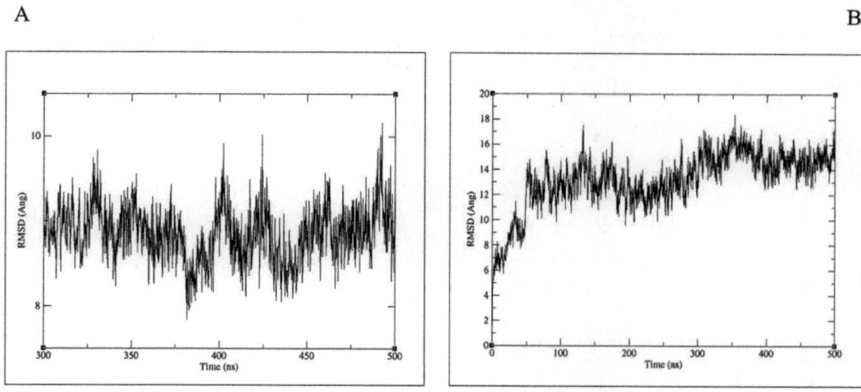

Fig. 11 RMSD relative to crystal structure of MITF **a** Model 1 and **b** Model 10

rigid structured regions, respectively. In model 10, the peaks reach amplitudes of ~ 12–14 Å, reflecting pronounced mobility across several segments, whereas model 1 displays reduced amplitudes at the interface, with deeper and broader troughs that signify local rigidification. This suppression of fluctuations is consistent with stabilization by persistent electrostatic and hydrogen-bonding contacts, most notably the ASP17-ARG366 salt bridge and adjacent polar residues. Importantly, the positions of the peaks remain aligned between the models, indicating that intrinsically flexible loops remain mobile in both cases, preserving adaptability for DNA binding. One of the RMSF plots employs a vertical offset to separate traces, so absolute values must be compared after correction; doing so will quantify the degree of fluctuation suppression [ΔRMSF = RMSF (Model 10)—RMSF (Model 1)] at the interface. Collectively, these results establish that Model 1 maintains a rigidified interface supported by stable contacts while allowing distal regions to remain flexible, providing a structural balance between stability of the heterodimeric core and adaptability at the periphery (ref. Fig. 12).

5 Results and Conclusion

The study of biomaterials in oncology provides crucial insights into structural & functional interactions of transcription factors, particularly the MITF-MAX complex, which plays a vital role in melanocyte biology and cancer progression. MITF, a master regulator of melanocyte development, differentiation, and survival, interacts with MAX, forming a b-HLH-Zip heterodimer that modulates gene expression related to cell proliferation, pigmentation, and apoptosis. Given the significance of this interaction in melanoma and other malignancies, investigating its structural and functional attributes post-docking offers valuable mechanistic understanding.

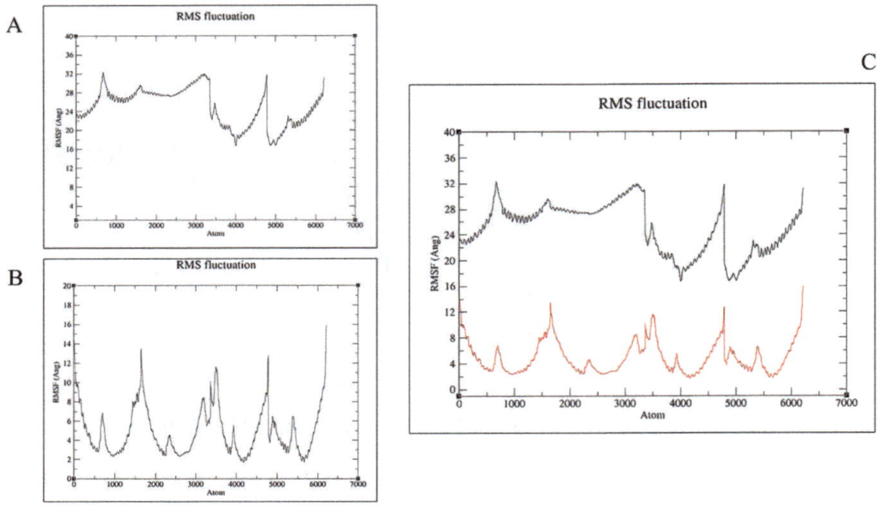

Fig. 12 RMSF analysis of **a** Model 1, **b** Model 10 and **c** comparative graph of the two models

Using GROMACS, we identified a stable MITF-MAX heterodimer (model 1) with: (i) a persistent hydrogen-bond network involving SER341/ASP345 and other polar contacts that anchor flexible loops, (ii) a prominent inter-protein salt bridge between **MITF-ASP17 and MAX-ARG366** that acts as a central electrostatic anchor, and (iii) RMSD/stability signatures that distinguish a structurally stable model (Model 1, RMSD $\approx$ 0.8–1.2 nm) from a more flexible alternative (Model 10, RMSD up to 1.7 nm). These observations identify residue-level determinants of MITF–MAX dimer stability and suggest specific sites that could mediate heterodimer assembly or be targeted to perturb the complex.

Deviation analyses relative to the crystal reference further emphasized this difference, as Model 10 explored broader conformational space with larger excursions, while Model 1 remained tightly clustered within a stable solution ensemble. Complementary residue-level fluctuation profiles (RMSF) reinforced this conclusion: Model 10 exhibited higher-amplitude peaks across multiple loops, whereas Model 1 showed suppressed fluctuations at the dimerization interface, consistent with stabilization by persistent electrostatic and hydrogen-bonding contacts. Importantly, distal loops in both models retained flexibility, suggesting an intrinsic adaptability that may facilitate DNA binding. Together, these observations identify residue-level determinants of MITF–MAX stability and highlight Model 1 as the more structurally robust heterodimeric configuration.

Biomaterial-based approaches hold potential in leveraging these structural insights for therapeutic applications. Engineered biomaterials, such as nanomaterials, hydrogels, and polymeric scaffolds, can serve as platforms for studying interactions under physiologically relevant conditions. Additionally, such systems can aid in targeted drug delivery, offering avenues for precision medicine strategies aimed at disrupting aberrant MITF signaling in melanoma and other cancers. The integration of biomaterials with computational modelling thus represents a promising interdisciplinary approach to bridge molecular insights with therapeutic innovations.

Molecular dynamics (MD) simulations serve as a powerful tool for comparative analysis, enabling the assessment of stability, conformational flexibility, and interaction energies of MITF-MAX complexes in different structural contexts. By performing MD simulations on docked MITF-MAX structures, the present study provides a comparative framework to evaluate dynamics changes in their binding interfaces, secondary structure stability, and residue-specific fluctuations. These simulations allow for a deeper understanding of how specific mutations or conformational variations affect dimerization and functional integrity, which can have direct implications for cancer research and targeted therapy.

Furthermore, mutations in MITF lead to distinct phenotypic consequences, including pigmentation disorders, deafness syndromes, and cancer susceptibility. The analysis of MD simulation trajectories enables the identification of key structural determinants that influence MITF-MAX complex stability, shedding light on potential therapeutic targets. By systematically evaluating these interactions in silico, researchers can predict the impact of specific mutations on protein function and develop targeted interventions.

Overall, this study highlights the significance of MD simulations in characterizing MITF-MAX interactions, providing a computational perspective to complement experimental findings. The synergy between biomaterials, molecular dynamics, and computational oncology paves the way for novel therapeutic strategies in melanoma and other MITF-associated malignancies. Future research should focus on integrating these computational insights with experimental validation, employing biomaterial-assisted drug delivery systems to enhance the precision and efficacy of cancer therapeutic.

Acknowledgements SS Thanks the ANRF for the Start-up research grant number SRG/2023/001007.

References

1. Ratner, B.D., Hoffman, A., Schoen, F.J., Lemons, J.E.: Front Matter. In: Biomaterials Science, p. iii. Elsevier, Amsterdam (2013). https://doi.org/10.1016/B978-0-08-087780-8.00148-0
2. Williams, D.F.: On the nature of biomaterials. Biomaterials **30**(30), 5897–5909 (2009). https://doi.org/10.1016/j.biomaterials.2009.07.027
3. Park, J.B., Lakes, R.S. (1992). Introduction to biomaterials. In: Biomaterials, pp. 1–6. Springer, Boston (1992). https://doi.org/10.1007/978-1-4757-2156-0_1
4. Ratner, B.D., Bryant, S.J.: Biomaterials: where we have been and where we are going (2004). https://doi.org/10.1146/annurev.bioeng.6.040803.140027
5. Ratner, B.D., Hoffman, A., Schoen, F.J., Lemons, J.E.: Biomaterials science: an introduction to materials in medicine. MRS Bull. **31**(1), 58–60 (2006). https://doi.org/10.1557/mrs2006.17

6. Barber, C.C., Burnham, M., Ojameruaye, O., McKee, M.D.: A systematic review of the use of titanium versus stainless steel implants for fracture fixation. OTA Int. **4**(3), e138 (2021). https://doi.org/10.1097/OI9.0000000000000138
7. Modjarrad, K., Ebnesajjad, S.: Handbook of polymer applications in medicine and medical devices. Elsevier (2014). https://doi.org/10.1016/C2012-0-07906-9
8. Höland, W., Schweiger, M., Watzke, R., Peschke, A., Kappert, H.: Ceramics as biomaterials for dental restoration. Expert Rev. Med. Devices **5**(6), 729–745 (2008). https://doi.org/10.1586/17434440.5.6.729
9. Cernencu, A.I., Vlasceanu, G.M., Serafim, A., Pircalabioru, G., Ionita, M.: 3D double-reinforced graphene oxide—nanocellulose biomaterial inks for tissue engineered constructs. RSC Adv. **13**(34), 24053–24063 (2023). https://doi.org/10.1039/D3RA02786D
10. Das, S., Ghosh, B., Sarkar, K.: Nanocellulose as sustainable biomaterials for drug delivery. Sens. Int. **3**, 100135 (2022). https://doi.org/10.1016/j.sintl.2021.100135
11. Adjei, I., Blanka, S.: Modulation of the tumor microenvironment for cancer treatment: a biomaterials approach. J Funct Biomater **6**(1), 81–103 (2015). https://doi.org/10.3390/jfb6010081
12. Hutmacher, D.W.: Biomaterials offer cancer research the third dimension. Nat. Mater. **9**(2), 90–93 (2010). https://doi.org/10.1038/nmat2619
13. Bo, Y., Wang, H.: Biomaterial-based in situ cancer vaccines. Adv. Mater. **36**(43) (2024). https://doi.org/10.1002/adma.202210452
14. Sousa, A.M., Amaro, A.M., Piedade, A.P.: Structural design optimization through finite element analysis of additive manufactured bioresorbable polymeric stents. Mater Today Chem **36**, 101972 (2024). https://doi.org/10.1016/j.mtchem.2024.101972
15. Goding, C.R., Arnheiter, H.: MITF—the first 25 years. Genes Dev. **33**(15–16), 983–1007 (2019). https://doi.org/10.1101/gad.324657.119
16. Pingault, V., Ente, D., Dastot-Le Moal, F., Goossens, M., Marlin, S., Bondurand, N.: Review and update of mutations causing Waardenburg syndrome. Hum. Mutat. **31**(4), 391–406 (2010). https://doi.org/10.1002/humu.21211
17. Steingrímsson, E., Copeland, N.G., Jenkins, N.A.: Melanocytes and the microphthalmia transcription factor network. Annu. Rev. Genet. **38**(1), 365–411 (2004). https://doi.org/10.1146/annurev.genet.38.072902.092717
18. García-Llorca, A., Eysteinsson, T.: The microphthalmia-associated transcription factor (MITF) and its role in the structure and function of the eye. Genes (Basel) **15**(10), 1258 (2024). https://doi.org/10.3390/genes15101258
19. Bharti, K., Liu, W., Csermely, T., Bertuzzi, S., Arnheiter, H.: Alternative promoter use in eye development: the complex role and regulation of the transcription factor MITF. Development **135**(6), 1169–1178 (2008). https://doi.org/10.1242/dev.014142
20. Hallsson, J.H., et al.: Genomic, transcriptional and mutational analysis of the Mouse microphthalmia locus. Genetics **155**(1), 291–300 (2000). https://doi.org/10.1093/genetics/155.1.291
21. Baranowska Körberg, I., et al.: A simple repeat polymorphism in the MITF-M promoter is a key regulator of white spotting in dogs. PLoS One **9**(8), e104363 (2014). https://doi.org/10.1371/journal.pone.0104363
22. Hauswirth, R., et al.: Mutations in MITF and PAX3 cause 'splashed white' and other white spotting phenotypes in horses. PLoS Genet. **8**(4), e1002653 (2012). https://doi.org/10.1371/journal.pgen.1002653
23. Zhang, C.-S., et al.: The lysosomal v-ATPase-ragulator complex is a common activator for AMPK and mTORC1, acting as a switch between catabolism and anabolism. Cell Metab. **20**(3), 526–540 (2014). https://doi.org/10.1016/j.cmet.2014.06.014
24. Hallsson, J.H., et al.: The basic helix-loop-helix leucine zipper transcription factor Mitf Is conserved in drosophila and functions in eye development. Genetics **167**(1), 233–241 (2004). https://doi.org/10.1534/genetics.167.1.233
25. Silvers, W.K.: The Coat Colors of Mice. Springer, New York (1979). https://doi.org/10.1007/978-1-4612-6164-3

26. Hollander, W.F.: Complementary alleles at the mi-locus in the mouse. Genetics **60**, 189 (1968)
27. Tachibana, M., et al.: Cloning of MITF, the human homolog of the mouse microphthalmia gene and assignment to chromosome 3p14. 1–p12.3. Hum. Mol. Genet. **3**(4), 553–557 (1994). https://doi.org/10.1093/hmg/3.4.553
28. Fuse, N., Yasumoto, K., Suzuki, H., Takahashi, K., Shibahara, S.: Identification of a melanocyte-type promoter of the microphthalmia-associated transcription factor gene. Biochem. Biophys. Res. Commun. **219**(3), 702–707 (1996). https://doi.org/10.1006/bbrc.1996.0298
29. George, A., et al.: Biallelic mutations in MITF cause coloboma, osteopetrosis, microphthalmia, macrocephaly, albinism, and deafness. Am. J. Human Genetics **99**(6), 1388–1394 (2016). https://doi.org/10.1016/j.ajhg.2016.11.004
30. Hartman, M.L., Czyz, M.: MITF in melanoma: mechanisms behind its expression and activity. Cell. Mol. Life Sci. **72**(7), 1249–1260 (2015). https://doi.org/10.1007/s00018-014-1791-0
31. Hemesath, T.J., et al.: Microphthalmia, a critical factor in melanocyte development, defines a discrete transcription factor family. Genes Dev. **8**(22), 2770–2780 (1994). https://doi.org/10.1101/gad.8.22.2770
32. Kuiper, R.P.: Regulation of the MiTF/TFE bHLH-LZ transcription factors through restricted spatial expression and alternative splicing of functional domains. Nucleic Acids Res. **32**(8), 2315–2322 (2004). https://doi.org/10.1093/nar/gkh571
33. Rehli, M., Lichanska, A., Cassady, A.I., Ostrowski, M.C., Hume, D.A.: TFEC is a macrophage-restricted member of the microphthalmia-TFE subfamily of basic helix-loop-helix leucine zipper transcription factors. J. Immunol. **162**(3), 1559–1565 (1999)
34. Mahony, C.B., Fish, R.J., Pasche, C., Bertrand, J.Y.: Tfec controls the hematopoietic stem cell vascular niche during zebrafish embryogenesis. Blood **128**(10), 1336–1345 (2016). https://doi.org/10.1182/blood-2016-04-710137
35. Yavuzer, U., Goding, C.R.: Melanocyte-specific gene expression: role of repression and identification of a melanocyte-specific factor, MSF. Mol. Cell Biol. **14**(5), 3494–3503 (1994). https://doi.org/10.1128/MCB.14.5.3494
36. Bentley, N.J., Eisen, T., Goding, C.R.: Melanocyte-specific expression of the human tyrosinase promoter: activation by the microphthalmia gene product and role of the initiator. Mol. Cell Biol. **14**(12), 7996–8006 (1994). https://doi.org/10.1128/MCB.14.12.7996
37. Pogenberg, V., et al.: Restricted leucine zipper dimerization and specificity of DNA recognition of the melanocyte master regulator MITF. Genes Dev. **26**(23), 2647–2658 (2012). https://doi.org/10.1101/gad.198192.112
38. Strub, T., et al.: Essential role of microphthalmia transcription factor for DNA replication, mitosis and genomic stability in melanoma. Oncogene **30**(20), 2319–2332 (2011). https://doi.org/10.1038/onc.2010.612
39. Hejna, M., Moon, W.M., Cheng, J., Kawakami, A., Fisher, D.E., Song, J.S.: Local genomic features predict the distinct and overlapping binding patterns of the bHLH-Zip family oncoproteins MITF and MYC-MAX. Pigment Cell Melanoma Res. **32**(4), 500–509 (2019). https://doi.org/10.1111/pcmr.12762
40. Aksan, I., Goding, C.R.: Targeting the microphthalmia basic helix-loop-helix–leucine zipper transcription factor to a subset of E-Box elements in vitro and in vivo. Mol. Cell Biol. **18**(12), 6930–6938 (1998). https://doi.org/10.1128/MCB.18.12.6930
41. Settembre, C., Medina, D.L.: TFEB and the CLEAR network, pp. 45–62 (2015). https://doi.org/10.1016/bs.mcb.2014.11.011
42. Solomon, D.L.C., Amati, B., Land, H.: Distinct DNA binding preferences for the c-Myc/Max and Max/Max dimers. Nucleic Acids Res. **21**(23), 5372–5376 (1993). https://doi.org/10.1093/nar/21.23.5372
43. David, C.J., Vanharanta, S.: Lineage-specific transcription factors in carcinogenesis. Annu. Rev. Cancer Biol. (2024). https://doi.org/10.1146/annurev-cancerbio-060524-105603
44. Wan, P., Hu, Y., He, L.: Regulation of melanocyte pivotal transcription factor MITF by some other transcription factors. Mol. Cell. Biochem. **354**(1–2), 241–246 (2011). https://doi.org/10.1007/s11010-011-0823-4

45. Blackwood, E.M., Eisenman, R.N.: Max: a helix-loop-helix zipper protein that forms a sequence-specific DNA-binding complex with myc. Science (1979) **251**(4998), 1211–1217 (1991). https://doi.org/10.1126/science.2006410
46. Li, J.-H., Liu, S., Zhou, H., Qu, L.-H., Yang, J.-H.: StarBase v2.0: decoding miRNA-ceRNA, miRNA-ncRNA and protein–RNA interaction networks from large-scale CLIP-Seq data. Nucleic Acids Res. **42**(D1), D92–D97 (2014). https://doi.org/10.1093/nar/gkt1248
47. Meyer, J.M., Eaves, L.J., Heath, A.C., Martin, N.G.: Estimating genetic influences on the age-at-menarche: a survival analysis approach. Am. J. Med. Genet. **39**(2), 148–154 (1991). https://doi.org/10.1002/ajmg.1320390207
48. Singh, A., Copeland, M.M., Kundrotas, P.J., Vakser, I.A.: GRAMM Web Server for Protein Docking, pp. 101–112 (2024). https://doi.org/10.1007/978-1-0716-3441-7_5
49. Chen, M., et al.: Hydrogel-embedded quantum dot-transcription factor sensors for quantitative progesterone detection. ACS Appl. Mater. Interfaces **12**(39), 43513–43521 (2020). https://doi.org/10.1021/acsami.0c13489
50. Richbourg, N.R., Peppas, N.A.: High-throughput FRAP analysis of solute diffusion in hydrogels. Macromolecules **54**(22), 10477–10486 (2021). https://doi.org/10.1021/acs.macromol.1c01752
51. Mundra, V., Li, W., Mahato, R.I.: Nanoparticle-mediated drug delivery for treating melanoma. Nanomedicine **10**(16), 2613–2633 (2015). https://doi.org/10.2217/nnm.15.111
52. Fattore, L., et al.: Oncosuppressive miRNAs loaded in lipid nanoparticles potentiate targeted therapies in BRAF-mutant melanoma by inhibiting core escape pathways of resistance. Oncogene **42**(4), 293–307 (2023). https://doi.org/10.1038/s41388-022-02547-9

Novel Molecules and Materials for Drug Discovery

Jai Singh, Nitish Kumar Singh, Saumya Dubey, Grijesh Jaiswal, Ananya Srivastava, Sandesh P. Telang, and Sandeep Kumar Srivastava

Abstract The advancement of cutting-edge materials and recently developed chemicals are drastically changing the dimension of the drug research environment worldwide. Small novel molecules are most significantly responsible for this shift because they target biological systems precisely. Their worthiness is demonstrated by their rapid utilization of automated high-throughput screening and rational drug design. At the same time, nanomaterials have increased the prospective for site-specific drug delivery, responsive release mechanisms and enhanced imaging. For instance, silica-based structures, inorganic and hybrid nanomaterials, and metal–organic frameworks (MOFs) have significantly improved medication, reliability and potency, thus bringing an evolution in pharmaceutical engineering. Intelligent and stimuli-responsive materials that respond to environmental cues like temperature or pH allow medications to function precisely. These have considerably impacted the drug development scenario and have opened the path for customized medicine. AI-powered modelling, predictive simulations and computational methods are uniformly meaningful as these technologies conserve time and money while completely modifying how we find, develop, and improve therapeutic candidates. The combination of bioengineering, chemistry and materials science has shown monumental promise in the contemporary success of CRISPR-loaded nanoparticles and multifunctional hydrogels. The future hinges on a more solution-based exceptional amalgamation of rational synthesis techniques, major cross-disciplinary associations, and real-world data. The regulatory complexity, long-term safety, and scalability of novel molecules and materials for drug discovery are still compelling issues. Machine learning, nanotechnology and systems biology will definitively run the progression of next-generation materials that remodel illness treatment, making the course of medication more exceptional and customized for each patient.

Keywords Drug discovery · High-throughput screening · Stimuli-responsive material · Metal-organic frameworks

---

J. Singh · N. K. Singh · S. Dubey · G. Jaiswal · A. Srivastava · S. P. Telang · S. K. Srivastava (✉)
Structural Biology and Bioinformatics Laboratory, Department of Biosciences, Manipal University Jaipur, Jaipur, Rajasthan, India
e-mail: sandeepkumar.srivastava@jaipur.manipal.edu

L. Ledwani and S. Srivastava (eds.), *Experimental and Computational Advances in Materials*, Engineering Materials, https://doi.org/10.1007/978-981-95-4180-5_16

1 Introduction

Drug development continues to be a key component of modern medicine's quest for novel and effective ways to treat, prevent, or manage illnesses. Historically, the process started with natural resources, where the first medicinal substances were found in plants, fungi, and minerals; well-known examples include penicillin from fungi, aspirin from willow bark, and morphine from opium poppies [1]. These findings demonstrated the enormous potential of nature while simultaneously highlighting its drawbacks, including its unpredictability, scarcity, and imprecision. Researchers shifted to laboratory synthesis as chemistry and technology improved, creating synthetic molecules that could be carefully engineered for increased efficacy, safety, and predictability [2]. This shift introduced engineered molecules capable of targeting diseases unresponsive to traditional therapies while also expanding the chemical space for drug development. With their own structural characteristics and focused methods of action, new molecules from tiny chemical compounds and peptides to nucleic acid-based therapeutics represent the forefront of innovation today. While nucleic acid-based therapeutics, like mRNA vaccines, show how molecular engineering can use the body's natural processes to produce protective proteins, as was the case during the COVID-19 pandemic, engineered peptides and proteins, for instance, can interact with specific signaling pathways, block malfunctioning genes, or stimulate immune responses against cancer cells [3]. But because of their complexity, these new compounds present problems with stability, solubility, and targeted distribution. This is where materials science comes into play. To get beyond pharmacokinetic and pharmacodynamic restrictions, advanced delivery technologies, including hydrogels, liposomes, nanoparticles, and biomimetic scaffolds are currently being developed [4]. For example, gold nanoparticles coated with chemotherapeutic agents enable precise delivery to tumors, lowering systemic toxicity, while lipid nanoparticles (LNPs) have demonstrated invaluable capabilities in stabilizing fragile mRNA molecules, guaranteeing their safe transport into cells, and promoting effective immune responses. Furthermore, smart, site-specific release systems have been made possible by stimuli-responsive materials that respond to environmental cues like pH, temperature, or enzyme activity, guaranteeing greater efficacy with fewer adverse effects [5]. Artificial intelligence (AI) has become a game-changing technology in contemporary drug discovery, along with developments in chemistry and materials [6]. It frequently took years before a single promising medication candidate made it to clinical trials due to the laborious and unaffordable nature of the old pipeline, which involved finding, screening, and validating molecules. By making it possible to analyze large datasets, such as chemical, proteomic, and genomic libraries, artificial intelligence (AI) has fundamentally changed this paradigm [6]. This has allowed for the more accurate creation of new molecules, the optimization of lead compounds, and the prediction of drug-target interactions. While technologies like AlphaFold have expedited biological understanding by precisely predicting protein structures, enabling logical and structure-based medication design, machine learning techniques lower the likelihood of late-stage failures. For instance, previously unthinkable, AI systems have discovered tiny compounds that can operate on "undruggable" cancer sites. AI broadens the chemical landscape in ways that are not achievable with manual techniques alone when paired with computational modeling and high-throughput screening. However, the synergistic fusion of biology, artificial intelligence, and materials science is what really makes modern drug development strong, not any one field [7]. Materials science makes sure that promising compounds can be delivered safely, effectively, and precisely to the right location in the body; artificial intelligence (AI) speeds up the discovery and optimization of drug candidates by sorting through data at a never-before-seen speed; and biology provides the fundamental knowledge of disease pathways, molecular targets, and biomarkers. For example, in cancer treatment, biology finds oncogenic mutations and signaling cascades, AI identifies the molecules that are most likely to interfere with them, and materials science-engineered nanocarriers deliver these molecules straight to tumor tissues, optimizing therapeutic effect while reducing harm to healthy cells. In addition to speeding up the medication development process, this interdisciplinary collaboration improves patient-centeredness, affordability, and personalization of medicines. Novel compounds and innovative materials are in a unique position to solve issues including medication resistance, poor solubility, and systemic toxicity as healthcare places an increasing emphasis on customized treatment plans [7]. Moreover, our perspective on the future of medicine is changing because of this integration. Scientists are increasingly using systematic, predictive, and adaptive methodologies rather than just traditional trial-and-error procedures, which allow for safer and more intelligent treatments. Personalized treatments that consider each patient's unique genetic and molecular characteristics are becoming increasingly possible, and intelligent delivery systems that adapt to biological settings are decreasing side effects and enhancing effectiveness. Treatment approaches for illnesses that were previously thought to be incurable have already changed because of the shift in technical growth, and further advancements are anticipated as it continues. A new age in drug development is being ushered in by the combination of innovative materials and novel compounds, aided by biological insights and AI-driven discoveries [8]. This age will guarantee greater accessibility, lower prices, and better health outcomes worldwide in addition to improving the effectiveness and precision of therapies. From natural cures to synthetic chemistry, from molecular engineering to AI-assisted design, and from unrefined delivery methods to highly designed nanomaterials, drug development has changed over time. We are now closer to safer, more individualized, and more effective treatments at every step of this process. Future medications will be more quickly developed, more precisely targeted, and more deeply patient-centered thanks to the framework for smarter, more systematic, and secure drug development provided by the convergence of biology, artificial intelligence, and materials science. This multidisciplinary approach ensures that the future of healthcare will be more inventive and hopeful as science and technology continue to progress, opening the door to a healthier world.

1.1 The Evolution of Drug Discovery

During the Renaissance and Enlightenment periods, the term drug came to have a better definite perception. The advancement in chemistry has changed the course of interpretation in studying medicinal molecules. Its progression reflects the rise of medicine, from herbal culture to molecular precision. In the current scenario, a drug is defined as any chemical that can influence biological function, involving both medicinal medicines and drugs of addiction. The scientific endeavour of drug discovery has undergone significant change over the past three centuries. It has evolved from an herbalism trial-and-error process based on basic chemistry to a complex technology-driven enterprise incorporating genomics, artificial intelligence, and personalized medicine [9]. The combination of changing healthcare perspectives, scientific discoveries, and technological advancements is reflected in the progression of drug discovery. Most of the drug development in the eighteenth century was experiential. Apothecaries and physicians made medications based on unscientific evidence and traditional knowledge systems rather than thoroughly mastering their chemical processes. Willow bark was used to cure fever and discomfort; in the late nineteenth century, its active component, salicylic acid, was isolated, and aspirin was made. Pharmacology as a medicinal subject was still immature, and drug development lacked scientific coordination and perfection [10]. In the nineteenth century, organic chemistry gained prominence, and pharmacology became a separate scientific field. An important turning point was the isolation and synthesis of pure substances. Morphine, which was extracted from opium in 1806 and quinine, which was isolated from cinchona bark in 1820, are the earliest instances of pure pharmaceuticals obtained from plant sources. The progress of the Industrial Revolution made possible large-scale pharmaceutical manufacturing, and the idea of synthetic pharmaceuticals was built when Bayer, a German multinational pharmaceutical company, created Trichloroacetaldehyde monohydrate ($C_2H_3Cl_3O_2$) and then aspirin (acetylsalicylic acid) in 1899 [11]. By the twentieth century, finding new drugs had evolved and become more methodical. The inventiveness of Paul Ehrlich's "magic bullet" in the 1900s changed the course of treatments designed for a disease [12]. His research created the concept of selective toxicity for arsphenamine, which is used against syphilis and is sold under the brand name Salvarsan. In the 1920s, the first wave of life-saving pharmaceuticals transpired with the discovery of insulin, and in the 1930s, with sulfa drugs like Prontosil, targeted therapies were made possible. In 1928, the discovery of penicillin by Alexander Fleming began the antibiotic era, which was mass-produced in the 1940s. This demonstrated the effectiveness of methodical drug research, which significantly declined the death rate from bacterial illnesses, and rational drug design emerged in the middle of the twentieth century [13]. Scientists were able to create compounds with biological targets, as the structure of DNA was discovered in 1953. Progress in biochemistry and molecular biology started to be used in medical research using methods like structure–activity relationship (SAR) models and high-throughput screening [14]. The production of popular medications such as oral contraceptives, beta-blockers, statins, and antihypertensives flourished in the pharmaceutical business. However, the thalidomide disaster in the 1950s called attention to the necessity of uncompromising drug testing and regulatory control, because of which the Food and Drug Administration (FDA) and other agencies were being given more control. 1990 The Human Genome Project (HGP) was launched, and genomics and proteomics became prominent. Drug discovery during this period was impacted by computational technologies, enabling molecular modeling and virtual screening of large compounds [15]. Using combinatorial chemistry, drug development businesses began swiftly building thousands of compounds for screening. The completion of HGP in 2003 marked the start of the twenty-first century's data-driven era by enabling researchers to identify new targets for illnesses and providing them with access to the full human genetic code. For complicated illnesses like cancer and neurological disorders, drug advancement began to be more informed by genomics, transcriptomics, and metabolomics data. Techniques like genomics, proteomics, and bioinformatics help locate possible biomolecular targets, such as proteins or genes, which perform serious roles in disease development pathways. Imatinib for chronic myeloid leukemia is one example of targeted therapy; thereby comprehending disease processes at the molecular level for treatment purposes. AI algorithms discover suitable drug contenders, forecast protein structures, enhance clinical trials, and utilize already-ratified medications [16]. Drug interactions can be simulated using in-silico modelling, virtual patients, and digital twins, allowing researchers to predict compounds' binding affinity to targets even before their synthesis. Structural biology is being transformed by businesses like DeepMind and its AlphaFold technology, making exceptionally precise predictions about protein folding and its 3D structure, which can be used to design molecules with optimal binding characteristics and open new routes for drug targeting [17]. Furthermore, a new period in quick vaccine development and treatment research was validated by mRNA technology, which was accelerated by the COVID-19 pandemic. In 2020, mRNA-based vaccine development technology demonstrated the platform's promise for future therapeutic development in several illnesses and the introduction of those products into the growing market by Moderna and Pfizer-BioNTech. Cell and gene treatments, which alter patients' cells to treat cancer, have become more popular. One example is chimeric antigen receptor T-cell therapy (CAR-T) [18]. By 2025, drug discovery will be at the intersection of data science, engineering, chemistry, and biology, each evolving a new route to better treatment strategies. AI-powered diagnostics decentralized clinical trials, and completely automated laboratories with an increased focus on equality and sustainability in global healthcare access are all signs of the future. Corrective therapies for congenital illnesses are being investigated using CRISPR gene editing technology [19]. The partnership between academia, biotech startups, and large pharmaceutical companies encourages an ecosystem of vibrant innovation. In conclusion, drug development has evolved into a demanding, multidisciplinary discipline from naturalist treatments and chance observations to combining technologies and knowledge to tackle ever-more complex health issues. The improvement from primitive preparations to customized therapies proves humanity's persistent search for more efficient, safer and fair remedies that

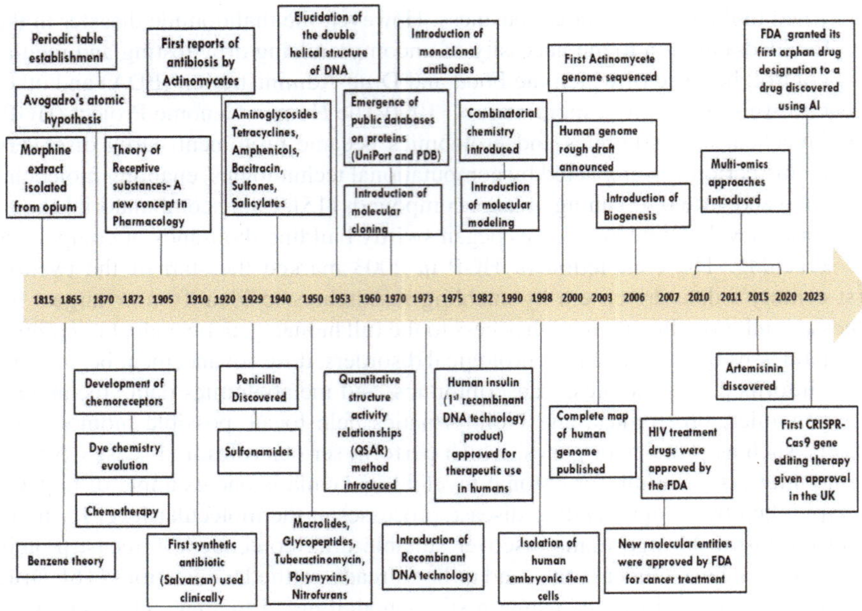

Fig. 1 Timeline of important events in drug discovery

will only deepen further. While traditional approaches are the foundational backbone for drug development and encourage natural product-based drug leads, modern techniques extend precision, accuracy, speed, and cost-efficiency. However, neither approach is unanimously exclusive (Fig. 1).

1.2 Need for Novel Molecules and Materials

With the evolving global health challenges, traditional drug molecules often fall short in productivity, protection, and particularity, which is a concern. A promising solution to bridge these gaps is offered by novel chemical entities (NCEs) and innovative biomaterials [20]. One prime rationality for probing novel molecules is the genesis of multidrug-resistant (MDR) pathogens and non-communicable diseases such as cancer, neurodegenerative disorders, and autoimmune conditions. The complexity of these diseases requires highly definite, multifunctional therapeutic molecules, which cannot be addressed by traditional small molecules or biologics alone. Advanced materials such as nanoparticles, lipid-based carriers, polymeric systems, and stimuli-responsive materials are revolutionizing drug delivery and developing new scaffolds, hybrid molecules, and molecular conjugates. This can overcome drug resistance and improve drug target selectivity. The advantages of utilizing these materials include controlled drug release, targeted delivery, increased bioavailability, minimizing undesirable effects and maximizing therapeutic outcomes [21]. Materials are being investigated for their potential to encapsulate drugs, respond to biological cues, and interact with biological systems more effectively and metal–organic frameworks (MOFs) and hydrogels are the best-suited options available in these conditions with customizable drug-loading capacities and biocompatibility. Furthermore, integrating AI and modern computational tools expands the authentication and refinement of novel molecules and materials, reducing the time and price associated with traditional trial-and-error routes. The requirement for novel molecules and materials is crucial to address urgent medical needs and personalized treatment and would lead the way in the next generation of precision medicine [22]. However, they have multiple limitations that hinder their overall effectiveness. One of the major challenges is sparse bioavailability, often due to very low solubility, quick metabolism, or substandard absorption. Many drugs degenerate before reaching their intended site of action, reducing much of their therapeutic impact. Especially in treatments like chemotherapy, non-specific diffusion is another drawback which leads to off-target effects and toxicity that can cause significant side effects in patients [23]. Cancer cells and pathogens often adopt certain mechanisms to dodge drug action, necessitating the use of higher doses or combination therapies, which further increases toxicity risks. These concerns are being solved by advanced materials recruited to improve drug delivery and its efficacy. To ensure that novel drugs reach the desired site while sparing healthy tissues, nanocarriers such as solid lipid nanoparticles, liposomes, dendrimers, and polymeric nanoparticles enable targeted drug delivery [24]. Thus, amalgamating advanced materials into drug design provides a much-needed solution to conventional limitations and has paved the path for safer, more accurate, and more effective treatments.

1.3 Emerging Trends in Drug Discovery

The prospect of drug discovery is quickly evolving, which is driven by advances in science, technology, and data analytics. As the worldwide healthcare environment faces rising drug resistance, complex diseases, researchers are unfolding innovative strategies to enhance drug development's speed, precision, and success. Several emerging trends are shifting the future of drug discovery and development (Table 1).

Table 1 Different types of emerging trends in drug discovery [24–27]

Serial No.	Trend	Description	Example
1	Artificial intelligence (AI) and machine learning (ML)	This predicts the drug-target interactions and accelerates drug discovery	Protein 3D structure prediction using AlphaFold based on primary amino acid sequence. The global database of protein structures was made public in 2021
2	CRISPR and gene editing	Precise gene editing and modification for genetic therapy and target validation	CRISPR-Cas9 made human cell resistant to HIV-I in 2013
3	Targeted protein degradation	Using bi-functional molecules such as Proteolysis targeting chimera (PROTACs) to degrade disease causing protein molecules	Bavdegalutamide, or ARV-110, is used to treat prostate cancer. ARV-110 is the first PROTAC to be evaluated in humans when Arvinas announced the start of the Phase I clinical trial in 2019
4	mRNA therapeutics	Delivers mRNA and inform cells to generate therapeutic proteins	Spikevax is the brand name of Moderna's COVID-19 vaccine (mRNA-1273). FDA Emergency Use Authorization (EUA) for the first time; 2021
5	Organs-on-chips	Devices for drug testing that are micro-engineered to mimic the activities of organs	In 2010, First microengineered liver-on-a-chip devices capable of mimicking hepatic tissue structure and metabolic activity were reported, showing relevance for drug metabolism and toxicity testing
6	Immunotherapy	Employs the immune system to target diseases, especially cancer	The first FDA-licensed CAR-T treatment for children and young adults with relapsed or refractory B-cell acute lymphoblastic leukemia (ALL) was Novartis' Kymriah (tisagenlecleucel), which was authorized in August 2017
7	Biomarker-driven drug discovery	Identifies biological indicators for precision therapies and predict response	One tumor biomarker that has become a landmark in biomarker-driven drug development and targeted therapy is the human epidermal growth factor receptor 2 (HER2). Herceptin (trastuzumab), the first HER2-targeted treatment for metastatic HER2-positive breast cancer, received FDA approval in 1998. This was among the first and most effective instances of cancer treatment based on biomarkers

(continued)

Table 1 (continued)

Serial No.	Trend	Description	Example
8	Multi-omics integration	Combination of genomic, proteomic, and metabolomic data to understand disease in detail	Since 2018, the use of multi-omics integration for the categorization of cancer subtypes, with transcriptomics and proteomics as key components, has grown significantly, with widespread acceptance starting in 2021. By 2025, multi-omics integration, which includes transcriptomics, proteomics, and occasionally other omics like epigenomics is widely used in academic and translational cancer research as a cutting-edge approach for subtype discovery
9	In silico drug design	Uses computational tools for virtual screening and modeling of compounds	Since 1990, Schrödinger software for computational chemistry research, simulation, molecular modeling, and virtual screening has been accessible
10	3D Bioprinting	Construction of 3D tissues/organs for drug testing and personalized medicine	3D bioprinted tumor models for drug testing were first reported in 2014, and it was shown that they more closely resembled the tumor microenvironment than 2D cultures
11	Microbiome-based drug discovery	Investigates the function of gut microbiota in health and illness to develop novel treatments	In microbiome-based medicine, SER-109 (marketed as Vowst) was the first FDA-approved oral microbiome therapy to restore a healthy gut microbiota and prevent recurrent *Clostridium difficile* infection (rCDI)
12	De novo drug design	AI-driven generation of novel chemical structures from scratch which can be a patentable drug candidate	The FDA authorized the use of Relenza for influenza prevention in adults and children five years of age on March 29, 2006
13	Epigenetic modulators	Targets enzymes that aid in modification of gene expression without altering DNA sequence	Suberoylanilide hydroxamic acid, (SAHA), is another name for Vorinostat, a histone deacetylase inhibitor (HDAC) that was used in 2006 to treat cutaneous T-cell lymphoma (CTCL)

(continued)

Table 1 (continued)

Serial No.	Trend	Description	Example
14	Antibody–drug conjugates (ADCs)	Combines antibodies with cytotoxic drugs for targeted therapy	Adcetris (Brentuximab), a monoclonal antibody Vedotin, is a powerful anti-tuberculin drug that was utilized in 2011 to treat certain CD30 lymphomas. Because of this, it was among the first antibody–drug conjugates in oncology to be licensed
15	Quantum computing in drug discovery	Simulates molecular interactions of drugs faster and more precisely than classical computers	In 2020, Qubit Pharmaceuticals was established, with the goal of drastically lowering the time, expense, and failure rate related to creating new medications. It operates at the nexus of quantum computing, artificial intelligence, and life sciences
16	Nanotechnology-based delivery	Use nanoparticles as a drug delivery system for precise delivery and reduce side effects	The first nanomedicine authorized for clinical use was Doxil (liposomal doxorubicin), which received FDA approval in 1995
17	Cell and gene therapies	Treats diseases by altering cells or genetic material	As a gene therapy for spinal muscular atrophy (SMA) in juvenile patients or in children younger than two years old who have bi-allelic mutations in the SMN1 gene, Zolgensma received FDA approval for the first time in 2019
18	Pharmacogenomics	Studies on how genes affect drug response, tailoring drug treatments based on a patient's genetic makeup	Genotyping of thiopurine methyltransferase (TPMT) to modify thiopurine dose. Clinical use of TPMT genotyping for thiopurine dosage adjustment began in the early 2000s, and in 2004 the FDA recognized it, and in 2011, Clinical Pharmacogenetics Implementation Consortium (CPIC) recommendations were established

2 Novel Small Molecules in Drug Discovery and Drug Design

The discovery of novel small molecule therapeutics remains fundamental to contemporary medicine, enhancing the therapeutic landscape across multiple domains of disease. Contemporary drug discovery mainly relies on rational design strategies and high-throughput technologies to identify and optimize lead compounds. The key methodologies that underpin small-molecule drug design include exploring structure-based and ligand-based approaches, high-throughput and fragment-based screening, and the critical role of computational techniques in transforming initial hits into promising drug candidates. Understanding these diverse methods is crucial to understanding the state of pharmaceutical innovation today and its future directions.

2.1 *Structure-Based Drug Design (SBDD)*

Structure-Based Drug Design (SBDD) uses the three-dimensional structures of biomolecules to guide the design of molecules that can bind to and modulate its activity [28]. Fundamental techniques in SBDD are X-ray crystallography and cryo-electron microscopy [29]. The co-crystallization of the protein–ligand or inhibitor complex illustrates the detailed mechanism of specific binding, where essential interactions such as hydrogen bonding, hydrophobic contacts, and electrostatic interactions are emphasized. This structural information can then be used to develop new compounds designed to affect these interactions and improve the affinity, selectivity, and efficacy. The HIV-1 protease inhibitors are among the most well-known and effective examples of SBDD for the treatment of HIV/AIDS. In 1995 FDA authorized the first protease inhibitor called Saquinavir, it selectively attaches itself to the HIV-1 protease enzyme's active site, blocking its function and stopping the virus from developing into its infectious form.

2.2 *Ligand-Based Drug Design (LBDD) and Pharmacophore Modeling*

Ligand-Based Drug Design (LBDD) provides an alternative strategy when the target protein's three-dimensional structure is unavailable or poorly defined [30]. To determine the fundamental structural and physicochemical characteristics necessary for binding, LBDD makes use of the knowledge of existing ligands that bind to the target. Novel molecules with comparable properties are then designed using this information. The essential molecular characteristics required for a molecule to interact with a particular biological target and trigger a biological response are represented abstractly by a pharmacophore. These characteristics usually consist of charged groups, hydrophobic areas, aromatic rings, and donors and acceptors of hydrogen bonds arranged in a particular three-dimensional space. The structures of a group of known active ligands for a specific target are analyzed to create pharmacophore models. The likelihood of finding new active ligands can then be increased by using this model as a query to screen virtual libraries of compounds or to direct the design of novel molecules that have the necessary features in the right spatial arrangement. Similarity search techniques, which can be applied to similarity-based virtual screening, operate on the premise that similar molecules typically have comparable characteristics. First, a reference active compound is used as input. Next, using the similarity principle, other compounds that are similar are found in the compound database.

2.3 High-Throughput Screening and Fragment-Based Drug Discovery

High-throughput screening (HTS), which remains one of the key elements in drug discovery today, offers efficiency in shifting through vast libraries of compounds to identify those that may exhibit biological effects [31]. Although leading diverse compounds can be achieved through HTS, the initial hits in the hit-to-lead phase often lack sufficient affinity and selectivity, necessitating heavy optimization, which could be viewed as a secondary deficit. This problem has augmented the popularity of complementary approaches like Fragment-Based Drug Discovery (FBDD) [32]. FBDD begins with collections of small chemical fragments rather than large, complex molecules, typically in the molecular weight range of 100–300 Daltons. These fragments are tested for binding to multiple sites on a target protein. The binding of these fragments is often accomplished and validated using sensitive protein-ligand interaction methods viz. surface plasmon resonance (SPR), nuclear magnetic resonance (NMR) spectroscopy, or X-ray crystallography.

2.4 Hit-to-Lead Optimization Using Computational Techniques

In drug discovery, initial hits from methods like fragment-based drug design (FBDD) or high-throughput screening (HTS) require refinement to improve potency, selectivity, and ADMET (Absorption, Distribution, Metabolism, Excretion, and Toxicity). ADMET evaluation is vital, as many compounds fail due to poor bioavailability, rapid clearance, adverse distribution, or high toxicity. Absorption ensures systemic circulation, distribution governs tissue targeting, metabolism determines stability, excretion influences clearance, and toxicity predicts harmful effects. Integrating in-silico, in-vitro, and in-vivo ADMET studies early reduces attrition and enhances success rates. Computational methods, such as molecular docking and molecular dynamics simulations, complement experimental assays by predicting protein–ligand interactions, designing structural analogs, and exploring drug–target dynamics [33]. These predictive tools, combined with screening data, provide a detailed chemical space

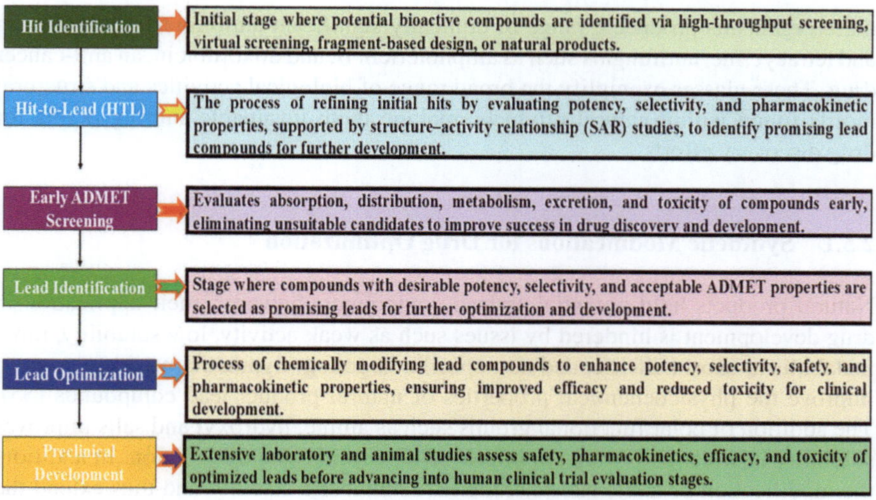

Fig. 2 General scheme showing hit discovery to lead optimization

outline, facilitating the selection of the most promising leads. Ultimately, this integration streamlines hit-to-lead optimization, improves efficiency, and accelerates the identification of drug candidates with strong therapeutic potential. (Fig. 2).

2.5 *Natural Products and Semi-Synthetic Derivatives in Drug Discovery*

Nature is rich in chemical diversity and is a source of secondary metabolites, which are of great importance in the field of drug discovery. These compounds have different ecological functions, such as defense, signaling, and competition, and are produced by organisms. The bioactivity of these compounds, alongside complex structures that took thousands of years to evolve, creates a treasure of leading therapeutic development compounds. Natural products represent a reservoir of structurally diverse and biologically active compounds, offering unique molecular scaffolds and mechanisms of action not typically found in synthetic compound collections. Several major classes of secondary metabolites have yielded numerous important drugs [34]. Flavonoids, which are known for their antioxidant and anti-inflammatory activities, display promising potential as anticancer and antiviral agents. Plants house a vast number of organic compounds that contain nitrogen called alkaloids, and they have multiple pharmacological properties. Some of these are resveratrol (antioxidant and potential anti-aging effects), quinine (an antimalarial), and vinblastine (an anticancer). A diverse group of natural products, known as polyketides, are synthesized by bacteria, fungi, and plants. Through the enzymatic polymerization of acyl units, these organisms produce a range of clinically significant antibiotics: erythromycin and tetracycline, antifungals such as amphotericin B, and doxorubicin, an anti-cancer drug. These classes exemplify the broad range of biological activities and structural motifs found within natural products, making them invaluable starting points for drug discovery efforts.

2.5.1 Synthetic Modifications for Drug Optimization

Natural products hold potential as lead compounds; however, their application in drug development is hindered by issues such as weak activity, low solubility, rapid metabolism, short half-life, and non-specific targeting. Synthetic modifications can improve the physicochemical properties of natural product lead compounds [35]. The addition of polar functional groups such as amine, hydroxyl and salts improves solubility in water and results in improved absorption and distribution. In addition, the modification of labile sites can increase metabolic stability and thus extend the drug's half-life, while structural changes increase the selectivity and affinity of the drug for the target protein and decrease the binding to off-target proteins, reducing side effects and increasing safety.

2.6 Peptide-Based Inhibitors

Peptide-based inhibitors have emerged as a compelling class of therapeutic molecules, especially protein–protein interactions (PPIs) [36]. Peptide-based inhibitors, which usually consist of short sequences of amino acids with molecular weights ranging from 0.5 to 5 kDa, address the difference between small molecule drugs and larger ones like antibodies. Because of their metabolism to amino acids, peptide-based inhibitors show lower toxicity and immunogenicity compared to small molecules and antibodies. Fast and cost-effective synthesis allows structural modifications and optimization. Peptide-based inhibitors primarily function by interfering with the binding of substrates to their target protein. This mechanism is mainly characterized by competitive and non-competitive inhibition [37]. In competitive inhibition, the peptide inhibitor competes at the active site, or binding interface, of a protein, resisting the natural substrate from binding. The binding of the inhibitor blocks the catalytic activity of the enzyme. For example, peptide-based angiotensin-converting enzyme (ACE) inhibitors compete with angiotensin I for binding at the active site of ACE [38], thus lowering the blood pressure. In cancer therapy, bound peptides that imitate the BH3 domain of pro-apoptotic proteins can competitively attach to anti-apoptotic BCL-2 family proteins [39], therefore triggering cell death. In non-competitive inhibition, the peptide inhibitor binds to a regulatory (allosteric) site rather than the active site. This binding induces a conformational change in the protein, which alters the catalytic activity and disrupts its interactions, regardless of substrate binding. For example, the peptide-based allosteric inhibitor FKC interacts with Tumor Necrosis Factor Receptor 1 (TNFR1) by attaching to a conformationally active site, consequently stabilizing the receptor in its inactive conformations and preventing signaling without directly obstructing ligand binding. This mechanism facilitates the regulation of protein function without direct competition at the primary binding site [40]. Although offering conspire potential, peptide-based inhibitors face many limitations and obstacles with stability and delivery [41]. Endogenous peptidases degrade natural peptides, leading to short in vivo half-lives and low bioavailability. Large size and polarity of the peptides are making their crossing cell membranes mobility poor and thus decreasing their overall potency against intracellular proteins. Additionally, in the GI tract extensive proteolysis and low permeability through the intestinal epithelium render peptides largely unsuitable for oral delivery and are primarily administered through injections. Currently, peptide design uses certain advanced techniques such as nanocarrier systems, cyclization, and PEGylation to strengthen and intensify therapeutic performance. PEGylation has an increasing effect on the peptides half-life and solubility, cyclization enhances reliability and receptor binding, whereas nanocarriers provide protection to peptides from deterioration. When combined, these strategies address issues such as fast enzymatic degradation and low bioavailability.

3 Nanomaterials in Drug Discovery

Tiny delivery systems that deliver their therapeutic payload at the ideal time and location or envision an appropriate treatment strategy that only hit the cells they are intended to kill, completely avoiding the healthy cells. Development in nanotechnology is opening doors for finding and delivering new medicines using nanomaterials. Their properties have transformed the drug delivery system in an unexpected ways due to nanoscale range. They have unique chemical and physical properties, like a massive surface area crammed into a tiny volume, surfaces we can tweak chemically almost at will, and often, an easier time getting along with our bodies' complex systems. These special traits make nanomaterials such powerful tools for medicine, especially for tasks like drug delivery, diagnostics, and therapies that need precision [42]. Nanotechnology offers a solution to many of these. Microscopic delivery vehicles, for example nanoparticles, liposomes, dendrimers, and specially designed polymer carriers are far more than mere passive carriers; they can protect medications, manage their timed release, and significantly amplify their effectiveness. One of the most exciting advancements is enabling these nanocarriers with address labels or targeting molecules, to enhance precision in delivering treatments as its guide them straight to disease hotspots, like tumor cells thus potentially delivering a knockout blow with far less collateral damage [43]. Nanotechnology goes beyond enhancing drug delivery. It is changing how we discover drugs in the first place. Finding new medicines is often a slow and expensive. Nanomaterials are lending a hand, for example using quantum dots or gold nanoparticles [44]. Researchers have shown how they can also be used in sophisticated sensors. These sensors in an incredibly

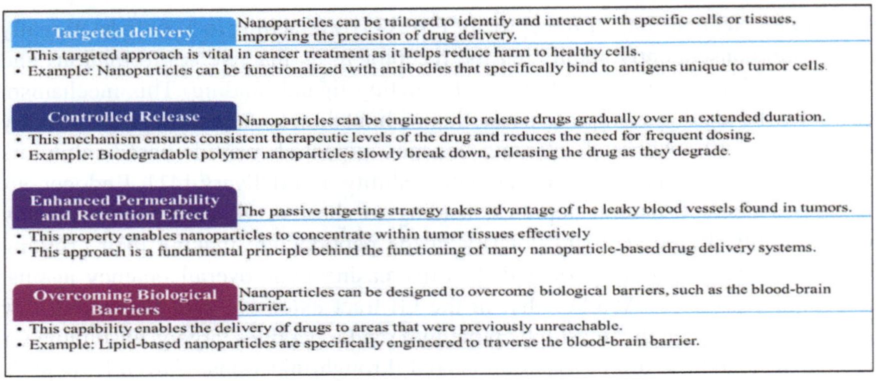

Fig. 3 Flow chart for the key concepts of nanoparticles

subtle ways can detect the interaction of potential drugs with targets in our bodies, helping scientists pick out the winners much faster. It's clear that nanotechnology is becoming a fundamental part of the pharmaceutical research toolkit. Working at this incredibly fine scale introduces its own set of puzzles. Progressing in this advanced field requires coordinated efforts from chemists, biologists, engineers, healthcare professionals, and policymakers, all working together to translate groundbreaking laboratory discoveries into practical benefits for patients (Fig. 3).

3.1 Nanoparticle-Based Drug Delivery Systems

Nanoparticle-Based Drug Delivery Systems (NBDDS) are transforming drug delivery by utilizing the unique characteristics of nanoparticles to improve therapy effectiveness and safety. They provide enhanced stability, solubility, and precise targeting, overcoming longstanding issues in conventional medicine [45] (Table 2).

3.1.1 Polymeric Nanoparticles

Polymeric nanoparticles are submicron-sized colloidal particles which are made up of natural or synthetic polymeric materials. They encase different range of drugs including proteins, peptides, and anticancer agents. Their biocompatibility and ability to modify surface properties make them a suitable candidate for site-specific delivery of the drug. Some of them are discussed for better understanding of polymeric nanoparticles. Polymeric micelles are spontaneously developed when placed in water, as water-loving and water-hating portions are both present in them. They immediately come together to form a core–shell structure, which is ideal for concealing medications in its core that are sensitive to water while giving the body a water-friendly

Table 2 Various types of (NBDDS)

	Nanoparticle-based drug delivery systems	Function
1	Liposomes	These microscopic vesicles from 10 to 200 nm which are made from phospholipid layers and resembling to cell membranes, can encapsulate water-soluble drugs in their interior or lipid-soluble drugs within their membranes. Their compatibility with biological systems makes them ideal for use in cancer therapies and vaccine development
2	Solid lipid nanoparticles (SLNs)	SLNs enhance the poorly water-soluble drugs bioavailability and are eligible for different routes of administration i.e., oral, topical, and parenteral. Curcumin-loaded SLNs enhance solubility and its therapeutic applications as an anti-inflammatory and antioxidant
3	Liquid nanoparticles (LNPs)	LNPs function as drug delivery systems by encapsulating therapeutic agents like mRNA, siRNA, or paclitaxel. For example, delivery of mRNA effectively into human cells in COVID-19 vaccines (Moderna and Pfizer)
4	Nano-emulsions	Nano-emulsions are colloidal dispersions, which are mixed with one another and result in formation of nanometer-sized droplets of (typically 20–200 nm). Surfactants or emulsifying agents stabilize these systems. Nano-emulsions are thermodynamically unstable but kinetically very stable hence offering unique properties such as high surface area, transparency, and enhanced bioavailability

exterior. Their small size makes it easier for them to enter tissues, which aids in the battle against cancer. Dendrimers are fascinating, highly branched molecules that look almost like nanoscale trees. This precise, layer-by-layer structure gives amazing control over size. Dendrimers can be engineered by many ways, such as incorporating drugs, targeting agents, or imaging molecules onto their surface, offering a versatile and adaptable platform for drug delivery [46]. Polymeric vesicles share structural similarities with fatty bubbles (also called liposome). They differ in their composition, being constructed from resilient synthetic polymers. The composition facilitates in various ways for example, it enhances their durability, allows for distinct release profiles, and offers an alternative method for delivering therapies into cells [47].

3.1.2 Inorganic Nanoparticles

Inorganic nanoparticles are usually 1–100 nm in size and these tiny particles are made of inorganic materials like metals, metal oxides, or ceramics. They have special physical and chemical characteristics such as quantum effects, their large surface area-to-volume ratio, their stability, adjustable optical characteristics, and magnetic activity make them perfect for therapeutic, imaging, and drug administration applications. Silicon Dioxide Nanoparticles(SiNPs) or Mesoporous Silica Nanoparticles (MSNs): They are porous in structure to facilitate various applications including drug delivery and molecular encapsulation. Made of silica (like glass), they're filled with tiny, ordered pores [48]. This structure gives them a huge internal surface area, ideal for soaking up significant amounts of drugs, from small molecules to larger proteins. MSNs are typically biocompatible and well-accepted by the body. Their silica-based framework offers robust protection for drug payloads, while researchers can extensively modify pore dimensions and surface chemistry to optimize their functionality. They are the gatekeepers to the pores which open only under specific conditions (like the acidity near a tumor) and decorate the outside with targeting molecules to ensure that the nanoparticle ends up in the right place [49]. This controlled and targeted delivery is basically used for powerful drugs to minimize the collateral damage. Researchers are even loading MSNs with multiple drugs to hit cancer cells [50]. By integrating imaging capabilities, these nano-sponges achieve theranostics, enabling simultaneous treatment and visualization, showcasing exceptional versatility in their applications. Gold Nanoparticles (AuNPs) have an undeniable charm, even at the stage of nanoscale it continues to captivate with its unique properties and appeal. Gold nanoparticles (AuNPs) are known for their biocompatibility, stability, and highly adaptable surface chemistry, which makes them exceptionally straightforward to modify. This enables scientists to connect drugs, targeting agents such as antibodies, or other functional components, effectively customizing the nanoparticle for its specific task [51]. The huge surface-area to volume ratio allows to carry decent amount of drugs. And with the right 'address label' attached, these can be directed towards cancer cells, thus delivering payloads more precisely. They can even help dissolve drugs that normally don't mix well with water. Cancer therapy is a major focus. AuNPs can ferry chemotherapy drugs right to the tumor. But they have another trick, they heat up when hit with certain kinds of light (near-infrared is often used as it goes through tissue). This allows for photothermal therapy—basically, using light to cook cancer cells tagged with AuNPs [52]. They're even being tested for delivering genes into cells. It seems like nano-gold has a bright future in medicine. Superparamagnetic Iron Oxide Nanoparticles (SPIONs) are tiny specks of rust that are magnetic in nature [53]. These particles are pulled up through an external magnet to deliver the drug cargo directly to a target site in the body. Importantly, these particles show up well on MRI scans, acting as contrast agents.

3.1.3 Inorganic and Hybrid Nanomaterials

Carbon based organic materials are workhorses, inorganic elements—metals, oxides, and semiconductors—offer different strengths (like useful optical or magnetic properties). Gold and silver nanoparticles, for instance, are valued for being stable and easy to function. Hybrid nanomaterials aim for the optimum integration, blending organic and inorganic components at the nanoscale. The aim is to achieve synergy by developing materials capable of performing tasks that neither component could accomplish independently. Researchers talk about 2-main classes which are based on their interactions. By mixing and matching, researchers hope to engineer materials perfectly suited for sensing, energy, or advanced medicine [54].

- **Class I**: Held together by weaker forces (van der Waals or hydrogen bonds).
- **Class II**: Linked by strong chemical bonds (covalent bond or ionic bond).

Quantum dots (QDs) are minuscule semiconductor crystals that glow brightly under light, and the color they glow depends precisely on their size, making them a bit bigger, the color shifts. This tunability, plus their intense brightness and resistance to fading, makes them fantastic tools for watching biology happen [55]. Scientists use QDs like tiny beacons to tag molecules or cells. Because they can have different sizes emitting different colors, they can track multiple things simultaneously in living tissue, in real time, over extended periods. It provides an opportunity to observe the intricate molecular processes underlying life or disease in real-time. Quantum Dots are being investigated as versatile drug carriers; likewise therapeutic delivery is integrated with tracking capabilities. Their higher sensitivity makes them well-suited for the highly precise biosensors development. These devices can detect disease markers at early stages or monitor treatment effectiveness, driving progress toward more personalized medical solutions [56].

3.2 *Smart and Stimuli-Responsive Materials*

Stimuli-responsive materials are an advanced innovation. These drug carriers are designed to actively respond to their environment, rather than merely functioning as passive transport systems. Engineered to undergo specific transformations such as altering their structure, releasing their therapeutic cargo, or becoming adhesive, they activate only upon encountering predetermined external triggers. The triggers could be any of them, a shift in acidity (pH), a change in temperature, certain enzymes, light, or the unique chemical environment inside a cell [57].

3.2.1 pH and Temperature Sensitive Nanocarriers

Nanocarriers have been developed as sophisticated platforms for drug delivery, enabling precise control over drug release customized to physiological conditions.

Engineered to respond to specific environmental stimuli, they offer multiple advantages, like enhancement of drug stability, improving bioavailability, and optimization of therapeutic effectiveness while minimizing undesirable side-effects. Researchers have created materials that remain stable at the body's neutral pH but undergo structural transformations, such as swelling, dissolving, or altering their configuration, in acidic environments. This innovative design allows drug carriers to move safely through the body and release their therapeutic cargo upon encountering acidic regions, including tumors or intracellular compartments with low pH levels [58]. Often using special polymers like poly-N-isopropylacrylamide (PNIPAM is a classic example), these carriers change drastically—like shrinking and squeezing out their contents when the temperature crosses a specific threshold. This threshold can be set near fever temperature, or doctors could use focused heat (from ultrasound or light) to trigger release only in a precisely targeted zone [59]. These environmentally sensitive systems promise site-specific, on-demand drug release, a major step towards more effective and less toxic treatments.

3.2.2 Light-Activated Drug Release System

Light offers a way to trigger drug release with amazing precision in both space and time. Using light (often near-infrared, which travels through tissue reasonably well), doctors could potentially 'switch on' drug activity exactly where needed, non-invasively [60]. There are several key mechanisms for light-activated drug release. Photothermal (Light into Heat) these light responsive nanocarriers (like gold nanoparticles and graphene oxide) absorb near-infrared light and convert it into heat. The conversion disrupts the carrier and releases the drug from carrier. This conversion method is particularly useful in cancer therapy. Photocleavage (Light-breakable links) – light-sensitive linkers are also known as photolabile protecting groups. They are integrated into drug delivery systems due to their photocleavage property. They interacted with drugs by those photolabile groups, and the bonds are cleaved after the light exposure to allow the controlled release of the encapsulated drug. Photoisomerization (Light-induced shape shifting) it is utilized in azobenzene based drug carriers, which have characteristics of light-induced conformational changes. Upon exposure to light, these carriers undergo structural transformations that modify their permeability or solubility, facilitating the controlled release of the encapsulated drug. Reactive Oxygen Species (ROS) they are generated by photosensitizers when exposed to the light irradiation. They enable drug delivery system due to their high chemical reactivity which causes the breakdown of drug carriers or causes disruption of cellular structures. Due to this breakage, the precise and localized release of drug in targeted regions occur very easily. This method enhances therapeutic effectiveness while minimizing damage to healthy tissues. These light-activated methods are efficient in treating localized problems viz. specific tumors, skin conditions thus offering precise control without surgery [61].

4 Computational Approaches for Novel Molecules and Materials

Quantum mechanics, simulation, and Artificial Intelligence (AI) have been the pillars of development of new molecules and materials via computational methods. Experiments are greatly reduced, and the process is speeded up by the application of these methods through high-throughput screening, prediction of properties, and mechanistic insight at atomic and electronic levels. Some key factors in the discovery and optimization of bioactive molecules are computational technologies such as molecular docking, virtual screening, and de novo design. Also, different types of materials modelling methods are employed in predicting drug-excipient interactions, ensuring physical and chemical stability, and designing better drug delivery systems. It is being heralded as a historic moment when AI algorithms are working with physics-based simulations to enable the forward-thinking movement in molecular and materials science research [62, 63]. Drug development has witnessed a complete evolution by means of artificial intelligence (AI) that has resulted in an optimal, quick, and accurate search for potential treatment drugs. The traditional methods of drug discovery, albeit successful, are frequently slow and resource intensive. The AI-driven tools to which molecular docking and virtual screening belong are very different in the way they function as they harness the power of big data and machine learning to predict ligand–receptor interactions very accurately [64]. These data mining approaches make it possible to swiftly go through huge chemical compound databases looking for the molecules that have the greatest chance to be the ligands of the newly introduced drugs; hence, they significantly speed up the discovery of new drugs at the very early stage [65]. Deep learning has been a very promising method for screening virtual de novo techniques and de-novo drug design. By understanding complex patterns from the data on biology and chemistry, deep learning models can create those molecular structures that are new to the world and at the same time have very good pharmaceutical characteristics [66]. The principle behind de-novo drug design is to design new molecules with the best biological activity from the beginning to the end, without using the set of existing chemical libraries. Traditionally, this was mainly realized by iterative synthesis and screening the valuable expertise of medicinal chemists. This has been vastly changed by deep learning, which now provides the use of data-driven methods that can generate new molecules with the necessary requirements indeed. Now, deep learning approaches have been the game-changer in the process of de-novo drug design. These models, particularly generative models, such as Recurrent Neural Networks (RNNs), Generative Adversarial Networks (GANs), and Variational Autoencoders (VAEs), have become a primary vehicle for drug discovery. By identifying the basic chemical features and attributes, these models could come up with new chemical entities that have not been seen before and extract knowledge from large collections of chemical compounds [67]. VAEs work by encoding molecules into a continuous latent space and then decoding them back into alternative structures, allowing researchers to explore chemical variations while optimizing important attributes such as solubility, binding affinity, or synthetic

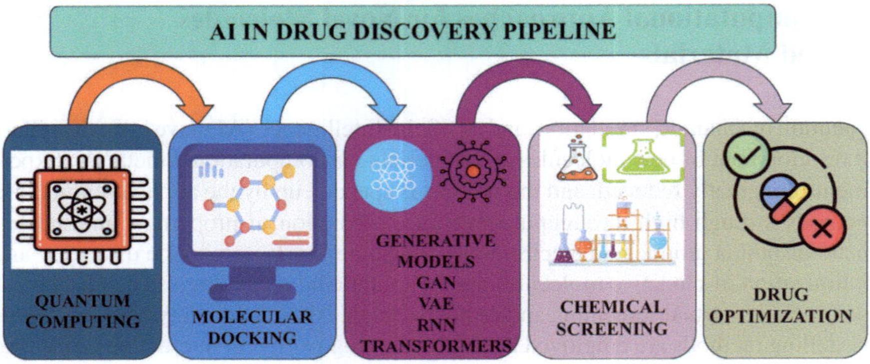

Fig. 4 Integration of AI in drug discovery

accessibility GANs, by contrast, rely on the interaction between two competing models—a generator that proposes new molecular structures and a discriminator that evaluates their validity, eventually producing realistic and chemically feasible compounds. RNNs are a class of neural networks designed to handle sequential data such as SMILES strings. They maintain a memory of previous inputs by passing information from one step to the next and can learn the chemical grammar needed to design novel molecules with specific biological properties. Transformer based models are inspired by natural language processing; this group of models represents molecules as sequences e.g. Simplified Molecular Input Line Entry System (SMILES strings) and learns the chemical syntax and semantics needed in creating new drug-like molecules with better properties [68] (Fig. 4).

4.1 Molecular Dynamics and Quantum Mechanics Simulations

Molecular Dynamics (MD) and Quantum Mechanics (QM) simulations serve as fundamental tools in computational chemistry. They provide deep insights at the atomic level, biomolecular interactions, conformational changes, and the electronic properties of drug molecules. Molecular docking characterizes the static state of ligand-receptor interactions; however, molecular dynamics and quantum mechanical simulations provide deep understanding of dynamic analysis at atomic and electronic functions, which are essential for determining drug activity and stability [69]. Molecular docking is a common technique in the preliminary phases of drug discovery to determine the binding orientation of a ligand with target protein. However, docking alone may not provide deep understanding of the movements of biomolecules or the dynamic nature of the process [70]. The movements of particles from their equilibrium positions during simulations can be deeply analyzed by molecular dynamics simulations at this step. The MD simulation includes force fields such as Assisted Model Building with Energy Refinement (AMBER), Chemistry at Harvard Advanced Region Molecular Mechanics (CHARMM), or GROningen MOlecular Simulation (GROMOS) to simulate inter- and intramolecular interactions. The system preparation stage includes solvation, ion addition, and energy minimization for making the system physiologically relevant during process. Simulations typically have an initial stabilization phase, followed by final runs by which analysis of trajectories for stability and interactions can be performed. The post-simulation analysis encompasses several elements, including root mean square deviation (RMSD), root mean square fluctuation (RMSF), hydrogen bonding, and the calculation of binding free energies, such as through interpolation of the solvent-accessible surface area or molecular mechanics-generalized born surface area methods. Using molecular dynamics, we can validate the docking poses, analyze conformational flexibility, and assess the stability of the interaction patterns between the ligand and the receptor in complex biological systems [71]. Quantum simulations utilizing quantum mechanics (QM) effectively elucidate drug–molecule interactions at a fundamental level by considering factors such as electronic structure, charge distribution, and atomic-level polarization. These techniques serve as precise models for covalent inhibition, metal ion coordination in metalloenzymatic active sites, and the delineation of reaction pathways, encompassing transition state geometries and activation barriers. Density Functional Theory (DFT) is a prominent quantum mechanical method due to its effective balance between computational efficiency and predictive accuracy for ground-state properties. In contrast, primarily techniques such as Hartree–Fock (HF) and post-HF methods, Møller-Plesset (MP2) perturbation theory and Coupled Cluster with Singles and Doubles (CCSD) provide greater accuracy but imply higher computational costs, making them more preferable for studies of small systems that necessitate electronic correlation. Using quantum mechanical approaches for chemically reactive sites and molecular mechanics for the remainder of the biomolecular system, Quantum Mechanics / Molecular Mechanics hybrid approaches coalesce quantum mechanics with classical mechanics. The integration of enzyme catalysis with ligand interactions in large-scale biological systems is shown using a combined scale of inference. It offers an extensive review of covalent drug binding, investigates the effect of protonation and tautomeric equilibrium on ligand affinity, and assesses physicochemical parameters such pKa and redox potential pertinent to the process of absorption, distribution, metabolism, excretion and toxicity (ADMET) profiling [72, 73].

4.1.1 Computational Material Science for Drug Formulations

Due to advancements in computational science and material science delivery systems in pharmaceutical formulation have notably been enhanced by the rational design of excipients. The identified routes allow for the prediction of molecular interactions at the surface level, thereby facilitating the incorporation of formulation components to improve stability, solubility, and bioavailability [74]. Molecular dynamics (MD) techniques provide a model that reveals atomistic details of the interaction between a drug and an excipient under physiologically relevant conditions. MD trajectories can provide insights into the movement of active pharmaceutical ingredients (APIs) within polymeric or lipid-based matrices, a key aspect of the development of controlled-release systems. Coarse-grained simulation improves the temporal and spatial resolutions while simplifying molecular concepts, and making it easier to represent mesoscale phenomena such as micelle formation and nanoparticle encapsulation. Free energy calculations are employed to evaluate the thermodynamic favorability of drug loading, miscibility, and release kinetics. Conversely, quantum mechanical calculations serve to investigate ion–excipient interactions and decomposition mechanisms at the electronic level [75]. Advance computing techniques have now become an indispensable tool for finding excipients with properties that go beyond simple carrier interaction. Molecular dynamics-based virtual screening and docking expedite the assessment of excipient libraries for compatibility with specific active pharmaceutical ingredients. Existing excipients may undergo in silico chemical alterations to improve performance attributes like hydrolysis resistance, permeability, or mucoadhesiveness [76]. The entire excipient design cycle can be accelerated by using advanced machine learning (ML) models to improve the prediction of crucial excipient features such as glass transition temperature (T_g), hygroscopicity, and chemical compatibility [77]. Modification include the derivatization of amphiphilic polymeric carriers to enhance the solubilization of poorly water-soluable medications, the use of modified cyclodextrins in the formation of inclusion complexes, and the computational prediction of antioxidant behavior to stabilize oxidation sensitive compounds [78]. When taken as a whole, these techniques represent a change in strategy toward knowledge-driven, mechanistic formulation design [79].

5 Bioorthogonal Chemistry Based Design Approaches

There is increasing focus for more efficient and targeted cancer treatments causing less harm to healthy tissues. Both targeted medicines and conventional chemotherapy frequently have poor specificity and off-target side effects. A promising approach to tumor-targeted drug development is bio-orthogonal chemistry, a potent technique for generating reactions that happen specifically in living things without interfering with regular biological functions. The application of bio-orthogonal processes is to enhance drug administration, imaging, and therapy in cancer treatment. The different important bio-orthogonal reactions are given as follows.

Click Chemistry is a group of processes that can be utilized to bind different molecules to medications that target cancer. The effective and selective reaction known as strain-promoted azide-alkyne cycloaddition (SPAAC) is used in medication administration and labeling. Tetrazine–Trans-Cyclooctene (TCO) reactions have been applied to imaging and medication delivery. Photo crosslinking reactions also allows for precise tumor targeting and those involving unnatural amino acids. Bio-orthogonal processes have also been utilized to design tumors activated prodrugs which are activated by the tumor microenvironment. Prodrugs can be tagged with bio-orthogonal chemical groups that are selectively reactive in the presence of tumor-associated enzymes or with other bioorthogonal pairs. These activation approaches ensure that the treatment is localized to the tumor thus minimizing damage to healthy cells. Researchers can monitor, image and diagnose tumor cells with high sensitivity by attaching bioorthogonal chemical groups to imaging agents (ranging from fluorophores to radiolabeled compounds). These imaging agents allow recording of location, size and evolution of tumors with high precision which enhances the specificity of cancer therapies by enabling more localized activation or release of therapeutics within the tumor. For example, bioorthogonal chemistry enables selective medication distribution by exploiting the receptors or enzymes specific to neoplasms. By conjugating drugs to bioorthogonal groups that react similarly with respect to the tumor environment, researchers can greatly enhance therapeutic efficacy while reducing systemic side effects.

5.1 In Drug Delivery, Metal–Organic Sustains Framework

Omar M. Yaghi's groundbreaking work in reticular chemistry led to the development of Metal-Organic Framework (MOFs), an innovation honored with the 2025 Nobel Prize in Chemistry. The intentional assembly of metal nodes and organic linkers into crystalline, permanently porous structures with precisely controllable characteristics was made possible by reticular chemistry. With its large surface area, pore size adjustment, chemical functionalization, and regulated release kinetics, MOFs have established themselves as cutting edge drug delivery systems, imaging, and tissue engineering platforms. Their biocompatibility and structural predictability serve as prime examples of how reticular chemistry converts molecular design concepts into therapeutic systems that are applicable to clinical settings. Recent studies focus on increasing the biocompatibility of MOFs by selecting appropriate metals and ligands, enhancing their stability in physiological environment, and promoting controlled release kinetics. There have also been some noteworthy investigations into using MOFs for combination therapies (i.e. for both drug delivery and, e.g., immunotherapy or photodynamic therapy for more efficient cancer treatment) [80]. The specificity and effectiveness of treatment are increased by these MOFs' ability to release their drug payloads in response to certain environmental cues within the tumor [81]. MOFs can work in concert with other therapeutic agents (such as immunotherapies or photodynamic treatments) to treat cancer more successfully [82]. The primary categories of environment-responsive MOFs are p^H responsive MOFs, redox responsive MOFs, thermally responsive MOFs, enzyme responsive MOFs.

5.1.1 DNA Origami in Drug Delivery and Vaccine Development

The process of DNA origami entails employing short "staple" strands that attach to the longer scaffold DNA strand to fold a long single-stranded DNA molecule into a certain shape. This technique makes it possible to create intricate and accurate nanoscale structures that have a wide range of uses in synthetic biology, molecular recognition, and nanomedicine. Because of its accuracy and adaptability, DNA origami is a perfect tool for researching and working with transcription factors. It has promise as a technique for assessing and utilizing transcription factors. It provides a very flexible environment for researching TF-DNA interactions, comprehending transcriptional control, and creating novel treatment approaches. Notwithstanding certain obstacles, further developments in synthetic biology and DNA nanotechnology may make DNA origami a potent tool for fundamental studies as well as therapeutic and gene regulatory applications in the clinic [83]. Drug delivery applications of DNA origami Protein–DNA interaction: DNA origami can replicate natural DNA sequences of interest, which can be used as a template to study the binding behaviour of transcription factors in a controlled manner. By presenting specific DNA motifs on the origami structures, the researchers can selectively control TF binding sites and thus explore the binding of TFs to their target DNA sequences with high spatial resolution. Such dynamic interactions of transcription factors can be studied by using DNA origami structures: They could be set up, for example, to have multiple binding sites for different TFs, which would allow the interaction of multiple TFs on a common region of DNA to be tested (either cooperatively or competitively). This is crucial for comprehending the intricate control of gene expression in response to various environmental factor and stimuli. Drug discovery and screening; In high-throughput screening procedures, DNA origami can be utilized to find substances that can alter transcription factor activity. A transcription factor binding site can also be displayed on the DNA origami so that researchers can then look for small chemicals, peptides, or other biomolecules that facilitate or inhibit TF binding. In fact, this use could be valuable for drug development efforts, particularly for diseases like cancer, in which aberrant TF activity is implicated.

6 Success Stories of Novel Molecules and Materials

Liposomal Doxorubicin (Doxil)—Doxorubicin or Adriamycin, is water soluble, photosensitive chemotherapeutic agent, isolated from *Streptomyces peucetius* var. *caesius*. Since its FDA approval in 1974. it has been used in combination therapies for treating a variety of cancers. Nano formulations of the Doxorubicin have been approved by FDA for its therapeutic use: Doxil for treating Kaposi's syndrome, ovarian cancer, breast cancer, and multiple myeloma (with Velcade); Myocet for breast cancer (with cyclophosphamide), Lipo-dox for breast and ovarian cancer [84]. Liposomes have evolved as a novel platform for a targeted drug delivery in cancer therapy, which addresses the side effects of traditional doxorubicin. Doxorubicin is a potent anthracycline chemotherapeutic agent widely used in the treatment of breast cancer, lymphomas, leukemias, sarcomas, and various other malignancies; however its major clinical limitation is dose-dependent cardiotoxicity, manifesting as acute, early-onset chronic, or late-onset cardiac dysfunction–which can progress to cardiomyopathy and heart failure. In addition to its cardiac effects, doxorubicin include multiple systemic adverse reactions, including myelosuppression, mucositis, alopecia, nausea, hepatotoxicity and extravasation injury and these cumulative toxicities significantly restricts its long term use and allowable lifetime dosing in clinical practice [85]. Liposomal encapsulation uplifts this approach by encapsulating the drug within lipid bilayers, allows control release of the drug, improve tumor targeting, and reduce off target toxicity [86]. PEGylated liposomal doxorubicin formulation such as Doxil, shows significant improvement in pharmacokinetic properties, increases bioavailability and decreases the cardiotoxicity and maintains the strong efficacy [86] The encapsulation of doxorubicin within liposome results in low peak plasma concentration of the drug, which makes it less available to cardiac tissues hence minimizing the risk of cardiac toxicity [86]. Liposomal formulation such as Doxil was formulated to reduce the toxicity and increase the therapeutic efficacy [87]. This formulation uses Poly (Ethylene glycol) to prolong the circulation time and facilitate accumulation in the tumor tissues through the enhance permeability and retention (EPR) effect [87]. Doxil® is the first FDA approved PEGylated liposomal drug. It was initially developed by Sequus Pharmaceuticals in 1955, for the treatment of multiple myeloma, ovarian cancer and HIV associated kaposi's sarcoma. Clinical studies have proven that Doxil is as effective as doxorubicin in action with lowering cardiotoxicity and myelo suppression [16]. Messenger RNA (mRNA) have been evolving as a novel group of therapeutic agents to treat and prevent various diseases like the development of mRNA-based nanoparticle infused COVID-19 vaccines such as those developed by Moderna (mRNA-1273) and Pfizer-BioNTech (BNT162b2) was remarkable scientific and medical breakthrough in recent history [88]. To be effective in vivo, mRNA require safe, effective and stable delivery systems that protect nucleic acid from degradation and allow efficient cellular uptake and release [89]. Several vaccines were developed against COVID-19, more than 30 vaccines approved and many reached phase 3 clinical trials [90]. The use of nanoparticles in the therapeutics reduces toxicity, enhances bioactivity, improves targeting and provides the means to control the release of the encapsulated substance [91]. The use of two mRNA-based vaccines developed using LNPs gained emergency approval during COVID-19 pandemic by United State -Food and Drug Administration was a unique mild-stone in the history of the mankind [92]. The basis of Moderna vaccine is stabilized mRNA of the viral spike protein. And Pfizer-BioNTech vaccine is based on a nucleoside modified RNA (modRNA) of the SARS-CoV-2 virus [93]. Both the vaccines Pfizer-BioNTech and Moderna have shown 95 and 94.5% efficacy in the clinical trial. Lipid nanoparticles have a potential to transport diversity of cargo molecules such as drugs and nucleic acids (mRNA, DNA, siRNA) which makes it versatile drug delivery system [94]. Both the vaccines have created global impact and the foundational mRNA work done by Katalin Kariko and Drew Weissman received global recognition in 2023 in the form of Nobel Prize in Physiology and

Medicine. The remarkable discovery of the mRNA-based vaccines results into the global investment into mRNA technology for future vaccine therapies development [95].

7 Regulatory Challenges and Future Trends for New Drug Materials

The regulatory framework for a new drug material is multifaceted, it requires constant updated frameworks to address the complexities of novel technologies and materials. The regulatory framework for novel drug materials is established to ensure safety, effectiveness, and quality while promoting innovation. The advancement in the field of nanotechnology leads to the development of nano pharmaceuticals and liposomal formulations, which has revolutionized the arena of drug discovery and therapeutics. However, their complex structures and small size are the challenges for regulatory bodies [96]. Key regulatory consideration includes safety assessment, means nanomaterials or any drug material should undergo intensive testing to evaluate the potential risk of cytotoxicity and long-term health impacts [97]. Efficacy and quality attributes such as the characterization of physicochemical properties of the material, the size, surface charge and drug release mechanism are essential factors which are considered by the regulatory [98]. Ensuring the production process which is continuous and scalable for manufacturing in industry is critical [99, 100]. Nanomedicines, which includes nanodrugs, nano vaccines and gene therapies, face unique hurdles and difficulties in the regulatory process due to their complexity and novelty. The lack of standardized protocols for testing and characterization of the nanomaterials complicates the assessment of safety and efficacy [101]. The regulatory bodies are constantly working towards making guidelines for the physiochemical characterization of the nanomedicines [102, 103]. The diversity in the nanomaterials such as dendrimers, liposomes, nanoparticles makes it difficult to follow and apply the same regulatory guideline to all the nanomaterials [104] the unique characteristics of the nanomaterials, such as their interactions with biological system, demand specialized safety assessments. The determination of the appropriate regulatory classification for the nanomedical products, such as devices, drugs or a combination product, is often critical. Regulatory framework and guidelines US FDA has issued specific guidelines for the drug products containing nanomaterials. The FDA generally adopts a more flexible and case-by-case strategy, emphasizing stability, quality attributes and physicochemical characterization. The FDA has a framework for the approval of complex injectable formulations, such as liposomes and nanoparticles. The European Medicines Agency (EMA) places greater focus on standardized evaluation methods, detailed comparability studies, and manufacturing consistency to ensure efficacy, safety and manufacturing processes ensuring across the European Union. International cooperation is required for harmonized regulatory practices and to address the global nanomedicine development. Key regulatory strategies—the detailed physicochemical characterization of nanomaterials is essential for regulatory submission. Measurements of zeta potential, particle size and drug release profiles are essential. Effective communication and active collaboration between industries and academia is crucial for navigating the regulatory landscape. The public–private partnership can facilitate the development of standardized testing methods and regulatory guidelines. Responsible drug development of novel medications is greatly defined by ethical considerations as well as by legal constraints. Issues regarding patient autonomy as well as informed consent are particularly central to areas as disparate as theragnostic as well as gene-based vaccines, where benefits and harms at inception may not be thoroughly known. Revelation of positive as well as negative consequences is vital to sustaining faith with the masses at large. The likelihood of residual health consequences with nanoparticles requires rigorous examination as well as ongoing post-marketing monitoring. Furthermore, fairness of access to advanced nanomedicines needs to be considered because their prohibitive cost as well as scientific complexity risk extending health inequities amongst many different groups. Lastly, environmental impacts of nanomaterials during production as well as disposal need to be considered with attention to sustainability as well as to advances in therapy. Prospects and opportunities–Innovative manufacturing technologies such as 3D printing, premix technologies and advances in nanofabrication are transforming drug manufacturing and developing complex formulations. These technologies offer new opportunities for patient, compliance and improving drug stability and bioavailability. Nanotechnology is well explored in areas such as cancer treatment, CRISPR, and gene therapy offer promising opportunities. The amalgamation of nanobots and advanced drug delivery systems holds potential for personalized and targeted medicine.

8 Conclusion

The landscape of drug discovery domain is being scientifically transformed by the evolution of innovative small molecules, smart materials and nanomaterials. Computational advancements, and futuristic manufacturing techniques like MOFs and DNA origami also provide new mechanisms to combat a wide range of diseases. Nanoparticles and novel nanomaterials have emerged as pivotal tools in modern pharmacology thereby improving drug stability, solubility, and targeted diffusion. Precise control over therapeutic release has been provided by smart stimuli-responsive materials that have further propelled the field by enabling drug delivery systems that respond intelligently to environmental triggers. Different types of computational methods, have significantly accelerated the pace of drug discovery and optimization of lead compounds and materials. Artificial intelligence (AI) plays an important role in drug discovery chemistry, where machine learning models predict biological activity, optimize synthesis pathways, and even design entirely new molecules improving synthesis routes and forecasting biological activity. Emerging technological horizons like quantum computing and next-generation sequencing will further augment drug discovery capabilities. However, the industry still faces challenges such as toxicity, biocompatibility, large-scale production, regulatory permissions, and ethical questions.

References

1. Hughes, J.P., Rees, S., Kalindjian, S.B., Philpott, K.L.: Principles of early drug discovery. Br. J. Pharmacol. **162**(6), 1239–1249 (2011)
2. Paul, S.M., Mytelka, D.S., Dunwiddie, C.T., Persinger, C.C., Munos, B.H., Lindborg, S.R., Schacht, A.L.: How to improve R&D productivity: the pharmaceutical industry's grand challenge. Nat. Rev. Drug Discovery **9**(3), 203–214 (2010)
3. Sliwoski, G., Kothiwale, S., Meiler, J., Lowe, E.W., Jr.: Computational methods in drug discovery. Pharmacol. Rev. **66**(1), 334–395 (2014)
4. Scannell, J.W., Blanckley, A., Boldon, H., Warrington, B.: Diagnosing the decline in pharmaceutical R&D efficiency. Nat. Rev. Drug Discovery **11**(3), 191–200 (2012)
5. Pushpakom, S., Iorio, F., Eyers, P.A., Escott, K.J., Hopper, S., Wells, A., Doig, A., Guilliams, T., Latimer, J., McNamee, C., Norris, A.: Drug repurposing: progress, challenges and recommendations. Nat. Rev. Drug Discovery **18**(1), 41–58 (2019)
6. Mak, K.K., Pichika, M.R.: Artificial intelligence in drug development: present status and future prospects. Drug Discov. Today **24**(3), 773–780 (2019)
7. Mullard, A.: 2017 FDA drug approvals. Nat. Rev. Drug Discovery **17**(2), 81–85 (2018)
8. Vanhaelen, Q., Lin, Y.C., Zhavoronkov, A.: The advent of generative chemistry. ACS Med. Chem. Lett. **11**(8), 1496–1505 (2020)
9. Brown, A.S., Patel, C.J.: A standard database for drug repositioning. Sci. Data **4**(1), 1–7 (2017)
10. Kaitin, K.I.: Deconstructing the drug development process: the new face of innovation. Clin. Pharmacol. Ther. **87**(3), 356–361 (2010)
11. Burslem, G.M., Crews, C.M.: Proteolysis-targeting chimeras as therapeutics and tools for biological discovery. Cell **181**(1), 102–114 (2020)
12. Cao, Y., Romero, J., Olson, J.P., Degroote, M., Johnson, P.D., Kieferová, M., Kivlichan, I.D., Menke, T., Peropadre, B., Sawaya, N.P., Sim, S.: Quantum chemistry in the age of quantum computing. Chem. Rev. **119**(19), 10856–10915 (2019)
13. Dobson, C.M., 2004. Chemical space and biology. Nature **432** (2019)
14. Driggers, E.M., Hale, S.P., Lee, J., Terrett, N.K.: The exploration of macrocycles for drug discovery—an underexploited structural class. Nat. Rev. Drug Discovery **7**(7), 608–624 (2008)
15. Fosgerau, K., Hoffmann, T.: Peptide therapeutics: current status and future directions. Drug Discov. Today **20**(1), 122–128 (2015)
16. Hou, X., Zaks, T., Langer, R., Dong, Y.: Lipid nanoparticles for mRNA delivery. Nat. Rev. Mater. **6**(12), 1078–1094 (2021)
17. Murphy, S.V., Atala, A.: 3D bioprinting of tissues and organs. Nat. Biotechnol. **32**(8), 773–785 (2014)
18. Roy, D., Cambre, J.N., Sumerlin, B.S.: Future perspectives and recent advances in stimuli-responsive materials. Prog. Polym. Sci. **35**(1–2), 278–301 (2010)
19. Keiser, M.J., Setola, V., Irwin, J.J., Laggner, C., Abbas, A.I., Hufeisen, S.J., Jensen, N.H., Kuijer, M.B., Matos, R.C., Tran, T.B., Whaley, R.: Predicting new molecular targets for known drugs. Nature **462**(7270), 175–181 (2009)

20. Wang, F., Zuroske, T., Watts, J.K.: RNA therapeutics on the rise. Nat. Rev. Drug Discov. **19**(7), 441–442 (2020)
21. Ventola, C.L.: The nanomedicine revolution: part 1: emerging concepts. Pharmacy Therap. **37**(9), 512 (2012)
22. Ledford, H.: CRISPR, the disruptor. Nature **522**(7554) (2015)
23. Arkin, M.R., Wells, J.A.: Small-molecule inhibitors of protein–protein interactions: progressing towards the dream. Nat. Rev. Drug Discov. **3**(4), 301–317 (2004)
24. Ginsburg, G.S., Willard, H.F.: Genomic and personalized medicine: foundations and applications. Transl. Res. **154**(6), 277–287 (2009)
25. Zhao, W., Meng, H.: Effects of genetic polymorphism of drug-metabolizing enzymes on the plasma concentrations of antiepileptic drugs in Chinese population. Bioengineered **13**(3), 7709–7745 (2022)
26. Schenone, M., Dančík, V., Wagner, B.K., Clemons, P.A.: Target identification and mechanism of action in chemical biology and drug discovery. Nat. Chem. Biol. **9**(4), 232–240 (2013)
27. Hopkins, A.L., Groom, C.R.: The druggable genome. Nat. Rev. Drug Discovery **1**(9), 727–730 (2002)
28. Anderson, A.C.: The process of structure-based drug design. Chem. Biol. **10**(9), 787–791 (2003)
29. Cheng, Y., Sali, A.: Integrative structural modeling with cryo-EM. Structure **27**(5), 717–724 (2019)
30. Langer, T., Hoffmann, R.D.: Pharmacophore modeling and virtual screening. Methods Mol. Biol. **1762**, 1–25 (2018)
31. Fox, S., et al.: High-throughput screening: update on practices and trends. Assay Drug Dev. Technol. **4**(6), 491–500 (2006)
32. Congreve, M., et al.: Fragment-based lead discovery. Drug Discov. Today **8**(19), 879–887 (2003)
33. Newman, D.J., Cragg, G.M.: Natural products as sources of new drugs from 1981 to 2014 and beyond. Bioorgan. Med. Chem. **24**(11), 2319–2335 (2016)
34. Butler, M.S.: The role of natural products in drug discovery. J. Nat. Prod. **67**(12), 2141–2153 (2004)
35. Cragg, G.M., Newman, D.J.: Natural products: a continuing source of novel drug leads. Biochimica et Biophysica Acta (BBA)—General Subjects **1830**(6), 3670–3695 (2013)
36. Wang, X., et al.: Rational design of peptide-based inhibitors disrupting protein-protein interactions. Front. Chem. **9**, 682675 (2021)
37. Wang, H., et al.: Peptide-based inhibitors of protein-protein interactions: biophysical, structural and cellular consequences of introducing a constraint. Chem. Sci. **12**(17), 5977–5993 (2021)
38. Kim, M., Kim, H., Lee, S.H.: Peptide-based ACE inhibitors: current status and future directions. Food Sci. Biotechnol. **29**(10), 1405–1414 (2020)
39. Gulati, N., Kumar, R., Reddy, B.S.R.: Peptide-based inhibitors for targeting BCL-2 family proteins in cancer therapy. Curr. Top. Med. Chem. **21**(12), 1096–1110 (2021)
40. Fischer, R., et al.: Discovery of a selective allosteric inhibitor targeting human TNF receptor 1. Sci. Rep. **4**, 6351 (2014)
41. Al Musaimi, O., et al.: Strategies for improving peptide stability and delivery. Pharmaceuticals **15**(10), 1283 (2022)
42. Kumar, A., Kashap, S., Mazahir, F., Sharma, R., Yadav, A.K.: Unveiling the potential of molecular imprinting polymer-based composites in the discovery of advanced drug delivery carriers. Drug Discov. Today 104164 (2024)
43. Filipczak, N., Yalamarty, S.S.K., Li, X., Parveen, F., Torchilin, V.: Developments in treatment methodologies using dendrimers for infectious diseases. Molecules **26**(11), 3304 (2021)
44. Zhong, H., Wang, Y.: Quantum dots on demand. Nat. Photonics **14**(2), 65–66 (2020)
45. Cheng, X., Xie, Q., Sun, Y.: Advances in nanomaterial-based targeted drug delivery systems. Front. Bioeng. Biotechnol. **11**, 1177151 (2023)
46. Boas, U., Heegaard, P.M.: Dendrimers in drug research. Chem. Soc. Rev. **33**(1), 43–63 (2004)

47. Yang, J., Jia, C., Yang, J.: Designing nanoparticle-based drug delivery systems for precision medicine. Int. J. Med. Sci. **18**(13), 2943 (2021)
48. Boppana, S.H., Kutikuppala, L.S., Sharma, S., Rangari, G., Misra, A.K., Kandi, V., Mishra, S., Singh, P.K., Rabaan, A.A., Mohapatra, R.K., Kudrat-E-Zahan, M.: Current approaches in smart nano-inspired drug delivery: a narrative review. Health Sci. Rep. **7**(4), e2065 (2024)
49. Vivero-Escoto, J.L., Huang, Y.T.: Inorganic-organic hybrid nanomaterials for therapeutic and diagnostic imaging applications. Int. J. Mol. Sci. **12**(6), 3888–3927 (2011)
50. Georgeous, J., AlSawaftah, N., Abuwatfa, W.H., Husseini, G.A.: Review of gold nanoparticles: synthesis, properties, shapes, cellular uptake, targeting, release mechanisms and applications in drug delivery and therapy. Pharmaceutics **16**(10), 1332 (2024)
51. Huang, H., Liu, R., Yang, J., Dai, J., Fan, S., Pi, J., Wei, Y., Guo, X.: Gold nanoparticles: construction for drug delivery and application in cancer immunotherapy. Pharmaceutics **15**(7), 1868 (2023)
52. Panja, A., Patra, P.: A review on quantum Dots (QDs) and their biomedical applications. 4open **6**, 1 (2023)
53. Ali, M.K., Javaid, S., Afzal, H., Zafar, I., Fayyaz, K., ul Ain, Q., Rather, M.A., Hossain, M.J., Rashid, S., Khan, K.A., Sharma, R.: Exploring the multifunctional roles of quantum dots for unlocking the future of biology and medicine. Environ. Res. **232**, 116290 (2023)
54. Slowing, I.I., Vivero-Escoto, J.L., Wu, C.W., Lin, V.S.Y.: Mesoporous silica nanoparticles as controlled release drug delivery and gene transfection carriers. Adv. Drug Deliv. Rev. **60**(11), 1278–1288 (2008)
55. Vallet-Regí, M., Colilla, M., González, B.: Medical applications of organic–inorganic hybrid materials within the field of silica-based bioceramics. Chem. Soc. Rev. **40**(2), 596–607 (2011)
56. Mamaeva, V., Sahlgren, C., Lindén, M.: Mesoporous silica nanoparticles in medicine—Recent advances. Adv. Drug Deliv. Rev. **65**(5), 689–702 (2013)
57. Stuart, M.A.C., Huck, W.T., Genzer, J., Müller, M., Ober, C., Stamm, M., Sukhorukov, G.B., Szleifer, I., Tsukruk, V.V., Urban, M., Winnik, F.: Emerging applications of stimuli-responsive polymer materials. Nat. Mater. **9**(2), 101–113 (2010)
58. Kalmouni, M.: pH-responsive nanocarriers for cancer therapy. Open University (United Kingdom) (2021)
59. Kucki, M., Cavelius, C., Kraegeloh, A.: Interference of silica nanoparticles with the traditional Limulus amebocyte lysate gel clot assay. Innate Immun. **20**(3), 327–336 (2014)
60. Linsley, C.S., Wu, B.M.: Recent advances in light-responsive on-demand drug-delivery systems. Ther. Deliv. **8**(2), 89–107 (2017)
61. Xing, Y., Zeng, B., Yang, W.: Light responsive hydrogels for controlled drug delivery. Front. Bioeng. Biotechnol. **10**, 1075670 (2022)
62. Joshi, R.P., Kumar, N.: Artificial intelligence for autonomous molecular design: a perspective. Molecules **26**(22), 6761 (2021)
63. Shahzad, K., Mardare, A.I., Hassel, A.W.: Accelerating materials discovery: combinatorial synthesis, high-throughput characterization, and computational advances. Sci. Technol. Adv. Mater. Methods **4**(1), 2292486 (2024)
64. Tiwari, P.C., Pal, R., Chaudhary, M.J., Nath, R.: Artificial intelligence revolutionizing drug development: exploring opportunities and challenges. Drug Dev. Res. **84**(8), 1652–1663 (2023)
65. Vazquez, M., Krallinger, M., Leitner, F., Valencia, A.: Text mining for drugs and chemical compounds: methods, tools and applications. Mol. Inf. **30**(6–7), 506–519 (2011)
66. Mouchlis, V.D., Afantitis, A., Serra, A., Fratello, M., Papadiamantis, A.G., Aidinis, V., Lynch, I., Greco, D., Melagraki, G.: Advances in de novo drug design: from conventional to machine learning methods. Int. J. Mol. Sci. **22**(4), 1676 (2021)
67. Tong, X., Liu, X., Tan, X., Li, X., Jiang, J., Xiong, Z., Xu, T., Jiang, H., Qiao, N., Zheng, M.: Generative models for de novo drug design. J. Med. Chem. **64**(19), 14011–14027 (2021)
68. Martinelli, D.D.: Generative machine learning for de novo drug discovery: a systematic review. Comput. Biol. Med. **145**, 105403 (2022)

69. Bengesi, S., El-Sayed, H., Sarker, M.K., Houkpati, Y., Irungu, J., Oladunni, T.: Advancements in generative AI: a comprehensive review of GANs, GPT, autoencoders, diffusion model, and transformers. IEEE Access (2024)
70. De Vivo, M., Masetti, M., Bottegoni, G., Cavalli, A.: Role of molecular dynamics and related methods in drug discovery. J. Med. Chem. **59**(9), 4035–4061 (2016)
71. Agarwal, S., Mehrotra, R.J.J.C.: An overview of molecular docking. JSM Chem. **4**(2), 1024–1028 (2016)
72. Hospital, A., Goñi, J.R., Orozco, M., Gelpí, J.L.: Molecular dynamics simulations: advances and applications. Adv. Appl. Bioinforma. Chem. **19**, 37–47 (2015)
73. Kotev, M., Sarrat, L., Gonzalez, C.D.: User-friendly quantum mechanics: applications for drug discovery. In: Quantum Mechanics in Drug Discovery, pp. 231–255 (2020)
74. Balasubramani, S.G., Chen, G.P., Coriani, S., Diedenhofen, M., Frank, M.S., Franzke, Y.J., Furche, F., Grotjahn, R., Harding, M.E., Hättig, C., Hellweg, A.: TURBOMOLE: modular program suite for ab initio quantum-chemical and condensed-matter simulations. J. Chem. Phys. **152**(18) (2020)
75. Krzyzaniak, J.F., Meenan, P.A., Doherty, C.L., Pencheva, K., Luthra, S., Cruz-Cabeza, A.: Integrating computational materials Science tools in form and formulation design. In: Computational Pharmaceutical Solid State Chemistry, pp. 117–144 (2016)
76. Sharma, A., Sagar, S., Alam, M.A., Kaur, M., Anwer, T., Sharma, P.K.: 11 Applications and molecular simulation strategies for excipient–excipient compatibility. In: Computational Drug Delivery: Molecular Simulation for Pharmaceutical Formulation, p. 247 (2024)
77. Djuris, J., Cvijic, S., Djekic, L.: Model-informed drug development: in silico assessment of drug bioperformance following oral and percutaneous administration. Pharmaceuticals **17**(2), 177 (2024)
78. Jiang, J., Lu, A., Ma, X., Ouyang, D., Williams, R.O., III.: The applications of machine learning to predict the forming of chemically stable amorphous solid dispersions prepared by hot-melt extrusion. Int. J. Pharmaceutics: X **5**, 100164 (2023)
79. Jiménez-Luna, J., Grisoni, F., Weskamp, N., Schneider, G.: Artificial intelligence in drug discovery: recent advances and future perspectives. Expert Opin. Drug Discov. **16**(9), 949–959 (2021)
80. Liu, G., Wold, E.A., Zhou, J.: Applications of bioorthogonal chemistry in tumor-targeted drug discovery. Curr. Top. Med. Chem. **19**(11), 892–897 (2019)
81. Ding, C., Tong, L., Feng, J., Fu, J.: Recent advances in stimuli-responsive release function drug delivery systems for tumor treatment. Molecules **21**(12), 1715 (2016)
82. Yan, C., Jin, Y., Zhao, C.: Environment responsive metal–organic frameworks as drug delivery system for tumor therapy. Nanoscale Res. Lett. **16**(1), 140 (2021)
83. Mentis, A.F.A., Papavassiliou, K.A., Papavassiliou, A.G.: DNA origami: a tool to evaluate and harness transcription factors. J. Mol. Med. **101**(12), 1493–1498 (2023)
84. Sritharan, S., Sivalingam, N.: A comprehensive review on time-tested anticancer drug doxorubicin. Life Sci. **278**, 119527 (2021)
85. Ahmad, M.Z., Mustafa, G., Abdel-Wahab, B.A., Pathak, K., Das, A., Sahariah, J.J., Kalita, P., Alam, A., Borthakur, P.P.: From bench to bedside: advancing liposomal doxorubicin for targeted cancer therapy. In: Results in Surfaces and Interfaces, p. 100473 (2025)
86. Gabizon, A.A., Gabizon-Peretz, S., Modaresahmadi, S., La-Beck, N.M.: Thirty years from FDA approval of pegylated liposomal doxorubicin (Doxil/Caelyx): an updated analysis and future perspective. BMJ Oncol. **4**(1), e000573 (2025)
87. Rios-Doria, J., Durham, N., Wetzel, L., Rothstein, R., Chesebrough, J., Holoweckyj, N., Zhao, W., Leow, C.C., Hollingsworth, R.: Doxil synergizes with cancer immunotherapies to enhance antitumor responses in syngeneic mouse models. Neoplasia **17**(8), 661–670 (2015)
88. Meo, S.A., Bukhari, I.A., Akram, J., Meo, A.S., Klonoff, D.C.: COVID-19 vaccines: comparison of biological, pharmacological characteristics and adverse effects of Pfizer/BioNTech and Moderna Vaccines. Europ. Rev. Med. Pharmacol. Sci. **25**(3) (2021)
89. Guerrini, G., Magrì, D., Gioria, S., Medaglini, D., Calzolai, L.: Characterization of nanoparticles-based vaccines for COVID-19. Nat. Nanotechnol. **17**(6), 570–576 (2022)

90. Khurana, A., Tekula, S., Saifi, M.A., Venkatesh, P., Godugu, C.: Therapeutic applications of selenium nanoparticles. Biomed. Pharmacother. **111**, 802–812 (2019)
91. Khurana, A., Allawadhi, P., Khurana, I., Allwadhi, S., Weiskirchen, R., Banothu, A.K., Chhabra, D., Joshi, K., Bharani, K.K.: Role of nanotechnology behind the success of mRNA vaccines for COVID-19. Nano Today **38**, 101142 (2021)
92. Patel, R., Kaki, M., Potluri, V.S., Kahar, P., Khanna, D.: A comprehensive review of SARS-CoV-2 vaccines: Pfizer, moderna & Johnson & Johnson. Hum. Vaccin. Immunother. **18**(1), 2002083 (2022)
93. Hagymási, K.: The breakthrough in vaccination: Nobel Prize in Physiology or Medicine 2023. Struct. Chem. **35**(2), 695–699 (2024)
94. Souto, E.B., Blanco-Llamero, C., Krambeck, K., Kiran, N.S., Yashaswini, C., Postwala, H., Severino, P., Priefer, R., Prajapati, B.G., Maheshwari, R.: Regulatory insights into nanomedicine and gene vaccine innovation: safety assessment, challenges, and regulatory perspectives. Acta biomaterialia (2024)
95. Ahmed, R., Maurya, V., Dwivedi, A., Mishra, M.K.: Regulation of nanomaterials and nanomedicines for clinical application. Appl. Drug Res. Clin. Trials Regulatory Affairs **10**(1), E050324227671 (2024)
96. Clogston, J.D., Foss, W., Harris, D., Oberoi, H., Pan, J., Pu, E., Guzmán, E.A.T., Walter, K., Brown, S., Soo, P.L.: Current state of nanomedicine drug products: an industry perspective. J. Pharmaceutical Sci. (2024)
97. Kapoor, M., Hughes, K., Tyner, K.M.: Regulatory considerations for cancer drug products containing nanomaterials. In: Nanotheranostics for Cancer Applications, pp. 277–299 (2018)
98. Soni, S.J.: Nanotechnology in pharmaceutical manufacturing: present innovations and future opportunities. J. Innov. Appl. Pharmaceutical Sci. (JIAPS) 1–6 (2024)
99. Kannan, B., Nandagawale, A., Karnati, P., Joga, R., Sabanis, C.D., Grover, P., Sindhu, R.K., Kumar, S.: Regulatory barriers to the marketing authorization of nanodrug delivery system. In: Nanotechnology and Drug Delivery, pp. 61–100 (2024)
100. Musazzi, U.M., Franzè, S., Condorelli, F., Minghetti, P., Caliceti, P.: Feeding Next-generation nanomedicines to Europe: regulatory and quality challenges. Adv. Healthcare Mater. **12**(30), 2301956 (2023)
101. Oualikene-Gonin, W., Sautou, V., Ezan, E., Bastos, H., Bellissant, E., Belgodère, L., Maison, P., Ankri, J.: Scientific Advisory Board of the ANSM, 2023. Regulatory assessment of nano-enabled health products in public health interest. Position of the scientific advisory board of the French National Agency for the Safety of Medicines and Health Products. Front. Public Health **11**, 1125577
102. De Jong, W.H., Geertsma, R.E., Borchard, G.: Regulatory safety evaluation of nanomedical products: key issues to refine. Drug Deliv. Transl. Res. **12**(9), 2042–2047 (2022)
103. Dri, D.A., Rinaldi, F., Carafa, M., Marianecci, C.: Nanomedicines and nanocarriers in clinical trials: surfing through regulatory requirements and physico-chemical critical quality attributes. Drug Deliv. Transl. Res. **13**(3), 757–769 (2023)
104. Ashokbhai, M.K., Sanjay, L.R., Sah, S.K., Roy, S., Kaity, S.: Premix technologies for drug delivery: manufacturing, applications, and opportunities in regulatory filing. Drug Discov. Today 104011 (2024)

Data Science and Artificial Intelligence in Materials

Enhancing Pulping System Performance: A Markov-Based Reliability and Availability Analysis in Paper Manufacturing Plants

Sumaira Rasool, Ashish Kumar, and Monika Saini

Abstract The main objective of the current investigation is to enhance the performance of pulping systems (PS) in paper manufacturing plants (PMP). The pulping system is a complex structure consisting of digester, decker, knotter, and opener in series configuration. To achieve the objective, a stochastic model is proposed and analyzed using Markovian approach. The concepts of exponentially distributed random variables, perfect repairs and switch devices incorporated in the development of stochastic models. In addition, by simple reliability arguments other system effectiveness measures also derived for PS as well as each subsystem. Reliability, availability, and maintainability are assessed at different points to predict system performance and repair efficiency. The steady-state behavior of PS is observed numerically and the impact of variation in random variables are explored in steady-state. From numerical results, it is revealed that the opener subsystem is highly sensitive in terms of reliability. The results derived can be utilized by PMP engineers and maintenance team to develop maintenance strategies.

Keywords Performance analysis · Pulping system · Markov approach · Exponential distribution

1 Introduction

Paper has played a key role in our daily life for many centuries. It was initially invented in China as a thin and flat material manufactured from pulp obtained from bamboo, bagasse and wood through chemical/mechanical procedures. The paper manufacturing process is very complex and involves many subsystems. Main subsystems of paper manufacturing plant are chipping, feeding, pulping, washing unit, bleaching unit, screening unit, stock preparation unit, and paper production unit arranged in a hybrid formation to optimize efficiency and productivity. Pulping system (PS) is a

S. Rasool · A. Kumar · M. Saini (✉)
Department of Mathematics and Statistics, Manipal University Jaipur, Jaipur, India
e-mail: monika.saini@jaipur.manipal.edu

L. Ledwani and S. Srivastava (eds.), *Experimental and Computational Advances in Materials*, Engineering Materials, https://doi.org/10.1007/978-981-95-4180-5_17

crucial system required for paper manufacturing. In paper manufacturing, the raw material, namely chips, fed into digester from the storage unit that is processed through knotter, decker and opener. Pulping is done mechanically or chemically or both. To maintain the performance of PMS, it becomes necessary that all the subsystems including PS perform with high efficiency and reliability. Several studies are performed by researchers to ensure high availability of industrial systems. However, Reliability, availability, maintainability and dependability (RAMD) investigation of PS not so extensively explored. RAMD analysis of complex systems is very helpful in recognition of possible modifications. The performance of the PS differs on the subsystem's performance. Therefore, it is essential to recognize the most sensitive subsystem. To implement a maintenance plan successfully it is important to examine each subsystem of the system.

Jo et al. [1] observed the failure rate and MTBF of an integrated converter using a Markov methodology. It evaluated the influence of standby on system trustworthiness and indicated significant improvement in the overall reliability of the converter. Fitrianingtyas et al. [2] focused on RAM measures of solar panels of hybrid power plants using the control chart method. This investigation is based on solar panel yield power statistics accumulated over the past four years. Choudhary et al. [3] investigated the RAMD measures of crystallization system. Iggi and Yusuf [4] derived RAMD measures of textile industry using the Markovian birth–death method in which five subsystems arranged in series with main, warm and cold standby units. Yusuf et al. [5] discussed RAMD measures of a hybrid system with various standby methodologies.

Ram et al. [6] investigated the RAMD indices of a paper mill using Markov birth–death processes, estimating key metrics such as MTBF, MTTF and dependability ratio for its six subsystems under steady-state conditions. Musa et al. [7] analyzed the performance of solar water pumping system through RAMD. Kabra [8] presented an illustration of RAMD measures for concentrated solar plants. Reliability, Availability, and Maintainability indices helped in identification of system improvement domains. They improve execution and maximize productivity of solar plants from a profit perspective. Maihulla and Yusuf [9] discussed RAMD indices of photovoltaic systems. Panwar and Kumar [10] suggested a probabilistic framework for performance evaluation of a pulping unit. The considered pulping system comprises five repairable subsystems and placed in series configuration. Sinwar et al. [11] predicted the physical processing unit's steady-state availability in a sewage treatment plant by developing a mathematical model.

Kumar et al. [12] estimated RAMD indices to enhance the operating execution of water supply plant under exponentially distributed random laws. Sayed et al. [13] analyzed the RAM indices for grid connected solar photovoltaic systems. Bahl et al. [14] proposed a procedure to estimate steady-state accessibility of distillery plants applying Petri nets. Zaidi and Goyal [15] discussed statistical properties and readiness evaluation of the pulping system. Sharma and Kumar [16] performed RAM investigation of repairable manufacturing structures exploiting certain data. Adhikary et al. [17] proposed RAM procedure for performance analysis of coal-fired thermal power plants. Pant et al. [18] performed the availability optimization of Kraft pulping system using particle swarm optimization. Tewari et al. [19] discussed the mathematical modeling and behavioral analysis of a refining system using Genetic Algorithm. Singh and Mahajan [20] determined the reliability and steady-state availability of a Utensils manufacturing plant.

After observing the above cited work, the present study is designed to improve the performance of a pulping system (PS) in paper manufacturing plant (PMP). The pulping system is a complex structure consisting of digester, decker, knotter, and opener in series configuration. To achieve the objective, a stochastic model is proposed and analyzed using Markovian approach. In addition, by simple reliability arguments other system effectiveness measures also derived for PS as well as each subsystem. The random variables associated with failure and repair rates of each subsystem are considered as exponentially distributed. The repairs and switch devices are perfect. The reliability, availability, and maintainability are estimated at different points to predict system performance and repair efficiency. The steady-state behavior of PS is observed numerically and the impact of variation in random variables are explored in steady-state. The results can be utilized by system designers and maintenance engineers to plan maintenance strategies.

2 Tools and Techniques

2.1 Symbolizations

The nomenclatures used for stochastic model development are as:

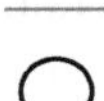	Structure is in working condition with full capacity
	Structure is in operational state but with reduced capacity
	Breakdown state of structure
P, Q, R, S	Completely functioning states of the subsystems
R_1, S_1	Reduced capacity state of the subsystems
p, q, r, s	Failed state of subsystems
φ_i and θ_j, i = 1, 2, 3, 4	Constant failure and repair rates of subsystems P, Q, R and S respectively
P_i; i = 1, 2, 3, 4	Steady state possibility of ith state of the system

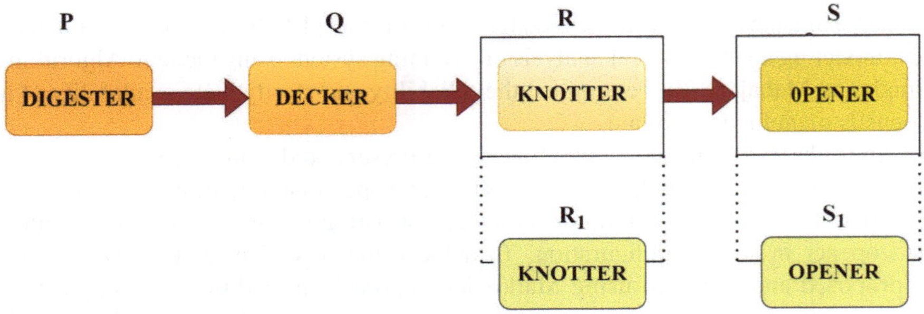

Fig. 1 Flow diagram of pulping system

2.2 Pulping System Explanation

The PS mainly comprises four subsystems: Digester, Decker, Knotter and Opener. All the subsystems of pulping system are arranged in series configuration. The Brief description of each subsystem is below (Fig. 1):

- **Subsystem P (Digester)**: In the pulping process, a mixture of wooden chips and chemicals (Na OH and Na_2S) is heated by steam in digester. It consists of one unit and if the digester fails, the cooking process stops and leads to a system failure.
- **Subsystem Q (Decker)**: The black liquor removal process is important in pulp production. It consists of one unit and failure in this unit leads to a complete halt in black liquor removal and disrupting the whole operation.
- **Subsystem R (Knotter)**: This subsystem includes a primary unit and a backup unit i.e. standby unit. The entire system causes complete failure if both units fail. This redundancy is vital for ensuring reliability.
- **Subsystem S (Opener)**: This subsystem includes a main unit and a standby unit. Complete failure occurs only if both units fail. The standby unit operates when the main unit fails, ensuring continuous operation.

2.3 Assumptions

The proposed PS works under the following assumptions:

- Repairs and breakdowns are unrelated to each other.
- No Concurrent breakdowns occur between the system.
- After getting repaired, the unit acts as new.

3 Transient Analysis of Pulping System Using RAMD Indices

In this section, transient behavior of PS and its subsystems are investigated using RAMD indices. For this purpose, mathematical models for each system derived according to the state transition diagrams as shown in Figs. 2, 3, 4, 5 and 6. The simple probabilistic arguments used for reliability measures evaluation and availability of subsystems evaluated by applying Markov birth–death process. The random variables associated with subsystems are considered as exponentially distributed. The Chapman-Kolmogorov differential-difference equations derived and later used for obtaining the steady-state equations. The detailed evaluation is performed as follows.

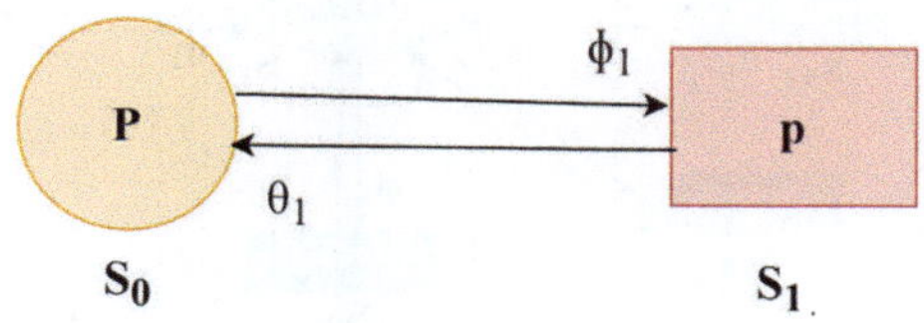

Fig. 2 STD of Digester

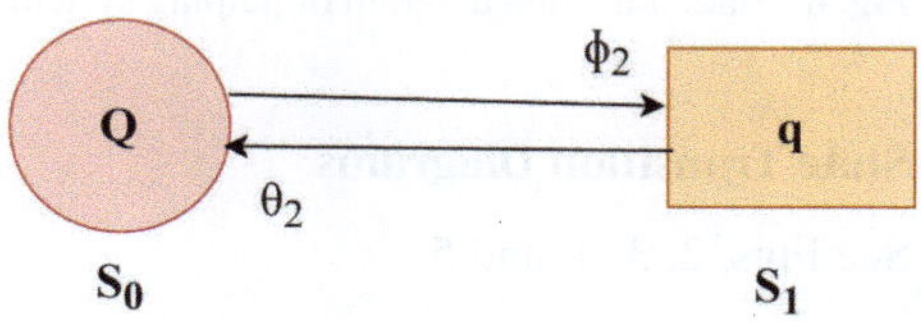

Fig. 3 STD of Decker

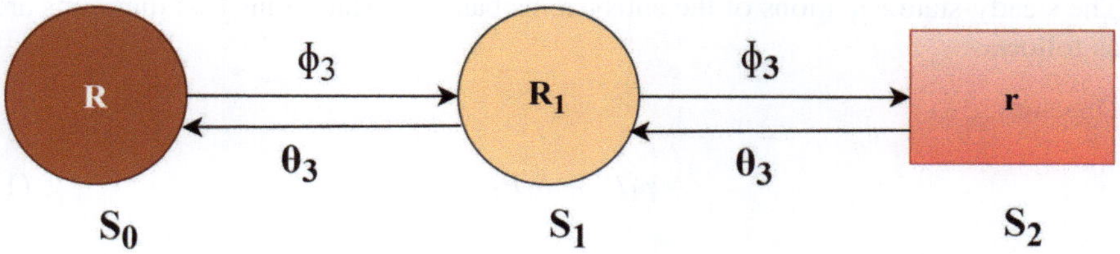

Fig. 4 STD of Knotter

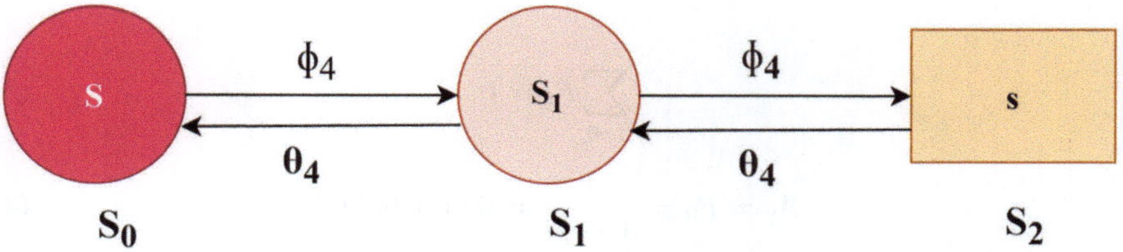

Fig. 5 STD of Opener

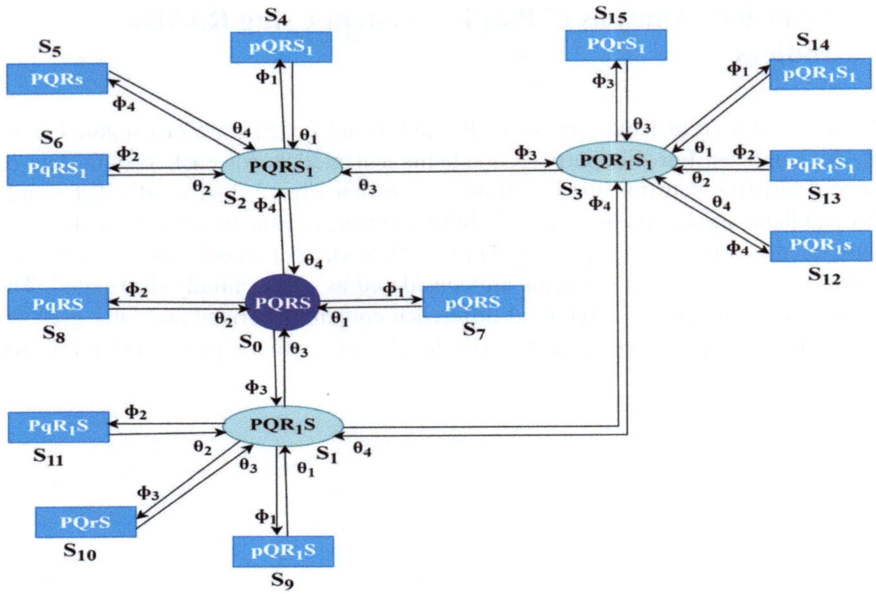

Fig. 6 State transition diagram of pulping system

State Transition Diagrams

See Figs. 2, 3, 4, and 5.

Steady State Equations and reliability Indices for Subsystems

The steady-state equations of the subsystems based on state transition diagrams are as follows:

Digester

$$-\varphi_1 P_0 + \theta_1 P_1 = 0 \tag{1}$$

$$-\theta_1 P_1 + \varphi_1 P_0 = 0 \tag{2}$$

Normalization condition is

$$\sum_{i=0}^{1} P_i = 1 \tag{3}$$

$$A_v = P_0 = \frac{1}{1 + \frac{\varphi_1}{\theta_1}} = 0.666666667 \tag{4}$$

Decker

$$-\varphi_2 P_0 + \theta_2 P_1 = 0 \tag{5}$$

$$-\theta_2 P_1 + \varphi_2 P_0 = 0 \tag{6}$$

Normalization condition

$$\sum_{i=0}^{1} P_i = 1 \tag{7}$$

$$A_v = P_0 = \frac{1}{1 + \frac{\varphi_2}{\theta_2}} = 0.909090909 \tag{8}$$

Knotter

$$-\varphi_3 P_0 + \theta_3 P_1 = 0 \tag{9}$$

$$-(\varphi_3 + \theta_3) P_1 + \varphi_3 P_0 + \theta_3 P_2 = 0 \tag{10}$$

$$-\theta_3 P_2 + \varphi_3 P_1 = 0 \tag{11}$$

$$\therefore \text{Availability } (A_v) = P_0 + P_1 = \left[1 + \frac{\varphi_3}{\theta_3} + \frac{\varphi_3^2}{\theta_3^2}\right] - 1\left[1 + \frac{\varphi_3}{\theta_3}\right] = 0.98630137 \tag{12}$$

Opener

$$-\varphi_4 P_0 + \theta_4 P_1 = 0 \tag{13}$$

$$-(\varphi_4 + \theta_4) P_1 + \varphi_4 P_0 + \theta_4 P_2 = 0 \tag{14}$$

$$-\theta_4 P_2 + \varphi_4 P_1 = 0 \tag{15}$$

$$\therefore \text{Availability } (A_v) = P_0 + P_1 = \left[1 + \frac{\varphi_4}{\theta_4} + \frac{\varphi_4^2}{\theta_4^2}\right] - 1\left[1 + \frac{\varphi_4}{\theta_4}\right] = 0.990990991 \tag{16}$$

Initial Conditions:

The initial conditions associated with system models are as follows:

$$P_i(0) = \begin{cases} 1 \ if \ i = 0 \\ 0 \ if \ i \neq 0 \end{cases} \quad (17)$$

Pulping System Reliability Indices:

$$\begin{aligned} R_{sys}(t) &= R_{ss_1}(t) \times R_{ss_2}(t) \times R_{ss_3}(t) \times R_{ss_4}(t) \\ &\Rightarrow R_{sys}(t) = e^{-0.001t} \times e^{-0.002t} \times e^{-0.003t} \times e^{-0.003t} = e^{-0.009t} \end{aligned} \quad (18)$$

$$\begin{aligned} A_{sys}(t) &= A_{ss_1}(t) \times A_{ss_2}(t) \times A_{ss_3}(t) \times A_{ss_4}(t) \\ &\Rightarrow A_{sys}(t) = 0.666666 \times 0.909090 \times 0.98630 \times 0.99099 \\ &= 0.592373 \end{aligned} \quad (19)$$

$$M_{Sys}(t) = M_{ss_1}(t) \times M_{ss_2}(t) \times M_{ss_3}(t) \times M_{ss_4}(t) = \left(1 - e^{-0.568001385t}\right) \quad (20)$$

$$D_{\min(sys)}(t) = D_{\min(ss1)} \times D_{\min(ss2)} \times D_{\min(ss3)} \times D_{\min(ss4)} = 0.205338389 \quad (21)$$

4 Steady-State Availability Evaluation of PS

The transient analysis of RAMD measures is very crucial for performance behavior exploration in short time domain and identification of most sensitive subsystem of PS. The performance of pulping system is prejudiced by the occurrence of subsystems' failures and repairs. Here, a stochastic model using the Markovian birth–death process developed and Chapman Kolmogorov differential-difference equations are derived by considering constant failure and repair rates. The numerical results of steady-state availability of the PS system obtained for a particular case and impact of variation in random variables observed. A state transition diagram of PS is developed as shown in Figure-6. By using probabilistic arguments, the differential-difference equations at each state of transition diagram derived. Let $P_i(t)$ be the probability that PS is in state 'i' at time "t" then the steady-state expression is derived as:

$$\begin{aligned} P_0(t + \Delta t) &= (1 - \varphi_1 \Delta t - \varphi_2 \Delta t - \varphi_3 \Delta t - \varphi_4 \Delta t) P_0(t) + \theta_1 \Delta t P_7(t) \\ &\quad + \theta_2 \Delta t P_8(t) + \theta_3 \Delta t P_1(t) + \theta_4 \Delta t P_2(t) \\ &\Rightarrow P_0(t + \Delta t) - P_0(t) \\ &= \Delta t \begin{pmatrix} -\varphi_1 P_0(t) - \varphi_2 P_0(t) - \varphi_3 P_0(t) - \varphi_4 P_0(t) + \theta_1 P_7(t) \\ + \theta_2 P_8(t) + \theta_3 P_1(t) + \theta_4 P_2(t) \end{pmatrix} \end{aligned} \quad (22)$$

Taking limit $\Delta t \to 0$, in Eq. (22), we get

$$\Rightarrow \lim_{\Delta t \to 0} \frac{P_0(t+\Delta t) - P_0(t)}{\Delta t} = -(\varphi_1 + \varphi_2 + \varphi_3 + \varphi_4)P_0(t) + \theta_1 P_7(t) + \theta_2 P_8(t) + \theta_3 P_1(t) + \theta_4 P_2(t)$$

After $\lim_{t \to \infty}$, we get

$$\lim_{t \to \infty} P_0'(t) = -(\varphi_1 + \varphi_2 + \varphi_3 + \varphi_4)P_0(t) + \theta_1 P_7(t) + \theta_2 P_8(t) + \theta_3 P_1(t) + \theta_4 P_2(t)$$

$$\Rightarrow (\varphi_1 + \varphi_2 + \varphi_3 + \varphi_4)P_0 = \theta_1 P_7 + \theta_2 P_8 + \theta_3 P_1 + \theta_4 P_2 \tag{23}$$

Similarly for other states, the steady state equations are as follows:

$$(\varphi_1 + \varphi_2 + \varphi_3 + \varphi_4 + \theta_3)P_1 = \theta_1 P_9 + \theta_2 P_{11} + \theta_3 P_{10} + \theta_4 P_3 + \varphi_3 P_0 \tag{24}$$

$$(\varphi_1 + \varphi_2 + \varphi_3 + \varphi_4 + \theta_4)P_2 = \theta_1 P_4 + \theta_4 P_5 + \theta_2 P_6 + \theta_3 P_3 + \varphi_4 P_0 \tag{25}$$

$$(\varphi_1 + \varphi_2 + \varphi_3 + \varphi_4 + \theta_3 + \theta_4)P_3 = \theta_1 P_{14} + \theta_2 P_{13} + \theta_3 P_{15} + \theta_4 P_{12} + \varphi_3 P_2 + \varphi_4 P_1 \tag{26}$$

$$\theta_1 P_4 = \varphi_1 P_2 \tag{27}$$

$$\theta_4 P_5 = \varphi_4 P_2 \tag{28}$$

$$\theta_2 P_6 = \varphi_2 P_2 \tag{29}$$

$$\theta_1 P_7 = \varphi_1 P_0 \tag{30}$$

$$\theta_2 P_8 = \varphi_2 P_0 \tag{31}$$

$$\theta_1 P_9 = \varphi_1 P_1 \tag{32}$$

$$\theta_3 P_{10} = \varphi_3 P_1 \tag{33}$$

$$\theta_2 P_{11} = \varphi_2 P_1 \tag{34}$$

$$\theta_4 P_{12} = \varphi_4 P_3 \tag{35}$$

$$\theta_2 P_{13} = \varphi_2 P_3 \tag{36}$$

$$\theta_1 P_{14} = \varphi_1 P_3 \tag{37}$$

$$\theta_3 P_{15} = \varphi_3 P_3 \tag{38}$$

After simplification of the above equations, the possibilities at each state are determined and stated below:

$$P_1 = \left(\frac{\theta_4 A5 + \varphi_4 A4}{A1A4}\right)P_0,\ P_2 = \left(\frac{\varphi_4 A4 + \theta_3 A5}{A2A4}\right)P_0,\ P_3 = \frac{A5}{A4}P_0,$$
$$P_4 = \frac{\varphi 1}{\theta_1}\left(\frac{\varphi_4 A4 + \theta_3 A5}{A2A4}\right)P_0,\ P_5 = \left(\frac{\varphi 4}{\theta_4}\right)\left(\frac{\varphi_4 A4 + \theta_3 A5}{A2A4}\right)P_0,\ P_6 = \left(\frac{\varphi 2}{\theta_2}\right)\left(\frac{\theta_4 A5 + \varphi_4 A4}{A1A4}\right)P_0,$$
$$P_7 = \left(\frac{\varphi 1}{\theta_1}\right)P_0,\ P_8 = \left(\frac{\varphi 1}{\theta_1}\right)P_0,\ P_9 = \left(\frac{\varphi 1}{\theta_1}\right)\left(\frac{\theta_4 A5 + \varphi_4 A4}{A1A4}\right)P_0,\ P_{10} = \left(\frac{\varphi 3}{\theta_3}\right)\left(\frac{\theta_4 A5 + \varphi_4 A4}{A1A4}\right)P_0,$$
$$P_{11} = \left(\frac{\varphi 2}{\theta_2}\right)\left(\frac{\theta_4 A5 + \varphi_4 A4}{A1A4}\right)P_0,\ P_{12} = \left(\frac{\varphi 4}{\theta_4}\right)\left(\frac{A5}{A4}\right)P_0,\ P_{13} = \left(\frac{\varphi 2}{\theta_2}\right)\left(\frac{A5}{A4}\right)P_0,$$
$$P_{14} = \left(\frac{\varphi 1}{\theta_1}\right)\left(\frac{A5}{A4}\right)P_0,\ P_{15} = \left(\frac{\varphi 3}{\theta_3}\right)\left(\frac{A5}{A4}\right)P_0.$$

And the initial conditions associated with state transition diagram is:

$$P_i(0) = \begin{cases} 1 \ \textit{if} \ i = 0 \\ 0 \ \textit{if} \ i \neq 0 \end{cases} \tag{39}$$

By using normalization conditions, i.e. sum of all probabilities is equal to one $\sum_{i=0}^{15} P_i = 1$, we get

$$P_0 = \frac{1}{1 + \left(1 + 2\frac{\varphi 2}{\theta_2} + \frac{\varphi 1}{\theta_1} + \frac{\varphi 3}{\theta_3}\right)\left(\frac{\theta_4 A5 + \varphi_4 A4}{A1A4}\right) + \left(1 + \frac{\varphi 1}{\theta_1} + \frac{\varphi 4}{\theta_4}\right)\left(\frac{\varphi_4 A4 + \theta_3 A5}{A2A4}\right) + \frac{\varphi 1}{\theta_1} + \frac{\varphi 2}{\theta_2} + \left(1 + \frac{\varphi 1}{\theta_1} + \frac{\varphi 2}{\theta_2} + \frac{\varphi 3}{\theta_3} + \frac{\varphi 4}{\theta_4}\right)\left(\frac{A5}{A4}\right)} \tag{40}$$

Now, the steady state availability of the PS is the sum of the probability of operating systems, which is given as:

$$Av = P_0 + P_1 + P_2 + P_3 \tag{41}$$

$$Av = \left(\frac{1}{\left(1 + \left(1 + \frac{\varphi 1}{\theta_1} + 2\frac{\varphi 2}{\theta_2} + \frac{\varphi 3}{\theta_3}\right)\left(\frac{\theta_4 A5 + \varphi_4 A4}{A1A4}\right) + \left(1 + \frac{\varphi 1}{\theta_1} + \frac{\varphi 4}{\theta_4}\right)\left(\frac{\varphi_4 A4 + \theta_3 A5}{A2A4}\right) + \frac{\varphi 1}{\theta_1} + \frac{\varphi 2}{\theta_2} + \left(1 + \frac{\varphi 1}{\theta_1} + \frac{\varphi 2}{\theta_2} + \frac{\varphi 3}{\theta_3} + \frac{\varphi 4}{\theta_4}\right)\left(\frac{A}{A}\right.}\right.$$
$$\times \left(1 + 1 + \left(\frac{\theta_4 A5 + \varphi_4 A4}{A1A4}\right) + \left(\frac{\varphi_4 A4 + \theta_3 A5}{A2A4}\right) + \frac{A5}{A4}\right) \tag{42}$$

where, $A1 = \theta_3 + \varphi_4$, $A2 = \varphi_4 + \theta_3$, $A3 = \varphi_4 + \theta_3$, $A4 = A3 - \frac{\varphi_3 \theta_3}{A2} - \frac{\varphi_4 \theta_4}{A1}$ and $A5 = \frac{\varphi_3 \varphi_4}{A2} - \frac{\varphi_3 \varphi_4}{A1}$.

5 Numerical Results

In this section, for a particular case numerical results of RAMD measures and steady-state availability are derived for pulping system. The arbitrary values of the random variables appended in Table 1. The RAMD indices for all subsystems of pulping system are presented in Table 2. It is observed that digester units are the least reliable and available subsystem in PS. Tables 3, 4, 5 and 6 revealed the impact of various failure rates of subsystems and time on pulping system reliability. It is observed that reliability of the pulping system declines sharply with respect to time and with the increase in the failure rate the reliability of the system decreases. Table 7 revealed that pulping systems reliability reached 0.637628152 after 50 days whereas Table 8 shows that maintenance chances for a failed subsystem is 99% after 30-time units. The study investigates the impact of changes in failure rates and observes how steady-state availability of the pulping system is influenced. After making 10% increment in each subsystem's failure rates the behavior of steady-state availability is analyzed with respect to pulping system's failure rate from 0.001 to 0.0015 as appended in Table 9. It is revealed that steady-state availability of the pulping system decreases with increment in the failure rates of subsystems. Table 10 examines the impact of changes in repair rates after making 10% increase in repair rates, and behavior of pulping system's availability studied with repair rate from 0.0021 to 0.0024. It is shown that the availability of pulping system increases with the increase in the repair rates of subsystems. It is also observed that subsystem opener is the most sensitive subsystem in the pulping system.

Table 1 Numerical values of random variables of PS

S. No.	Subsystem of PS	Breakdown rate (φ_i)	Restore rate (θ_j)
1	Digester (P)	$\varphi_1 = 0.001$	$\theta_1 = 0.002$
2	Decker (Q)	$\varphi_2 = 0.002$	$\theta_2 = 0.02$
3	Knotter (R)	$\varphi_3 = 0.0015$	$\theta_3 = 0.012$
4	Opener (S)	$\varphi_4 = 0.0015$	$\theta_4 = 0.015$

Table 2 RAMD indices for subsystems of pulping system

RAMD indices of subsystems	SS_1	SS_2	SS_3	SS_4
Reliability	$e^{-0.001t}$	$e^{-0.002t}$	$e^{-0.003t}$	$e^{-0.003t}$
Availability	0.666666667	0.909090909	0.98630137	0.990990991
Maintainability	$(1 - e^{-0.002t})$	$(1 - e^{-0.02t})$	$(1 - e^{-0.21600t})$	$(1 - e^{-0.33000t})$
Dependability (D_{min})	0.30685282	0.744157212	0.939765561	0.956876327
MTBF (hours)	1000	500	333.333333	333.333333
MTTR (hours)	500	50	4.62963	3.030303
Dependability ratio (d)	2	10	72.0004607	110.000001

Table 3 Effect of change in failure rate of Digester on system's reliability

Time (days)	Pulping system			
	$\phi_1 = 0.0008$	$\phi_1 = 0.001$	$\phi_1 = 0.0012$	$\phi_1 = 0.0014$
10	0.915760877	0.913931185	0.91210515	0.910282762
20	0.838617983	0.835270211	0.831935804	0.828614707
30	0.76797354	0.763379494	0.758812931	0.754273685
40	0.703280122	0.697676326	0.692117182	0.686602333
50	0.644036421	0.637628152	0.631283646	0.625002268

Table 4 Effect of change in failure rate of Decker on system's reliability

Time (days)	System			
	$\phi_2 = 0.0018$	$\phi_2 = 0.002$	$\phi_2 = 0.0022$	$\phi_2 = 0.0024$
10	0.906648904	0.913931185	0.903029552	0.901225297
20	0.822012235	0.835270211	0.815462371	0.812207037
30	0.745276491	0.763379494	0.736386619	0.731981528
40	0.675704114	0.697676326	0.664978879	0.65968027
50	0.612626394	0.637628152	0.600495579	0.594520548

Table 5 Effect of change in failure rate of Knotter on system's reliability

Time (days)	System			
	$\phi_3 = 0.0013$	$\phi_3 = 0.0015$	$\phi_3 = 0.0017$	$\phi_3 = 0.0019$
10	0.917594231	0.900324523	0.894938749	0.889585193
20	0.841979173	0.810584246	0.800915364	0.791361816
30	0.772595232	0.729788874	0.716770194	0.703983754
40	0.708928928	0.65704682	0.641465421	0.626253524
50	0.650509095	0.591555364	0.574072261	0.557105862

Table 6 Effect of change in failure rate of Opener on system's reliability

Time (days)	System			
	$\phi_4 = 0.0013$	$\phi_4 = 0.0015$	$\phi_4 = 0.0017$	$\phi_4 = 0.0019$
10	0.917594231	0.900324523	0.894938749	0.889585193
20	0.841979173	0.810584246	0.800915364	0.791361816
30	0.772595232	0.729788874	0.716770194	0.703983754
40	0.708928928	0.65704682	0.641465421	0.626253524
50	0.650509095	0.591555364	0.574072261	0.557105862

Table 7 Impact of time variation on reliability of PS and its subsystems

Time (days)	R_{SS1}	R_{SS2}	R_{SS3}	R_{SS4}	R_{SYS}
10	0.990049834	0.980198673	0.970445534	0.970445534	0.913931185
20	0.980198673	0.960789439	0.941764534	0.941764534	0.835270211
30	0.970445534	0.941764534	0.913931185	0.913931185	0.763379494
40	0.960789439	0.923116346	0.886920437	0.886920437	0.697676326
50	0.951229425	0.904837418	0.860707976	0.860707976	0.637628152

Table 8 Impact of time variation on maintainability of PS and its subsystems

Time (days)	M_{SS1}	M_{SS2}	M_{SS3}	M_{SS4}
10	0.009950166	0.019801327	0.181269247	0.884676473
20	0.019801327	0.039210561	0.329679954	0.986700484
30	0.029554466	0.058235466	0.451188364	0.998466253
40	0.039210561	0.076883654	0.550671036	0.999823123
50	0.048770575	0.095162582	0.632120559	0.999979602

Table 9 Effect of variation in failure rates on availability

φ_1	Base Values	φ_1 + 10% of φ_1	φ_2 + 10% of φ_2	φ_3 + 10% of φ_3	φ_4 + 10% of φ_4
0.001	0.622596675	0.603800467	0.61874438	0.622463699	0.622490196
0.0012	0.586105916	0.566194905	0.582690719	0.585988069	0.586011552
0.0013	0.569418926	0.54909567	0.566194894	0.569307693	0.569329858
0.0014	0.553655821	0.532998962	0.55060734	0.553550661	0.553571616
0.0015	0.53874194	0.517819124	0.535855054	0.536787276	0.53866221

Table 10 Effect of variation in repair rates on availability

θ_1	Base Values	θ_1 + 10% of θ_1	θ_2 + 10% of θ_2	θ_3 + 10% of θ_3	θ_4 + 10% of θ_4
0.0021	0.63196475	0.649740201	0.635616465	0.632089355	0.632064386
0.0022	0.640729215	0.658154794	0.644483221	0.6408573	0.640831634
0.0023	0.648946588	0.666030307	0.652797791	0.649077979	0.64905165
0.0024	0.656666542	0.672764425	0.660610198	0.656801079	0.65677412
0.0025	0.663932918	0.680358832	0.667964602	0.664070449	0.66404289

6 Conclusion

Reliability indices are essential in performance analysis of industrial systems as it directly influences operational efficiency. Here, a stochastic model for pulping system is proposed under the concepts of redundancy and numerical results derived for a particular case by considering random variables as exponentially distributed.

It is observed that the pulping system attains the lowest reliability 0.592373 that is sufficiently less in comparison to least reliable subsystem digester. The transient reliability of pulping system and its subsystems sharply decline with the time as well as increment in failure rate. The numerical values of steady-state availability revealed that failure of digester and opener units influence the availability of whole pulping system and needs extra attention. The subsystem opener is the most sensitive subsystem of the pulping system in paper manufacturing plants. To enhance the reliability of the entire system, it is suggested operating the opener subsystem with utmost care. The findings of the study may be helpful for system designers and maintenance engineers to plan the maintenance strategies.

References

1. Jo, J.S., Kim, S.P., Gyu Oh, S., Kim, T.J., Kang, F.S., Park, S.J.: Markov model-based reliability analysis considering the redundancy effect of modular converters. IEEE Access (2024)
2. Fitrianingtyas, P., Asyari, M.K., Elyawati, N., Musyafa, A.: Reliability, maintainability and availability analysis of solar power plant in Pantai Baru using voltage measurement data. In: E3S Web of Conferences, vol. 479, p. 05001. EDP Sciences (2024)
3. Choudhary, R., Maan, V.S., Kumar, A., Saini, M.: Performance modeling of crystallization system in sugar plant using RAMD approach. Reliab. Theory Appl. **19**(4(80)), 301–312 (2024)
4. Iggi, A.M., Yusuf, İ: RAMD analysis of mixed standby serial manufacturing system. Sigma J. Eng. Nat. Sci. **42**(4), 1116–1132 (2024)
5. Yusuf, I., Sanusi, A., Lawan, Z.D.: A comparative study of hybrid systems using RAMD analysis. Risk Assess. Manage. Decis. **1**(1), 75–87 (2024)
6. Ram, M., Goyal, N., Kumar, A.: Dependability analysis of paper mill plant under multi-state repairable systems execution. Nonlinear Stud. **31**(2) (2024)
7. Musa, M.A., Yusuf, I., Dakingari, A.U.: Performance analysis of solar water pumping system through RAMD. Life Cycle Reliab. Saf. Eng. **12**(4), 309–321 (2023)
8. Kabra, R.: Reliability availability and maintainability study for concentrated solar plants and how to account for current limitations and challenges. In: 2023 Middle East and North Africa Solar Conference (MENA-SC), pp. 1–3. IEEE (2023)
9. Maihulla, A.S., Yusuf, I.: Reliability, availability, maintainability, and dependability analysis of photovoltaic systems. Life Cycle Reliab. Saf. Eng. **11**(1), 19–26 (2022)
10. Panwar, N., Kumar, S.: Availability analysis using Markovian approach and maintenance planning of a process industry. Int. J. Ind. Syst. Eng. **41**(3), 277–294 (2022)
11. Sinwar, D., Saini, M., Singh, D., Goyal, D., Kumar, A.: Availability and performance optimization of physical processing unit in sewage treatment plant using genetic algorithm and particle swarm optimization. Int. J. Syst. Assur. Eng. Manage. **12**(6), 1235–1246 (2021)
12. Kumar, A., Singh, R., Saini, M., Dahiya, O.: Reliability, availability and maintainability analysis to improve the operational performance of soft water treatment and supply plant. J. Eng. Sci. Technol. Rev. **13**(5) (2020)
13. Sayed, A., El-Shimy, M., El-Metwally, M., Elshahed, M.: Reliability, availability and maintainability analysis for grid-connected solar photovoltaic systems. Energies **12**(7), 1213 (2019)
14. Bahl, A., Sachdeva, A., Garg, R.K.: Availability analysis of distillery plant using petri nets. Int. J. Qual. Reliab. Manage. **35**(10), 2373–2387 (2018)
15. Zaidi, Z., Goyal, Y.K.: Mathematical analysis and availability of the pulping system in the paper industry. Int. J. Model. Optim. **4**(1), 31 (2014)

16. Sharma, S.P., Kumar, D.: RAM analysis of repairable industrial systems utilizing uncertain data. Appl. Soft Comput. **10**(4), 1208–1221 (2010)
17. Bose, D., Chattopadhyay, S., Bose, G., Adhikary, D., Mitra, S.: RAM investigation of coal-fired thermal power plants: a case study. Int. J. Ind. Eng. Comput. **3**(3), 423–434 (2012)
18. Pant, M., Thangaraj, R., Abraham, A.: Optimization of a Kraft pulping system: using particle swarm optimization and differential evolution. In: 2008 Second Asia International Conference on Modelling & Simulation (AMS), pp. 637–641. IEEE (2008)
19. Tewari, P.C., Joshi, D., Rao, M.S.: Mathematical modeling and behavioral analysis of a refining system using genetic algorithm. In: Proceedings of National Conference on Competitive Manufacturing Technology and Management for Global Marketing. Chennai (2005)
20. Singh, J., Mahajan, P.: Reliability of utensils manufacturing plant—a case study. Opsearch **36**, 260–269 (1999)

Artificial Intelligence and Nanotechnology in the Comprehensive Management of Colon Cancer: From Risk Prediction to Precision Therapy

Renu Khangarot and Abhijeet Singh

Abstract The chapter provides an overview of the escalating problem of antimicrobial resistance (AMR) and the promising potential of artificial intelligence (AI) and machine learning (ML) in addressing AMR. As a threat to global health, AMR reduces the effectiveness of the drugs we already have (despite established AMR definitions and guidelines) and challenges treatment outcomes. Traditional diagnostics were effective, but they are also relatively time consuming, resource intensive, and slow down the clinical decision-making process. We demonstrated how AI and ML approaches are speeding up microbiology diagnostics, predicting drug resistance, and personalizing drug treatments by allowing rapid, accurate, and most importantly, cost-effective solutions. The chapter presented several applications of AI and ML to AMR including identification of bacteria, detection of resistance genes, and predictive analytics to optimize treatment strategies while also providing a sampling of the foundational algorithmic and computational models that have underpinned these advances. We highlighted two key elements that enabled these advances: leveraging big data and bioinformatics, and integrating clinical datasets to provide predictive analytics applicable to the field of microbiology. Finally, there was critical reflection upon the challenges, limitations, and ethical considerations of applying AI in the field of clinical healthcare along with future research and policy possibilities. All of this information becomes part of the body of knowledge as we show how AI and ML can enhance the global response to AMR and improve the clinical outcomes for patients.

Keywords Antimicrobial resistance (AMR) · Artificial intelligence (AI) · Machine learning (ML) · Microbial diagnostics · Drug resistance prediction · Personalized medicine · Bioinformatics · Predictive analytics · Clinical microbiology · Healthcare innovation

---

R. Khangarot · A. Singh (✉)
Cancer Research Laboratory, Department of Biosciences, Manipal University Jaipur, Jaipur, Rajasthan, India
e-mail: abhijeet.singh@jaipur.manipal.edu

L. Ledwani and S. Srivastava (eds.), *Experimental and Computational Advances in Materials*, Engineering Materials, https://doi.org/10.1007/978-981-95-4180-5_18

1 Introduction

Colon cancer, or cancer that arises from the large bowel, is still one of the leading health burdens globally. In 2020 alone, over 1.9 million new colorectal cancer (CRC) cases and approximately 935,000 deaths resulted from CRC, making it the third most commonly diagnosed cancer and the second leading cause of cancer deaths. In high-income countries, the incidence rates are higher than globally, and in India, even with the rising lifestyle risk of Westernized food choices, CRC diagnoses have been increasing even in younger adults (age < 50) challenging long established screening recommendations. Although there have been improvements in colonoscopy-led screening and adjuvant therapies, the clinical outcomes are still lacking. In particular, the 5-year survival for localized tumors may reach greater than 90%, but drops to anywhere between 10 and 15% in metastatic disease. Conventional chemotherapy regimens such as 5 FU and irinotecan are limited by their systemic toxicity, non-selectivity for tumors, and varying efficacy related to tumor heterogeneity, drug resistance, and relapse [1].

Emerging developments in Artificial Intelligence (AI) and Nanotechnology have revealed significant opportunities to change the current approach to colon cancer care. Examples of AI include machine learning (ML), deep learning (DL), reinforcement learning (RL), and predictive analytics, all of which are able to recognize patterns in manifold data from genomics, histology, radiology, and real-world clinical data [2]. Examples of nanotechnology include engineered nanoscale carriers (e.g., liposomes, dendrimers, mesoporous silica nanoparticles, quantum dots, etc.) able to accomplish targeted drug delivery, controlled release, enhanced imaging of disease, and targeted therapy of the tumor microenvironment.

Why an integrated AI–nanotech strategy is essential:

- AI can predict the optimum properties of nanoparticles (e.g. size, surface chemistry, targeting ligands) that is best matched to the nano-carrier design's tumor microenvironment to enhance uptake and loss of target tissue effect.
- In diagnostics, nanobiosensors can significantly increase sensitivity for early detection of biomarkers (e.g. ctDNA, CEA, miRNAs) from easy-to-obtain samples (e.g., stool, blood), which is potentially scalable for screening applications.
- There are theranostic nanoplatforms with imaging and therapeutic properties (e.g., SPIONs, gold nanoparticles), which can be integrated with AI based imaging analyses and feedback loops to adjust doses or assess residual disease in real time [3].

At the population health level, AI-based polygenic risk scores (PRS) and lifestyle modeling contribute to personalizing the start of screening and timing intervals of screening potentially allowing recommendations to start screening up to a decade sooner for those high genetic risk individuals without clinical features of the disease. Translating this blend into health care practice involves overcoming significant obstacles: biological heterogeneity and variability of human tumor vasculature versus rodent models which subsequently affects nanoparticle delivery to the tumors, current regulatory frameworks involve no precedent for regulatory oversight on combined AI, medical device and nanomedicine systems; challenging scalability and reproducibility in manufacturing; algorithm bias and data privacy issues [4].

This chapter will pool together current evidence and emerging innovations through the entire continuum of care for colon cancer from risk prediction and early detection through diagnosis, targeted therapy, recurrence surveillance, and personalized follow-up. Each section will focus on:

1. Scientific foundations of AI and nanotechnology,
2. Examples from the world and clinical trials (e.g. GI Genius, liposomal irinotecan),
3. Synergistic integration planning,
4. Limits and ethical/regulatory considerations, and
5. Future research directions including multi-omics integration, transparency in AI explainability, regulatory sandboxes, and scalable production.

This chapter connects notions from oncology, computer science, and nano-engineering to develop a roadmap to precision, adaptive, and minimally invasive colon cancer management where early detection in faster, treatments are smarter, and outcomes are considerably more predictable [5].

2 Colon Cancer: Background and Clinical Landscape

2.1 *Epidemiology*

In 2022, colorectal cancer (CRC), including colon cancer specifically, was the third most common diagnosed cancer in the world, with over 1.9 million new cases and approximately 904,000 deaths, representing about 10% of deaths from cancer. It has a higher incidence in high-income areas of the world such as Europe, North America, and Oceania (ASIR > 30 per 100,000), while low- and middle-income countries are on the lower end of the spectrum but have rapidly increasing incidence (ASIR 8–17).

Rising incidence among young adults (age < 50) is concerning worldwide: someone born in 1990 will have double the risk of being diagnosed with colon cancer as someone born in 1950, and CRC is the leading cause of cancer mortality in men under the age of 50 in the United States. Projections in Europe for individuals aged 25–49 years show significantly higher early-onset CRC deaths related to lifestyle risk factors from obesity, alcohol, and other risk factors [6].

2.2 Risk Factors and Pathophysiology

CRC follows the multistep sequence of adenoma–carcinoma and includes genetic mutations of APC, KRAS, TP53 and other oncogenes, as well reaffirmation of microsatellite instability (MSI) pathways (Lynch's syndrome, which accounts for 15–20% of sporadic CRC), in a small fraction of cases of CRC. Risk consistently increases with diets rich in red/processed meats, saturated fat, obesity, diabetes and sedentary lifestyle, with hazard ratios reported to be 1.25–1.57. Approximately 5–10% of CRC cases are attributed to heritable syndromes (FAP, Lynch), while the vast majority of cases (> 90%) are sporadic [7]. In countries like India, CRC's incidence is growing particularly among urban populations as they shift from traditional diets to Western-style foods and lifestyle factors reduce physical activity.

2.3 Clinical Presentation and Staging

Symptoms emerge late and often include changes in bowel habits, occult and visible bleeding from the rectum, unexplained weight loss, fatigue, and so forth. Thus, it is not surprising that almost 90% of CRC cases are diagnosed following clinical symptoms reported by the patient or an incidental diagnosis made by a health care professional. Most cases are detected during a routine colonoscopy (~ 80%) or as a patient presented to the emergency department as part of an emergency surgery (~ 10%).

Staging is often done following the AJCC TNM staging system (8th edition): T1–T4 (depth of invasion), N0–N2 lymph node involvement, and M0–M1 distance pertaining to metastasis. Stage grouping I–IV directly is closely associated with prognosis; Stage I is predominantly surgical treatment (with ~ 90% 5 year survival); and in Stage IV, survival drops to only 13–15% [8].

2.4 Current Treatment Paradigm

The main treatment modality for early-stage (I–II) colon cancer is surgical resection. For stage III colon cancer and high-risk stage II colon cancer, adjuvant chemotherapy which may include FOLFOX (5-FU + oxaliplatin) or CAPOX (oxaliplatin + capecitabine) is recommended for these patients. For low-risk stage II patients, the treatment duration may reduce from the conventional 6 months of treatment to 3 months with no loss of efficacy and a lower risk of peripheral neuropathy. Immunotherapy (e.g., PD-1 inhibitors such as pembrolizumab) is reserved for patients with MSI-H/MMR-deficiency. A random phase II trial for rectal MSI H cancers showed 100% complete clinical response with dostarlimab, but data specific to colon cancers are still accumulating [9]. Other therapeutic modalities in colon cancers would include radiation (more often used in the case of rectal cancer), targeted agents (EGFR inhibitors), and precision agents which are rapidly emerging.

2.5 *Prognosis and Global Disparity*

The 5-year relative survival for CRC in general has been reported as > 90% for localized, 70% for regional and declines to 13–14% for distant metastatic CRC. Global survival disparities also demonstrate the level of global HDI countries with high HDI report significantly higher rates of early detection and better management compared to countries with low HDI or low-middle HDI. Countries on their way to being a low-middle HDI report increasing cancer incidence, mortality benefits from so called clinical trials, but often in the absence of any formal screening. For example, India reported 64,863 CRC new cases and approximately 38,367 deaths in 2022 with dismal survival compared to developed countries [10]. The background information establishes our need for early identification and detection, personalized assessment of risk and adoption of precision therapeutics, where AI is a potential solution and nanotechnology for therapeutics.

3 Challenges to Conventional Management

Despite major advances in screening and treatment options, management of colon cancer remains poor owing to a variety of clinical, systemic, and biological barriers and limitations.

3.1 *Late Presentation and Poor Screening Uptake*

Given the fact that early stages of CRC are often asymptomatic, screening uptake and access are poor (specifically for young adults and in low and middle income countries [LMICs]). A colonoscopy is the gold-standard test, however, it will miss lesions, particularly flat or proximal adenomas, and there are not enough resources. Up to 90% of cases are still diagnosed after symptom onset, ultimately leading to a higher stage at presentation and subsequently worse outcomes. The fecal occult blood test can reduce CRC mortality by ~ 20–30%, but it does not possess good sensitivity and does not significantly reduce overall mortality [11]. This indicates there is a use for procedures like, but are not definitive or broad in scope.

3.2 Biological Heterogeneity and Resistance to Treatment

CRC is also marked by significant inter-patient and intra-tumor biological heterogeneity with multiple molecular subtypes (i.e., CMS1–4; MSI vs MSS). Standard chemotherapy regimens address populations of patients instead of individual patients, which means there will be variations in efficacy. Drug resistance (either de novo or induced) significantly limits treatment response, as it develops at various stages of disease, especially in patients with Stage III or IV disease [12]. Additionally, combination therapy methods, though often used, will usually require more severe side effects systemically and those strategies often fail because the genetic resistance for obviously established patient is entirely different.

3.3 Toxicity and Lack of Precise Delivery

Conventional treatment options such as 5-FU, irinotecan, oxaliplatin, etc. will often result in adverse event related toxicities such as neutropenia, neuropathy, mucositis, and organ toxicity, presenting challenges to escalation and lengthy duration of therapy. In addition, there is no precision when it comes to systematic chemotherapy, leading to insufficient tumor concentration, while also exposing normal and healthy organs to drugs.

3.4 Resource and Infrastructure Deficiencies

Healthcare infrastructure is insufficient in most areas (especially in LMICs like India) leading to lack of colonoscopy, genetic evaluation and treatment options. Areas of higher colon cancer incidence rates will typically have better access (urban facilities), while rural, underserved and uninformed populations suffer from limited access. All of this contributes to late stage presentations and sub-optimal disease trajectories.

3.5 Sub-optimal Risk Stratification

Current guidelines rely on age (> 50) and family history for initiating screening. The guidelines/methods are woefully insufficient at identifying many individuals with high genetic risk or those with early-onset disease. Moreover, lifestyle factors and influences of the microbiome are poorly integrated into standard screening pathways, thereby reducing any potential yield for prevention and delays diagnosis [13].

3.6 *Inconvenient Follow-Up*

Post-resection surveillance pathways using CEA monitoring and imaging are inconvenient and costly. Often, by the time a recurrence is detected through imaging, it is at a late stage and therefore too late for effective intervention.

3.7 *Call for Innovation*

In light of the aforementioned growing complexities, innovation is needed in:

- Predictive risk algorithms that consider more than just age or family history
- Automated high-accuracy diagnostics (e.g., AI-assisted colonoscopy, digital pathology)
- Targeted treatment systems that offer reduced toxicity and improved efficacy
- Dynamic surveillance systems utilizing biomarkers and real-time analytics.

Integration of AI and nanotechnology opens potential pathways towards bridging the gaps identified above to facilitate precision diagnostics, personalized risk assessment, targeted therapeutic delivery, and adaptive monitoring systems. The subsequent sections will identify advances in each of these areas.

4 Artificial Intelligence in Colon Cancer Management

4.1 *Risk Prediction and Population Screening*

Conventional screening protocols, including age-appropriate colonoscopy screens and standard fecal immunochemical testing (FIT), typically overlook high-risk individuals who do not possess clinical signals (ex: a family history of CRC or clinical symptoms). This gap is even more troubling with respect to early-onset colon cancer in younger populations. Thus, PRS combined with machine learning (ML) improves the capabilities of risk stratification and delivers more sophisticated personalization. Recent population-based analyses and segmented research projects using the UK Biobank have demonstrated the utility of PRS: in one case–control study (2585 CRC cases vs 9362 controls), individuals in the top quintile of PRS had an increased CRC risk of 2.70 times vs median risk, despite accounting for both family history of CRC and lifestyle factors (aOR 2.70; 95% CI: 2.27–3.19). Moreover, this significant PRS calculation remained elevated in younger individuals (< 50 years of age), where no traditional risk factors for CRC were present (aOR 2.87; CI: 1.65–5.00) [14].

AI-driven risk prediction frameworks integrate genomics alongside lifestyle, demographic, and clinical data moving beyond static genetic frameworks. For example, multiple studies have sought to combine polygenic risk scores (PRS), with non-genetic models (e.g. QCancer 10) and have demonstrated improved model performance than either PRS or the non-genetic model alone. Machine learning algorithms, such as XGBoost, LDpred2, have been effectively employed in the context of UK Biobank to improve risk estimation knowing dozens to hundreds of SNPs and other features of phenotype [15].

A specific example of a public health application is Bupa's Health Insights tool (Powered by Genomics) being used in NHS trials in the UK. Individuals who, for example, score in the top decile of the PRS have an average CRC risk of the average 50 year old at around 37.6. This individual is afforded the opportunity to access targeted screening, rather than the entire population normative screening program, over a decade earlier [16].

Taken together, these data also suggest that ML PRS might provide actionable improvements for early identification, personalized screening intervals, and resource targeting. By moving away from population thresholds towards personalized schedules, AI-driven risk prediction could lead to fewer late-stage diagnoses, reduced barriers to preventive measures, and improve survival outcomes concerning colon cancer.

4.2 Diagnosis: Colonoscopy and Histopathology

Artificial intelligence (AI) has greatly improved the precision and reliability of colon cancer diagnosis in two distinct areas: real time colonoscopy (AI-assisted colonoscopy) and digital histopathology.

AI-Assisted Colonoscopy

GI Genius intelligent endoscopy module (Medtronic) was approved as the first FDA-cleared AI-based tool to support real-time polyp detection. In a large multicenter randomized tandem colonoscopy trial (NCT03954548), GI Genius reduced the adenoma miss rate (AMR) by half (AI 15.5% vs. no AI 32.4%, adjusted OR 0.38; 95% CI: 0.23–0.62; $p < 0.001$). Subsequent Cochrane meta-analyses confirmed consistent reductions in AMR (~twofold improvement) in small (< 5 mm), flat, proximal, and distal lesions. The real-world clinical use also increased the overall adenoma detection rate (ADR) from ~ 48 to ~ 56.6% (sessile serrated lesion detection also improved from ~ 8.3 to ~ 11.6%) [17]. The AI systems actively mark suspicion in real-time, particularly in metastatic liaisons that human endoscopists may miss due to fatigue or subtle morphology.

AI in Histopathology

In pathology, deep learning models trained on whole-slide images (WSI) have been shown to achieve diagnostic performance comparable to expert pathologists. A large-scale system called "Chief" developed at Harvard Medical School was able to achieve overall accuracy of ~ 96% in detecting cancer in whole slides of colon tissue and improved on previous methods by 36% [18].

Another interpretable AI model trained on ~ 10,500 WSIs (with external validation on TCGA and PAIP datasets) internally achieved 93.44% accuracy and 99.6% sensitivity for dysplasia detection and ~ 84.9% accuracy on an external cohort of cases presenting diagnostic challenges. Another multi-institutional model designed to classify the four major types of colorectal polyps (tubular adenoma, villous adenoma, hyperplastic polyp, sessile serrated adenoma) reported 93.5% accuracy internally and 87.0% accuracy externally, which approximates the level of performance achieved by local pathologists [19].

AI tools have also been evaluated for their ability to predict microsatellite instability-high (MSI-H) status directly from H&E images. A 2025 meta analysis of 19 studies (N = 33,383 samples) with pooled internal sensitivity/specificity of ~ 0.88/0.86 and an external sensitivity of 0.93, specificity 0.71, with an AUC ~ 0.92, provides strong evidence for molecular inference without the need for costly genomic testing [20].

These advanced AI systems automate feature extraction (e.g. gland morphology, nuclear architecture, tumor–stroma ratio) and decrease inter-observer variability and diagnostic time in resource-limited settings.

4.3 Treatment Planning and Decision Support

AI based decision-support tools are playing an increasingly larger role when determining complex treatment decisions for colon cancer patients. In practice, these tools pull from multimodal patient data such as cancer genomics, comorbidities, and treatment history to recommend differential treatment plans.

For example, IBM Watson for oncology (WFO) is informed by a comprehensive list of all the available evidence and expert consensus of the literature. In a retrospective study of 656 colon cancer patients (stage II–IV) in Korea, the absolute concordance between WFO recommended regimens and actual choices made was 48.9% with a general concordance (acceptable "for consideration" included) of 65.8%. Concordance varied by cancer stage reported from 32.5% with stage II to 58.8% with stage III. Notably, younger patients (~ 63.8% concordance) had much greater concordance with WFO recommendations than older patients (~ 20.2%) [21].

A growing body of patients are not enough that diagnosis and prognosis can be incorporated in AI-supported treatment pathways Instead, a prospective study that involved Watson for Oncology for colorectal cancer cases at a multidisciplinary treatment center has reported 91% concordance on AI recommended treatment plans across stages II–IV. In addition, concordance across treatment stages including, neoadjuvant, surgery, adjuvant, first-line or second-line ranged from 87 to 100% concordance across cases highlighting the adaptability of AI and how it can learn from feedback loops [21].

More recent frameworks go beyond broad recommendation systems. One of the models for ML is CRCNet based on deep learning from very large slide images that can predict patient survival outcomes and whether or not they will benefit from adjuvant chemotherapy in Stage II/III CRC. With this model, we could potentially classify patients in high- and low-risk subgroups—including which of them could benefit from chemotherapy.

Related work in colorectal cancer patients has also been established with ensemble ML models using logistic regression, random forests, SVM, and XGBoost to predict adverse events with chemotherapy (e.g., cardiotoxicity) and prognosis outcomes, with AUC values of 0.78–0.90, which could influence optimizing pre-emptive regimens (BMC AI, Palestine cohort) [22]. Overall, although platform-based cognitive computing models like Watson have potential in matching treatment options with evidence-based guidelines, it appears that ensemble ML models offer more detailed possibilities for predictions of toxicity, outcomes, and risks of surgery. Using these data-driven approaches with the supervising ability of the clinician might help in providing a more personalized, safer, and effective treatment plan for colon cancer.

4.4 Monitoring and Recurrence Detection

Post-operative surveillance for colon cancer involves detecting recurrence accurately and as soon as possible while limiting unnecessary treatment. With the recent development of AI (artificial intelligence) models to predict relapse by evaluating longitudinal clinical data, clinical and pathological data, biomarker data, and more into the model, surveillance for recurrence can now be more timely and precise.

AI Models Using Clinical and Biomarker Data

AI Models Using Clinical and Biomarker Data (which also uses Machine Learning (ML)) used a retrospective cohort study of 396 colon cancer patients using CatBoost classifiers using clinicopathological and laboratory variables such as, CEA, albumin, neutrophils, lymph node involvement; this model gave an $AUC = 0.92$ and accuracy of 88%, so this model has excellent prognostic value (from routine clinical data).

Other models whose predictive power and accuracy is valuable for patient clinicians, are ensemble ML (machine learning) models (XGBoost, Random Forest, SVM) which used Brazilian datasets to predict overall and cancer-specific survival outcomes with accuracy of ~ 80% across the 1, 3, and 5 year follow-ups, contributing value to clinical decision support in resource-limited environments [23].

Digital Pathology and Histology Based Prognostic

Deep learning approaches utilizing whole slide images (WSIs) from thousands of colon tumor specimens were able to stratify patients by risk of recurrence. The QuantCRC algorithm by the Mayo Clinic segments multiple tumor regions to extract quantitative histologic features that are prognostic of recurrence. QuantCRC delineates patients at very low risk allowing them to possibly avoid unnecessary adjuvant chemotherapy from high risk groups that require intense follow up.

A larger deep-learning survival model developed by Wulczyn et al. (NPJ Digital Medicine, 2021) utilized over 3,652 CRC cases from multiple cohorts and produced a comparable 5-year disease-specific survival AUC $\approx$ 0.70. This model provided meaningful prognostic value that complemented standard clinical staging (T and N category, grade). Interpretable features derived from clustering accounted for up to 80% of model variance. Additionally, the use of a multimodal deep learning framework (combination of histology with clinical or genomic data) provided an ËtAUC score that was significantly higher than those we reported for WSI n ~ 0.743 to n ~ 0.771–0.773 for 5-year survival predictions [24].

Tumor-Infiltrating Lymphocytes (TILs)

Automated deep learning quantification of TILs in H&E stains demonstrated that low density of TILs ($\leq$ 17 per mm^2) in stage II/III proficient MMR tumors was independently associated with decreased relapse-free survival (HR multivariate $\approx$ 1.37–2.38), thus confirming TILs as a valid prognostic biomarker.

5 Nanotechnology in Colon Cancer Management

5.1 *Nanocarriers for Targeted Drug Delivery in Colon Cancer*

Nanocarriers, including liposomes, polymeric nanoparticles, dendrimers and mesoporous silica nanoparticles (MSNs) have the potential to revolutionize colon cancer strategies by providing colon specificity, controlled release, and specific targeting resulting in decreased systemic toxicity.

Liposomes: Ligand functionalized liposomes improve tumor-specific delivery of chemotherapeutics. For example, hyaluronic acid coated liposomes loaded with 5-FU achieved over 70% reduction in tumor volume in HT29 xenografts with minimal off-target effects when compared with free drug. Surface carboxylated PEGylation also enhanced serum stability and circulation time. A systematic review (2012–2022) found liposomal 5-FU or irinotecan (folic acid or transferrin targeted) consistently resulted in significant tumor burden reduction and improved survival in murine models with reduced hematological and gastrointestinal toxicity [25].

Mesoporous Silica Nanoparticles (MSNs): MSNs with tumor-triggered gatekeeper had colon-specific payload release properties. HA and galactosyl-chitosan functionalized MSNs loaded with Doxorubicin exhibited pH/enzyme activated release at colon simulated conditions. MSNs released more than 85% of drug at pH6.8 and only 15% at pH7.4 in vitro; in vivo this targeting resulted in ~ 60% tumor burden reduction without weight loss [26]. Magnetic-MSNs decorated with EpCAM aptamers allowed targeted delivery and provided MRI contrast. This confirmed the imaging and delivery potential.

PAMAM Dendrimers Highly branched polymers were used for simultaneous delivery and imaging. PEGylated dendrimers loaded with curcumin and conjugated with gold nanoparticles exhibited anti-cancer activity and CT contrast enhancement in HT29 and C26 colon cancer xenografts resulting in more than 70% tumor regression. Capecitabine conjugated G4 PAMAM constructs produced similar effects with added benefits of reduced toxicities, such as hepatotoxicity and myelosuppression, compared to free drug [26].

Hydroxyapatite Nanoparticles (HANs) are emerging as colon-targeting delivery systems, which exhibit site specific accuracy with oxaliplatin consumed at a slow release of 48 h, good amounts of cellular uptake in Caco2 colon cancer cells, and low cytotoxicity toward fibroblasts. The early phases of in vivo work on rodents demonstrated superior shrinkage of tumor size, with less toxicity in systemic release than traditional use.

They continue to combine active targeting (ligand-association), passive targeting (via the EPR effect), and stimuli-responsiveness (pH and enzymatic triggers), providing a higher degree of precision. In pharmacokinetics, drug release through these agents can show prolonged half-lives, high AUC levels in tumor tissue, and low systemic exposures, which is good clinical practice in terms of improved therapeutic intervention outcomes. The integration of these technologies could offer a multi-faceted solution to improving the inherited challenges associated with drug resistance, treatment enveloping, changing tumor microenvironments, and the complication of treatment-associated toxicity in colon cancer [27].

5.2 Theranostic Nanotechnology for Colon Cancer

Theranostics is referred to the use of a single nanoplatform that provides both therapeutic and diagnostic capabilities, resulting in personalized medicine that allows for continuous monitoring of treatment effectiveness while administering targeted therapy. This dual-processing capability is attractive as a method in colon cancer because it may address several issues, including tumor heterogeneity, optimized imaging accuracy, and lessened systemic toxicity.

One potential example is that of gold nanoparticles (AuNPs) functionalized with tumor-specific ligands such as folate, or EpCAM aptamers, used simultaneously as photothermal therapy (PTT) as well as contrast enhanced CT imaging. The gold nanoparticles preferentially aggregate at tumor locations because of the enhanced permeability and retention (EPR) effect and receptor mediated endocytosis. As the AuNPs are subjected to near infrared (NIR) laser illumination, heating will lead to tumor ablation, and CT imaging will display the location and distribution of nanoparticles [28].

Superparamagnetic iron oxide nanoparticles (SPIONs), another theranostic agent, have demonstrated to be efficacious in an MRI-guided delivery of 5-fluorouracil (5-FU). A combination of PEG coated SPIONs conjugated with 5-FU and hyaluronic acid led to targeted drug delivery with a high T2-weighted imaging contrast in colon cancer-bearing mice and greater than 60% reduction in tumor volume, while confirming spatial tracking of the drug release [28].

Additionally, quantum dots (QDs) doped with therapeutic agents can also be visualized with high resolution fluorescence imaging and have controlled drug release utitlization. Despite concerns about toxicity, the more recent evolution of biocompatible QDs like graphene QDs have potential for also providing real-time tumor monitoring, and can also provide pH-responsive drug release mechanisms in acidic tumor microenvironments.

These theranostic approaches provide clinicians with the unique opportunity to monitor a therapeutic response, make real-time dosing adjustments, and potentially predict tumor recurrence earlier than allowed with traditional diagnostics, all of which are preferential in the complex progression of colon cancer. Nonetheless, the most prevalent translational barriers still exist in the form of toxicity, clearance, and regulatory approval that must be overcome before clinical applicability [29].

5.3 Nanobiosensors for Colon Cancer Early Detection

Detection of colorectal cancer at an early stage results in better prognostic outcomes, but standard non-invasive options currently available to detect colorectal cancer, such as colonoscopy and fecal occult blood testing, are generally invasive and sensitive, or have insufficient sensitivity. Nanobiosensors, which combine biological recognition elements that bind to target biomarkers with nanomaterials that can amplify detection signals, provide a sensitive, non-invasive, and cost-effective option for the detection of early-stage colorectal cancer.

Electrochemical nanobiosensors developed using gold nanoparticles (AuNPs) and graphene oxide (GO) have proved to be powerful tools for the detection of specific biomarkers for colorectal cancer, such as carcinoembryonic antigen (CEA), KRAS mutations, and microRNAs (miRNAs). For example, in developing the electrochemical nanobiosensor based on AuNPs and functionalized with anti-CEA antibodies, Zhao et al. developed a detection limit of 0.02 ng/mL, which is well below physiologically relevant serum levels in early cancer patients (Biosensors and Bioelectronics). GO, or even better, in combination with carbon nanotubes as a nanomaterial, can increase the electron transfer rate and provide stability in electrochemical biosensors [30].

Surface Plasmon Resonance (SPR) nanobiosensors can provide label-free detection in clinical samples. For example, in a recent study, researchers created SPR sensors that were functionalized with a circular, single stranded nucleic acid aptamer that recognizes miRNA-21, an important early marker of colorectal carcinogenesis. Attomolar sensitivity was achieved in serum-derived samples from patients.

Biosensors based on magnetic nanoparticles (MNPs) in combination with microfluidics and PCR amplification can isolate and detect mutated DNA like KRAS G12D from liquid biopsies. These have enabled point of care testing (POCT) with results available in less than an hour, and real-time monitoring for high-risk patients [31].

The multiplexing and miniaturization capabilities of nanobiosensors provides simultaneous detection of multiple cancer markers, improving specificity and sensitivity for early detection. Integration of nanobiosensors with smartphones and artificially intelligent (AI) algorithms are currently being explored for real-time interpretability and remote diagnostics.

Although the potential of nanobiosensors is great, there are limitations to be considered including reproducibility, long-term stability and regulatory standards. Nanobiosensors can fundamentally alter early screening strategies by advancing diagnostics for colon cancer to democratize cancer detection, to be more personalized and predictive.

5.4 AI-in-Integrative Nanoplatforms for Precision Oncology in Colon Cancer

By converging artificial intelligence (AI) with nano-technology, we have experienced a new shift in precision oncology in the management of colon cancer, which has many challenges related to inter patient variability, heterogeneity and treatment resistance. AI in integrative nano-platforms utilize machine learning algorithms to design nanoparticles, predict drug responses, and potentially undertake personalized therapy treatments.

Recent developments indicate how deep learning algorithms can learn from enormous image databases (e.g. histopathology or colonoscopy), identify tumor features and provide protocols for formulation of targeted nanoparticles. The power of AI in precision oncology can also be exemplified in colon cancer by studying how convolutional neural networks have outperformed other classifiers by over 30% with respect to identifying colon cancer molecular subtypes and better understand development of the ligands for functionalization of nanoparticles [32].

AI is crucial for predicting biodistribution and pharmacokinetics in the design of nanoparticles. Tools such as Bayesian optimization and genetic algorithms contribute significantly to understanding the potential optimal characteristics, e.g. size, shape and surface charge, of biodistribution and pharmacokinetics for colon targeted delivery. A study reported the impact of AI in customizing PEGylated nanoparticles to penetrate more effectively across the intestinal mucosa of colon tumors where it achieved higher tumor to normal tissue ratios.

Additionally, AI offers real-time feedback monitoring and control in theranostics. Nanoparticles with biosensors relay information (such as pH, oxygen, and drug release) to the cloud-based AI where doses can be adjusted dynamically or external stimuli (such as light or magnetic fields) can be triggered. This closed-loop system facilitates maximum efficacy with minimal side effects.

AI is also being investigated in the creation of nanocarrier libraries, where machine learning can cluster molecular features via unsupervised learning and match them with patient omics data to achieve hyper personalized therapy.

In summary, with coalescing of nanotechnology and AI, a new pathway towards smart, adaptive, and individualized treatments for colon cancer is now possible, transforming the concept of precision oncology into clinical reality.

6 Convergence of AI and Nanotechnology in Colon Cancer

6.1 Introduction: The Rationales for Integration

The management of colon cancer is on the cusp of a major change where an intersection of two emerging technologies (Artificial Intelligence [AI] and nanotechnology) will create a paradigm shift in diagnosis, treatment, and monitoring. Colon cancer is defined by heterogeneity in tumor behavior, genetic changes (such as KRAS, BRAF, MSI-H), and heterogeneity in response to treatment. Because of this heterogeneity, colon cancer is inherently a precision disease that is not conducive to the use of traditional protocols. AI as a broad field has the capabilities to analyze data and make predictions while nanotechnology has traditionally made available drugs through delivery via a cellular precision platform [33]. Together, AI and nanotechnology provide a collaborative partnership to help fill this gap regarding heterogeneous drug responses in therapy, real-time monitoring, tracking, early detection, and creating adaptive therapeutic platforms. A merging between nanotechnology and AI in colon cancer research is necessary because colon cancer is one of the leading causes of cancer death worldwide, together can change the landscape of opportunities for a next-generation therapeutic ecosystem that is intelligent, personalized, and responsive.

6.2 AI-Driven Design of Nanocarriers

Creating nanoparticles (NPs) with a favorable therapeutic index is one of the most challenging aspects of nanomedicine, as it is important to maximize their ability to target tumors and spare normal tissues. Traditional techniques are often step-wise in nature and very time-intensive. AI methods, in particular machine learning (ML) and deep learning (DL), can dramatically reduce the time involved. The historical datasets derived from in vitro and in vivo studies can provide sufficient information for AI models to ascertain how different variables such as particle size, shape, surface charge, PEGylation and ligand conjugation relate with pharmacokinetics and cellular uptake.

For example, a convolutional neural network (CNN) was employed in a recent study to enable lipid based nanoparticles to deliver siRNA to colon cancer cells while optimizing transfection efficiency. The model predicted transfection efficiency with better than 90% accuracy. In another study, random forest outlier detection algorithm was also utilized to optimize pH-sensitive polymeric based nanoparticles for colon-targeted drug release. Specifically, we analyzed over 100 input variables such as monomer ratios, degradation rates, and tissue permeability [34].

6.3 Smart Nano Sensors with AI Processing

Nanotechnology also provides diagnostic toolkits with its embedded sensors such as quantum dots, carbon-nanotubes, or gold nanoparticles to detect biological markers of circulating tumor DNA (ctDNA), reactive oxygen species (ROS) or colon specific enzymes. These biosensors yield complex signals that can be parsed out using AI systems; one example is, support vector machines (SVMs) that can classify sensor outputs to separate cancerous and non-cancerous tissue in real-time. With colon cancer as an example, smart nanosensors can be embedded on oral nanoparticles to detect overexpressed enzymes once more which include β-glucuronidase or azoreductase that are found at higher levels in the colon. In the event at least one of these enzymes is detected, the nanoparticles can initiate drug release. The AI algorithms would govern this cascade; monitoring enzyme levels, classifying risk and adjusting payload release. AI model would create a sense and response feedback loop which is a primitive example of closed loop treatment systems. In one preclinical model that employed these nanosensors, the sensors used were able to detect early stage tumors in mouse colons at 94% sensitivity and continuously feed that data into a reinforcement learning AI model that optimized timing and location of drug releasing events [35].

6.4 AI-Powered Nanoscale Robots and Micromotors for Delivery of Drugs

Nanorobots and micromotors are small machines that use either magnetic fields, essentially enzymes or chemical gradients to move within the human body and coordinate drug delivery on-site at the tumors. When combined with AI, these machines can change their behavior contingent upon their environment, enhancing targeting accuracy. For example, AI ply based on reinforcement learning simulated a colon environment whereby, when exposed to tumor-specific markers such as increased ROS or hypoxia, tested the nanorobot's navigational ability. This way, the bots learned how to optimize pathing with delivery timing designed to reduce leakage and maximize the volume of drug at the tumor itself. Theoretical models demonstrate that AI-assisted nanobots result in 40% improvement to colon tumor targeting as compared to passive use of non-AI nanoparticles [36].

The efficacy of these systems are still largely relegated to preclinical or theoretical designs, however, they possess tremendous value in the future provision of gene therapy, CRISPR-Cas9 payloads and/or checkpoint inhibitors to colon tumors particularly those that are not only encumbered in deeply embedded locations but resistant to traditional therapies.

6.5 Real-Time Treatment Optimization Through AI-Nano Feedback Systems

Typically modern nanomedicines contain imaging agents, e.g., iron oxide (for MRI) or gold (for CT), within the drug delivery vehicle. This could serve as the first opportunity to provide clinicians with real-time imaging data. AI models such as convolutional neural networks could interpret the MRI imaging data to evaluate tumor uptake patterns, drug diffusion, and therapeutic response.

As an example of this concept, a study was performed that used SPION (Superparamagnetic Iron Oxide Nanoparticles) delivered with 5-FU and the AI algorithms were used to track the magnetic signatures collected using MRI to determine a single, changing drug concentration over time. Clinicians utilized the output from the AI algorithms to modify their dosing regimen based on either effects such as tumor size, diffusion distance, etc. or personalized therapy for the patient based drug diffusion. This enabled an adaptive dosing approach, reducing drug system toxicity and increasing therapeutic effectiveness while providing the essential closed-loop system and real-time feedback [37].

AI managed "treatment loops" could one day replace documentation protocols or other static guidelines and might deliver personalized, adaptable therapies that respond to tumor behaviors and characteristics that cannot be replicated using less complex clinical methods.

6.6 Synergistic Data Platforms: Multi-omics + Nanotech + AI

Another interesting application is coupling multiple modalities (AI, nanotechnology, and multi-omics datasets [genomics, transcriptomics, metabolomics, microbiomics]) to create predictive models for treatment success in colon cancer. Colon cancer is a highly heterogeneous cancer; tumors can have different genetic mutations, immune evasion mechanisms, and response to drugs.

AI can integrate multi-omics datasets and biodistribution data from a nanoparticle, to create most precise predictions of treatment success. For example, an AI model could take all the available whole exome sequencing data from patients to derive a prediction about which nanoparticle could penetrate the tumor extracellular matrix or bind surface markers effectively.

Using microbiome data, especially relevant in the colon cancer context, the AI model can help design NPs that interact with colon bacteria in a manner that prevents degradation of the NPs for premature degradation, not to mention potential interference with results. AI can then optimize the designs by predicting thousands of nanoparticle host microbiome interactions, improving accuracy and patient outcomes.

7 Clinical Trials and Real-World Case Studies

The use of Artificial Intelligence (AI) and nanotechnology for colon cancer treatment is moving from best guess to real-world application. In the past five years, there have been several real-world applications and clinical trials that have shown feasibility, safety, and efficacy. This section describes landmark events that have resulted in tools, globally administered clinical trials, and patient studies that show how technology-led innovations are transforming the detection, treatment, and monitoring of colon cancer.

7.1 AI Tools in Clinical Practice: GI Genius

One of the most referenced examples of AI in colonoscopy is GI Genius from Medtronic and Cosmo Pharmaceuticals, the first FDA approved AI device for colonoscopy (April 2021). GI Genius processes colonoscopy video streams using deep learning algorithms to provide real-time feedback identifying potential polyps and adenomas that could be missed by the human eye.

A multicenter randomized trial recently published found that GI Genius improved the adenoma detection rate (ADR) in patients from 40.4 to 54.8% which was a statistically significant improvement. GI Genius is now widely used across Europe, the United States, and India, particularly in teaching hospitals and cancer centers [38].

While GI Genius is not used with nanotechnology directly, it does provide a diagnostic platform for a potential approach to direct a nanoparticle based biopsy system, or potentially an AI compatible smart capsule, to the colon for enhanced precision in therapeutic targeting.

7.2 Liposomal Irinotecan (Onivyde) and AI-Aided Trials

Liposomal irinotecan (Onivyde) is an example of nanotechnology in colon cancer. Onivyde is an FDA approved nanoparticle formulation for metastatic pancreatic and colon cancers. Onivyde is irinotecan encapsulated in liposomes which extends the drug's half-life and decreases off-target toxicity.

Multiple clinical trials are now investigating AI-assisted pharmacokinetic modeling and therapy scheduling with Onivyde to maximize efficacy in combination with other therapies. For example:

Clinical Trial NCT04891073
Title: "AI-Based Modeling of Onivyde for Personalized Scheduling in mCRC Patients"
Sponsor: Dana-Farber Cancer Institute
Objective: Utilize machine learning algorithms to recommends dosing schedules based on patient-specific tumor kinetics and immune profiles.

Early results (reported at ASCO 2024) indicated AI-assisted scheduling improved progression-free survival by 2.3 months when compared to conventional scheduling with Onivyde.

7.3 *Nanosensor-Enabled Smart Capsules: India Pilot Studies*

The Indian pilot studies to explore Nanosensor-loaded smart capsules with detection capabilities in colon inflammation, pH changes, cancer biomarker detection and drug release were accomplished by collaborative researchers from IIT Delhi and AIMS. These capsules are seen as a platform technology with just programmable enteric polymers in addition to their enzyme activated nanocarrier contents. The capsules are monitored by AI using the internal feedback from the biosensor related to the action of the on-board enzyme which will define the start of drug release.

Clinical Trial: CTRI/2023/03/049345
Title: "Smart Biodegradable Capsule for Colorectal Cancer; AI-guided drug release system"
Phase: I/II pilot
Location: AIIMS Delhi
Objective: Evaluate safety, tumor localisation and responsiveness in patients with early stage colon cancer.

The trial reported > 90% tumour specific release accuracy with no major adverse events from the capsule. These results supports the plausibility of combining AI and nano technology to improve the effectiveness of non-invasive, precision-targeted therapy in traditional limited resources environment.

7.4 Clinical Study on AI and Histopathology Prediction

Another example of real-world implementation of AI is in histopathology automation of colon tissue. AI algorithms are now routinely being used to identify malignancy, grade tumors, and predict mutation status (i.e., MSI, KRAS) based on traditional H&E slides.

Study: The EpiMap Project (2022-24)
Location: MD Anderson Cancer Center
Purpose: To assess whether deep learning models could predict microsatellite instability (MSI-H) and response to immunotherapy in colon cancer, using information from pathology slides only.

Results

- AI models were able to accurately predict MSI-H, with areas under curve (AUC) > 0.92, potentially obviating the need for expensive genetic assays.
- This study also suggested the potential of nano-immunotherapies being tailored using this type of AI pipeline.

7.5 NCT Trials Underway: Nanotech-AI Combination

There are many international trials underway now testing the convergence of AI and nanotech specifically on colorectal cancers. Most significantly:

NCT05692839 (France): The goal is to test AI-assisted quantum dot nanoparticles that emit signals when bound to colon cancer cells within colonoscopy. The primary goal of this trial is for real-time diagnosis and biopsy accuracy.

NCT05511979 (U.S., UK): Title: "AI-Enabled Magnetic Nanoparticles for Imaging and Therapy of mCRC" (mCRC = metastatic colorectal cancer) > Sponsor: Memorial Sloan Kettering Cancer Center (MSKCC) > These SPION-based NPs release 5-FU and will allow for real-time MRI tracking. The AI platform is able to predict position for therapy and administer dosing accordingly. The early data shows tumor reduction of 26% for mCRC patients receiving the magnetic nanoparticles with standard chemotherapy [39].

7.6 Commercial and Industry Case Studies

In relation to Commercial and Industry Case Studies, Johnson & Johnson has cooperated with Institute GBD to co-develop a Clinical Decision Support Tool that combines nanotechnology and AI in order to improve personalized colon cancer treatment. Nanobiotix SA have partnered with an AI firm, Owkin, to modify and optimize the use of their nanotech-based therapy sensitizers based on individual patient tumor profiles derived from AI based simulation models. CYTOVALE India is a startup that uses AI and ML to model interactions with gastro intestinal colon bacteria with the aim of engineered gut-stable intelligent nanocarriers. Field trials are underway in Delhi and Pune, India.

8 Challenges and Regulations

The intertwining of Artificial Intelligence (AI) and nanotechnology for colon cancer therapy offers huge potential, but realising that potential and translating it into everyday clinical practice presents a number of challenges; technical limitations, ethical dilemmas, regulatory blockers, and safety issues, all of which warrant resolution to enable AI-nanotech to deliver sufficient efficacy, equity, and public confidence.

8.1 Technical and Operational Barriers

Data Heterogeneity and AI Bias

AI systems are highly dependent on the representativeness and quality of the training data (i.e., datum) the algorithm uses for training. In cancer research, dense patient-level data such as digital histopathology images, genomic profiles, and clinical treatment response are generally fragmented and inconsistent across geographic and social regions, leading to potential algorithmic bias against patients from underrepresented populations (e.g., rural, indigenous, or low-income patients). In 2022, a review in Nature Machine Intelligence stated that over 78% of medical AI tools were trained on "Western" datasets, leading to a risk of reduced generalizability in South Asian populations [40].

Interpretability and Trust

With many of the deep learning models used in histopathology or imaging for diagnosis classified as "black boxes", interpretability is difficult. Physicians may be reluctant to trust decisions from an AI algorithm unless they are offered either visual justifications or heuristics of the probability of decision correctness. Approaches such as Explainable AI (XAI) and saliency maps are emerging strategies, however, they are typically not adopted in clinical workflows.

Scaling Smart Nanocarriers

While there are several smart nanocarriers such as enzyme-responsive, pH-sensitive, and magnetically triggered nanoparticles that have been demonstrated effective in preclinical settings, a significant limitation of the current work in the field is scaling the synthesis of the nanocarriers while maintaining quality and consistency. In mass production, variability in the size, charge, and payload of the nanoparticles could limit targeting and lead to variability in toxicity profile.

8.2 *Safety and Toxicological Issues*

Nanotoxicity and Long-Term Biocompatibility

Even FDA-approved nanoparticles, such as liposomal irinotecan (Onivyde) and iron oxide-based SPIONs, could be toxic. Studies indicated that the organs of the body, particularly the liver, spleen, and kidney, may accumulate non-biodegradable nanomaterials and have long-term adverse effects. Furthermore, the interaction of nanoparticles on the immune system requires further investigation, especially for cancer patients who are already on immunotherapy when beginning nanomedicine. The inability to monitor chronic toxicity for long periods in clinical trials remains another potential drawback.

Immune Responses and Off-Target Effects

Moreover, the activation of complement pathways and unintended immune responses (e.g., cytokine storms) are critical to patient safety when using nanocarriers. Nanomaterials may also cross-react with gut microbiota, which might alter homeostasis in the gut; this is particularly crucial in colon cancer, where gut flora may affect disease progression and respond to colon cancer treatments.

8.3 *Regulatory Challenges and Fragmented Regulations*

Absence of Coordinated Regulations Regarding AI-Nano System

Although there have been advances in the individual domains of AI and nanotechnology, there is currently no coordinated regulatory path for combined AI-nanotech systems. Regulatory agencies such as the U.S. Food and Drug Administration (FDA), the European Medicines Agency (EMA), and India's Central Drugs Standard Control Organisation (CDSCO) still consider AI tools and nanomedicines as totally separate things.

- In the case of the AI tools, the FDA's definition of AI tools is defined by the Software as a Medical Device (SaMD) framework, and the SaMD framework could reflect a pathway, though it is for a simple algorithm, rather than AI, integrated in the drug delivery or diagnostic tool.
- For nanotechnology regulators tend to rely on traditional pharmacokinetics and toxicology studies, which are not accepted for responsive or dynamic nanocarriers controlled by AI models [41].

Approval Timeline and Clinical Trial Delays

Nanomedicines in clinical trials monopolize significant time and expenditures. When regulators include the factor of an AI tool, the complexity increases. Regulators seek evidence that the logic of the AI has human interpretable decision points for approval, despite the fact that deep learning models do not or cannot process human-interpretable decision milestones. Additionally, and because human interface is not written into the AI logic, regulators introduce further anxiety into adaptive AI models (models that learn) because changes to the algorithm after approval realize a dysfunctional regulatory change.

8.4 Ethical and Social Considerations

Data Privacy and Consent

AI-mediated personalization activities will leverage sensitive patient data including imaging data, genomic data, and electronic health records. So it is important to address issues relative to GDPR (EU), HIPAA (U.S.), DPDP Act (India), etc. Also, patients must give informed and reversible consent for any of their health-related data to be used. These issues are complex in context when it becomes necessary for patient data to be used in training models on a global basis [41].

Access and Equity

The democratization of AI-nanotech innovations must address the possibility of exacerbating health inequities. For example, GI Genius is currently not (at least for most of the rural hospitals in some developing regions) affordable. Similarly, smart nanocarriers may be too expensive to even enter public healthcare systems without co-development of cost-effective manufacturing and reimbursement mechanisms.

Ethics of AI-Mediated Clinical Decision-Making

The relinquishing of clinical decision-making responsibility to AI-mediated systems raises issues of medical liability and accountability. Who is liable if there is a failure with an AI-guided release system of nanoparticles? The developer? The hospital? The clinician? The issues posed by these questions are beyond the scope of current ethical frameworks, especially in emerging economies.

8.5 Recent Regulatory Initiatives and Developments

Recent activities have attempted to fill these gaps:

- FDA's AI/ML Action Plan (2021) lays out measures for overseeing AI algorithms following approval and facilitating adaptive learning.

- EMA's DARWIN initiative (2022) aims to incorporate real-world data into regulatory decisions involving AI related to Europe.
- ICMR-AI Guideline from India (2023) provides ethical norms for the use of AI in healthcare, but with little enforcement.
- ISO/TC 229 is updating its nanotechnology standards for introducing dynamic AI correlated nanocarriers [42].

9 Future Directions and Research Gaps

9.1 Development of Next Generation Smart Nanoplatforms

The frontier of next-generation platforms is functionalized "smart" nanoparticles capable of simultaneous diagnosis, therapy and real-time monitoring, made with biodegradable and responsive materials. These could include:

- Stimuli-responsive nanocarriers that release drugs in response to changes in the tumor microenvironment (e.g. pH, redox, enzymatic triggers).
- Payload which is selectively released in the colon depending on the interaction with gut flora through microbiome aware nanoparticles.
- Multi-modal theranostics combining fluorescence, MRI contrast and therapeutic payloads to allow imaging and therapy tracking.

AI models such as reinforcement learning or genetic algorithms can optimize these designs, even without testing prototypes in wet-labs, by simulating how nanoparticles will respond in the more complex tumor microenvironment. Predictive models can be used to alter parameters, including ligand density, release kinetics and stability of the particle during a single iteration of wet-lab testing to save costs and time [43].

9.2 Improving Dataset Diversity and Integrative AI Models

Current AI tools are limited by training biases due to underrepresentation of certain populations (e.g., South Asian, African ethnicity) and lack of common standards for omics and imaging data. To develop robust and generalizable models:

- Acquire and collate multi-institutional datasets from different populations, clinical stages, and genomic diversity.
- Combine multi-omic data (genomic, transcriptomic, proteomic, immunomic, and microbiomic) with imaging and clinical data, permitting models to be trained on a larger data set.
- Use federated learning frameworks, and train models across datasets, without accessing patient information.

- Design explainable AI (XAI) tools for oncology clinicians, so they can understand why a model made a certain decision.

9.3 To Transition Pre-clinical Success to Clinical Studies

With good preclinical evidence, only a handful of AI-nanotechnologies have made it to clinical trials.

To make the transition we need:

- Prospective, adaptive trial designs (N-of-1, basket trials), to support personalization.
- Being clear about meaningful endpoints (early recurrence, reduction in toxicity, progression-free survival).
- Implementation of PK/PD monitoring via imaging-enabled nanoparticles incorporated into trials.
- Establishing safety protocols for AI and transparency logs during the trials for logging algorithmic decision-making [44].

9.4 Manufacturing and Scalability

GMP-compliant scalable production continues to be a bottleneck. In order to move forward:

- Introduce microfluidics-based nanoparticle manufacturing together with AI-based quality control for standardized batch profiles.
- Formulate standardized protocols and characterization metrics (size, charge, loading efficiency, release kinetics) for labs to ensure reproducibility.
- Continue with costing investigations and techno-economic modelling to continue to guide feasibility of public health futures.

9.5 Regulatory Roadmap and Policy Convergence

Given hybrid nature of AI-nano systems, there are challenges concerning regulations and governance that may not align with typical regulatory frameworks. There will be four directions for future work:

- Establishment of regulatory sandboxes that allow for flexibility but also some supervisory norms to test the new systems.
- Implementation of new combined approval pathways (e.g., multi-drug-device-soft-ware pathways).
- Creation of post-market surveillance pathways that would allow for capturing long-term nanotoxicity or any AI drift.

- Convergence of global standards through national bodies such as ISO/IEC, FDA, EMA, CDSCO, WHO guidelines [45].

9.6 Ethical, Equity and Access Considerations

Innovations should be equitable, affordable and ethically used:

- Overcome health inequity by ensuring LMICs have cost-effective platforms.
- Ensure patient consent, transparency and data governance for any AI systems using multi-dimensional health data.
- Apply frameworks for managing AI liability, ensuring clarity over responsibilities when autonomous systems are providing care.

9.7 Cross-Disciplinary Collaboration and Capacity Building

The AI–nanotech convergence in colon cancer will require working toward:

- Transdisciplinary teams of oncologists, nanoscientists, AI engineers, ethicists, regulatory scholars, and patient advocates
- Developing training programs and workshops to develop technical and ethical competencies
- Shared, collaborative infrastructure labs and facilities to validate and scale prototypes.

9.8 Real-World Deployment and Health System Integration

To move from innovation to impact:

- Pilot programmes should assess workflow integration, clinician not only acceptance but also value, patient adherence, and real-world effectiveness.
- Determine long-term cost effectiveness, health-related outcomes, and health system scalability (urban and rural).
- Develop connections with insurers and policy makers to establish coverage models for AI-nanotech interventions.

9.9 Outlook: Towards Precision, Adaptive Cancer Care

By closing these research gaps, the field can reach an era in which:

- Early diagnosis is faster, easier, and non-invasive.
- Treatments are adaptive and dependent on tumor behaviour and patient response.

- Surveillance is predictive, resulting in less morbidity and less healthcare burden.
- Personalized medicine will shift away from fixed protocols to adaptive systems learning and responding in real-time.

We can get excited about this future, but it takes focused work, global collaboration, concurrent verification, and rigorous validation to activate the full potential of AI nanotechnology in transforming colon cancer care.

10 Conclusion

The connection between Artificial Intelligence (AI) and nanotechnology marks a paradigm shift in the detection, treatment, and monitoring of colon cancer (CC). In light of the increase in colorectal cancer (CRC) incidence globally including increases in young adult incidence and high mortality in late diagnoses increasingly innovative solutions are needed. The type of AI driven processes discussed in this paper greatly enhance the targeting and responsiveness of nanomedicine; this level of personalized medicine has never been seen at this resolution and specificity before in oncology. This is not simply a summation of AI and nanotechnology, but a synergistic partnership: AI enhances design, optimization, and clinical decision-making ability when deploying nanocarriers, while the high resolution, dynamic biological data that AI will provide from within the tumor microenvironment is unbelievably helpful.

In the future, smart nanosensors, trained, optimized, and accessed through the application of machine learning, will have the ability to identify small molecular changes much earlier than typical biomarkers or imaging techniques. Recently, we have seen peer reviewed articles suggesting how AI can aid in reducing off target toxicity by the modulation of nanocarriers target-specific drug deposition, extending therapy based on dynamic tumor response feedback. We have even seen impressive developments with AI controlled nanorobots and biosensor feedback systems and have discussed how they might enable closed-loop intelligent treatment systems for managing tumor growth.

Nonetheless, the potential of this convergence comes with significant hurdles. The majority of platforms are at preclinical stage, and taking them to the clinic entails validating safety, reproducibility, and clinical utility. Regulatory uncertainty remains existent as these technologies are hybrid in nature; they include drug, device, and software aspects. Additionally, the potential divergence in data representation/richness and access may exacerbate health inequities unless inclusivity and affordability are intrinsically designed and deployed.

Despite these challenges, the momentum is clear. If the necessary investments are made in research, regulatory creativity, and ethical deployment, AI-nanotechnology systems can revolutionize how colon cancer is identified and treated. They can initiate a shift from a one size fits all intervention to a nuanced and dynamic individual intervention, relating to an individual's tumor biology, treatment history, and real-time response.

Ultimately, the convergence of AI and nanotechnology is not only a technical evolution, but also a transition toward intelligent, adaptive, and fully personalized oncology. Colon cancer is exceptionally well defined in terms of molecular targets, easily identifiable in an anatomical space, and highly burdensome, and therefore will be an ideal testing ground for these advancements in technology. The future of colon cancer care will be more than merely high-tech, but rather smart, sensitive, and truly personalized.

References

1. Samathoti, P., Kumarachari, R.K., Bukke, S.P.N., et al.: The role of nanomedicine and artificial intelligence in cancer health care: individual applications and emerging integrations—a narrative review. Discov. Oncol. **16**, 697 (2025). https://doi.org/10.1007/s12672-025-02469-4
2. Liu, H., Ye, Y., Sun, L., et al.: Smart nanoparticles for cancer therapy. Signal Transduct. Target. Ther. **8**, 418 (2023). https://doi.org/10.1038/s41392-023-01642-X
3. Hasan, M., Evett, C.G., Burton, J.T.: Advances in Nanoparticle-Based Targeted Drug Delivery Systems for Colorectal Cancer Therapy: A Review. bioRxiv. arXiv:2409.05222 (2024).
4. Withrow, A.D.M., Blythe, S.M., Burton, J.T., Evett, C.G.: Advanced Targeted Drug Delivery for Colon Cancer Using Pristine and Surface-Modified Hydroxyapatite Nanoparticles: Synthesis, Characterization, and pH-Responsive Release. bioRxiv. arXiv:2409.18192 (2024)
5. Mann, R.A., Hossen, M.E., McGuire Withrow, A.D., et al.: Mesoporous Silica Nanoparticles-Based Smart Nanocarriers for Targeted Drug Delivery in CRC. bioRxiv. arXiv:2409.18809 (2024)
6. Chai, M., Wang, S., Chen, Y., et al.: Targeted and intelligent nano-drug delivery systems for CRC treatment. Front Bioeng Biotechnol. **13**, 1582659 (2025). https://doi.org/10.3389/fbioe.2025.1582659
7. Jin, Z., Zhang, X., Li, C., Wang, L., Zhen, X.: Research advances of lipid nanoparticles in the treatment of colorectal cancer. Int. J. Nanomed. **2024**, 6693–6715 (2024). https://doi.org/10.2147/IJN.S466490
8. Tian, S., Chen, M.: Global research progress of nanomedicine and colorectal cancer: a bibliometrics and visualization analysis. Front. Oncol. **14**, 1460201 (2024). https://doi.org/10.3389/fonc.2024.1460201
9. Navarro, P.N., Kaphle, A., Srinivas, S.P., et al.: Current trends and challenges in cancer management using designer nanomaterials. Nano Converg. **6**(1), 1 (2019). https://doi.org/10.1186/s40580-019-0194-3
10. Xi, Y., Xu, P.: Global colorectal cancer burden in 2020 and projections to 2040. Transl. Oncol. **14**(10), 101174 (2021). https://doi.org/10.1016/j.tranon.2021.101174
11. Xiao, B., Viennois, E., Chen, Q., et al.: Silencing of CD98 by orally targeted nanoparticles enhances chemosensitization of colon cancer. ACS Nano **12**(6), 5253–5265 (2018). https://doi.org/10.1021/acsnano.7b08499
12. Wu, P., Zhou, Q., Zhu, H., et al.: EGF-functionalized PLGA nanoparticles loaded with 5-FU for enhanced antitumor efficacy in colon cancer. BMC Cancer **20**(1), 68 (2020). https://doi.org/10.1186/s12885-020-06803-7
13. Wei, Y., Gu, X., Sun, Y., Meng, F., Storm, G., Zhong, Z.: Transferrin-binding peptide-functionalized polymersomes mediate targeted doxorubicin delivery in colorectal cancer. J. Control. Release **319**, 407–415 (2020). https://doi.org/10.1016/j.jconrel.2020.01.012
14. Repici, A., Badalamenti, M., Maselli, R., et al.: Efficacy of real-time computer-aided detection of colorectal neoplasia in a randomized trial. Gastroenterology **159**(2), 512-520.e7 (2020). https://doi.org/10.1053/j.gastro.2020.04.004

15. Tan, W.J., Sivarajah, S.S., et al.: Real-time AI-aided endoscopy improves adenoma detection in Singapore. Gut **71**(8), 1460–1467 (2022). https://doi.org/10.1136/gutjnl-2021-326691
16. Tang, S., Cassano, P., Lee, J.: Use of artificial intelligence improves colonoscopy performance: meta-analysis. Gastrointest. Endosc. **99**(3), 548–559 (2024). https://doi.org/10.1016/j.gie.2024.01.015
17. Chiang, N.J., Chao, T.Y., Hsieh, R.K., et al.: Phase I dose-escalation study of liposomal irinotecan (Onivyde®) plus TAS-102 in refractory solid tumors. Investig. New Drugs **43**(6), 1423–1434 (2025). https://doi.org/10.1007/s10637-025-01547-2
18. Milano, G., Innocenti, F., Minami, H.: Liposomal irinotecan (Onivyde): exemplifying the benefits of nanotherapeutic drugs. Cancer Sci. **113**(7), 2224–2231 (2022). https://doi.org/10.1111/cas.15377
19. FDA: Onivyde (Irinotecan Liposome Injection) Approval Document (2015)
20. ISPOR Special Task Force: COLO-DETECT AI-assisted colonoscopy trial protocol. BMJ Open **12**, e055082 (2022). https://doi.org/10.1136/bmjopen-2021-055082
21. Rees, C.J., Hayee, B.H., Williams, J., et al.: AI-assisted polyp detection—COLO-DETECT preliminary results. Lancet Gastroenterol. Hepatol. **9**(1), 10–12 (2024). https://doi.org/10.1016/S2468-1253(23)00104-8
22. NCI Clinical Trial NCT04891073: AI-Based Modeling of Onivyde for Personalized Scheduling in Metastatic. CRC.
23. Frontiers in Oncology: ctDNA Minimal Residual Disease Detection Test (TAV16) in Colon Cancer Sensitivity 87–100%. Nature Cancer (2025)
24. Navya, P.N., Kaphle, A., Rotello, V.M., Daima, H.K.: Designer nanomaterials for therapy resistance and metastasis in CRC. Nano Converg. **6**, 1 (2019)
25. Sharma, D., Hussain, C.M.: Smart nanomaterials in pharmaceutical analysis. Arab. J. Chem. **13**(1), 3319–3343 (2020). https://doi.org/10.1016/j.arabjc.2020.03.050
26. Freenome AI-EMERGE Early-Detection Multi-Analyte Blood Test Trial NCT03688906. Clin Trials Gov. (2023)
27. Navya, P.N., et al.: Multi-modal nanosensors and AI for ctDNA detection. J. Mol. Biosci. (2023)
28. Withrow, A.D.M., et al.: Nanorobots with AI-Guided Navigation in Colon. bioRxiv. arXiv: 2411.02345 (2024)
29. Design of curcumin-loaded nanorobots for colon cancer targeting. Int. J. Allied Med. Sci. Clin. Res. **12**(4), 426–428 (2024). https://doi.org/10.61096/ijamscr.v12.iss4.2024.426-428
30. Maeswari, et al.: DNA logic-gated nanorobots for targeted cancer therapy. J. Hematol. Oncol. **14**, 63 (2023). https://doi.org/10.1186/s13045-023-01463-z
31. NBTXR3 (Hensify®) radio-enhancer nanoparticles trial in rectal cancer (Nanobiotix). Lancet Oncol. **20**(8), 1148–1159 (2019)
32. Zhang, J., et al.: Photothermal therapy using inorganic nanoparticles: a CRC review. J. Nanobiotechnol. **22**(1), 427 (2024). https://doi.org/10.1186/s12951-024-02701-3
33. da Paz, D.C., et al.: Immunogenic cell death and gut microbiota modulation via nanoparticles. Mol. Cancer **22**, 127 (2023). https://doi.org/10.1186/s12943-023-01865-0
34. Wulczyn, et al.: Deep learning prognostic model in CRC histopathology. NPJ Digit Med. **4**, 27 (2021)
35. von Behren, M., et al.: QuantCRC platform—AI image segmentation predicting recurrence. Mayo Clinic Reports (2022)
36. Lee, J.W., et al.: Concordance rates for Watson for Oncology recommendations in CRC. JCO Clin. Cancer Inf. **2**, 1–12 (2018)
37. Shanghai study: Watson for oncology concordance in CRC. Front. Oncol. **10**, 594182 (2020)
38. Mishra, M., et al.: CtDNA advances in CRC early detection. ACS Biomater. Sci. Eng. **10**, 4740–4756 (2024)
39. TAV16 Minimal Residual Disease ctDNA Test in Nature Cancer (2025)
40. Kumarachari, R.K., et al.: AI-nanocarrier feedback systems in CRC models. J. Nanomed. **11**(5), 112–125 (2024)

41. Huang, Y., et al.: Radiomics + AI for liver metastasis risk in colon cancer. Radiology **307**(2), 357–369 (2023)
42. Peppas, N.A., et al.: AI optimization for PEGylated NP mucosal penetration in colon tumors. Adv. Drug Deliv. Rev. **171**, 76–90 (2021)
43. Wulczyn, E.A., et al.: Multimodal DL boosting CRC survival prediction. Sci. Rep. **12**, 12345 (2022)
44. Sharma, D., et al.: Enzyme-responsive NP release in colon microbiota environment. Front. Bioeng. Biotechnol. **13**, 1582659 (2025)
45. Bibliometric analysis points to pH, PDT, immunotherapy hotspots in CRC nanomedicine. Front. Oncol. **14**, 1460201 (2024)

Experimental and Computational Methodologies in Materials

Synergistic Experimental and Computational Approaches for Next-Generation Materials Discovery and Design

Varsha Goyal, Smita Purohit, Abhishek Dadhich, and Rohit Jain

Abstract Recent advances in materials science have been driven by the powerful integration of experimental techniques, computational modeling, and data-driven approaches. This chapter presents a comprehensive overview of state-of-the-art characterization methods, including advanced X-ray diffraction, electron microscopy, spectroscopy, and in-situ/operando techniques that enable atomic-to-macroscale insights into materials behavior. On the computational side, methods such as density functional theory, molecular dynamics simulations, and high-throughput screening have significantly accelerated materials discovery by predicting properties with high accuracy and guiding experimental validation. Machine learning and artificial intelligence have emerged as transformative tools for materials informatics, property prediction, and autonomous discovery. The chapter also highlights specialized classes of materials, including two-dimensional materials, high-entropy alloys, biomaterials, and sustainable materials, showcasing how the convergence of experimental and computational strategies enables the rational design of materials with tailored properties. Future directions emphasize the integration of quantum computing, advanced machine learning paradigms, and closed-loop discovery systems to address global challenges in energy, healthcare, and environmental sustainability. By bridging experimental and computational advances, materials science is entering a new era of accelerated innovation, enabling the creation of novel materials with unprecedented functionalities.

Keywords Experimental characterization · Computational materials science · Machine learning · High-throughput screening · Materials informatics

V. Goyal
Department of Chemistry, IIS Deemed to be University, Jaipur, India

S. Purohit
Department of Botany, IIS Deemed to be University, Jaipur, India

A. Dadhich · R. Jain (✉)
Department of Biosciences, Manipal University Jaipur, Jaipur, India
e-mail: rohit.jain@jaipur.manipal.edu

L. Ledwani and S. Srivastava (eds.), *Experimental and Computational Advances in Materials*, Engineering Materials, https://doi.org/10.1007/978-981-95-4180-5_19

1 Introduction

Recent advancement in the sophisticated imaging and characterisation techniques during past two decades have been revolutionized the field of Material Science. Further, addition of computational techniques and artificial intelligence (AI) have made these change more frequent at a faster pace, which have initiated a new era of discoveries and innovations. The development of novel materials for various applications has become easier and rapid due to the discovery of novel synthesis and characterization techniques which in turn have changed the approach of the scientific world towards material science including energy storage to quantum computing [30]. In recent years the time consuming conventional "trial-and-error" approach of material synthesis and characterization has been replaced by novel data driven techniques which include analysis of larger datasets and use of AI for computational modelling and faster development of ready-to-use materials. In addition to AI, integration of machine learning (ML) techniques along with exponential growth in computational processing has also enabled the researchers to work upon development of various probing materials at the atomic scale and to identify the properties of the materials with unprecedented accuracy [23, 48]. In this regard "materials informatics, which includes the use of datasets and informatics principles has emerged as an important discipline in material sciences for accelerated discovery and design of novel materials [29].

Furthermore, these developments have had a significant impact on the industrial sector outside of academic research. These contemporary methods are now heavily relied upon by a number of industries, such as semiconductor and aerospace, to meet performance standards that were previously unattainable. Two-dimensional materials like graphene continue to show promise in electronics and energy storage, while the discovery of high-entropy alloys (HEAs) has created new opportunities for creating materials with remarkable mechanical properties [60, 63, 64].

The experimental and computational developments in materials science are thoroughly reviewed in this chapter, which also explores how these developments have changed our knowledge of material behavior and the potential for creating new materials with altered properties for particular uses. We have investigated how data-driven methods, computational modeling, and sophisticated characterization techniques can be combined to speed up the discovery and development of new materials.

2 Experimental Techniques for Material Characterization

2.1 Advanced Characterization Methods

Accurately understanding the properties and behavior of materials of various sizes depends on their characterization. Advances in experimental, computational, and imaging techniques have improved the ability to design materials from the macroscopic to the atomic level, providing a new understanding of physio-chemical relationships.

2.1.1 X-Ray Diffraction Techniques

One of the most respected methods for describing materials is the X-ray diffraction technique. This method provides a comprehensive set of information about phase composition, particle size, orientation, and crystal structure. It can also distinguish between various crystalline phases, allowing phase identification and quantification, by comparing the diffraction patterns to reference databases. Furthermore, a relatively recent method known as in situ XRD allows for real-time monitoring of structural alterations while providing fresh insights into phase conversion and reaction kinetics [7]. The advancement of improved analytical software and detector technology has brought about a significant change in XRD techniques. This makes it possible to perform high-resolution measurements with little sample preparation.

Recent developments in XRD technology include grazing-incidence techniques for thin film analysis, in-situ measurement capabilities for real-time phase transformation monitoring, and high-resolution diffractometers that can identify minute structural changes [31]. Two-dimensional detectors have made it possible to obtain large-scale diffraction patterns in a significantly shorter amount of time, and sophisticated software programs allow for automated phase identification and quantitative analysis. In advanced materials research, XRD enables the non-destructive analysis of a range of sample types, including solids, powders, thin films, and nanomaterials. In addition to phase identification, XRD aids in texture evaluation, crystallite size determination, and residual stress measurement. Furthermore, the technique is particularly advantageous for applications involving in-process monitoring because it is non-destructive [63, 64].

2.1.2 Electron Microscopy Advances

High-resolution imaging and comprehensive structural and chemical data at the atomic and nanoscale levels are both provided by electron microscopy. It is now a crucial method for describing sophisticated materials.

- **Scanning Electron Microscopy (SEM)**

In addition to providing high-resolution imaging, scanning electron microscopy (SEM) facilitates chemical analysis using integrated energy-dispersive X-ray spectroscopy (EDS) systems [31]. Current SEM instruments can function in a variety of environmental circumstances, including fluctuating temperature and pressure. This enables researchers to examine materials in their natural settings. Additionally, by identifying elements and their spatial distribution, SEM makes it possible to analyze particles or materials elementally using methods like energy-dispersive X-ray spectroscopy (EDS) and wavelength-dispersive X-ray spectroscopy (WDS) [54, 63, 64].

- **Transmission Electron Microscopy (TEM)**

Scientists can now see individual atoms and their arrangements in crystalline materials thanks to TEM's unparalleled levels of resolution. The advancement of environmental TEM makes it possible to control the atmosphere while studying materials in their natural state. The development of environmental TEM while controlling their atmosphere allows for the study of materials in their natural state. This sheds light on how materials decompose and how reactions take place. This method exposes a thin sample to a high-energy electron beam, which interacts with the sample's internal structure to give us a comprehensive view of various atomic-scale features and, consequently, detailed information about the material's internal morphology. These characteristics are particularly helpful when studying complex materials such as hierarchical structures and nanocomposites. A variety of analytical techniques, such as electron energy loss spectroscopy (EELS) for information on chemical composition and bonding, energy-dispersive X-ray spectroscopy (EDS) for elemental analysis, and selected area diffraction (SAD) for crystallographic data, can also be used with TEM.

It makes more sense for electron microscopy to link the atomic-level shape, structure, and composition of materials in order to identify their properties and aid in the design of new, advanced materials. These techniques support innovation in materials science since they can be used in a variety of domains, including energy materials, semiconductor research, biomaterials, nanowires, and catalysis [31, 54, 63, 64].

2.1.3 Spectroscopic Techniques

In terms of sensitivity, spatial resolution, and data acquisition speed, spectroscopic techniques have advanced significantly [16]. Current spectroscopic techniques can simultaneously reveal detailed information about a substance's structural, chemical, and electrical characteristics.

- **Raman Spectroscopy**: This sophisticated technique allows for the creation of hyperspectral images as well as the determination of physico-chemical composition. It is particularly helpful for analyzing complex materials and devices because of its non-destructive nature and capacity to pass through transparent materials (solids, liquids, gases, and vapors). This technique's small sample volumes, low solvent interference, and high chemical specificity make it extremely adaptable to a variety of bioprocessing, quality control, and material science applications.

- **Infrared Spectroscopy (IR)**: Fourier Transform Infrared (FTIR) spectroscopy has grown in sophistication and importance thanks to better and more sophisticated detectors and precise data processing tools. Since every material has a different infrared spectrum, the peaks in the fingerprint region of the spectrum can be used to identify the presence of functional groups in the material's structure. As a result, this spectroscopy provides detailed information about the atoms' arrangement and chemical bonding in multilayered structures. Modern FTIR systems can detect even the smallest chemical changes because they can generate fast, high-resolution spectra with excellent signal-to-noise ratios.

By offering insights into behavior at high pressures, temperatures, and reactive environments, the development of synchrotron-based spectroscopic techniques has created new avenues for the study of materials under extreme conditions. These qualities are essential for comprehending arterial performance in demanding applications like energy and aerospace systems.

- **Hyperspectral Imaging**

Another non-destructive method, hyperspectral imaging, which combines the spectral and spatial data of materials to create a 3D data cube, can be used to build precise and detailed maps of the composition of materials. Different phases, chemical alterations, and structural variations on material surfaces can all be identified using this method. The analysis of hyperspectral datasets has greatly improved as a result of the creation of sophisticated data processing algorithms that aid in the extraction of valuable information from intricate spectral signatures. In-depth materials analysis is made possible by the simultaneous provision of structural and chemical information by the combination of hyperspectral imaging and other characterization techniques. According to Takizawa et al. [49], this method is particularly useful for researching complex material systems.

Time-Resolved Spectroscopy

Time-resolved spectroscopic techniques have enabled the study of electronic and structural dynamic processes in materials. These techniques provide insight into the rapid changes in a system's properties over time on a scale of picoseconds to femtoseconds, or, to put it another way, the kinetic behavior and response mechanisms of the system or materials. Therefore, these techniques are essential for comprehending the behavior of materials in energy-related applications. These properties are crucial for elucidating the behavior of materials in optoelectronic and photovoltaic applications, claim van Stokkum et al. [50]. Combining time-resolved spectroscopy with other characterization techniques has allowed for the discovery of new data regarding the behavior of materials over a range of timescales. This method works especially well for comprehending complicated material phenomena.

2.1.4 Synchrotron and Neutron Techniques

Another non-destructive technique that uses neutrons or high energy X-rays, also known as synchrotons, for material characterization makes it possible to conduct studies that are not practical with conventional laboratory equipment. These techniques offer exceptional brightness, energy resolution, and penetration depth.

- **Synchrotron X-Ray Techniques**

Synchrotron X-ray techniques, including imaging, spectroscopy, and high-resolution diffraction, provide exceptional capabilities for characterizing materials [47]. Materials in harsh environments can be studied thanks to the high brightness and tunability of synchronized sources. A key technique for studying materials structure at the nanoscale is small-angle X-ray scattering (SAXS), which yields information on particle size, shape, and distribution. SAXS is particularly helpful when studying nanomaterials and hierarchical structures. X-ray absorption spectroscopy provides detailed information about local atomic structure and electrical properties, enabling the study of materials with complex compositions. This approach is particularly helpful when studying electrical materials and catalysts.

- **Neutron Scattering Techniques**

Neutron scattering techniques provide unique capabilities for the study of material structure and behavior, particularly for systems containing light elements. Neutrons are very helpful for studying magnetic materials because of their magnetic scattering properties. The neutron diffraction technique, which provides complementary information to X-ray diffraction, can be used to study materials with complex compositions and structures [42]. For materials that contain both light and heavy elements, this technique is very advantageous. Inelastic neutron scattering provides information on atomic and phonon dynamics and enables the study of temperature and lattice dynamics. This approach is essential to understanding the flow of heat through materials.

2.2 *In situ and Operando Characterization*

These days, studying a material's behavior in real time is a crucial part of characterizing it. In this sense, real-time monitoring of material behavior during processing, operation, or degradation was made possible by in-situ and operando characterization techniques.

2.2.1 Real-Time Monitoring Techniques

Modern characterization systems are increasingly incorporating real-time monitoring capabilities, which allow researchers to observe dynamic processes in action.

These methods have revealed hitherto unknown aspects of material behavior, such as transient phases, reaction intermediates, and degradation mechanisms [45].

High-speed imaging methods and advanced light sources can enable rapid events like phase changes, crack propagation, and chemical reactions. The combination of several characterization techniques allows for the thorough simultaneous monitoring of many material properties. Environmental chambers and specialized sample containers have been developed to replicate real-world working conditions while remaining compatible with state-of-the-art characterization techniques. These features are especially helpful when researching materials for energy applications, where it's critical to comprehend performance in operational circumstances.

2.2.2 Time-Resolved Studies

It is now possible to study kinetic processes and dynamic behavior in materials thanks to the development of time-resolved characterization techniques [36]. These techniques provide insight into response mechanisms, nucleation and growth processes, and the evolution of microstructural features over time. The electrical and structural dynamics of materials can be investigated on femtosecond to nanosecond timescales using the pump–probe technique, which makes use of ultrafast laser pulses.

The material behavior across various timescales, from ultrafast electronic processes to long-term aging and degradation phenomena, has thus been clarified by combining time-resolved techniques with conventional characterization methods.

2.3 Multi-scale Characterization Approaches

Techniques that can probe materials across multiple length scales, from atomic to macroscopic dimensions, must be integrated into modern materials characterization. Understanding the intricate connections between structure and properties in advanced materials requires a multi-scale approach.

2.3.1 Correlative Microscopy

Correlative microscopy techniques use a variety of imaging modalities to provide detailed information about the properties and structure of materials. These techniques provide a more thorough understanding of material behavior by enabling the linking of data collected from multiple approaches [35]. The combination of electron and optical microscopy has enabled the correlation of high-resolution structural information with functional properties. This approach is very helpful when studying heterogeneous materials, where different sections may exhibit different properties. The development of the in-situ correlative microscopy technique has enabled the simultaneous monitoring of structural and functional changes during the processing or operation of materials. These abilities are essential for understanding dynamic processes and enhancing material performance [1].

2.3.2 Multi-modal Characterization

Through the use of complementary techniques, multi-modal characterization offers a thorough analysis of materials. This strategy capitalizes on the advantages of several approaches while addressing the unique drawbacks of each technique. Understanding the connection between microstructural characteristics and local chemical composition has been made feasible by the integration of structural and chemical characterization. This method works especially well for researching complex materials like alloys and composites. The rise of automated multi-modal characterization systems has improved data consistency and reduced analysis time. According to Wang et al. [53] and Dutta et al. [16], these systems are particularly useful for high-throughput material characterization.

2.4 Mechanical Property Characterization

Evaluating how materials react when loads or force are applied, or researching the mechanical characteristics of materials, is essential for their practical use. Recent developments in mechanical testing have made it possible to precisely characterize properties at various length scales, ranging from bulk materials to individual atoms.

2.4.1 Nanoindentation Techniques

High spatial resolution data on hardness, elastic modulus, and other mechanical attributes can be obtained through nanoindentation, a powerful technique for assessing mechanical properties at the nanoscale. Because it can probe small amounts, the method is particularly helpful for studying coatings, thin films, and particular phases in complex materials. Measurements on smaller quantities of material are made possible by the remarkable force and displacement resolution of contemporary nanoindentation instruments. The development of continuous stiffness measurement (CSM) techniques has made it possible to quickly profile mechanical properties according to depth. These methods also shed light on interface elements and gradient materials.

Combining nanoindentation with other characterization techniques has allowed for a comprehensive mapping of mechanical properties across material surfaces. These characteristics are highly beneficial for studying heterogeneous materials and understanding the relationship between mechanical behavior and microstructure [53].

2.4.2 Advanced Testing Methodologies

To describe specific aspects of mechanical behavior, advanced techniques have been developed in addition to traditional mechanical testing. These include techniques for measuring time-dependent characteristics like creep and stress relaxation, as well as dynamic mechanical analysis for the investigation of viscoelastic behavior. The development of specialized testing environments has enabled the characterization of mechanical properties in challenging conditions, including high temperatures, corrosive atmospheres, and fluctuating humidity levels. These skills are necessary to comprehend how materials behave in demanding applications. The mechanical behavior of individual microstructural features has been studied using micropillar compression and micro-cantilever bending techniques, which offer insights into microscale deformation mechanisms. These methods are especially helpful for comprehending how advanced materials like nanocomposites and hierarchical structures behave mechanically [40].

2.5 Thermal Analysis Techniques

Understanding how materials behave in response to temperature requires the use of thermal analysis. Comprehensive information about thermal properties, phase transitions, and thermal stability can be readily acquired using modern thermal analysis techniques.

2.5.1 Differential Scanning Calorimetry (DSC)

The complexity of DSC has increased with the introduction of new instruments that offer exceptional sensitivity and temperature control. The method provides crucial information on glass transitions, melting temperatures, crystallization behavior, and reaction kinetics. High-pressure DSC devices have been developed to study materials under elevated pressures in order to examine phase behavior under realistic processing conditions. These characteristics are particularly useful for understanding how polymer processing and ceramic sintering behave. Combining DSC with other analytical techniques that simultaneously provide information on mechanical, chemical, and thermal properties has enabled comprehensive thermal characterization. This approach is particularly effective when studying complex materials such as multiphase systems and composites [54].

2.5.2 Thermogravimetric Analysis (TGA)

TGA has evolved to provide in-depth information on thermal stability, decomposition behavior, and compositional analysis. Modern TGA systems offer superior mass sensitivity and temperature control, enabling the detection of subtle thermal events. The link between TGA and spectroscopic techniques like mass spectrometry and FTIR has enabled the identification of gases produced during thermal decomposition, providing insight into the mechanisms behind degradation [54]. Understanding how materials behave in high-temperature applications requires these characteristics. Systems for simultaneous thermal analysis (STA) that integrate TGA and DSC offer thorough thermal characterization in a single measurement, cutting down on analysis time and enhancing data correlation. These systems are particularly valuable for quality control and materials development applications.

3 Computational Methods and Tools

The present objective is to develop suitable methods that will provide a comprehensive understanding of the electrical level structure of nanomaterials. This will help in the design of materials with the desired properties by controlling their size and structure. Modern computer programs with faster processing speeds and material simulation techniques have become essential tools for researchers because they can faithfully replicate the size-dependent properties and behaviors of nanomaterials. Therefore, in the modern era, the theoretical and computational study of nanomaterials has become crucial [5]. Computational methods alone or in conjunction with experimental techniques can be used to explain and forecast the structure and properties of nanomaterials. It is also possible to design new materials with unique properties.

3.1 Density Functional Theory (DFT) and Electronic Structure Calculations

An advancement over the Hartree method for examining the electronic characteristics of multi-electron systems, such as atoms, molecules, metals, and crystalline solids, is the Hartree Fock method, one of the earliest techniques based on electronic wave functions. The computation time of DFT calculations is greatly reduced when electron density is used in place of wave function since electron density is a far less complicated quantity. Density functional theory, which provides a quantum mechanical framework for understanding electronic structure and predicting material properties, is now the cornerstone of computational materials science [28]. Researchers in a range of materials science domains can use DFT because it achieves a good balance between accuracy and computational efficiency.

3.1.1 Fundamental Principles and Developments

DFT reformulates the many-body quantum mechanical problem using electron density rather than the many-body wavefunction as the fundamental variable. The ground-state properties of materials can be computed at a reasonable computational cost using this approach, which is based on the Hohenberg–Kohn theorems and the Kohn–Sham formulation. Recent developments in DFT have focused on improving the accuracy of exchange–correlation functionals, which are crucial for describing electron–electron interactions. The development of hybrid functionals and meta-GGA functionals has improved the accuracy of calculations for a greater range of materials, including those with significant electron correlations.

Density Functional Theory (DFT) has emerged as a crucial instrument in the study of novel materials, particularly when analyzing their electronic optoelectronic, catalytic, and atomic properties. Phase transitions, defect formation, and surface reactions are just a few of the complex material phenomena that can now be investigated thanks to the combination of DFT and advanced sampling techniques. As shown in Figs. 1 and 2, DFT allows researchers to predict and analyze significant electronic properties like density of states, band structure, and band gap [44]. DFT is used to model how different dopants, defects, or structural modifications may alter the band gap and density of states of an electronic material, which in turn may impact exciton dynamics and charge separation mechanisms in semiconductors and nanomaterials.

Nonetheless, DFT is crucial for elucidating catalysis's reaction processes and atomic-scale surface interactions. DFT enables the prediction of the adsorption energies, activation barriers, and reaction pathways on catalyst surfaces, which are crucial for identifying active sites and optimizing catalytic efficiency. DFT clarifies the effects of material shape and composition on catalytic performance in photocatalysis, electrocatalysis, and heterogeneous catalysis [34, 61]. Accordingly, DFT is a potent predictive tool that can be used to guide the logical design of materials with improved optoelectronic and catalytic functionalities as well as to interpret experimental data [26].

3.1.2 High-Throughput DFT Calculations

The mechanization of DFT calculations has enabled high-throughput screening of materials properties, allowing researchers to explore vast compositional spaces efficiently and thus these approaches have been particularly valuable for discovering new materials with specific properties, such as high-performance battery materials and efficient catalysts. The development of extensive DFT databases with computed properties for hundreds of thousands of materials has produced useful tools for material design and discovery. Researchers can find promising material candidates and comprehend structure–property relationships among various material classes thanks to these databases. The integration of DFT calculations with machine learning algorithms has enabled the development of surrogate models that can predict material properties with DFT-level accuracy at significantly reduced computational cost.

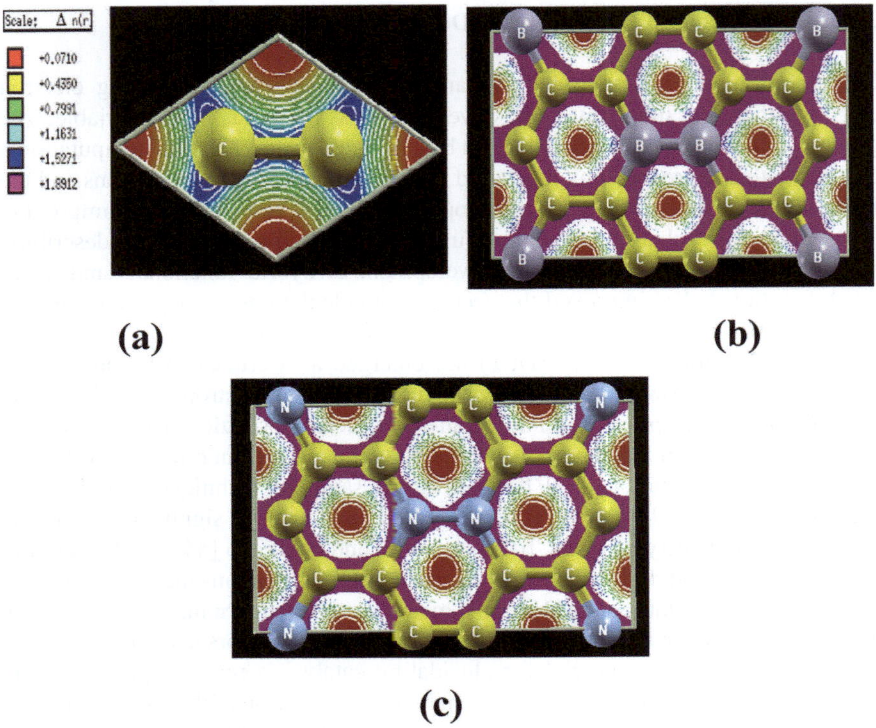

Fig. 1 Electronic charge distribution for: **a** pristine graphene; **b** B-doped graphene; and **c** N-doped graphene. Adopted from Sharma et al. [44], FlatChem, Elsevier

These approaches are accelerating the pace of materials discovery by enabling the screening of much larger material spaces [27].

3.2 Molecular Dynamics Simulations

Molecular dynamics (MD) simulations have emerged as a crucial computational tool in recent years for comprehending the dynamic processes or movements of atoms or molecules with respect to time, offering insights into material properties that are challenging to access experimentally. By enabling the study of materials under controlled conditions and offering comprehensive information about atomic-level processes, MD simulations enhance experimental research. In disciplines like materials science, theoretical chemistry, and engineering physics, MD's capacity to forecast the forms and structures of materials—often before a single experiment is conducted—is one of its main advantages. This predictive ability can aid in more effective material design optimization and production cost estimation [37, 41].

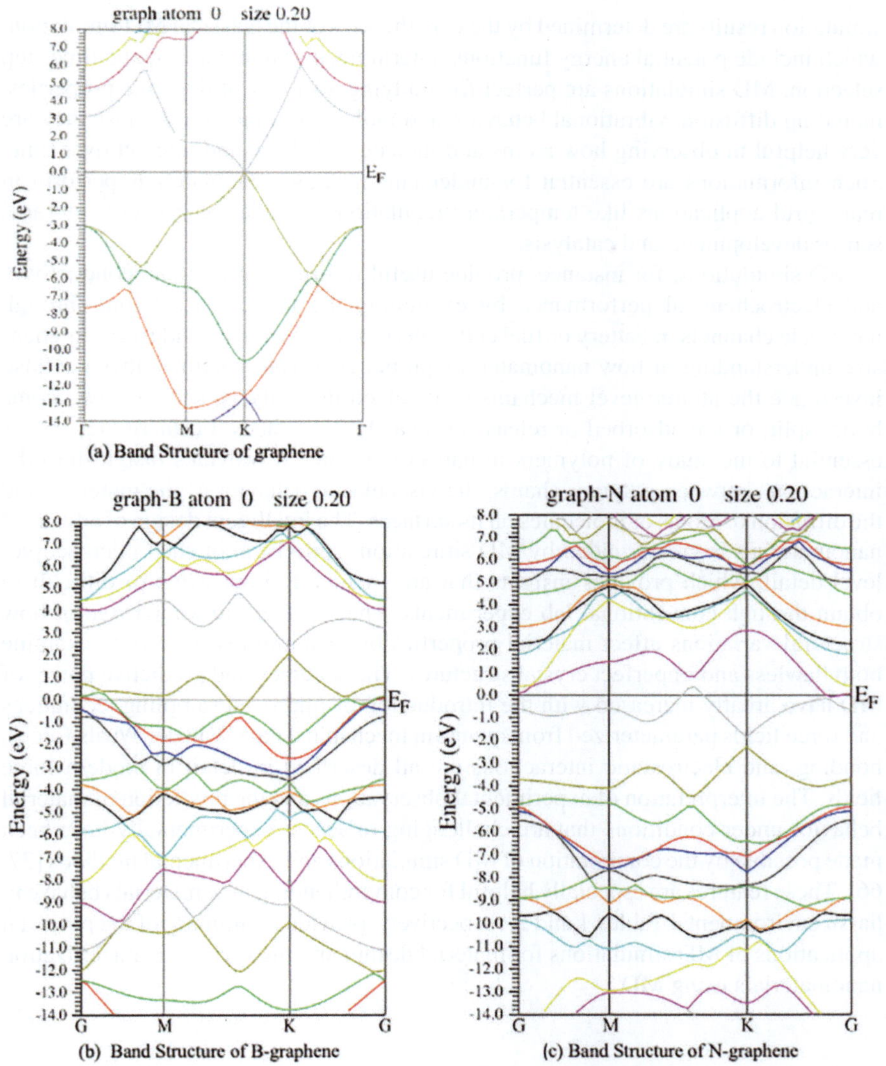

Fig. 2 Band structure of: **a** pristine graphene; **b** B-doped graphene; and **c** N-doped graphene. Adopted from Sharma et al. [44], FlatChem, Elsevier

3.2.1 Classical Molecular Dynamics

MD simulations allow researchers to capture the time-dependent evolution of materials under a variety of conditions, including changes in temperature, pressure, mechanical stress, or chemical environment, by numerically solving Newton's equations of motion for systems of interacting particles. The accuracy and stability of the simulation results are determined by the core theories of the classical MD simulation, which include potential energy functions, intermolecular interactions, and time step selection. MD simulations are perfect for studying a variety of dynamic properties, including diffusion, vibrational behavior, and molecular transport, because they are very helpful in observing how atoms and molecules behave and interact over time. Such informations are essential for understanding how nanomaterials perform in real-world applications like temperature regulation, drug delivery, energy storage, sensor development, and catalysis.

MD simulations, for instance, provide useful information on ionic conductivity and electrochemical performance by examining the movement of ions through nanoscale channels in battery or fuel cell materials. In order to provide a comprehensive understanding of how nanomaterials propel chemical reactions, they can also investigate the atomic-level mechanisms involved in catalysis, such as how atoms bond, split, or are adsorbed or released on catalytic surfaces. Furthermore, MD is essential to the study of polymers in nanocomposites. It provides insight into the interactions between polymer chains, the viscoelastic behavior of the material, and the diffusion of atoms or molecules on its surfaces. The intelligent design of advanced nanomaterials is greatly aided by MD simulations' provision of such microscopic-level detail, which provides insights that are frequently impossible or difficult to obtain through conventional lab experiments. They further our knowledge of how structural variations affect material properties by enabling researchers to examine both flawless and imperfect crystal structures. The accuracy and predictive power of MD have greatly increased with the introduction of improved sampling techniques and force fields parameterized from quantum mechanical data. Van der Waals forces, bonding, and electrostatic interactions are all described in detail in modern force fields. The interpretation of experimental observations and the prediction of material behavior under conditions that are challenging to access experimentally have been made possible by the combination of MD simulations and experimental methods [27, 66]. These features are especially helpful for comprehending how materials behave in harsh environments. Tables 1 and 2, respectively, provide a summary of the potential applications of MD simulations for material design and the various characterization nanomaterials using MD.

3.2.2 Ab Initio Molecular Dynamics

By fusing quantum mechanical computations with classical molecular dynamics, ab initio molecular dynamics (AIMD) enables extremely precise simulations of atomic interactions using first-principles techniques, most commonly density functional theory (DFT). AIMD is essential for comprehending intricate atomic-level phenomena in the field of advanced materials that are otherwise difficult to observe experimentally.

Ab initio MD simulations allow the study of materials where bond formation and breaking are significant by combining the precision of quantum mechanical calculations with molecular dynamics. These simulations are especially useful for

Table 1 Key areas where MD simulations are important to enhance material design

Aspect	The function of MD simulations	Outcome	References
Material optimization	Examine the mechanical, electrical, thermal, and structural characteristics	Customizing nanomaterials for particular uses (e.g., thermal stability, strength, conductivity)	Gajewicz et al. [18]
Atomic-scale insights	Model the dynamics, bonding, and interactions of atoms in nanomaterials	Enhanced comprehension of the behavior of materials at the nanoscale	Kedir et al. [25]
Surface modifications	Use ligands, acids, or other modifiers to simulate surface functionalization	Improved command over characteristics like as stability, reactivity, and wettability	Bo et al. [4]
Phase transitions	Forecast solid, liquid, and amorphous state changes at the nanoscale	Comprehending the phase behavior and stability of nanoparticles	Aliko et al. [2]
Interface studies	Analyze interactions at material interfaces, such as liquid–solid or heterostructures	Creation of improved composite materials, coatings, or adhesives	Ding et al. [13]
Mechanical properties	Learn the deformation, stress–strain relationship, and fracture behaviour	Creating superior high-quality nanomaterials with high durability	Zhao et al. [67]
Transport phenomena	Investigate diffusion, heat transfer, and charge mobility	Material optimization for temperature control, catalysis, and energy storage	Shang et al. [43]

researching materials synthesis, catalysis, and chemical reactions. AIMD simulations are readily available for ever-more complex systems because of the advancement of effective algorithms and the availability of high-performance computing resources. It is especially useful for researching how materials behave in harsh environments like high pressures, temperatures, and electric fields. For instance, in materials such as perovskites, battery electrodes, solid electrolytes, and catalysts, AIMD aids in the prediction of phase transitions, diffusion mechanisms, reaction mechanisms, defect dynamics, and chemical reactions at surfaces or interfaces. Its predictive power speeds up the search and development of next-generation functional materials by allowing the design of materials with specific qualities for electronic devices, energy storage, and catalytic applications. Additionally, the development of more effective simulation techniques that preserve accuracy while cutting down on computational costs has been made possible by the integration of ab initio MD with

Table 2 Characterization of nanomaterials using MD

Property	Method	Insights gained	Applications	References
Mechanical properties	Shear deformation, fracture mechanics, tensile/ compressive testing, and stress–strain simulations	Tensile strength, yield stress, elasticity, and deformation processes	Nanomaterials in automotive, aerospace, and construction industries	Nanoth et al. [33]
Thermal conductivity	Green–Kubo method (equilibrium MD) or NEMD for temperature gradient simulations	Heat transfer, thermal conductivity, phonon transport behavior	Thermal management materials, electronics cooling, thermoelectric	Dionne et al. [14]
Structural characterization	Bond analysis	The atomic arrangement, crystal structure, bonding patterns	Material design, nanostructure stability	Zhang et al. [65]
Dynamic properties	MD simulates atomic diffusion, molecular transport, and phonon behavior	Diffusion rates, molecular interactions, phonon scattering	Catalysis, sensors, drug delivery, gas storage, and filtration systems	Sibanda et al. [46]
Electronic properties	MD combined with DFT or other quantum mechanical methods	Band structure, charge density, and electron transport	Nanoelectronics, semiconductors, sensors, photovoltaics	Nandee et al. [32]
Optical properties	MD simulations paired with quantum methods to study light interaction (e.g., absorption, reflection)	Optical behavior, refractive index, absorption spectra	Photonic devices, optical coatings, and solar energy materials	Weinbub and Kosik [58]
Surface interactions	Simulate interactions with functional groups, adsorption, desorption, and surface modifications	Surface adsorption, interfacial strength, surface energy	Drug delivery systems, catalysis, nanocomposite interfaces	Van Tran et al. [51]
Chemical stability	Reactive MD simulations (ReaxFF)	Reaction pathways, bond dissociation, stability	Corrosion resistance, catalysis, material degradation	Chen et al. [9]

machine learning techniques [21]. These methods are broadening the range of issues that atomistic simulations can solve. The ability of ab initio molecular dynamics (AIMD) to capture the influence of a material's electronic structure on its behavior is unique. As a result, it can precisely forecast how atoms will bond, move charges, and vibrate in response to temperature, particularly in small, nanoscale structures. However, larger, more traditional systems where the material properties are already well understood through experimentation are better suited for classical simulations. AIMD provides a more thorough and accurate look at what's occurring at the atomic level, even though classical methods are quicker [59].

4 Machine Learning and Artificial Intelligence

Artificial intelligence (AI) and machine learning (ML), two quickly evolving technologies, have the potential to drastically alter the materials processing industry. An introduction to machine learning and artificial intelligence is given in this section, along with an analysis of their potential uses in materials processing and a list of their benefits over conventional techniques [12].

A subfield of artificial intelligence called machine learning (ML) focuses on developing statistical models and algorithms that let computers learn from data and make decisions or predictions on their own without the need for explicit programming. According to Butler et al. [6], these models are made to identify and use patterns in data to guide their forecasts and choices. Data preparation, model training, and model evaluation are the three primary steps of a typical machine learning workflow. The following are some examples of possible uses of AI and ML in materials processing:

1. **Materials Design and Discovery**: The discovery of new materials with desirable properties can be accelerated by ML and AI. The time and expense needed for the discovery of new materials can be greatly decreased by using machine learning and artificial intelligence (ML) algorithms to analyze vast datasets of material properties and structures and find patterns that may be useful in predicting the properties of new materials.
2. **Materials Characterization and Analysis**: Large datasets of materials characterization data, including diffraction patterns, spectroscopy data, and microscopy images, can be effectively analyzed using machine learning and artificial intelligence. Researchers can learn important details about the composition and characteristics of materials by spotting trends and characteristics in these datasets.
3. **Materials Processing Optimization and Control**: ML and AI can be used to optimize and control a number of materials processing parameters, including composition, temperature, and pressure. Additionally, ML and AI algorithms can modify processing parameters to increase efficiency, quality, and consistency by evaluating real-time data from sensors and feedback loops.

4.1 Supervised Learning Algorithms

Using experimental and computational data to train models that can then predict properties for new materials, supervised learning algorithms have been widely used to predict mechanical, electronic, and thermal properties [57]. Because these algorithms provide a label or output for every input, they are able to predict an output for new data by learning a mapping function from input to output. By adding significant descriptors, ranging from basic compositional features to intricate structural descriptors, advanced featured engineering techniques have increased the accuracy of machine learning models. The creation of more reliable and accurate predictive models has been made possible by the integration of various data sources, such as computational computations, experimental measurements, and literature data. These approaches are particularly valuable for materials informatics applications.

4.2 Unsupervised Learning and Pattern Recognition

Self-driven, unsupervised learning employs algorithms to uncover hidden patterns and relationships in unlabeled data. Its most crucial feature is that the training data does not contain any pre-defined output. These methods have been used to find new material classes, analyze structure–property relationships, and spot patterns in materials data. These methods are especially useful for discovering unexpected correlations and conducting exploratory data analysis.

Materials can be categorized into discrete groups with comparable attributes by using clustering algorithms to find similar materials based on their structures or properties. These methods have proven useful for comprehending the variety of materials and locating promising fields for additional study. By visualizing high-dimensional materials data using dimensionality reduction techniques, researchers can spot patterns and trends that aren't visible in the original data [3]. These methods are especially helpful for comprehending intricate relationships between structure and properties.

4.3 Deep Learning and Neural Networks

Deep learning is a well-known machine learning technique that makes use of multi-layered artificial neural networks. These methods have demonstrated great promise for applications in materials science, especially when dealing with complex, high-dimensional data. Layers (input, hidden, and output) of interconnected nodes or neurons make up a neural network model. Properties, materials design, characterization data analysis, microscopy image analysis, crystal structure prediction, and defect identification have all benefited from the successful application of neural networks.

In order to create new materials with desired properties, generative models such as generative adversarial networks and variational auto-encoders have been developed for materials design. The automated materials discovery process has advanced significantly as a result of these methods [10].

4.4 Autonomous Materials Discovery

The future of materials science is represented by autonomous materials discovery systems, which enable completely automated materials discovery by fusing robotic synthesis and characterization with artificial intelligence. According to Häse et al. [22], these systems have the potential to significantly quicken the rate of materials discovery.

- **Robotic Synthesis Systems**

Systems for robotic synthesis have been created that can create materials with predetermined compositions and processing parameters on their own. These systems have advanced monitoring and control features. Adaptive synthesis protocols that can optimize synthesis conditions based on feedback from characterization have been made possible by the integration of artificial intelligence with synthesis robots [22]. For effective materials discovery, these skills are essential. Flexible synthesis capabilities have been made possible by the development of modular robotic systems, enabling researchers to investigate various synthesis conditions and routes. These systems are especially useful for learning new synthesis techniques.

- **Closed-Loop Discovery Systems**

Autonomous materials discovery is made possible by closed-loop discovery systems, which combine experimental capabilities and artificial intelligence. These systems have the ability to suggest new materials, synthesize them, describe their characteristics, and gain knowledge from the outcomes. By concentrating experimental efforts on the most promising candidates, the development of active learning algorithms has made it possible to explore the materials space efficiently. For autonomous discovery systems to operate as efficiently as possible, these algorithms are especially crucial. Comprehensive property evaluation in autonomous systems has been made possible by the integration of various characterization techniques. Finding materials with ideal qualities requires these skills [38].

4.5 Quantum Computing Applications

A new area of computational materials science called quantum computing holds promise for resolving issues that traditional computers are unable to handle. Because of their special characteristics, quantum systems are especially well-suited to modeling quantum mechanical processes in materials.

- **Quantum Algorithms for Materials Science**

The electronic structure problem has been solved using quantum algorithms, which may provide exponential speedups over classical approaches for specific problem classes. When studying strongly correlated materials, where traditional methods are ineffective, these algorithms show great promise. Proof-of-concept demonstrations of quantum advantage for materials science problems have been made possible by the development of variational quantum eigensolvers, which allow the simulation of small molecular systems on existing quantum hardware. As quantum hardware advances, it is anticipated that these methods will scale to larger systems. For materials property prediction and materials design, quantum machine learning algorithms have been developed; in some cases, they may be superior to classical machine learning [19, 20]. Although these methods are still in their infancy, they have potential for use in the future.

Research is currently underway to develop hardware for quantum computing that is specifically tailored for materials science applications. The applicability of various quantum computing architectures, such as trapped ions and superconducting qubits, to materials science issues is being assessed. In materials science, software frameworks for quantum computing are being created, offering researchers easily navigable interfaces for putting quantum algorithms into practice. For the larger materials science community to have access to quantum computing, these frameworks are essential. The potential for hybrid approaches that combine the advantages of both paradigms is being investigated through the integration of quantum computing with classical computing resources. Even before completely fault-tolerant quantum computers are developed, these methods might offer useful benefits.

5 High-Throughput Materials Discovery

5.1 Experimental High-Throughput Approaches

Because high-throughput experimental techniques allow for the quick synthesis and screening of numerous material candidates, they have completely changed the materials discovery process. These methods have proven especially effective in fields like drug discovery, electronic materials, and catalysis [39].

5.1.1 Combinatorial Synthesis

Large arrays of materials with systematically different compositions can be prepared thanks to combinatorial synthesis techniques. These methods have proven especially effective in finding novel materials with ideal characteristics. The creation of materials libraries with thousands of compositions has been made possible by automated synthesis systems, which has significantly sped up the discovery process. These systems have advanced monitoring and control features. Rapid screening of material properties made possible by the combination of synthesis and high-throughput characterization has made it possible for researchers to quickly identify promising candidates. For catalyst discovery, this strategy has proven especially effective [8].

5.1.2 Automated Characterization Systems

In order to facilitate the quick screening of material properties across large libraries, automated characterization systems have been developed. These systems have sophisticated data management and robotics features. High-throughput screening protocols have made it possible to quickly assess a material's performance for a given application. These procedures are especially useful for locating materials with ideal characteristics. Comprehensive materials screening has been made possible by the integration of various characterization techniques, which has yielded detailed information about the relationships between structure and property. This method is especially useful for comprehending the behavior of complex materials [39].

5.2 Computational High-Throughput Methods

Computational high-throughput methods have enabled the screening of millions of materials candidates, identifying promising materials for further study. These approaches have been particularly successful for discovering new materials with specific properties.

5.2.1 Materials Databases and Informatics

Calculated properties for hundreds of thousands of materials are included in extensive materials databases. These databases facilitate the identification of structure–property relationships and offer useful resources for materials discovery. Techniques in materials informatics have been created to evaluate sizable datasets and find promising material candidates. These methods speed up the discovery of materials by combining data mining, machine learning, and domain expertise. Comprehensive materials informatics techniques that provide both predicted and measured properties for materials have been made possible by the integration of experimental and computational databases. Finding promising materials and confirming computational predictions depend on this integration [8, 24, 39].

5.2.2 Automated Workflows

High-throughput computations with little human involvement are now possible thanks to automated computational workflows. These workflows make use of advanced data processing and job management tools. High-throughput computations are now more consistent and reliable thanks to the creation of standardized computational protocols. In large-scale studies, these procedures are especially crucial for guaranteeing data quality. Comprehensive materials discovery pipelines have been made possible by the combination of computational workflows and experimental validation [24]. These pipelines combine the accuracy of experimental validation with the speed of computational screening.

5.3 Data Management and Analysis

There are many difficulties in managing and analyzing the massive datasets produced by high-throughput methods. To overcome these obstacles, sophisticated data management and analysis tools have been created.

5.3.1 Data Infrastructure

Managing the massive amounts of data produced by high-throughput techniques requires a strong data infrastructure. Scalable data analysis, retrieval, and storage must be supported by this infrastructure. Data interoperability has been enhanced and the integration of data from various sources made possible by the creation of standardized data formats and metadata schemas. In order to create thorough materials databases, these standards are essential. The purpose of cloud-based data management systems is to offer scalable and easily accessible data processing and storage capabilities. Researchers from various institutions can work together thanks to these systems.

5.3.2 Data Analysis Tools

To extract valuable information from massive materials datasets, sophisticated data analysis tools have been developed. These tools use statistical techniques and machine learning algorithms to find trends and correlations. To help researchers examine and comprehend complex materials datasets, visualization tools have been developed. These tools are especially crucial for spotting patterns and trends in high-dimensional data. Automated analysis and the identification of materials with desired properties have been made possible by the integration of data analysis tools with materials databases. The rate of materials discovery is increasing as a result of these capabilities [56, 57].

6 Specialized Materials and Applications

6.1 Two-Dimensional Materials

Due to their special qualities and potential uses in electronics, energy storage, and sensing, two-dimensional materials have become one of the most fascinating fields of materials science research. A whole family of 2D materials with a wide range of characteristics and uses were made possible by the discovery of graphene in 2004.

Because of its remarkable mechanical, thermal, and electrical characteristics, graphene—the first isolated 2D material—is still being thoroughly researched. It holds promise for next-generation electronic devices due to its distinct band structure and high carrier mobility. In addition to graphene, many other 2D materials have been found, such as MXenes, hexagonal boron nitride (h-BN), and transition metal dichalcogenides (TMDs). Because of their distinct qualities, these materials can be used in a variety of ways. Designer materials with specific properties have been made possible by the development of heterostructures, which are made up of various 2D materials stacked on top of one another. The best qualities of various 2D materials can be combined in these heterostructures to produce new functions.

High-quality 2D material synthesis is still a major challenge, and several methods, such as liquid-phase exfoliation, chemical vapor deposition, and mechanical exfoliation, are being developed [15]. Depending on the intended use, each approach has pros and cons. For 2D materials to be used in practice, scalable synthesis techniques must be developed. Particularly promising for creating high-quality, large-area 2D materials is chemical vapor deposition. Because 2D materials are atomically thin and sensitive to environmental factors, processing them poses special difficulties. For the handling, transfer, and incorporation of 2D materials into devices, specialized methods have been developed [11].

6.2 High-Entropy Alloys

A paradigm shift in alloy design is represented by high-entropy alloys (HEAs), which combine several principal elements to produce materials with special qualities. These materials are being researched for a number of applications because they have demonstrated remarkable mechanical properties. According to Xu et al. [60], the design of HEAs is predicated on the idea of optimizing configurational entropy by combining several elements in nearly equal amounts. This method stops brittle intermetallic compounds from forming and stabilizes single-phase solid solutions. For HEAs, a number of synthesis techniques have been developed, such as additive manufacturing, mechanical alloying, and arc melting. Regarding composition control, microstructure, and processing scalability, each approach has unique benefits. Phase formation and property predictions have been made possible by the development of computational tools for HEA design, which has directed experimental efforts toward compositions that show promise. These tools forecast alloy behavior by taking into account kinetic and thermodynamic factors.

High strength, superior corrosion resistance, and good stability at high temperatures are just a few of the remarkable qualities that HEAs have proven to possess. These characteristics make them appealing for demanding chemical processing, energy, and aerospace applications. Many compositions of HEAs exhibit exceptional strength and ductility combinations, and their mechanical qualities are frequently better than those of conventional alloys. HEAs' remarkable mechanical performance is a result of their distinct microstructures. Applications of HEAs are being investigated in a number of domains, such as catalysis, wear-resistant coatings, and structural components. Because of their adaptability, HEAs can be used in a variety of applications [63, 64].

6.3 Biomaterials and Bioengineering

With the development of sophisticated materials for medical applications ranging from drug delivery systems to implants, the field of biomaterials has undergone significant change. Strict specifications for functionality, long-term stability, and biocompatibility must be met by these materials. Advanced designs that can interact with biological systems in controlled ways are incorporated into modern biomaterials. These materials may be bioactive, biodegradable, or designed to promote specific cellular responses. The development of biomimetic materials that mimic natural biological structures has led to improved performance in medical applications. These materials can promote tissue regeneration and integration with existing biological systems. Nanomaterials have found extensive applications in biomedicine, offering unique properties for drug delivery, imaging, and therapeutic applications. The small size and high surface area of nanomaterials enable them to interact with biological systems at the cellular level.

Advanced biomaterials facilitate tissue growth and regeneration by acting as scaffolds. The development of smart biomaterials that can respond to biological signals has enabled more sophisticated tissue engineering approaches [55]. These materials can deliver growth factors, respond to pH changes, or degrade at controlled rates. 3D printing has revolutionized tissue engineering by enabling the fabrication of complex, patient-specific scaffolds. This technology allows for precise control of scaffold architecture and the incorporation of multiple materials [52].

6.4 Sustainable Materials

As society works to address resource scarcity and environmental issues, the development of sustainable materials has grown in importance. Throughout their lifecycle, these materials are made to have as little of an impact on the environment as possible. Traditional petroleum-based materials can be replaced with bio-based materials made from renewable resources. These materials can lessen their impact on the environment and their reliance on fossil fuels. High-performance materials made from renewable resources are now possible thanks to the development of sophisticated bio-based composites. These materials offer advantages for the environment and can perform on par with conventional materials. Applications where end-of-life disposal is an issue have led to the development of biodegradable materials. These substances can break down organically without producing toxic byproducts.

Designing materials with recycling and circular economy principles in mind has grown in significance. These strategies seek to optimize resource use and reduce waste. In order to recover valuable materials from complex waste streams, sophisticated recycling techniques have been developed. These techniques can enable the recovery of high-value materials that would otherwise be lost. The development of materials that can be easily disassembled and recycled has become an important design consideration. These materials can help create more sustainable product lifecycles [17, 62].

7 Future Directions

7.1 Integration of Experimental and Computational Approaches

To address the complex issues that materials science faces, synergistic workflows that leverage the benefits of both computational and experimental approaches are essential. However, certain tests are still necessary for verifying computational predictions in order to boost confidence in computational methods. This validation process exposes the assumptions and limitations of computational models. Rapid confirmation of computer predictions is now possible thanks to the development of automated experimental validation techniques. These systems can provide feedback to improve computational models and systematically test predicted qualities. Incorporating uncertainty quantification into computational and experimental methods has improved the reliability of materials discovery. These techniques help identify which areas need better models or more data.

Closed-Loop Materials Discovery

The future of materials science lies in closed-loop materials discovery systems that combine synthesis, characterization, and computation. These systems are able to optimize the properties of materials and independently explore their space. By concentrating experimental efforts on the most promising candidates, the development of active learning algorithms has made it possible to explore the materials space efficiently. These algorithms are especially crucial for maximizing discovery systems' effectiveness. The integration of multiple characterization techniques has enabled comprehensive property evaluation in closed-loop systems. These capabilities are crucial for identifying materials with optimized properties.

7.2 Emerging Computational Paradigms

Emerging computational paradigms have the potential to transform materials science by making it possible to investigate issues that were previously unsolvable. These paradigms include advanced machine learning techniques, neuromorphic computing, and quantum computing.

7.2.1 Quantum Computing Applications

Quantum computing holds particular promise for materials science because of the quantum mechanical nature of materials. Quantum computers may be able to solve a number of materials science issues ten times faster than conventional computers. Among the current areas of study in the field of quantum algorithms for materials science are algorithms for electronic structure computations, materials optimization, and property prediction. These algorithms are evaluated using current quantum equipment. It is being investigated how to integrate quantum computing with traditional computing resources, which could lead to hybrid strategies that incorporate the advantages of both paradigms. These methods could offer useful benefits even before completely fault-tolerant quantum computers are developed.

7.2.2 Advanced Machine Learning Approaches

Deep learning and reinforcement learning are two examples of advanced machine learning techniques being used to tackle more challenging materials science issues. Complex relationships and high-dimensional data can be handled by these methods. Physical constraints and prior knowledge can now be incorporated into models thanks to the development of machine learning techniques informed by physics. These methods aid in guaranteeing that models produce predictions that are physically plausible.

The creation of hybrid approaches that combine the precision of physical models with the effectiveness of machine learning has been made possible by the integration of machine learning with conventional computational techniques.

8 Challenges and Opportunities

To fully utilize advanced materials, a number of materials science issues still need to be resolved despite tremendous advancements. These difficulties also present chances for ground-breaking discoveries.

- **Data Challenges**: Accessibility, quantity, and quality of materials data continue to be major obstacles. For materials science to advance, data infrastructure and standards must be improved. To maximize the value of research data, materials scientists must develop FAIR (Findable, Accessible, Interoperable, Reusable) data principles. These guidelines make it possible for data to be shared and reused efficiently. Complex data management systems and analysis tools are needed to integrate experimental and computational data. These systems have to manage a variety of data scales and types.
- **Scalability and Translation**: Converting lab findings into real-world uses is still very difficult. Multidisciplinary cooperation is necessary to close the gap between basic research and real-world applications. In order to convert novel materials into useful applications, scalable synthesis and processing techniques must be developed. These techniques need to be both environmentally friendly and economically feasible. To solve complex problems, materials science must be integrated with other fields like computer science, biology, and engineering. New teaching strategies and cooperative frameworks are needed for this integration.
- **Societal Impact and Sustainability**: Because advanced materials enable technologies that address global challenges like energy, healthcare, and environmental sustainability, materials science has a significant impact on society. Future studies in materials science must take sustainability and societal impact into account.

9 Conclusion

The field of materials science has undergone a major revolution as a result of improvements in experimental techniques and computational power. Because of the unprecedented opportunities this convergence has created for materials discovery, design, and optimization, it has fundamentally changed the way we approach materials research and development.

The experimental advancements discussed in this chapter, ranging from atomic-resolution microscopy to high-throughput synthesis, have provided scientists with powerful tools for understanding and working with materials at different length scales. These capabilities have facilitated the production of materials with specialized features for certain purposes as well as the discovery of new material phenomena. Combining in-situ and operando characterisation methods has helped close the gap between lab research and real-world applications by offering vital insights into how materials behave in realistic settings.

Computational approaches have evolved over time, from simple tools for property prediction to sophisticated platforms for materials design and discovery. With the help of machine learning, molecular dynamics simulations, and density functional theory, researchers can now quickly investigate vast material domains and predict properties with unprecedented accuracy. The emergence of materials informatics as a distinct field provided the framework for integrating experimental and computational data to expedite materials discovery.

The application of artificial intelligence and machine learning to materials research represents a paradigm shift that is still in its early stages. The full potential of these techniques has not yet been realized, despite their impressive prior performance in materials design and property prediction. The pace of materials discovery could be further accelerated by the creation of autonomous discovery systems that can combine synthesis, computation, and characterization.

Examples of areas where the combination of state-of-the-art experimental and computational methods has yielded groundbreaking results include high-entropy alloys, biomaterials, sustainable materials, two-dimensional materials, and other specialty materials discussed in this chapter. These materials are excellent illustrations of how modern materials science can be applied to challenging issues in energy, healthcare, and environmental sustainability.

Many important trends that will impact materials science are emerging as we look to the future. The continued integration of computational and experimental methods will enable more targeted and efficient material discovery. As new computer paradigms like quantum computing and advanced machine learning are developed, the scope of problems that can be resolved will expand. The increased focus on sustainability and societal impact will spur the development of materials that can address global issues.

However, significant problems still need to be solved before advanced materials science can be used to its full potential. These include improving data quality and standardization, developing scalable synthesis and processing methods, and translating laboratory research into practical applications. To overcome these challenges, sustained investments in interdisciplinary collaboration, education, and research infrastructure are required.

Materials science has a very promising future because it has the potential to create materials that were previously unimaginable. Artificial intelligence, sophisticated computational methods, and contemporary experimental techniques are coming together to create a new paradigm for materials research. This paradigm promises to accelerate research and enable the creation of materials with unprecedented properties and capabilities. As we push the boundaries of what is practical to address the main issues facing humanity and create a more sustainable and prosperous future, materials science will become increasingly important.

The transformation of individual atoms into useful materials is one of the most fascinating and challenging issues in physics. Through the techniques and tools discussed in this chapter, scientists are gaining unprecedented capabilities to understand and control this process, opening up new possibilities for materials that could transform society and technology. The potential to unleash the full potential of matter and create materials that will impact human civilization in the future is presented by the combination of experimental and computational advancements at the start of a new era in materials science.

References

1. Aleman, A.M., Wang, X., Becker, T., Slater, C.E., Zhang, D., Chen, J., Brien, G.L., Zuo, J.-M., Rodriguez, J.A.: Multimodal in situ characterization uncovers unexpected selectivity pathways in high-entropy alloy electrocatalysts. J. Am. Chem. Soc. **147**(12), 10309–10319 (2025). https://doi.org/10.1021/jacs.4c16707
2. Aliko, V., Vasjari, L., Ibrahimi, E., Impellitteri, F., Karaj, A., Gjonaj, G., Piccione, G., Arfuso, F., Faggio, C., Istifli, E.S.: From shadows to shores—quantitative analysis of CuO nanoparticle-induced apoptosis and DNA damage in fish erythrocytes: a multimodal approach combining experimental, image-based quantification, docking and molecular dynamics. Sci. Total. Environ. **906**, 167698 (2023). https://doi.org/10.1016/j.scitotenv.2023.167698
3. Bhat, N., Nguyen, M., Lange, L.W., Oses, C., Curtarolo, S., Brenner, D.W., Toher, C., Vecchio, K.S., Levy, O.: Unsupervised learning and pattern recognition in alloy design. Digit. Discov. **3**, 1792–1807 (2024). https://doi.org/10.1039/D4DD00282B
4. Bo, Z., Li, C., Yang, H., Ostrikov, K., Yan, J., Cen, K.: Design of supercapacitor electrodes using molecular dynamics simulations. Nanomicro Lett. **10**(2) (2018). https://doi.org/10.1007/s40820-018-0188-2
5. Bordonhos, M., Galvão, T.L.P., Gomes, J.R.B., Gouveia, J.D., Jorge, M., Lourenço, M.A.O., Pereira, J.M., Pérez-Sánchez, G., Pinto, G., Silva, M.L., Tedim, C.M., Zêzere, J., Bruno, B.: Multiscale computational approaches toward the understanding of materials. Adv. Theory Simul. **6**(11), 2200628 (2023). https://doi.org/10.1002/adts.202200628
6. Butler, K.T., Davies, D.W., Cartwright, H., Isayev, O., Walsh, A.: Machine learning for molecular and materials science. Nature **559**(7715), 547–555 (2018). https://doi.org/10.1038/s41586-018-0337-2
7. Chambers, M.S., Hunter, R.D., Hollamby, M.J., Pauw, B.R., Smith, A.J., Snow, T., Danks, A.E., Schnepp, Z.: In situ and ex situ X-ray diffraction and small-angle X-ray scattering investigations of the sol–gel synthesis of Fe_3N and Fe_3C. Inorg. Chem. **61**, 6742–6749 (2022)
8. Chen, C., Nguyen, D.T., Lee, S.J., Baker, N.A., Karakoti, A.S., Lauw, L., Owen, C., Mueller, K.T., Bilodeau, B.A., Murugesan, V., Troyer, M.: Accelerating computational materials discovery with machine learning and cloud high-performance computing: from large-scale screening to experimental validation. J. Am. Chem. Soc. **146**(29), 20009–20018 (2024). https://doi.org/10.1021/jacs.4c03849
9. Chen, J., Jiang, X., Sun, H., Shao, Z., Fang, Y., Shu, R.: Phase transformation and strengthening mechanisms of nanostructured high-entropy alloys. Nanotechnol. Rev. **10**(1), 1116–1139 (2021). https://doi.org/10.1515/ntrev-2021-0071
10. Choudhary, K., DeCost, B., Wang, Y., Chen, C., Jain, A., Tavazza, F., Agrawal, A., Choudhary, A., Goyal, S.: Recent advances and applications of deep learning methods in materials science. npj Comput. Mater. **8**, 59 (2022). https://doi.org/10.1038/s41524-022-00734-6
11. Deng, Y., Wang, J., Che, J., Peng, Y., Su, Y., Tian, X., Xu, Z., Zhou, S., Zhao, Y., Zhang, H.: Precision chemistry for two-dimensional materials. Precis. Chem. **2**(4), 27797–27800 (2024). https://doi.org/10.1021/prechem.4c00065

12. Dimiduk, D.M., Holm, E.A., Niezgoda, S.R.: Innovation: perspectives on the impact of machine learning, deep learning, and artificial intelligence on materials, processes, and structures engineering. Integrating Mater. Manuf. **7**, 157–172 (2018)
13. Ding, H., Tian, W., Ma, Y.: Designing nanoparticle translocation through membranes by computer simulations. ACS Nano **6**(2), 1230–1238 (2012)
14. Dionne, C.J., Thakur, S., Scholz, N., Hopkins, P., Giri, A.: Enhancing the thermal conductivity of semiconductor thin films via phonon funneling. NPJ Comput. Mater. **10**(1) (2024). https://doi.org/10.1038/s41524-024-01364-w
15. Duan, X., Iamprasertkun, P., Wang, X., Jiang, S., Liu, W., Kiriya, D., Wang, C., Xu, M., Zhou, D., Li, Z., Chen, G.: Advances in liquid-phase exfoliation for scalable processing of two-dimensional materials. Precis. Chem. **2**(4), 27801–27816 (2024)
16. Dutta, P.K., Gupta, M.C., Manivannan, V., Chakrabarti, C.L.: Synthesis and characterization of advanced materials. In: ACS symposium series, pp 249–261. American Chemical Society, Washington, DC (1997). https://doi.org/10.1021/bk-1997-0680.ch015
17. Fratzl, P.: Introduction: sustainable materials. Chem. Rev. **123**(5), 3097–3101 (2023). https://doi.org/10.1021/acs.chemrev.3c00091
18. Gajewicz, A., Rasulev, B., Dinadayalane, T.C., Urbaszek, P., Puzyn, T., Leszczynska, D., Leszczynski, J.: Advancing risk assessment of engineered nanomaterials: application of computational approaches. Adv. Drug Deliv. Rev. **64**(15), 1663–1693 (2012)
19. Govoni, M., Ma, H., Galli, G.: Solving Materials Problems with a Quantum Computer: Quantum Embedding Simulations for Spin Defects in Solids. Argonne National Laboratory Newsroom (2020). https://www.anl.gov/article/solving-materials-problems-with-a-quantum-computer
20. Guo, Z., Li, R., He, X., Guo, J., Ju, S.: Harnessing quantum power: revolutionizing materials design through advanced quantum computation. Mater. Genome Eng. Adv. **2**(4), e73 (2024). https://doi.org/10.1002/mgea.73
21. Hafner, J.: Ab-initio simulations of materials using VASP: density-functional theory and beyond. J. Comput. Chem. **29**(13), 2044–2078 (2008). https://doi.org/10.1002/jcc.21057
22. Häse, F., Roch, L.M., Aspuru-Guzik, A.: Autonomous intelligent agents for accelerated materials discovery. Chem. Sci. **11**, 11864–11876 (2020). https://doi.org/10.1039/D0SC01101K
23. Hauke, B.M., Poland, M., Zimmermann, L., Döring, T., Adamcik, J., Zamponi, M., Egry, I., Schneider, H., de Ligny, D., Behrens, H.: Experimental and computational methods for studying the relaxation behavior of high-temperature oxide glasses. J. Am. Ceram. Soc. **108**(9), 4768–4781 (2025). https://doi.org/10.1111/jace.20440
24. Helge, S.S., Bhowmik, A., Gregoire, J.M.: Advancing energy materials through high-throughput experiments and computation. J. Mater. Chem. A **12**, 28 (2024). https://doi.org/10.1039/D3TA05860C
25. Kedir, W.M., Li, L., Tan, Y.S., Bajalovic, N., Loke, D.K.: Nanomaterials and methods for cancer therapy: 2D materials, biomolecules, and molecular dynamics simulations. J. Mater. Chem. B (2024). https://doi.org/10.1039/d4tb01667j
26. Ko, H.Y., Lam, S., Kulik, H.J.: High-throughput condensed-phase hybrid density functional theory: SeA implementation in quantum ESPRESSO. J. Chem. Theory Comput. **19**(13), 3610–3625 (2023). https://doi.org/10.1021/acs.jctc.2c00827
27. Kohlmeyer, A., Luo, T., Barreto, D., Cornelius, C.J.: Molecular dynamics simulations of advanced polymers for gas separation applications. Macromolecules **46**(11), 4707–4714 (2013). https://doi.org/10.1021/ma4005236
28. Kohn, W., Sham, L.J.: Self-consistent equations including exchange and correlation effects. Phys. Rev. **140**(4A), A1133–A1138 (1965). https://doi.org/10.1103/PhysRev.140.A1133
29. Krzywanski, J., Sosnowski, M., Grabowska, K., Zylka, A., Lasek, L., Kijo-Kleczkowska, A.: Advanced computational methods for modeling, prediction and optimization—a review. Materials **17**(14), 3521 (2024). https://doi.org/10.3390/ma17143521
30. LeSar, R.: Materials informatics: an emerging technology for materials development. Stat. Anal. Data Min.: ASA Data Sci. J. **1**, 372 (2009).https://doi.org/10.1002/sam.10034

31. Mohan, S., Kumar, S., Ramakrishnan, T., Palani, I.A., Parthasarathy, G.: Mastering material insights: advanced characterization techniques. Ind. Eng. Chem. Res. **54**(28), 6931–6944 (2015). https://doi.org/10.1021/acs.iecr.5c00447
32. Nandee, R., Chowdhury, M.A., Hossain, N., Rana, M.M., Mobarak, M.H., Islam, M.A., Aoyon, H.: Experimental characterization of defect-induced Raman spectroscopy in graphene with BN, ZnO, Al_2O_3, and TiO_2 dopants. Results Eng. **21**, 101738 (2023). https://doi.org/10.1016/j.rineng.2023.101738
33. Nanoth, R., Jayanarayanan, K., Kumar, P.S., Balachandran, M., Pegoretti, A.: Static and dynamic mechanical properties of hybrid polymer composites: a comprehensive review of experimental, micromechanical and simulation approaches. Compos. Part A Appl. Sci. Manuf. **174**, 107741 (2023). https://doi.org/10.1016/j.compositesa.2023.107741
34. Nørskov, J.K., Bligaard, T., Rossmeisl, J., Christensen, C.H.: Towards the computational design of solid catalysts. Nat. Chem. **1**, 37–46 (2009). https://doi.org/10.1038/nchem.121
35. Odermatt, P.D., Shivanandan, A., Deschout, H., Jankele, R., Nievergelt, A.P., Feletti, L., Davidson, M.W., Radenovic, A., Fantner, G.E.: High-resolution correlative microscopy: bridging the gap between single molecule localization microscopy and atomic force microscopy. Nano Lett. **15**(8), 4896–4904 (2015). https://doi.org/10.1021/acs.nanolett.5b00572
36. Osakabe, N., Tonomura, A., Fukuhara, A., Otake, Y., Shinagawa, K., Fujita, J., Haraguchi, S.: Characterization of dynamics in materials using time-resolved electron microscopy. Adv. Imaging Electron Phys. **109**, 167–207 (1999). https://doi.org/10.1016/S1044-5803(99)00022-4
37. Plimpton, S.: Fast parallel algorithms for short-range molecular dynamics. J. Comput. Phys. **117**(1), 1–19 (1995). https://doi.org/10.1006/jcph.1995.1039
38. Pogue, E.A., New, A., McElroy, K., Le, N.Q., Pekala, M.J., McCue, I., Gienger, E., Domenico, J., Hedrick, E., McQueen, T.M., Wilfong, B., Piatko, C.D., Ratto, C.R., Lennon, A., Chung, C., Montalbano, T., Bass, G., Stiles, C.D.: Closed-loop superconducting materials discovery. npj Comput. Mater. **9**, 190 (2023). https://doi.org/10.1038/s41524-023-01131-3
39. Potyrailo, R., Rajan, K., Stoewe, K., Takeuchi, I., Chisholm, B., Lam, H.: Combinatorial and high-throughput screening of materials libraries: review of state of the art. ACS Comb. Sci. **13**(6), 579–633 (2011). https://doi.org/10.1021/co200007w
40. Prouster, D.J., Trelewicz, J.R., Snead, L.L., Hu, X., Morrall, D., Koyanagi, T., Parish, C.M., Tan, L., Katoh, Y., Wirth, B.D.: Advanced synchrotron characterization techniques for fusion materials science. ACS Appl. Mater. Interfaces **11**(50), 46692–46709 (2019). https://doi.org/10.1021/ammi.9b12647
41. Rapaport, D.C.: The Art of Molecular Dynamics Simulation, 2nd edn. Cambridge University Press, Cambridge, UK (2004). https://doi.org/10.1017/CBO9780511813467
42. Scalambra, F., Parker, S.F., Fernandez-Alonso, F., Silverwood, I.P.: Neutron scattering techniques for studying metal complexes in aqueous and nonaqueous solutions. ACS Omega **6**(1), 215–226 (2021). https://doi.org/10.1021/acsomega.0c05030
43. Shang, Y., Dehkordi, R.B., Chupradit, S., Toghraie, D., Sevbitov, A., Hekmatifar, M., Suksatan, W., Sabetvand, R.: The computational study of microchannel thickness effects on H_2O/CuO nanofluid flow with molecular dynamics simulations. J. Mol. Liq. **345**, 118240 (2021). https://doi.org/10.1016/j.molliq.2021.118240
44. Sharma, R., Khan, S., Goyal, V., Sharma, V., Sharma, K.S.: Investigation on effect of boron and nitrogen substitution on electronic structure of graphene. FlatChem **1**, 20–33 (2017). https://doi.org/10.1016/j.flatc.2016.10.001
45. Shi, K.-Y., Xia, Z.-N., Gan, T.-T., Jiang, X.-M., Xia, C.: Applications of real-time monitoring techniques in chemical synthetic ingredients. Guang Pu Xue Yu Guang Pu Fen Xi **30**(2), 499–503 (2010). PMID: 20384154
46. Sibanda, D., Oyinbo, S.T., Jen, T.: A review of atomic layer deposition modelling and simulation methodologies: density functional theory and MD. Nanotechnol. Rev. **11**(1), 1332–1363 (2022). https://doi.org/10.1515/ntrev-2022-0084
47. Sprouster, D.J., Trelewicz, J.R., Snead, L.L., Hu, X., Morrall, D., Koyanagi, T., Parish, C.M., Tan, L., Katoh, Y., Wirth, B.D.: Advanced synchrotron characterization techniques for fusion

materials science. ACS Appl. Mater. Interfaces **11**(50), 46692–46709 (2019). https://doi.org/10.1021/ammi.9b12647
48. Tahmasebi, P., Ghanbarian, B., Bai, H.: A state-of-the-art review of experimental and computational methods towards bridging different scales for modeling porous materials. Prog. Solid State Chem. **65**, 100415 (2023). https://doi.org/10.1016/j.progsolidstchem.2023.100415
49. Takizawa, S., Hiramatsu, K., Lindley, M., Gala de Pablo, J., Ono, S., Goda, K.: High-speed hyperspectral imaging enabled by compressed sensing in time domain. Adv. Photonics Nexus **2**(2), 026008 (2023). https://doi.org/10.1117/1.APN.2.2.026008
50. van Stokkum, I.H.M., Larsen, D.S., van Grondelle, R.: Global and target analysis of time-resolved spectra. Biochim. Biophys. Acta **1657**(2–3), 82–104 (2004). https://doi.org/10.1016/j.bbabio.2004.04.011
51. Van Tran, T., Nguyen, D.T.C., Le, H.T., Vo, D.N., Nanda, S., Nguyen, T.D.: Optimization, equilibrium, adsorption behavior and role of surface functional groups on graphene oxide-based nanocomposite towards diclofenac drug. J. Environ. Sci. **93**, 137–150 (2020). https://doi.org/10.1016/j.jes.2020.02.007
52. Walters-Shumka, J.P., Cheng, C., Jiang, F., Willerth, S.M.: Recent advances in modeling tissues using 3D bioprinted nanocellulose bioinks. ACS Biomater. Sci. Eng. **11**(4), 1882–1896 (2025). https://doi.org/10.1021/acsbiomaterials.5c00123
53. Wang, J., Zhou, Q., Li, Z., Liu, C., Wu, X., Liu, J.: Using nanoindentation to characterize the mechanical and creep properties of shale. ACS Omega **7**(15), 12927–12940 (2022). https://doi.org/10.1021/acsomega.2c01190
54. Wang, X., Zhang, H., Liu, Y., Zhao, F., Li, S., Yu, J.: A teaching-learning framework for materials characterization. J. Chem. Educ. **101**(2), 580–586 (2024). https://doi.org/10.1021/acs.jchemed.3c00974
55. Wang, Y., Wang, Z., Dong, Y.: Collagen-based biomaterials for tissue engineering. ACS Biomater. Sci. Eng. **9**(3), 1132–1150 (2023). https://doi.org/10.1021/acsbiomaterials.2c00730
56. Ward, L., Dunn, A., Faghaninia, A., Zimmermann, N.E.R., Bajaj, S., Wang, Q., Montoya, J., Chen, J., Bystrom, K., Dylla, M., Chard, K., Asta, M., Persson, K.A., Snyder, G.J., Foster, I., Jain, A.: Matminer: an open source toolkit for materials data mining. Comput. Mater. Sci. **152**, 60–69 (2018). https://doi.org/10.1016/j.commatsci.2018.05.018
57. Ward, L., Liu, R., Krishna, A., Hegde, V.I., Agrawal, A., Choudhary, A., Wolverton, C.: Including crystal structure attributes in machine learning models of formation energies via Voronoi tessellations. Phys. Rev. B **96**(2), 024104 (2017). https://doi.org/10.1103/PhysRevB.96.024104
58. Weinbub, J., Kosik, R.: Computational perspective on recent advances in quantum electronics: from electron quantum optics to nanoelectronic devices and systems. J. Phys. Condens. Matter **34**(16), 163001 (2022). https://doi.org/10.1088/1361-648x/ac49c6
59. Wimmer, E., Najafabadi, R., Young, G.A., Jr., Ballard, J.D., Angeliu, T.M., Vollmer, J., Chambers, J.J., Niimi, H., Shaw, J.B., Freeman, C., Christensen, M., Wolf, W., Saxe, P.: Ab initio calculations for industrial materials engineering: successes and challenges. J. Phys. Condens. Matter **22**(38), 384215 (2010). https://doi.org/10.1088/0953-8984/22/38/384215
60. Xu, H., Shahbazian-Yassar, R.: High-entropy alloys in catalysis: progress, challenges, and prospects. ACS Mater. Au **4**(10), 3630–3644 (2024). https://doi.org/10.1021/acsmaterialsau.4c00080
61. Yan, Q., Yu, L., Truhlar, D.G., Gagliardi, L.: A combined theoretical and experimental study of the photophysical properties of organic photovoltaic materials. J. Phys. Chem. C **121**(17), 8939–8949 (2017). https://doi.org/10.1021/acs.jpcc.7b02019
62. Zeng, X.: A review on design of sustainable advanced materials by using artificial intelligence. Adv. Mater. Sustain. Manuf. **1**, 10006 (2024). https://doi.org/10.35534/amsm.2024.10006
63. Zhang, J., Tian, Y., Shi, Y., Liu, X., Liu, X., Zeng, Q., Wang, Y., Liu, J., Gu, L., Liu, Y., Ma, S., Yu, R., Liu, J.: Design and development of high-entropy alloys with a tailored composition and phase structure based on thermodynamic parameters and film thickness using a novel combinatorial target. ACS Omega **8**(31), 28333–28343 (2023). https://doi.org/10.1021/acsomega.3c02222

64. Zhang, Q., Bai, S., Snaith, H.J., Meng, Q., Cao, G., Ran, R., Shao, Z., Wang, D.: Advanced nanomaterials and characterization techniques for photon-enabled applications. Acc. Mater. Res. **4**(5), 3752–3765 (2023). https://doi.org/10.1021/accountsmr.3c00012
65. Zhang, X., Zheng, P., Ma, Y., Jiang, Y., Li, H.: Atomic-scale understanding of oxidation mechanisms of materials by computational approaches: a review. Mater. Des. **217**, 110605 (2022). https://doi.org/10.1016/j.matdes.2022.110605
66. Zhang, Y., Remsing, R.C., Xi, Y., Patel, S., Weeks, J.D., Voth, G.A.: Molecular dynamics study of ion transport in solid-state electrolytes. J. Am. Chem. Soc. **138**(19), 5975–5985 (2016). https://doi.org/10.1021/jacs.6b02192
67. Zhao, C., Tao, Y., Yu, Y.: Molecular dynamics simulation of thermal and phonon transport characteristics of nanocomposite phase change material. J. Mol. Liq. **329**, 115448 (2021). https://doi.org/10.1016/j.molliq.2021.115448

MIX
Papier aus verantwortungsvollen Quellen
Paper from responsible sources
FSC® C105338
FSC
www.fsc.org

If you have any concerns about our products,
you can contact us on
ProductSafety@springernature.com

In case Publisher is established outside the EU,
the EU authorized representative is:
Springer Nature Customer Service Center GmbH
Europaplatz 3, 69115 Heidelberg, Germany

Printed by Libri Plureos GmbH
in Hamburg, Germany